Molecular
Basis of
Insulin Action

Molecular Basis of Insulin Action

Edited by

Michael P. Czech
University of Massachusetts Medical School
Worcester, Massachusetts

Plenum Press • New York and London

Library of Congress Cataloging in Publication Data

Main entry under title:

Molecular basis of insulin action.

Includes bibliographies and index.
1. Insulin — Physiological effect. I. Czech, Michael P. [DNLM: 1. Insulin — pharma-
codynamics. WK 820 M718]
QP572.I5M65   1985   615'.365   84-26423
ISBN 0-306-41843-6

QP
572
,I5
M65
1985

©1985 Plenum Press, New York
A Division of Plenum Publishing Corporation
233 Spring Street, New York, N.Y. 10013

Printed in the United States of America

Contributors

Joseph Avruch, Howard Hughes Medical Institute Laboratory, Harvard Medical School, the Diabetes Unit and Medical Services, Massachusetts General Hospital; and the Department of Medicine, Harvard Medical School, Boston, Massachusetts 02114

Perry J. Blackshear, Howard Hughes Medical Institute Laboratory, Harvard Medical School, the Diabetes Unit and Medical Services, Massachusetts General Hospital; and the Department of Medicine, Harvard Medical School, Boston, Massachusetts 02114

Roger W. Brownsey, Department of Biochemistry, University of Bristol Medical School, Bristol, BS8 1TD, England. *Present address:* Department of Biochemistry, University of British Columbia, Vancouver, Canada

K. Cheng, Department of Pharmacology, University of Virginia School of Medicine, Charlottesville, Virginia 22908

Michele A. Cimbala, Department of Biochemistry, University of Massachusetts Medical Center, Worcester, Massachusetts 01605

G. A. Clawson, Department of Pathology, University of California, San Francisco, California 94143

Philip Cohen, Department of Biochemistry, University of Dundee, Dundee DD1 4HN, Scotland

J. Craig, Department of Internal Medicine, University of Virginia School of Medicine, Charlottesville, Virginia 22908

Marco Crettaz, Research Division, Joslin Diabetes Center and Department of Medicine, Brigham and Women's Hospital, Harvard Medical School, Boston, Massachusetts 02215

Samuel W. Cushman, Cellular Metabolism and Obesity Section, National Institute of Arthritis, Diabetes, and Digestive and Kidney Diseases, National Institutes of Health, Bethesda, Maryland 20205

Michael P. Czech, Department of Biochemistry, University of Massachusetts Medical School, Worcester, Massachusetts 01605

Joseph F. Daigneault, Department of Biochemistry, University of Massachusetts Medical Center, Worcester, Massachusetts 01605

Roger J. Davis, Department of Biochemistry, University of Massachusetts Medical Center, Worcester, Massachusetts 01605

Richard M. Denton, Department of Biochemistry, University of Bristol Medical School, Bristol, BS8 1TD, England

I. D. Goldfine, Cell Biology Research Laboratory, Harold Brunn Institute, Mount Zion Hospital and Medical Center, San Francisco, California 94120; and the Departments of Medicine and Physiology, University of California, San Francisco, California 94143

Alan G. Goodridge, Departments of Pharmacology and Biochemistry, Case Western Reserve University, Cleveland, Ohio 44106

Florin Grigorescu, Research Division, Joslin Diabetes Center and Department of Medicine, Brigham and Women's Hospital, Harvard Medical School, Boston, Massachusetts 02215

Richard W. Hanson, Department of Biochemistry, Case Western Reserve University School of Medicine, Cleveland, Ohio 44106

Hans U. Häring, Research Division, Joslin Diabetes Center and Department of Medicine, Brigham and Women's Hospital, Harvard Medical School, Boston, Massachusetts 02215

Kim A. Heidenreich, Department of Medicine, University of California at San Diego, La Jolla, California 92093

Yaacov Hod, Department of Biochemistry, Case Western Reserve University School of Medicine, Cleveland, Ohio 44106

Steven Jacobs, The Wellcome Research Laboratories, Research Triangle Park, North Carolina 27709

Leonard Jarett, Department of Pathology and Laboratory Medicine, University of Pennsylvania School of Medicine, Philadelphia, Pennsylvania 19104

C. Ronald Kahn, Research Division, Joslin Diabetes Center and Department of Medicine, Brigham and Women's Hospital, Harvard Medical School, Boston, Massachusetts 02215

Kathleen L. Kelly, Department of Pathology and Laboratory Medicine, University of Pennsylvania School of Medicine, Philadelphia, Pennsylvania 19104

Frederick L. Kiechle, Department of Clinical Pathology, William Beaumont Hospital, Royal Oak, Michigan 48072

Tetsuro Kono, Department of Physiology, School of Medicine, Vanderbilt University, Nashville, Tennessee 37232

Yan C. Kwok, Howard Hughes Medical Institute Laboratory, Harvard Medical School, the Diabetes Unit and Medical Services, Massachusetts General Hospital; and the Department of Medicine, Harvard Medical School, Boston, Massachusetts 02114

J. Larner, Department of Pharmacology, University of Virginia School of Medicine, Charlottesville, Virginia 22908

David Lau, Department of Biochemistry, University of Massachusetts Medical Center, Worcester, Massachusetts 01605

E. Locher, Department of Internal Medicine, University of Virginia School of Medicine, Charlottesville, Virginia 22908

David S. Loose, Department of Biochemistry, Case Western Reserve University School of Medicine, Cleveland, Ohio 44106

S. Lance Macaulay, Department of Pathology and Laboratory Medicine, University of Pennsylvania School of Medicine, Philadelphia, Pennsylvania 19104

C. Malchoff, Departments of Pharmacology and Internal Medicine, University of Virginia School of Medicine, Charlottesville, Virginia 22908

B. Richard Martin, Department of Biochemistry, University of Cambridge, Cambridge, CB2 IOW, England

Jay M. McDonald, Division of Laboratory Medicine, Departments of Pathology and Medicine, Washington University School of Medicine, St. Louis, Missouri 63110

Herman M. Meisner, Department of Biochemistry, Case Western Reserve University School of Medicine, Cleveland, Ohio 44106

Thomas B. Miller, Jr., The Department of Biochemistry, University of Massachusetts Medical School, Worcester, Massachusetts 01605

Richard D. Moore, Biophysics Laboratory, State University of New York, Plattsburgh, New York 12901; and the Department of Physiology and Biophysics, College of Medicine, University of Vermont, Burlington, Vermont 05405

Cristina Mottola, Department of Biochemistry, University of Massachusetts Medical School, Worcester, Massachusetts 01605

Raphael A. Nemenoff, Howard Hughes Medical Institute Laboratory, Harvard Medical School, the Diabetes Unit and Medical Services, Massachusetts General Hospital; and the Department of Medicine, Harvard Medical School, Boston, Massachusetts 02114

Yoshitomo Oka, Department of Biochemistry, University of Massachusetts Medical School, Worcester, Massachusetts 01605

Jerrold M. Olefsky, Department of Medicine, University of California at San Diego, La Jolla, California 92093

Janice C. Parker, Department of Pathology and Laboratory Medicine, University of Pennsylvania School of Medicine, Philadelphia, Pennsylvania 19104

Peter J. Parker, Imperial Cancer Research Fund Laboratories, London, England

Harrihar A. Pershadsingh, Division of Laboratory Medicine, Departments of Pathology and Medicine, Washington University School of Medicine, St. Louis, Missouri 63110

Jeffrey E. Pessin, Department of Physiology and Biophysics, The University of Iowa, Iowa City, Iowa 52242

Mark Pierce, Howard Hughes Medical Institute Laboratory, Harvard Medical School, the Diabetes Unit and Medical Services, Massachusetts General Hospital; and the Department of Medicine, Harvard Medical School, Boston, Massachusetts 02114

Paul F. Pilch, Department of Biochemistry, Boston University School of Medicine, Boston, Massachusetts 02118

F. Purrello, Cell Biology Research Laboratory, Harold Brunn Institute, Mount Zion Hospital and Medical Center, San Francisco, California 94120; and the Department of Physiology, University of California, San Francisco, California 94143

Marilyn D. Resh, The Biological Laboratories, Harvard University, Cambridge, Massachusetts 02138

Joshua B. Rubin, Department of Biochemistry, Boston University School of Medicine, Boston, Massachusetts 02118

C. Schwartz, Department of Pharmacology, University of Virginia School of Medicine, Charlottesville, Virginia 22908

Jonathan R. Seals, Department of Biochemistry, University of Massachusetts Medical School, Worcester, Massachusetts 01605

Virender S. Sheorain, Howard Hughes Medical Institute and Department of Physiology, Vanderbilt Medical Center, Nashville, Tennessee 37232

Michael A. Shia, Department of Biochemistry, Boston University School of Medicine, Boston, Massachusetts 02118

Ian A. Simpson, Cellular Metabolism and Obesity Section, National Institute of Arthritis, Diabetes, and Digestive and Kidney Diseases, National Institutes of Health, Bethesda, Maryland 20205

Thomas R. Soderling, Howard Hughes Medical Institute and Department of Physiology, Vanderbilt Medical Center, Nashville, Tennessee 37232

Sumiko Takayama, Research Division, Joslin Diabetes Center and Department of Medicine, Brigham and Women's Hospital, Harvard Medical School, Boston, Massachusetts 02215

S. Tamura, Department of Pharmacology, University of Virginia School of Medicine, Charlottesville, Virginia 22908

M. Thompson, Department of Pharmacology, University of Virginia School of Medicine, Charlottesville, Virginia 22908

R. Vigneri, Cell Biology Research Laboratory, Harold Brunn Institute, Mount Zion Hospital and Medical Center, San Francisco, California 94120; and the Department of Physiology, University of California, San Francisco, California 94143

Morris F. White, Research Division, Joslin Diabetes Center and Department of Medicine, Brigham and Women's Hospital, Harvard Medical School, Boston, Massachusetts 02215

Lee A. Witters, Diabetes Unit and Medical Services, Massachusetts General Hospital; and the Department of Medicine, Harvard Medical School, Boston, Massachusetts 02114

James R. Woodgett, Salk Institute, La Jolla, California 92037

Anthony Wynshaw-Boris, Department of Biochemistry, Case Western Reserve University School of Medicine, Cleveland, Ohio 44106

Kin-Tak Yu, Department of Biochemistry, University of Massachusetts Medical School, Worcester, Massachusetts 01605

Kenneth Zierler, Department of Physiology, The Johns Hopkins University School of Medicine, Baltimore, Maryland 21205

Preface

One day, in a moment of weakness, I fell prey to the temptation to organize and edit this volume on the mechanism of insulin action. The major reason for attempting to resist, of course, is the amazing speed at which advances are being made in this field. The usefulness of books such as this is often quickly compromised by new findings obtained during and just after publication. Happily for the contributors to this volume and myself, this unfortunate fate does not appear to be in store for us. New and important findings will undoubtedly continue to flow in this field during the next few years, but I believe this will increase rather than decrease the usefulness of this volume. As a matter of fact, as we go to press, I am delighted both that I was tempted and that I failed to resist.

There are two basic reasons for my enthusiasm about this book, and they both relate to this issue of timeliness. First, each of the contributors has had an opportunity to update the scientific content of the various chapters only a few months before actual publication of this volume. The material presented in this volume is, at publication, contemporary with the current original literature. This volume thus provides an excellent framework for assessing new discoveries in this field for some time to come. The second reason for my enthusiasm derives from the current status and stage of maturity of investigative activity in the field of insulin action. As demonstrated by the wide representation of topics discussed in this volume, a large number of hypotheses and approaches have been employed in attempting to solve the central mysteries underlying cellular signaling by insulin. Despite this diversity and intensity in current investigative effort, the molecular basis of insulin action remains unsolved. It is hoped that at least one of the hypotheses presented in this volume will turn out to be a key element of signal transduction by insulin. It is conceivable that together, several of the hypotheses discussed here may be part of the final answer. However, it is also possible that a completely new, currently unknown pathway of cellular signaling is utilized by insulin and that we are in store for new and unexpected twists on the trail to a final solution. The point to be made here is that whichever of these possibilities turns out to be correct, the work left to be done is almost certain to be extensive and difficult. It is likely that new technology will have to be developed to achieve this goal. I am hopeful that this book will serve as a bridge leading from where we have been to where we must go.

One rather unique aspect of this volume is the multiplicity of treatments of several common topics by various involved laboratory groups. The strategy in these cases was to provide a forum for multiple viewpoints about focal issues and to allow development of subtle differences in the perspectives and approaches employed by several contributing laboratories. Although this means there is a small amount of repetition of introductory material in the chapters on similar topics, we have tried to minimize redundancy. The gain is that important diversity of thinking is presented. Although it was not possible, because of space limitations, to include chapters from all major laboratories in this field, broad representation of divergent experimental approaches has been achieved. It is indeed gratifying that so many scientific leaders have enthusiastically agreed to contribute to this book. The reader can expect to obtain from this volume information on the action of insulin which truly reflects both the depth and breadth of this field.

One of the striking features of the extensive work in the field of insulin action is the profound and direct effect it has had on other fundamental biological processes. As is clearly illustrated in this volume, a myriad of new technologies have been developed to deal rigorously with key questions about the many steps in the pathways of insulin action. In many cases, these new techniques have been employed in other fields to attack similar biological questions about different biological structures and molecules. For example, the more general fields of receptor biology and enzyme regulation have clearly drawn upon and benefited from technology initially developed in the field of insulin action. Moreover, investigations on this topic have often led to fundamental knowledge about cell biology and biochemistry quite independent of insulin action. It is interesting, in this regard, to recall that in previous decades the insulin molecule itself was repeatedly used as a prototype for technology development and to gain insights into general protein chemistry. Similarly, the direct relevance of many of the contributions in this volume to much more general biological questions makes this book important reading for scientists in multiple disciplines.

With the above considerations in mind, I thank Kirk Jensen from Plenum Publishing Corporation for providing the temptation that sparked the genesis of this volume. Special thanks go to all my colleagues, both near and far, who have generously contributed their knowledge and insights to this effort. This book for me will serve as a reminder of the happy and cooperative interactions I have had with many great investigators around the world. It will also serve as a testament to the exciting scientific period that we have just passed through. Indeed, these past few years of work on insulin action have given us all the ingredients of truly vibrant science, including explosive methodological development, rapid advances in insight on several fronts, and even controversy. I believe this book makes it clear that much more of these will be coming our way.

Michael P. Czech

Contents

1

Signal Initiation by the Insulin Receptor

Subunit Structure and Regulation of the Insulin-Receptor Complex

Jeffrey E. Pessin, Cristina Mottola, Kin-Tak Yu, and Michael P. Czech

I. INTRODUCTION

Methods and reagents developed during the late 1970s have led to significant new insights into the protein structure of the insulin receptor over the past five years. The importance of this information resides in the presumption that the insulin receptor initiates all the biological responses mediated by the hormone and therefore that its structure must contain key information related to cellular signaling mechanisms. Because the mechanism of insulin action is still not understood, this assumption has not yet been proved. However, the data available are fully consistent with this view. Furthermore, it is probable that the insulin receptor, like many other receptor systems, contains structural information that catalyzes other functions such as desensitization of the insulin response and transport of insulin to or among various cellular compartments, as well as its own recycling from intracellular compartments to the cell-surface membrane. It seems reasonable to assume that various receptor domains participate both specifically and directly in these various biological functions. It follows that an important future objective in this field is to obtain the primary sequence and general protein structure of this receptor in order to better define the biochemical mechanisms by which the insulin receptor fulfills its various functions.

Although we may be a long way from this ultimate goal, we now have a very useful framework to be used as a springboard for such future studies. The use of methods such as affinity labeling and affinity chromatography as well as the availability

Jeffrey E. Pessin • Department of Physiology and Biophysics, The University of Iowa, Iowa City, Iowa. *Christina Mottola, Kin-Tak Yu, and Michael P. Czech* • Department of Biochemistry, University of Massachusetts Medical School, Worcester, Massachusetts.

of specific anti-insulin receptor antibody preparations have provided definitive iden-
tifications of the insulin-receptor subunits and have formed a sound basis for hypotheses
related to subunit stoichiometry and structure within the receptor complex. Furthermore, the finding that the insulin receptor is an enzyme expressing tyrosine kinase
activity has led to detailed characterization of its enzymology. In this chapter, we wish
to discuss two basic issues related to the structure of the insulin receptor. The first is
a model for the structure of the insulin-receptor subunit proposed by our laboratory
group[1,2] and independently by Jacobs and colleagues.[3] Basically, this model proposes
that the predominant minimum covalent subunit structure consists of a heterotetrameric
complex of disulfide-linked subunits of two types, as illustrated in Fig. 4. Although
recent results in our laboratory and others suggests that other receptor subunit configurations may exist, we believe that this heterotetrameric model for the minimum
covalent insulin-receptor subunit structure remains most consistent with the data now
available on this issue. Second, we discuss and present new data relating to the
regulation of the insulin-receptor tyrosine kinase activity. Such regulation occurs upon
tyrosine phosphorylation of the receptor itself. In addition, we recently observed that
a cyclic adenosine monophosphate (cAMP)-dependent mechanism, initiated by hormones that activate adenylate cyclase, inhibits the binding of insulin to its receptor
and uncouples insulin binding and receptor tyrosine kinase activation in isolated fat
cells. These findings may provide a useful basis or future studies designed to probe
the detailed relationships between the structure of the insulin receptor and its biological
functions.

II. THE SUBUNITS OF THE INSULIN RECEPTOR STRUCTURE

There now appears to be a general consensus that insulin receptors contain at
least one subunit of 125,000–135,000 daltons and another subunit type of approximately 95,000 daltons. The larger subunit was first identified by affinity purification
and electrophoresis on sodium dodecylsulfate (SDS) polyacrylamide gels.[4] That this
subunit is indeed associated with high-affinity insulin-receptor complexes was demonstrated by its specific labeling using photoactive insulin analogs[5–7] and affinity
crosslinking techniques.[1,2,8–9] As shown in Fig. 1, [125I]insulin is covalently crosslinked to this 125,000-dalton receptor subunit upon incubation with receptor-containing
membranes or with cells in the presence of the chemical crosslinker disuccinimidylsuberate. Affinity labeling of this subunit is increased with increasing concentrations
of [125I]insulin and is abolished when the incubation is carried out in the presence of
a saturating concentration of unlabeled insulin (Fig. 1). This receptor subunit is also
immunoprecipitated using anti-insulin receptor immunoglobulin preparations that are
(1) prepared in rabbits injected with purified insulin receptor,[10] (2) derived from
patients diagnosed as having acanthosis nigricans who are resistant to insulin,[9,11,12]
and (3) prepared by hybridoma techniques.[12,14] This subunit is a glycoprotein in that
it can be observed in immunoprecipitates of receptor derived from cells incubated with
labeled amino acids or saccharides.[11,12] This subunit appears to be quite similar in a
variety of tissues,[6–22] although in brain it may have a somewhat different molecular
weight.[18] We denoted this subunit as the α subunit of the insulin receptor.[1,2]

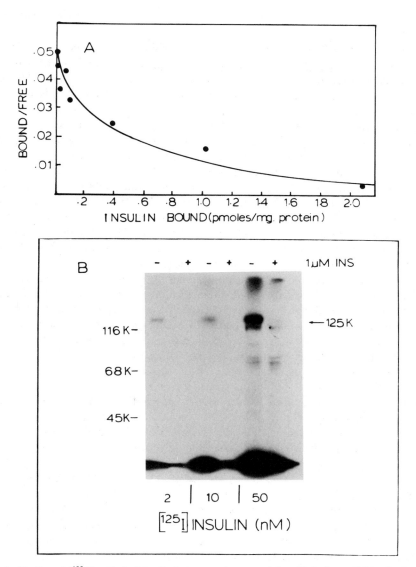

Figure 1. Binding of [^{125}I]insulin to fat cell plasma membranes and the effect of crosslinking increasing concentrations of [^{125}I]insulin to fat cell plasma membranes. (A) Purified fat cell plasma membranes in Krebs–Ringer phosphate buffer were incubated with buffer containing 1% bovine serum albumin, 2×10^{-10}M [^{125}I]insulin, and various concentrations of unlabeled hormone for 30 min at 23°C. Incubation was terminated by the addition of 3.0 ml of ice cold Krebs–Ringer phosphate buffer containing 0.1% albumin. Specific binding was measured by a rapid filtration assay. (B) [^{125}I]Insulin at the concentrations indicated was incubated for $\frac{1}{2}$ hr at 23°C with fat cell plasma membranes. Half of the incubation mixtures also contained 1 μM unlabeled insulin ($+$ 1μM INS). The incubation was terminated by cooling the mixture with 0°C, and the subsequent steps of crosslinking, electrophoresis, staining, drying, and autoradiography were carried out. The autoradiograph of the dried 7.5% acrylamide gel shows the molecular-weight markers indicated on the left and an arrow on the right pointing to the specifically labeled $M_r = 125,000$ protein. (From Pilch and Czech.[37])

The 95,000-dalton insulin-receptor subunit is readily identified by affinity chromatography,[3,20,21] immunoprecipitation,[10,12–15] or photoaffinity labeling with certain photoactive insulin analogs.[22] Interestingly, this receptor subunit is affinity-labeled using [125I]insulin and chemical crosslinkers much less intensely than is the α subunit.[1,2,9] For example, in Fig. 1 this subunit is not visualized under conditions in which very significant labeling of the α subunit is achieved. That the 95,000-dalton insulin receptor subunit is also a glycoprotein is demonstrated by its labeling with labeled saccharides.[11,12] Staining of this subunit by either Coomassie brilliant blue or silver staining techniques is much less intense than that observed for the α subunit.[21] This receptor subunit is exquisitely sensitive to elastaselike proteases and exhibits higher sensitivity to proteases in general compared with the α subunit.[23] The 95,000-dalton subunit is phosphorylated in ^{32}P-labeled intact cells upon addition of insulin[24,25] and appears to contain a site(s) that bind adenosine triphosphate (ATP) with high affinity.[26–28] Thus, this receptor subunit may contain the tyrosine kinase activity associated with insulin receptor.[24–29] This receptor subunit we denoted as the β subunit.[2]

Recent data produced by Yip and colleagues[30–33] have led this group to propose that 40,000-dalton subunits also reside in the insulin-receptor complex. These workers found that a 40,000-dalton band was photolabeled upon incubation of B29-monoazidobenzoyl-insulin or B1-monoazidobenzoyl-insulin with isolated adipocytes (but not plasma membranes). Because the appearance of this band was not affected by the presence of proteolytic inhibitors in the incubation system and could not itself be generated by digestion of either the α or β subunit bands with proteases or cell homogenate, it was suggested that one or more native subunits reside in this 40,000-dalton band.[31,32] Furthermore, it was found that the 40,000-dalton band labeled by B1-monoazidobenzoyl-insulin differs in its sensitivity to trypsin compared with the 40,000-dalton band labeled by B29-monoazidobenzoyl-insulin.[32] Yip and colleagues therefore proposed that two 40,000-dalton subunits reside in the insulin-receptor complex.[31]

These data on the 40,000-dalton species are obviously of considerable interest, but there are several reasons to suspect that this protein is not a native insulin-receptor subunit. Native receptor subunits associated with the receptor complex would be expected to be present in affinity-purified as well as immunoprecipitated preparations of insulin receptors. However, highly purified insulin receptors isolated on immobilized insulin appear to be devoid of this 40,000-dalton species.[20,21] In addition, analysis of immunoprecipitated insulin-receptor subunits on dodecylsulfate gels again indicates the absence of this band.[9,10,12–15] The lack of confirmation of the putative 40,000-dalton subunit by these alternate rigorous methods has prompted us to consider other explanations for this detection by photoaffinity labeling. Yip and colleagues have correctly argued that the 40,000-dalton species cannot be a fragment of the β subunit because the B1-monoazidobenzoyl-insulin analog that labels the 40,000-dalton species is not capable of photolabeling the 95,000-dalton β subunit.[33] Therefore, we suggest the possibility that the 40,000-dalton band may be derived from the α subunit by a proteolytic cleavage. In this regard, we have previously noted a 43,000-dalton receptor fragment derived from the α subunit apparently by proteolysis.[23] We consider this alternative more likely than the hypothesis that the 40,000-dalton species is a distinct native subunit of the insulin receptor.[33] Clearly, further data are required to elucidate the origin of this 40,000-dalton receptor species.

Another receptor subunit that has been recently observed is a 210,000-dalton species.[14] This protein is observed upon immunoprecipitation of insulin receptors from IM-9 lymphocytes and 3T3L1 adipocytes using anti-insulin receptor immunoglobulin.[34–36] A subunit with similar mass and properties has recently been labeled in intact cells with [^{35}S]methionine, [^{3}H]leucine, or [1,6-^{3}H]glucosamine.[34,35] However, this species, like the 40,000-dalton protein described above, is not observed upon purifying insulin-receptor complex using immobilized insulin.[21] Furthermore, this species has not been observed in any of the affinity-labeling experiments performed with a number of insulin analogs or with the combination of chemical crosslinker and [^{125}I]insulin. These negative data suggest that the 210,000-dalton receptor subunit may not bind [^{125}I]insulin or that it has very low affinity for the hormone. Interestingly, the 210,000-dalton receptor protein appears to contain peptides similar to both α and β subunits upon exhaustive tryptic hydrolysis.[14] Kasuga *et al.* proposed that this component may therefore represent a proreceptor species.[14] Recent data compiled by independent laboratories have indeed suggested that labeled amino acids incorporated into this receptor subunit could be chased into α and β subunits.[34–36] Taken together, these results suggest that the 210,000-dalton receptor subunit represents a precursor polypeptide containing both α and β subunits and that it may not bind insulin with high affinity before its conversion to the mature subunits.

III. SUBUNIT STOICHIOMETRY WITHIN THE INSULIN-RECEPTOR COMPLEX

A major advance that ultimately led to a viable hypothesis regarding the structure of the insulin-receptor subunit was the discovery that the α and β insulin-receptor subunits are linked by disulfide bonds in a high-molecular-weight complex.[1,6,37–39] The molecular weight of the native disulfide-linked insulin-receptor structure was estimated to be 350,000–400,000,[1,6,32] although this range represents only a rough approximation because of the difficulties in assessing molecular weights on SDS gels. Of further interest was the fact that affinity-purified insulin receptor or insulin receptor that was affinity crosslinked with [^{125}I]insulin and disuccinimidylsuberate could be partially reduced with low levels of reductant to yield partially reduced receptor fragments of intermediate molecular weights. For example, Fig. 2 depicts an experiment in which rat adipocyte membranes are treated with increasing concentrations of dithiothreitol followed by incubation with N-ethylmaleimide and affinity labeling of insulin receptor (lanes A–E). Incubation of the membranes with 10 mM dithiothreitol results in the formation of 210,000-dalton fragments (lane E) that can be further dissociated in the presence of SDS into α subunits (lane J) and β subunits (not visualized on the gels in this experiment due to less intense labeling of the α subunit). Thus the 210,000-dalton partially reduced receptor fragment was hypothesized to be α-S-S-β, and to compose one-half of the native receptor complex.

Closer examination of the high-molecular-weight native nonreduced receptor com-

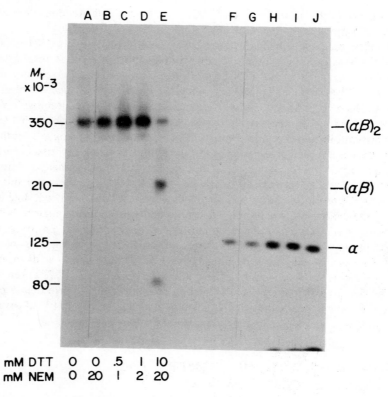

Figure 2. Effect of dithiothreitol on the reduction state of insulin receptors in membranes from rat adipocyte. Rat adipocyte membranes were incubated for 20 min at 23°C in Krebs–Ringer–phosphate buffer, pH 7.4 (1.2 mg/ml of membrane protein) in the presence of the indicated concentrations of dithiothreitol. Membranes were then pelleted at 10,000 × g for 5 min and resuspended in Krebs–Ringer–phosphate buffer, pH 7.4, containing the indicated concentrations of N-ethylmaleimide. After 20-min incubation under these conditions at 23°C, membranes were pelleted again, resuspended in Krebs–Ringer–phosphate buffer, pH 7.4, containing 1% bovine serum albumin. Affinity labeling of these membranes was performed by successive incubation with 10 nM [^{125}I]insulin and 0.2 mM disuccinimidyl suberate. (Lanes A–E) Affinity-labeled membranes were boiled in the presence of 1% sodium dodecylsulfate and electrophoresed (100 μg of membrane protein per lane) on 5% polyacrylamide gels. (Lanes F–J) Counterparts of samples on lanes A–E were boiled in the presence of 1% sodium dodecylsulfate and 10 mM dithiothreitol before electrophoresis. The autoradiogram was obtained from a fixed, dried gel. (From Massague and Czech.[38])

plex suggested that at least three forms could be resolved. As shown in Fig. 3B, affinity labeling of a crude membrane fraction with [^{125}I]insulin and crosslinker leads to visualization of distinct bands of 350,000, 320,000, and 290,000 daltons, respectively. The 350,000-dalton receptor species is often the major species labeled when intact cells are employed or when extreme caution is taken to prevent proteolysis during preparation of membranes. Figure 3 also demonstrates that the two lower-molecular-weight forms (320,000 and 290,000) can be generated from the highest-molecular-weight form (350,000) by proteolysis with elastase *in vitro*, suggesting that the lower-molecular-weight forms are not native insulin receptors.[2,23] Careful analysis of the highest-molecular-weight insulin-receptor complex affinity crosslinked by [125I]insulin and disuccinimidylsuberate using partial and complete reduction and sodium dodecylsulfate polyacrylamide gel electrophoresis (SDS-PAGE) reinforced the concept that (1) the insulin receptor consists of two receptor halves, each containing one α disulfide linked to one β subunit and (2) the two halves were also disulfide linked.[1,39] Jacobs and colleagues came to a similar conclusion using affinity-labeled insulin receptor subjected to a similar analysis.[19] Furthermore, we were able to deduce that the β subunit was the site of endogenous or elastase-mediated proteolysis of insulin receptor.[23] This hypothesis is depicted in Fig. 4. The detailed experiments that led to the

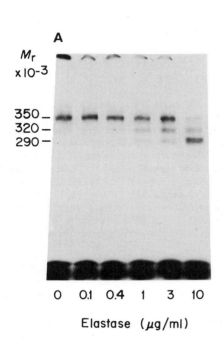 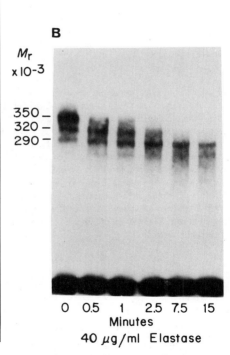

Figure 3. Generation of receptor heterogeneity by elastase. Intact adipocytes were labeled with [125I]insulin. (A) Affinity-crosslinked cells were treated with the concentrations of elastase indicated for 2 min, and proteolysis was stopped by diluting and washing the cells. A crude membrane fraction was prepared; this material was subjected to electrophoresis on a 5.0% acrylamide gel. (B) A crude membrane fraction was prepared directly after the cells were affinity-labeled. Portions of the membrane were boiled for 1 min in sample buffer and then treated with elastase (40 μg/ml) for the times indicated prior to a second 1-min boil. Electrophoresis was performed on a 5.0% acrylamide gel; autoradiograms show the stained, dried gels. (From Massague *et al.*[23])

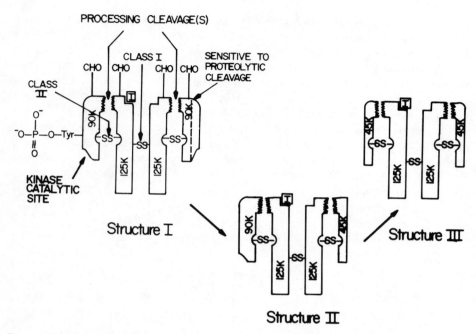

Figure 4. Proposed minimum covalent subunit structure of the native insulin receptor and two major proteolytically cleaved forms. Structure I presents the hypothetical structure of the native insulin receptor. This structure consists of α and β subunits disulfide-linked in a heterotetrameric complex. The proposed location of class I disulfides links the two symmetrical halves of the receptor; these disulfide(s) are extremely sensitive to reductant.[1,39] The α and β subunits are believed to be synthesized as one continuous polypeptide chain, which is then cleaved during processing to yield the mature α and β subunits.[34–36] The β subunit is phosphorylated upon insulin activation in intact cells or in detergent extracts[24,25] and appears to contain one or more tyrosine kinase catalytic sites.[29] The β subunit contains a region at about the center of its amino acid sequence that is extremely sensitive to elastaselike proteases.[23] Cleavage of one β subunit within the structure yields structure II, while cleavage of both β subunits yields structure III. These latter proteolytically cleaved structures are often major constituents of the total receptor population in isolated membrane fractions.

deduction of the stoichiometry and structure of this subunit were previously reviewed.[38]

Recent improvements in methodology for affinity-chromatography of insulin receptor, particularly in respect to elution of purified insulin receptor from immobilized insulin, have permitted more rigorous testing of the hypothesis represented in Fig. 4, using nanogram to microgram quanitities of highly purified receptor.[20,21] With such purified insulin-receptor preparations, insulin-receptor subunits can be analyzed and visualized by direct staining techniques rather than by reliance on labeling. Such preparations of highly purified insulin receptor retain their ability to bind [^{125}I]insulin with high affinity and exhibit insulin-activated tyrosine kinase activity.[29] Electrophoretic analysis of the highly purified insulin-receptor complex on nonreduced SDS-PAGE demonstrated the presence of three major high-molecular-weight species of

320,000, 300,000, and 275,000 M_r as well as some lower-molecular-weight minor species.[21] Upon electrophoresing these three major receptor species in a second dimension in the presence of reductant, Fujita-Yamaguchi demonstrated that the 320,000-M_r receptor is composed of α and β subunits, the 300,000-M_r receptor is composed of α, β, and a 45,000-M_r protein, and the 275,000-M_r species contains α subunit and the 45,000-M_r species but no β subunit.[21] Upon peptide mapping, the 45,000-dalton receptor protein was shown to be homologous to the β subunit, as would be expected for β_1, while the α and β subunits are quite different in structure. Fujita-Yamaguchi[21] concluded that these results were fully compatible with the model depicted in Fig. 4 and with previous results from our laboratory using affinity crosslinking methodology.[1,23,39]

In recently performed experiments similar to those described above, we have observed similar results. Highly purified insulin receptor was prepared according to the method described by Fujita-Yamaguchi,[21] and the preparation was electrophoresed on a 5% polyacrylamide gel in the absence of reductant. Three major high-molecular-weight silver-stained bands were visualized, corresponding almost identically to the three receptor forms we previously identified by affinity labeling. When each of these three nonreduced major receptor forms is analyzed in the presence of reductant, the predicted subunit compositions are observed: (1) the highest-molecular-weight, native receptor structure is composed of only α and β subunits; (2) the lowest-molecular-weight receptor species is composed of only α and β_1 subunits; and (3) the intermediate, 320,000-M_r receptor form is composed of all three subunit types, consistent with its proposed asymmetric structure (receptor form II in Fig. 4). At this point, we cannot eliminate the possibility that other noncovalently linked subunits also exist.

We and others[3,14,21,39] have observed other oligomeric forms of the insulin receptor when it is electrophoresed in nonreducing gels, usually in minor amounts. For example, free α-S-S-β and α-S-S-α and α and β subunits have all been observed in nonreduced gels. It is important to note that these apparent receptor fragments may not reside as free units within the membrane. Experiments in our laboratory have demonstrated that reduction of the class I disulfides in the native receptor complex (yielding α-S-S-β fragments when analyzed by SDS-PAGE) does not actually dissociate the receptor complex in mild detergent solution.[38,39] Thus, the native intact heterotetrameric receptor complex remains associated with noncovalent bonds upon receptor reduction in the native membrane. In intact cells or membranes, a small but significant fraction of receptor complexes may be in a partially reduced state, although the receptor complex remains intact. Alternatively, it is possible that small amounts of receptor forms other than the I form shown in Fig. 4 reside in cell membranes, but this hypothesis remains to be demonstrated.

Recent experiments from several laboratories have convincingly documented a major receptor conformation change in response to insulin binding.[40,41] Experiments in our laboratory demonstrated a marked increase in sensitivity of the α insulin-receptor subunit to exogenous protease action upon binding hormone.[40] This increased protease digestion of the α subunit occurred in intact cells as well as isolated membranes. More recently, Maturo et al.[42] were able to detect a dramatic change in the chromatographic profile of the insulin receptor in detergent solution when bound to insulin. The insulin-activated structural change confers an apparent lower Stokes radius to the receptor, an observation previously reported by Ginsberg et al.[43] Krupp and Livingston have also reported apparent changes in electrophoretic mobility of the insulin receptor upon its binding to hormone.[44] Taken together, these data suggest that a marked alteration in receptor structure is elicited by insulin binding; it seems possible that this conformational change may be related to the activation of tyrosine kinase activity. This latter point remains to be documented.

It is important to note that the insulin-receptor structure appears to be homologous with at least one other receptor species—the high-affinity receptor for insulinlike growth factor I. Several laboratories have demonstrated that this growth factor receptor species contains subunits remarkably similar to those of the insulin receptor in respect to apparent molecular weight[45,46] and that this receptor has a similar apparent subunit stoichiometry.[45–49] This receptor species exhibits highest affinity for insulinlike growth factor I but also binds insulin with low affinity. Like the insulin receptor, the insulinlike growth factor I receptor also appears to contain two classes of intramolecular disulfides.[47] In addition, its β subunit, which is autophosphorylated in response to ligand binding, as is the insulin β-receptor subunit,[49,50] contains a site sensitive to proteases about midway into the amino acid sequence.[47] Interestingly, small variations in affinity of the insulinlike growth factor I receptor among various tissues suggest the possibility that multiple such receptors exist with subtle structural differences. An important future objective will be to sort out the structural basis of the apparent homology between the insulin receptors and the insulinlike growth factor I receptors.

IV. REGULATION OF INSULIN-RECEPTOR KINASE BY TYROSINE PHOSPHORYLATION

The discovery by Kasuga, Kahn, and colleagues that insulin receptor is phosphorylated in intact cells in response to insulin binding[24,25] and in purified form is associated with insulin-activated tyrosine kinase activity[29] has prompted intense study of the insulin receptor as an enzyme. Clearly, a most important issue for future work relates to the physiological role of this property of the insulin receptor. In our laboratory, we have focused our attention on mechanisms that might regulate the insulin receptor tyrosine kinase activity and that therefore may be important in modulating at least some of the biological functions of the insulin receptor. This section reviews results related to two such receptor regulatory mechanisms: (1) the ability of tyrosine phosphorylation of the insulin receptor β subunit to activate the receptor-associated tyrosine kinase activity, and (2) the ability of cAMP in intact fat cells to desensitize or uncouple the insulin-activated tyrosine kinase activity.

Studies by Zick et al.[51] and Rosen et al.[52] have suggested that insulin receptor autophosphorylated by incubation with ATP is more active in catalyzing phosphorylation of exogeneous substrates on tyrosine residues. Studies in our laboratory show a close association between ^{32}P incorporated into the receptor β subunit during an autophosphorylation period using $[\gamma\text{-}^{32}P]$-ATP and the subsequent activity of the insulin receptor kinase using histone as substrate (Fig. 5). Rosen et al.[52] further discovered

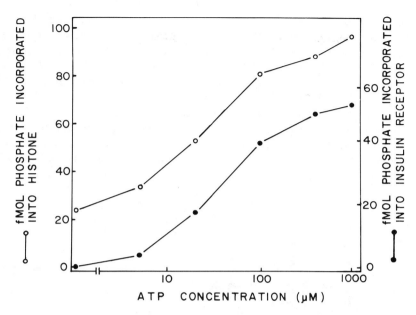

Figure 5. Relationship between the phosphorylation of the insulin receptor β subunit and its associated kinase activity. Insulin receptor-bearing insulin-agarose preparations were incubated with different concentrations of unlabeled or [γ-^{32}P]-ATP for 1 hr at 22°C. The preparations were washed extensively to remove the unreacted ATP. Samples incubated with [γ-^{32}P]-ATP were boiled in 0.2 ml of electrophoresis buffer containing sodium dodecylsulfate (SDS) and dithiothreitol DTT and electrophoresed on an 8% gel. The ^{32}P-labeled band corresponding to the β subunit of the insulin receptor was excised, and radioactivity was determined by liquid scintillation counting. Samples incubated with unlabeled ATP were used to assay the receptor-associated kinase activity using histone as substrate and 5 μM [γ-^{32}P]-ATP. The assay period was 1 hr. (From Yu and Czech.[53])

that phosphorylated insulin receptor, in the absence of insulin, appears to remain fully active in its ability to phosphorylate exogeneous protein substrates. Thus, dissociation of insulin from the phosphorylated receptor does not decrease its associated kinase activity, nor does a second addition of insulin further stimulate the tyrosine kinase activity. The insulin dependence of the receptor kinase activity could be restored upon dephosphorylation of insulin receptor with alkaline phosphatase. These results demonstrated that phosphorylated insulin receptor exhibits markedly elevated tyrosine kinase activity independent of the presence of insulin. This may be an important mechanism in intact cells for amplifying the activation of insulin receptor kinase by insulin. Furthermore, the data suggest that deactivation of insulin receptor kinase activity may involve both dissociation of insulin and dephosphorylation of insulin receptor by cellular phosphatases.

Studies in our laboratory were designed to determine whether specific phosphorylation sites on the insulin receptor are linked to the activation of receptor tyrosine kinase activity. In order to determine this, receptors phosphorylated with various concentrations of [γ-^{32}P]-ATP were exhaustively trypsinized and the resulting phosphopeptides were resolved on high-pressure liquid chromatography (HPLC) using a propanol gradient.[52] Three major peaks of receptor phosphopeptides are resolved by HPLC, and the corresponding fractions can be analyzed for phosphoamino acids after hydrolysis in HCl. Figure 6A depicts the amino acids phosphorylated in three HPLC fractions containing receptor phosphopeptides after the trypsinization. Insulin receptor phosphorylated with 5, 20, and 400 μM [γ-^{32}P]-ATP exhibited serine and threonine as well as tyrosine [^{32}P]phosphate, although the relative distribution of these phosphoamino acids among the three receptor phosphopeptide fractions was quite different (Fig. 6A). Phosphorylation of serine, threonine, and tyrosine in the receptor phosphopeptides of fractions 2 and 3 all increased with increasing concentrations of [γ-^{32}P]-

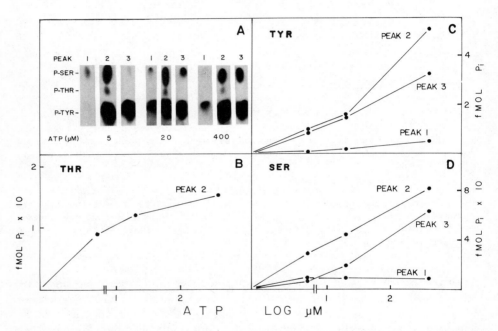

Figure 6. Analyses of the phosphoamino acid contents in HPLC peaks 1, 2, and 3 and the dose-dependent response of each phosphoamino acid residue to increasing concentrations of ATP. Insulin–agarose preparations containing adsorbed insulin receptor were prepared. The receptor preparations were phosphorylated at 22°C for 1 hr with 5, 20, and 400 μM of [γ-^{32}P]-ATP with specific activities of 1600, 400, and 200 μCi/nmole, respectively. The samples were then washed, reduced, and alkylated and were electrophoresed on dodecylsulfate gels. The gel pieces containing the ^{32}P-labeled β subunit were incubated with N-tosyl-L-phenylalanyl chloromethyl ketone-treated trypsin for a total of 20 hr at 37°C. The tryptic digest was loaded on a reverse-phase HPLC C-18 column, and the ^{32}P-labeled tryptic peptides were eluted in a linear 0–40% 1-propanol gradient. The radioactivity in each fraction was determined by Cherenkov counting. Three labeled peaks were obtained. Only one-half the sample at 5 μM ATP was used for the analysis. After hydrolysis in HCl as well as electrophoresis and autoradiography, the phosphoamino acid spots were excised and their corresponding radioactivity determined by Cherenkov counting. (A) Autoradiogram of the ^{32}P-labeled phosphoamino acids in peaks, 1, 2, and 3. (B–D) Respective dose-dependent phosphate incorporation into the threonine, tyrosine, and serine residues in each peak to increasing concentrations of ATP. (From Yu and Czech.[53])

ATP (Fig. 6). Although incorporation of ^{32}P into phosphotyrosine is 5 to 50 times higher than incorporation into serine or threonine, detectable levels of the latter two phosphoamino acids are clearly observed under our experimental conditions. Thus, it was possible to conclude from this experiment that phosphorylation of serine, threonine, and tyrosine in phosphopeptides of peaks 2 and 3 from the insulin receptor correlate with activation of the receptor tyrosine kinase activity. Nevertheless, it is possible that serine and threonine phosphorylation is mediated by the presence of contaminating kinases in the receptor preparation.

A more definitive correlation between receptor phosphorylation at specific sites and activation of the receptor kinase was achieved by treatment of phosphorylated receptor with alkaline phosphatase at 15°C.[53] Insulin receptor immobilized on insulin agarose was phosphorylated with 400 μM [γ-^{32}P]-ATP at 22°C for 1 hr and washed free of the labeled ATP. In order to achieve partial dephosphorylation, bovine alkaline phosphatase was added and, after an appropriate incubation period, was also washed away from the immobilized insulin receptor. The insulin-receptor preparation was then assayed for its ability to catalyze tyrosine kinase activity using histone as substrate. The receptor phosphorylation sites that were dephosphorylated by the bacterial alkaline phosphatase under these conditions were also assayed, using HPLC. Dephosphorylation of the insulin receptor with alkaline phosphatase resulted in a 60–70% decrease in receptor kinase activity as well as a similar decrease in the receptor [^{32}P]tyrosine content.[53] When the partially dephosphorylated insulin receptor preparation that exhibited the markedly decreased kinase activity was exhaustively trypsinized and the phosphopeptides resolved on HPLC, approximately 90% of the dephosphorylation due to alkaline phosphatase treatment was found to occur in peak 2 (Table 1). Dephosphorylation of the insulin receptor under these conditions could be completely attributed

Table 1. Quantitation of the Phosphoamino Acid Contents in the Tryptic Domains of the β Subunit of the Insulin Receptor with or without Alkaline Phosphatase Treatment[a,b]

Alkaline phos- phatase	Phospho- amino acid	Peak 1 (cpm)	Decrease due to phos- phatase (cpm)	Peak 2 (cpm)	Decrease due to phos- phatase (cpm)	Peak 3 (cpm)	Decrease due to phos- phatase (cpm)
−	Ser	35	—	425	—	83	—
+	Ser	24	11	421	4	80	3
−	Thr	—	—	139	—	—	—
+	Thr	—	—	133	6	—	—
−	Tyr	173	—	2387	—	545	—
+	Tyr	121	52	960	1427	417	138

[a] The pooled fractions of HPLC peaks 1, 2, and 3 from autophosphorylated receptor or autophosphorylated receptor treated with alkaline phosphatase were lyophilized, and their phosphoamino acid contents were analyzed. Following the excision of the phosphoamino acid spots from the chromatogram, the radioactivity in each was determined by Cherenkov counting. Similar results were obtained when the analysis was performed on the peak fractions from another similar experiment with or without alkaline phosphatase treatment.
[b] Reprinted with permission from Yu and Czech.[53]

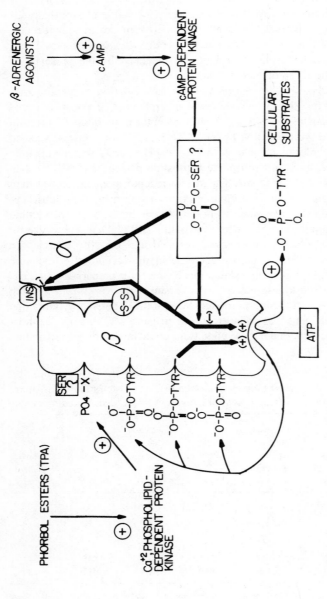

Figure 7. Regulatory mechanisms that modulate insulin-receptor binding and tyrosine kinase activities. For the sake of simplicity in this schematic representation, the insulin receptor is depicted as one α and one β subunit. Insulin binding to this complex activates an associated tyrosine kinase activity that may be present on the β subunit; this activity catalyzes phosphorylation of β subunit as well as exogenous substrates. Autophosphorylation of specific tyrosine residue(s) further activates the tyrosine kinase activity and releases its dependence on insulin binding. Dephosphorylation of these tyrosine phosphate residue(s) deactivates the receptor tyrosine kinase activity. Increased cAMP levels in response to isoproterenol or other hormones mediate, presumably through activation of the cAMP-dependent protein kinase, rapid inhibition of insulin binding to the receptor and an uncoupling of the insulin activation of receptor tyrosine kinase activity in isolated fat cells. Phorbol diesters mediate the phosphorylation of insulin receptor in intact cells by a mechanism that may involve activation of C-kinase. This protein kinase may either directly (as shown) or indirectly catalyze phosphorylation of the insulin receptor. The physiological effect of phorbol diester-mediated receptor phosphorylation is unknown, but it may have effects on either insulin binding or tyrosine kinase activity, or both.

to the loss of tyrosine phosphate with little or no effect on serine or threonine phosphates (Table 1). Dephosphorylated receptor was capable of autophosphorylation upon addition of $[\gamma^{-32}P]$-ATP. These data indicate that in a specific receptor phosphopeptide fraction the tyrosine phosphate sites are closely associated with activation of the receptor-associated tyrosine kinase activity.

We cannot at present eliminate the possibilities that other tyrosine phosphorylation sites (e.g., those present in HPLC peaks 1 and 3) are also involved in receptor kinase activation or that a combination of phosphorylations may play a role in modulating the kinase activity. However, the data suggest that one or more phosphotyrosine site on the receptor that fractionate with the phosphopeptide(s) in peak 2 are the most likely candidates for a regulatory role in insulin receptor kinase activation. Thus, as depicted schematically in Fig. 7, our current hypothesis is that binding of insulin to its receptor results in an activation of the associated tyrosine kinase activity, which leads to rapid autophosphorylation of the insulin receptor β subunit. Autophosphorylation of specific tyrosine residues on this subunit leads to marked further activation of the insulin receptor kinase and a release from its dependence on insulin binding. Dephosphorylation of these tyrosine phosphate sites causes deactivation of the receptor kinase and renews its dependence on insulin binding.

V. INSULIN RECEPTOR REGULATION BY OTHER PROTEIN KINASES

A. Metabolic Antagonism between cAMP and Insulin Action

Since its discovery 25 years ago as a glycogenolytic intracellular mediator in liver,[54,55] cAMP has been recognized as a ubiquitous regulatory molecule controlling diverse metabolic processes in both prokaryotic and eukaryotic cells. In animals, its synthesis is catalyzed almost exclusively by the hormone-sensitive adenylate cyclase system, the activity of which is regulated by a variety of hormones and ligands.[56] The principal role of cAMP as an intracellular second messenger in the hormonal regulation of cellular metabolism is via the activation of the cAMP-dependent protein kinases.[57,58] All the diverse effects of cAMP on intermediary metabolism are thought to be a direct consequence of the phosphorylation of specific regulatory enzymes in these pathways.[59,60]

It has been well documented that the action of agents that elevate cAMP levels (e.g., catecholamines and glucagon) have an antagonistic influence on several insulin-responsive cellular enzyme systems such as glycogen synthase and glycogen phosphorylase in muscle,[61,62] liver,[63,64] and adipocytes[65,66] and hormone-sensitive lipase in adipocytes.[67,68] It therefore appeared appropriate to investigate whether insulin might act by altering the intracellular levels of this cyclic nucleotide. In this regard, insulin was found to inhibit the rise in cAMP levels due to submaximal concentrations of hormones opposing insulin action, such as catecholamines and glucagon in liver[69–71] and adipose tissue.[72,73] Soderling[74] reported that insulin markedly inhibited the epinephrine activation of the cAMP-dependent protein kinase in adipose tissue which was associated with a decrease in the relative amount of catalytic subunit. Insulin has also been reported to inhibit the hormone-stimulated adenylate cyclase activity in liver plasma membranes and adipocyte homogenates,[75,76] although several groups have failed to support this claim.[77–80] Others have demonstrated that the ability of insulin to lower cAMP levels may be a result of an increase in the insulin-sensitive low K_m cAMP phosphodiesterase activity.[81–85] It is interesting to note that in fat cells the stimulatory effect of insulin on phosphodiesterase activity is dependent on the presence of extracellular Ca^{2+},[86] whereas the antilipolytic action of insulin is not. More recently, Weber and Appelman[87] have presented evidence for a specific form of insulin-sensitive cyclic nucleotide phosphodiesterase distinguishable from three other insulin-insensitive forms of the enzyme. The activation of the cAMP phosphodiesterase has been correlated with the conversion of a dephosphorylated membrane form to a phosphorylated one.[88] Several groups have reported that insulin treatment of rat adipocytes or liver plasma membranes releases some low-molecular-weight material(s) (1000–1500 M_r) that are capable of inhibiting adenylate cyclase activity,[89] of activating the cAMP phosphodiesterase,[90] and of lowering cAMP levels in intact rat adipocytes.[91,92]

Although appealing, the concept that the decrease in intracellular cAMP levels mediates most actions of insulin in extrahepatic tissues has not been generally demonstrated.[93] Insulin has been shown to have no effect[94–98] or to even increase[99,100] cAMP levels in muscle or adipocytes while still activating glycogen synthase in muscle[98,99] or inhibiting lipolysis in adipocytes.[94–98] Claus and Pilkis[101] found that in the absence of added free Ca^{2+} in the medium, insulin action on gluconeogenesis was unaffected, while its effect on cAMP levels was abolished. The concept that intracellular levels of cAMP mediate all the actions of insulin, even in liver, therefore seems untenable, and unequivocal evidence that cAMP functions solely in mediating any of the effects of insulin is lacking. Nevertheless, substantial evidence has accumulated that the mutual antagonistic actions of insulin versus hormones that elevate intracellular cAMP levels ultimately result in alterations in the phosphorylation state of key intermediary metabolism enzymes.[59–61] For example, cAMP-dependent protein kinase phosphorylation of glycogen synthase and phosphorylase kinase results in the inhibition of the former and activation of the latter enzyme. In contrast, insulin causes dephosphorylation of these enzymes, resulting in the activation of glycogen synthase and inhibition of phosphorylase kinase. The mechanism by which insulin induces the decrease in the phosphorylation state (e.g., decreased kinase activity or increased phosphatase activity) of these and several other regulatory enzymes is under intense investigation and has not been established to date.

B. cAMP Action on [^{125}I]Insulin Binding to Insulin Receptor

Although numerous experiments over the past 20 years have been carried out to examine the effects of insulin on the intracellular levels of cAMP, it was surprising to discover that until recently no studies on the effects of elevated cAMP levels on the insulin receptor had been reported.[102,103] If one considers the binding of ligand to a cell-surface receptor as the commitment step in a metabolic pathway, the regulation of receptor binding activity should be expected. The known antagonism between the actions of cAMP and insulin, as described in the previous section, leads to the hypothesis that cAMP might act to desensitize or inactivate the insulin receptor. Regulation of hormone receptors by heterologous hormones has been previously described for a number of different hormone receptor systems.[104–107] Glucocorticoids have been recently shown to increase β-adrenergic receptors in human lung,[108] epidermal growth factor (EGF) receptors in human foreskin fibroblasts,[109] insulin receptors in IM-9 lymphocytes,[110] and 3T3-C2 fibroblasts[111] and to inhibit the expression of Fc reulocyte cell line HL-60,[112] as well as insulin receptors in 3T3-L1 fibroblasts[113] and adipocytes.[114] All these effects of steroid hormones have been shown to require relatively long periods of exposure to hormone and are dependent on protein synthesis. Other receptor–receptor regulations can occur extremely rapidly, within several minutes at 37°C, and are apparently independent of protein synthesis. Examples include insulin activation of the IGF-II receptor,[115,116] thyrotropin-releasing hormone (TRH) activation of somatostatin C binding,[117] and relaxin activation of insulin binding.[118] Rapid inhibitions of a receptor function by a second receptor system have been demonstrated for inhibition by platelet-derived growth factor (PDGF), fibroblast-derived growth factor, and vasopressin of EGF binding[119–123] as well as phorbol ester inhibition of both insulin[124,125] and EGF[126–130] binding. On the basis of these considerations, we reasoned that, in a "negative" control sense, catecholamines may rapidly regulate the physiological responses of cells to insulin (or other antagonistic growth factors) by inhibiting insulin-receptor binding or insulin receptor signaling properties, or both.

We[102] and others[103] have recently documented that indeed incubation of isolated rat adipocytes with hormones known to elevate cAMP levels (isoproterenol, norepinephrine, epinephrine, and soterenol) rapidly inhibit the binding of [^{125}I]insulin and [^{125}I]-EGF by approximately 50% (Table 2). Isoproterenol and soterenol have a relatively high specificity for the β-adrenergic receptor,[131,132] whereas norepinephrine and epinephrine have affinity for both α- and β-adrenergic receptor systems.[131] In order to demonstrate that these inhibitions of receptor binding activities are specifically related to elevated intracellular cAMP levels induced via the β-adrenergic receptor, the effects of a variety of pharmacological agents were investigated. Propranolol and alprenolol, which are specific β-receptor antagonists,[131] had no effect on [^{125}I]insulin and [^{125}I]-EGF binding and completely blocked the isoproterenol-induced inhibition of binding. Phenylephrine (an α-receptor agonist) and phentolamine (an α-receptor antagonist) also had no effect on peptide hormone binding, and phentolamine was unable to block the isoproterenol-induced inhibition of binding (Table 2). A cAMP phosphodiesterase inhibitor, 3-isobutyl-1-methylxanthine,[133] was as potent as the β-receptor agonists in promoting the binding inhibition. Adenosine is a potent inhibitor of adenylate cyclase in rat adipocytes[134] and was found capable of completely antagonizing the isoproterenol-induced decrease in binding. Furthermore, dibutyryl cyclic adenosine monophosphate (Bt$_2$cAMP) mimicked the decrease in insulin and EGF binding, whereas dibutyryl cyclic guanosine monophosphate (Bt$_2$cGMP) did not.

Lipolytic agents that raise cAMP levels in fat cells have also been shown to either decrease[135,136] or have no effect[95] on the intracellular concentration of ATP. To exclude the possibility that the inhibition of binding was a result of decreased ATP levels, adipocytes were incubated both in the presence and in the absence of isopro-

Table 2. Effect of Various Agents on the Binding of Insulin and
EGF to the Rat Adipocyte[a,b]

	Percent inhibition of binding	
Addition	Insulin	EGF
None	0	0
Isoproterenol (1 μM)	48 ± 15	46 ± 9
Norepinephrine (1 μM)	51 ± 17	48 ± 4
Epinephrine (1 μM)	47 ± 17	44 ± 2
Soterenol (1 μM)	53 ± 15	52 ± 10
Propranolol (10 μM)	0 ± 6	12 ± 5
Propranolol (10 μM) + isoproterenol (1 μM)	16 ± 12	16 ± 2
Alprenolol (10 μM)	0 ± 6	3 ± 3
Alprenolol (10 μM) + isoproterenol (1 μM)	2 ± 3	5 ± 4
Phenylephrine (10 μM)	6 ± 3	1 ± 1
Phentolamine (50 μM)	3 ± 3	7 ± 12
Phentolamine (50 μM) + isoproterenol (1 μM)	50 ± 12	54 ± 8
Mix (200 μM)	66 ± 3	66 ± 5
Mix (200 μM) + isoproterenol (1 μM)	72 ± 6	70 ± 6
Adenosine (1 mM)	0 ± 15	5 ± 7
Adenosine (1 mM) + isoproterenol (1 μM)	10 ± 5	15 ± 16
Bt_2cAMP	44 ± 8	25 ± 6
Bt_2GMP	10 ± 1	9 ± 9
Sodium butyrate (1 mM)	1 ± 14	2 ± 12
Sodium oleate (1 mM)	0 ± 10	6 ± 6

[a] Isolated rat adipocytes were pretreated with various concentrations of agents for 30 min at 37°C and the percentage inhibition of binding at 1.0 nM [^{125}I]insulin and [^{125}I]-EGF was determined as previously described.[104]
[b] Reprinted with permission from Pessin et al.[102]

terenol in medium containing 0.5% bovine serum albumin (BSA), or 4% BSA plus 20 mM D-glucose or D-fructose. These latter conditions prevented the associated decrease in ATP levels[137,138] and had no effect on the catecholamine-regulated loss of insulin and EGF receptor binding activities.[102] These results clearly demonstrate that in rat adipocytes, β-adrenergic receptor agonists acting via elevated levels of cAMP specifically inhibit [^{125}I]insulin and [^{125}I]-EGF binding as a direct action that regulates the peptide hormone receptors in parallel with other cellular functions.

The catecholamine-induced inhibition of insulin binding would be expected to manifest itself physiologically by decreasing the sensitivity of the cell to insulin action. In fact, catecholamines have been shown to have marked diabetogenic properties not only by decreasing insulin secretion,[139] but by inducing a peripheral resistance to insulin as well.[140,141] Figure 8 shows the effects of isoproterenol on insulin action in vitro under conditions in which insulin-receptor activity is decreased. As has been previously established, insulin stimulates insulinlike growth factor II (IGF-II) binding[115,116] and D-glucose transport[142] in isolated rat adipocytes. The insulin

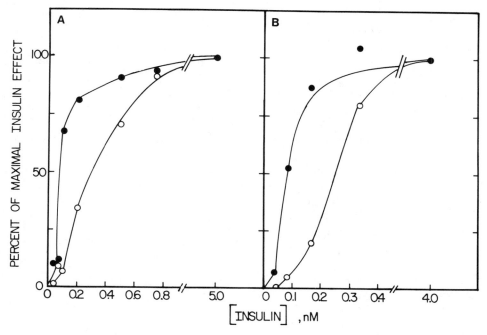

Figure 8. Effect of isoproterenol treatment on the insulin stimulation of IGF-II binding and D-glucose transport. Adipocytes were incubated for 30 min at 37°C in the absence (●) or presence (○) of 1.0 μM isoproterenol. The cells were then incubated for an additional 15 min with the indicated concentration of insulin. The binding of 1.0 nM IGF-II (A) and uptake of 3-O-methylglucose (B) were determined as previously described.[102] The data are normalized to the maximum stimulation found at saturating insulin concentrations in the control cells. (From Pessin *et al.*[102])

dose–response relationship with respect to IGF-II binding and glucose transport indicates that half-maximal activation occurs at ~0.1–0.3 nM insulin (Fig. 8). Isoproterenol treatment of cells shifted the dose–response curve approximately threefold to the right, as compared with untreated cells without any alteration in the basal or maximally stimulated level of transport or [^{125}I]-IGF-II binding activity. Thus, under these conditions, catecholamines appear to desensitize the adipocyte to the metabolic actions of insulin at submaximal concentrations, while at maximal insulin concentrations normal responsiveness is observed.

C. cAMP-Mediated Regulation of Insulin Receptor Tyrosine Kinase Activity

We recently extended our observations that isoproterenol induces inhibition of insulin binding in isolated rat adipocytes by examining the effects of elevated cAMP levels on the autophosphorylating activity of the insulin receptor. We first developed a methodology to preserve the isoproterenol-induced inhibition of insulin-binding activity in cell-free, partially purified insulin-receptor preparations.[143] Thus, insulin receptors, partially purified from membrane extracts by immobilization and elution from wheat germ agglutinin-agarose, retain decreased [125I]insulin-binding activity when prepared from isoproterenol-treated adipocytes. This inhibition of insulin binding does not appear to occur as a result of a change in the number of insulin receptors based on estimates using specific anti-insulin receptor antibody.[143] This conclusion is also consistent with the fact that at very high [125I]insulin concentrations, no change in insulin binding is observed due to isoproterenol action. Therefore, the effect of cAMP on the insulin receptor appears to reflect an alteration in either some property of the structure of the insulin receptor itself or in a regulatory component.

Under conditions in which the relative quantity of insulin receptors isolated from control and isoproterenol-treated cells was the same, we examined the basal (in the absence of insulin) and the hormone-stimulated (in the presence of saturating insulin concentrations) autophosphorylating activity of insulin receptors (Fig. 9). This experiment demonstrates that isoproterenol treatment of the intact cells induces an inhibition of the insulin-stimulated receptor autophosphorylation reaction in partially purified insulin receptor in vitro when incubated with [γ-^{32}P]-ATP. When insulin is not added to the insulin receptor preparation, no significant effect on the basal phosphorylating activity of the receptor due to isoproterenol is observed (Fig. 9, left four lanes). In the experiment depicted in Fig. 9, the autoradiograms showing basal phosphorylations (left four lanes) were exposed for longer periods than the insulin-induced phosphorylations and thus appear more intense. In fact, insulin stimulated receptor autophosphorylation by 20- to 30-fold in these experiments. The inhibition of insulin-stimulated autophosphorylation of insulin receptor β subunit in vitro by isoproterenol treatment of intact cells was about 50%. Thus, isoproterenol treatment appears to desensitize the cells to insulin action by inhibiting insulin binding and by partiallly uncoupling insulin binding from the activation of the intrinsic kinase activity of the insulin receptor.

The isoproterenol-induced inhibition of insulin-stimulated receptor kinase activity is analogous to the heterologous catecholamine-induced desensitization of the β-adrenergic receptor-linked adenylate cyclase system.[144-148] However, unlike the hormone-sensitive adenylate cyclase system, no evidence as yet exists demonstrating that insulin-stimulated kinase activity of the insulin receptor is a necessary event in biological responsiveness to insulin action.

The above results demonstrating the dramatic effects of isoproterenol on the insulin receptor in fat cells suggest the hypothesis, depicted in Fig. 7 that one or more cAMP-mediated serine phosphorylations lead to at least two actions. First, insulin receptors are inhibited in their binding to insulin by an inactivation process that greatly decreases receptor affinity for the hormone. The fact that [125I]insulin binding to receptors from isoproterenol-treated cells is the same as that in control cells when [125I]insulin is present at very high concentrations suggests an affinity change in the insulin receptor caused by isoproterenol. Interestingly, even when this apparent affinity change is overcome by very high concentrations of insulin such that no difference in binding is observed due to the isoproterenol, the activation of tyrosine kinase activity is markedly depressed. Thus, the putative serine phosphorylation or phosphorylations due to activation of cAMP-dependent protein kinase appears to alter insulin receptor structure

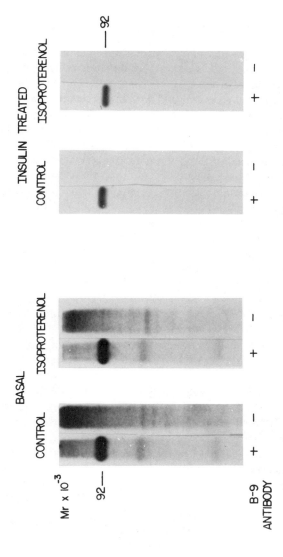

Figure 9. Effect of isoproterenol treatment on the autophosphorylating activity of the insulin receptor. Isolated rat adipocytes were incubated in the presence and absence of isoproterenol. The cells were washed and homogenized, and a crude plasma membrane fraction was prepared by differential centrifugation. The membranes were solubilized and the insulin receptors were partially purified by wheat germ agglutinin–Sepharose affinity chromatography as described by Harrison and Itin.[15] The insulin receptors were then incubated for 30 min at 20°C in the absence or in the presence of 200 nM insulin. The *in vitro* autophosphorylation reaction was then initiated by the addition of 5 μM [γ-^{32}P]-ATP (50 μCi/nmole) for 10 min at 4°C, and the insulin receptors were immunoprecipitated as described. The autoradiogram of the receptor phosphorylations in the absence of insulin (basal) was exposed longer than that for the plus insulin conditions. In fact, insulin stimulated the autophosphorylation 20- to 30-fold in these experiments.

such that its response to insulin binding is dampened and subsequent tyrosine kinase activation is decreased. Whether the hypothetical serine phosphate site(s) might reside on the insulin receptor itself or are present on other regulatory proteins is not known.

D. Phorbol Ester Actions on Insulin Receptors

The tumor-promoting phorbol diesters are agents that are nontumorigenic by themselves but that will induce tumors when administered after subthreshold doses of carcinogens.[149] The most potent phorbol diester, 12-0-tetradecanoyl phorbol-13-acetate (TPA), has been studied in many systems and has been shown to induce a myriad of *in vitro* biochemical changes. In the context of this discussion, a most notable effect is the inhibition of insulin-receptor binding activity.[124,125] Many cells contain a high-affinity phorbol ester receptor with a K_d in the nanomolar range.[150,151] It was recently demonstrated that TPA can directly activate a partially purified preparation of the Ca^{2+}, phospholipid-dependent protein kinase[152] and that the phorbol receptor co-purifies with the kinase activity during partial purification by ammonium sulfate precipitation, DEAE-cellulose, and gel-filtration chromatography.[150] This kinase was also shown to have an absolute requirement for divalent cations, preferentially Ca^{2+}.[153] Diacylglycerol dramatically increases the apparent affinity of the enzyme for Ca^{2+} and phospholipid, causing a marked activation of the kinase at micromolar Ca^{2+} concentrations.[154] TPA appears to substitute for diacylglycerol, thereby increasing the affinity of the enzyme for Ca^{2+} and phospholipid.

Treatment of several human tissue culture cell lines (IM-9 lymphocytes, U-937 macrophagelike cells, and HL-60 promyelocytic leukemia cells) with TPA has been shown to cause rapid inhibition of insulin-receptor binding activity.[124,125] This inhibition of binding appears to occur by a decrease in receptor affinity and not by an alteration in the number of cell-surface insulin receptors. The characterization (i.e., time, temperature, reversibility) of the TPA-induced inhibition of insulin binding in these tissue culture cells is similar to that which we have reported for the catecholamine-induced inhibition of insulin binding in isolated rat adipocytes. Recently, Jacobs *et al.*[155] demonstrated that TPA treatment of IM-9 lymphocytes labeled with ^{32}P enhances the phosphorylation of the insulin receptor β subunit approximately fourfold.

Although no experimental evidence has been presented demonstrating that the TPA-induced increase in receptor phosphorylation correlates directly with the decrease in receptor-binding activity previously reported,[124,125] such a hypothesis is quite appealing on the basis of the current evidence. Recent studies have indicated that EGF receptor is a target of phorbol diester-mediated phosphorylation and can serve as a substrate for the C-kinase *in vitro*.[156–158] EGF receptor phosphorylation by phorbol diesters is associated with decreased binding to EGF and decreased EGF receptor tyrosine kinase activity.

One of the simplest interpretations of the results described above is depicted in Fig. 7. This hypothesis suggests that phorbol diesters act directly to activate cellular C-kinase, which catalyzes phosphorylation of the insulin-receptor β subunit on serine or threonine residues. It is possible, and perhaps likely, that a more complex mechanism mediates this process and that other cellular components are involved. Whether the receptor phosphorylation site(s) mediated by phorbol diester action in intact cells might affect the receptor tyrosine kinase activity, as appears to be the case with EGF receptor[159] or the possible functions of the tyrosine phosphates on the receptor is unknown.

VI. CONCLUSIONS

During the past 5 years, very significant structural information has been obtained about the insulin receptor. In addition, powerful new tools for further work are available. We believe we can make the following overall conclusions:

1. The heterotetrameric, disulfide-linked structure proposed for the insulin receptor complex (see Fig. 4) remains consistent with most of the data available and should serve as a useful framework for further experimentation.
2. The insulin-receptor-associated tyrosine kinase activity is markedly activated by tyrosine phosphorylation of its β subunit, which may be an important regulatory mechanism in intact cells.
3. The structure of the insulin receptor serves as a target for the actions of other cellular signaling systems (e.g., hormones that stimulate adenylate cyclase), and therefore its biological functions are probably tightly controlled by a variety of factors.

Regulation of the insulin receptor may result from a highly integrated network of serine and tyrosine phosphorylations and dephosphorylations that modulate both insulin-binding activity and the coupling of insulin binding to tyrosine kinase activation. A most important question that remains for future investigation is the precise mechanism(s) whereby the insulin receptor structure with its associated tyrosine kinase mediates its many biological effects.

VII. REFERENCES

1. Massague, J., Pilch, P.F., and Czech, M.P., 1980, *Proc. Natl. Acad. Sci. USA* **77**:7137–7141.
2. Czech, M.P., Massague, J., and Pilch, P.F., 1981, *Trends Biochem. Sci.* **6**:222–225.
3. Jacobs, S., Hazum, E., and Cuatrecasas, P., 1980, *J. Biol. Chem.* **255**:6937–6940.
4. Jacobs, S., Shechter, Y., Bissell, K., and Cuatrecasas, P., 1977, *Biochem. Biophys. Res. Commun.* **77**:981–988.
5. Yip, C.C., Yeung, C.W.T., and Moule, M.L., 1978, *J. Biol. Chem.* **253**:1743–1745.
6. Jacobs, S., Hazum, E., Shechter, Y., and Cuatrecasas, P., 1970, *Proc. Natl. Acad. Sci. USA* **76**:4918–4921.
7. Wisher, M.H., Baron, M.D., Jones, R.H., and Sonksen, P.H., 1980, *Biochem. Biophys. Res. Commun.* **92**:492–498.
8. Pilch, P.F., and Czech, M.P., 1979, *J. Biol. Chem.* **254**:3375–3381.
9. Kasuga, M., van Obberghen, E., Yamada, K.M., and Harrison, L.C., 1981, *Diabetes* **30**:354–357.
10. Kull, F.C., Jr., Jacobs, S., Su, Y.-F., Svoboda, M.E., Van Wyk, J.J., and Cuatrecasas, P., 1983, *J. Biol. Chem.* **258**:6561–6566.
11. Van Obberghen, E., Kasuga, M., LeCam, A., Hedo, J.A., Itin, A., and Harrison, L.C., 1981, *Proc. Natl. Acad. Sci. USA* **78**:1052–1056.

12. Hedo, J.A., Kasuga, M., van Obberghen, E., Roth, J., and Kahn, C.R., 1981, *Proc. Natl. Acad. Sci. USA* **78:**4791–4795.
13. Roth, R.A., Cassell, D.Y., Wong, K.Y., Maddolux, B.A., and Goldfine, I.D., 1982, *Proc. Natl. Acad. Sci. USA* **79:**7312–7316.
14. Kasuga, M., Hedo, J.A., Yamada, K.M., and Kahn, C.R., 1982, *J. Biol. Chem.* **257:**10392–10399.
15. Harrison, L.C., and Itin, A., 1980, *J. Biol. Chem.* **255:**12066–12072.
16. Im, J.H., Frangakis, C.Y., Meezan, E., Di Bona, D.R., and Kim, H.D., 1982, *J. Biol. Chem.* **257:**11128–11134.
17. Im, J.H., Meezan, E., Rackley, C.E., and Kim, H.D., 1983, *J. Biol. Chem.* **258:**5021–5026.
18. Heidenreich, K.A., Zahniser, N.R., Berhanu, P., Brandenburg, D., and Olefsky, J.M., 1983, *J. Biol. Chem.* **258:**8527–8530.
19. Jacobs, S., and Cuatrecasas, P., 1981, *Endocrinol. Rev.* **2:**251–263.
20. Fujita-Yamaguchi, Y., Choi, S., Sakamoto, Y., and Itakura, K., 1983, *J. Biol. Chem.* **258:**5045–5049.
21. Fujita-Yamaguchi, Y., 1984, *J. Biol. Chem.* **259:**1206–1211.
22. Yip, C.C., Moule, M.L., and Yeung, C.W.T., 1982, *Biochemistry* **21:**2940–2945.
23. Massague, J., Pilch, P.F., and Czech, M.P., 1981, *J. Biol. Chem.* **256:**3182–3190.
24. Kasuga, M., Karlsson, F.A., and Kahn, C.R., 1982, *Science* **215:**185–187.
25. Kasuga, M., Zich, Y., Blithe, D.L., Karlsson, F.A., Haring, H.U., and Kahn, C.R., 1982, *J. Biol. Chem.* **257:**9891–9894.
26. Roth, R.A., and Cassell, D.J., 1983, *Science* **219:**299–301.
27. Shia, M.A., and Pilch, P.F., 1983, *Biochemistry* **22:**717–721.
28. van Obberghen, E., Rossi, B., Kowalski, A., Gazzano, H., and Ponzio, G., 1983, *Proc. Natl. Acad. Sci. USA* **80:**945–949.
29. Kasuga, M., Fujita-Yamaguchi, Y., Blithe, D.L., and Kahn, C.R., 1983, *Proc. Natl. Acad. Sci. USA* **80:**2137–2141.
30. Yip, C.C., Moule, M.L., and Yeung, C.W.T., 1982, *Biochemistry* **21:**2940–2945.
31. Yip, C.C., and Moule, M.L., 1983, *Fed. Proc.* **42:**2842–2845.
32. Yip, C.C., and Moule, M.L., 1983, *Diabetes* **32:**760–767.
33. Yip, C.C., Yeung, C.W.T., and Moule, M.L., 1980, *Biochemistry* **19:**70–76.
34. Hedo, J.A., Kahn, C.R., Hayashi, M. Yamad, K.M., and Kasuga, M., 1983, *J. Biol. Chem.* **258:**10020–10026.
35. Jacobs, S., Kull, F.C., Jr., and Cuatrecasas, P., 1983, *Proc. Natl. Acad. Sci. USA* **80:**1228–1231.
36. Deutsch, P.J., Wan, C.F., Rosen, O.M., and Rubin, C.S., 1983, *Proc. Natl. Acad. Sci.* **80:**133–136.
37. Pilch, P.F., and Czech, M.P., 1980, *J. Biol. Chem.* **255:**1722–1731.
38. Massague, J., and Czech, M.P., 1982, *J. Biol. Chem.* **257:**6729–6738.
39. Massague, J., and Czech, M.P., 1980, *Diabetes* **29:**945–947.
40. Pilch, P.F., and Czech, M.P., 1980, *Science* **210:**1152–1153.
41. Donner, D.B., and Yonkers, K., 1983, *J. Biol. Chem.* **258:**9413–9418.
42. Maturo, J.M., Hollenberg, M.D., and Aglio, L.S., 1983, *Biochemistry* **22:**2579–2586.
43. Ginsberg, B.H., Kahn, C.R., and De Meyts, P., 1976, *Biochem. Biophys. Res. Commun.* **73:**1068–1074.
44. Krupp, M.N., and Livingston, J.N., 1978, *Proc. Natl. Acad. Sci. USA* **75:**2593–2598.
45. Bhaumick, B., Bala, R.M., and Hollenberg, M.D., 1981, *Proc. Natl. Acad. Sci. USA* **78:**4279–4283.
46. Chernausek, S.D., Jacobs, S., and van Wyk, J.J., 1981, *Biochemistry* **20:**7345–7350.
47. Massague, J., and Czech, M.P., 1982, *J. Biol. Chem.* **257:**5038–5045.
48. Kasuga, M., van Obberghen, E., Nissley, S.P., and Rechler, M.M., 1981, *J. Biol. Chem.* **256:**5305–5308.
49. Rubin, J.B., Shia, M.A., and Pilch, P.F., 1983, *Nature* **305:**438–440.
50. Jacobs, S., Kull, F.C., Jr., Earp, H.S. Inobode, M.E., Van Wyk, J.J., and Cuatrecasas, P., 1983, *J. Biol. Chem.* **258:**9581–9584.
51. Zick, Y., Whittaker, J., and Roth, J., 1982, *J. Biol. Chem.* **258:**3431–3434.
52. Rosen, O.M., Herrera, R., Yetunde, O., Petruzzelli, L.M., and Cobb, M.H., 1983, *Proc. Natl. Acad. Sci. USA* **80:**3237–3240.
53. Yu, K.T., and Czech, M.P., 1984, *J. Biol. Chem.* **259:**5277–5286.
54. Rall, T.W., and Sutherland, E.W., 1958, *J. Biol. Chem.* **232:**1065–1076.

55. Sutherland, E.W., and Rall, T.W., 1960, *Pharmacol. Rev.* **12:**265–299.
56. Ross, E.M., and Gilman, A.G., 1980, *Annu. Rev. Biochem.* **49:**533–564.
57. Krebs, E.G., and Beavo, J.A., 1979, *Annu. Rev. Biochem.* **48:**923–929.
58. Nimmo, H.G., and Cohen, P., 1977, *Adv. Cyclic Nucleotide Res.* **8:**145–166.
59. Ingebritsen, T.S., and Cohen, P., 1983, *Science* **221:**331–338.
60. Cohen, P., 1982, *Nature* **296:**613–619.
61. Larner, J., Villar-Palasi, C., Goldberg, N.D., Bishop, J.S., Huijing, F., Wenger, J.I., Sasko, H., and Brown, N.E., 1968, in *Control of Glycogen Metabolism* (Whelan, W.J., ed.), Academic Press, London, pp. 1–18.
62. Schlender, K.K., Wei, S.H., and Villar-Palasi, C., 1969, *Biochim. Biophys. Acta* **191:**272–278.
63. Bishop, J.S., and Larner, J., 1969, *Biochim. Biophys. Acta* **171:**374–377.
64. Sutherland, E.W., and Robinson, G.A., 1966, *Pharmacol. Rev.* **18:**145–161.
65. Miller, E., Fredholm, B., Miller, R.E., Steinberg, D., and Mayer, S.E., 1975, *Biochemistry* **14:**2470–2480.
66. Miller, R.E., Miller, E.A., Fredholm, B., Yellin, J.B., Eichner, R.D., Mayer, S.E., and Steinberg, D., 1975, *Biochemistry* **14:**2481–2488.
67. Corbin, J.D., Reimann, E.M., Walsh, D.A., and Krebs, E.G., 1970, *J. Biol. Chem.* **245:**4849–4851.
68. Khoo, J.C., Aquino, A.A., and Steinberg, D., 1974, *J. Clin. Invest.* **53:**1124–1131.
69. Jefferson, L.S., Exton, J.H., Butcher, R.W., Sutherland, E.W., and Park, C.R., 1968, *J. Biol. Chem.* **243:**1031–1038.
70. Exton, J.H., Lewis, S.B., Ho, R.J., Robison, G.A., and Park, C.R., 1971, *Ann. NY Acad. Sci.* **185:**85–100.
71. Siddle, K., Kane-Maguire, B., and Campbell, A.K., 1973, *Biochem. J.* **132:**765–773.
72. Butcher, R.W., Baird, C.E., and Sutherland, E.W., 1968, *J. Biol. Chem.* **243:**1705–1712.
73. Jungas, R.L., 1966, *Proc. Natl. Acad. Sci. USA* **56:**757–763.
74. Soderling, T.R., Corbin, J.D., and Park, C.R., 1973, *J. Biol. Chem.* **248:**1822–1829.
75. Illiano, G., and Cuatrecasas, P., 1972, *Science* **175:**906–908.
76. Hepp, D.K., and Renner, R., 1972, *FEBS Lett.* **20:**191–194.
77. Pilkis, S.J., Exton, J.H., Johnson, R.A., and Park, C.R., 1974, *Biochim. Biophys. Acta* **343:**250–267.
78. Pohl, S.L., Birnbaumer, L., and Rodbell, M., 1971, *J. Biol. Chem.* **246:**1849–1856.
79. Thompson, W.J., Williams, R.H., and Little, S.A., 1973, *Biochim. Biophys. Acta* **302:**329–337.
80. Rosselin, G., and Freychet, P., 1973 *Biochim. Biophys. Acta* **304:**541–551.
81. House, P.D.R., Poulis, P., and Weidemann, M.J., 1972, *Eur. J. Biochem.* **24:**429–437.
82. Loten, E.G., and Sneyd, J.G.T., 1970, *Biochem. J.* **120:**187–193.
83. Manganiello, V., and Vaughan, M., 1973, *J. Biol. Chem.* **248:**7164–7170.
84. Sinman, B., and Hollenberg, C.H., 1974, *J. Biol. Chem.* **249:**2182–2187.
85. Kono, T., Robinson, F.W., and Sarver, J., 1976, *J. Biol. Chem.* **250:**7826–7835.
86. Desai, K., and Hollenberg, C.H., 1975, *Israel J. Med. Sci.* **11:**540–550.
87. Weber, H.W., and Appleman, M.M., 1982, *J. Biol. Chem.* **257:**5339–5341.
88. Marchmont, R.J., and Houslay, M.D., 1981, *Biochem. J.* **195:**653–660.
89. Kiechle, F.L., and Jarett, L., 1981, *FEBS Lett.* **133:**279–282.
90. Saltiel, A.R., Siegel, M.I., Jacobs, S., and Cuatrecasas, P., 1982, *Proc. Natl. Acad. Sci. USA* **79:**3513–3517.
91. Zhang, S-R., Shi, Q-H., and Ho, R.J., 1983, *J. Biol. Chem.* **258:**6471–6476.
92. Kiechle, F.L., Jarett, L., Kotagal, N., and Popp, D.A., 1981, *J. Biol. Chem.* **256:**2945–2951.
93. Czech, M.P., 1977, *Annu. Rev. Biochem.* **46:**359–384.
94. Jarett, L., Steiner, A.L., Smith, R.M., and Kipnis, D.M., 1972, *Endocrinology* **90:**1277–1284.
95. Fain, J.N., and Rosenberg, L., 1972, *Diabetes* **21**(Suppl. 2):414–425.
96. Fain, J.N., 1971, *Mol. Pharmacol.* **7:**465–479.
97. Siddle, K., and Hales, C.N., 1974, *Biochem. J.* **143:**97–103.
98. Craig, J.W., Rall, T.W., and Larner, J., 1969, *Biochim. Biophys. Acta* **177:**213–219.
99. Fain, J.N., 1975, *J. Cyclic Nucleotide Res.* **1:**359–366.
100. Goldberg, N.D., Villar-Palasi, C., Sasko, H., and Larner, J., 1967, *Biochim. Biophys. Acta* **148:**665–672.

101. Claus, T.H., and Pilkis, S.J., 1976, *Biochim. Biophys. Acta* **421:**246–262.
102. Pessin, J.E., Gitomer, W., Oka, Y., Oppenheimer, C.L., and Czech, M.P., 1983, *J. Biol. Chem.* **258:**7386–7394.
103. Lonnroth, P., and Smith, U., 1983, *Biochem. Biophys. Res. Commun.* **112:**972–979.
104. Kahn, C.R., 1976, *J. Cell Biol.* **70:**261–286.
105. Zeleznik, A.J., Midgely, A.R., and Reichert, L.E., 1974, *Endocrinology* **95:**818–825.
106. McGuire, W.L., Horwitz, K.B., and Zava, D.T., 1978, *Metabolism* **27:**487–501.
107. Lippman, M.E., and Allegra, J.C., 1978, *N. Engl. J. Med.* **299:**930–933.
108. Fraser, C.M., and Venter, J.C., 1979, *Fed. Proc.* **38:**362.
109. Baker, J.B., Barsh, G.S., Carney. D.H., and Cunningham, D.D., 1978, *Proc. Natl. Acad. Sci. USA* **75:**1882–1886.
110. Fantus, I.G., Saviolakis, G.A., Hedo, J.A., and Gorden, P., 1982, *J. Biol. Chem.* **257:**8277–8283.
111. Knutson, V.P., Ronnett, G.V., and Lane, M.D., 1982, *Proc. Natl. Acad. Sci. USA* **79:**2822–2826.
112. Crabtree, G.R., Munck, A., and Smith, K.A., 1979, *Nature* **279:**338–339.
113. Grunfeld, C., Baird, K., van Obberghen, E., and Kahn, C.R., 1981, *Endocrinology* **109:**1723–1730.
114. Cigolini, M., and Smith, L., 1979, *Metabolism* **28:**502–510.
115. King, G.L., Rechler, M.M., and Kahn, C.R., 1982, *J. Biol. Chem.* **257:**10001–10006.
116. Oppenheimer, C.L., Pessin, J.E., Massague, J., Gitomer, W., and Czech, M.P., 1983, *J. Biol. Chem.* **258:**4824–4830.
117. Schonbrunn, A., and Tashjian, A.H., Jr., 1980, *J. Biol. Chem.* **255:**190–198.
118. Olefsky, J.M., Saekow, M., and Kroc, R.L., 1982, *Ann. NY Acad. Sci.* **380:**200–216.
119. Wrann, M., Fox, C.F., and Ross, R., 1980, *Science* **210:**1363–1365.
120. Heldin, C-H., Wasteson, A., and Westermark, B., 1982, *J. Biol. Chem.* **257:**4216–4221.
121. Wharton, W., Leof, E., Pledger, W.J., and O'Keefe, E.J., 1982, *Proc. Natl. Acad. Sci. USA* **79:**5567–5571.
122. Rozengurt, E., Brown, K.D., and Pettican, P., 1981, *J. Biol. Chem.* **256:**716–722.
123. Rozengurt, E., Collins, M., Brown, K.D., and Pettican, P., 1982, *J. Biol. Chem.* **257:**3680–3686.
124. Grunberger, G., and Gorden, P., 1982, *Am. J. Physiol.* **243:**319–324.
125. Thomopoulos, P., Testa, U., Gourdin, M.-F., Hervy, C., Titeux, M., and Vainchenker, W., 1982, *Eur. J. Biochem.* **129:**389–393.
126. Lee, L.S., and Weinstein, B., 1978, *Science* **202:**313–315.
127. Lee, L.S., and Weinstein, B., 1979, *Proc. Natl. Acad. Sci. USA* **76:**5168–5172.
128. Brown, K.D., Dicker, P., and Rozengurt, E., 1979, *Biochem. Biophys. Res. Commun.* **86:**1037–1043.
129. Shoyab, M., DeLarco, J.E., and Todaro, G.J., 1979, *Nature* **279:**387–391.
130. King, A.C., and Cuatrecasas, P., 1982, *J. Biol. Chem.* **257:**3053–3060.
131. Minneman, K.P., Pittman, R.N., and Molinoff, P.B., 1981, *Annu. Rev. Neurosci.* **4:**419–461.
132. Dugan, K.W., Cho, Y.W., Gomoll, A.W., Aviado, D.M., and Lish, P.M., 1968, *J. Pharmacol. Exp. Ther.* **164:**290–301.
133. Butcher, R.W., and Sutherland, E.W., 1962, *J. Biol. Chem.* **237:**1244–1250.
134. Fain, J.N., and Malbon, C.C., 1979, *Mol. Cell. Biochem.* **25:**143–169.
135. Hepp, D., Challoner, D.R., and Williams, R.H., 1968, *J. Biol. Chem.* **243:**4020–4026.
136. Schwabe, U., Schonhofer, P.S., and Ebert, R., 1974, *Eur. J. Biochem.* **46:**537–545.
137. Angel, A., Desai, K., and Halperin, M.L., 1971, *Metabolism* **20:**87–99.
138. Bihler, I., and Jeanrenaud, B., 1970, *Biochim. Biophys. Acta* **202:**496–506.
139. Robertson, R.P., and Porte, D., 1973, *Diabetes* **22:**1–8.
140. Deibert, D.C., and DeFronzo, R.A., 1980, *J. Clin. Invest.* **67:**717–721.
141. Sacca, L., Vigorito, C., Cicala, M., and Ungaro, B., 1982, *J. Clin. Invest.* **69:**284–293.
142. Czech, M.P., 1980, *Diabetes* **29:**399–409.
143. Pessin, J.E., Oka, Y., Gitomer, W., and Czech, M.P., (submitted).
144. Hoffman, B.B., and Lefkowitz, R.J., 1980, *Annu. Rev. Pharmacol. Toxicol.* **20:**581–608.
145. Hoffman, B.B., Kilpatrick, M.D., and Lefkowitz, R.J., 1980, *J. Cyclic Nucleotide Res.* **5:**355–366.
146. Stadel, J.M., DeLean, A., Kilpatrick, M.D., Sawyer, D.D., and Lefkowitz, R.J., 1981, *J. Cyclic Nucleotide Res.* **7:**37–47.
147. Simpson, I.A., and Pfeuffer, T., 1980, *Eur. J. Biochem.* **111:**111–116.

148. Hudson, T.H., and Johnson, G.L., 1981, *Mol. Pharmacol.* **20**:694–703.
149. Slaga, T.J., Sivak, A., and Boutwell, R.K., 1978, *Mechanisms of Tumor Promotion and Carcinogenesis,* Raven Press, New York.
150. Niedel, J.E., Kuhn, L.J., and Vandenbark, G.R., 1983, *Proc. Natl. Acad. Sci. USA* **80**:36–40.
151. Kikkawa, U., Tatrai, Y., Minakuchi, R., Inohara, S., and Nishizuka, Y., 1982, *J. Biol. Chem.* **257**:13341–13348.
152. Castagna, M., Takai, Y., Kaibuchi, K., Sano, K. Kikkawa, U., and Nishizuka, Y., 1982, *J. Biol. Chem.* **257**:7847–7851.
153. Wise, C.B., Raynor, R.L., and Kuo, J.F., 1982, *J. Biol. Chem.* **257**:8481–8488.
154. Kishimoto, A., Takai, Y., Mori, T., Kikkawa, U., and Nishizuka, Y., 1980, *J. Biol. Chem.* **255**:2273–2276.
155. Jacobs, S., Sahyoun, N.E., Saltiel, A.R., and Cuatrecasas, P., 1983, *Proc. Natl. Acad. Sci. USA* **80**:6211–6213.
156. Cochet, C., Gill, G.N., Meisenhelder, J., Cooper, J.A., and Hunter, T., 1983, *J. Biol. Chem.* **259**:2553–2558.
157. Iwashita, S., and Fox, C.F., 1983, *J. Biol. Chem.* **259**:2559–2567.
158. Davis, R., and Czech, M.P., 1984, *J. Biol. Chem.* **259**:8545–8549.
159. Davis, R., and Czech, M.P., 1984, *J. Biol. Chem.,* in press.

2

Immunochemical Characterization of Receptors for Insulin and Insulinlike Growth Factor-I

Steven Jacobs

Receptors for insulin and insulinlike growth factors are present in target cells in extremely low quantities. They comprise only a small fraction of a percentage of the total membrane protein. Methods useful for their detection and characterization must therefore be exquisitely sensitive and specific. The high affinity and specificity of the hormones themselves, which are requisite for their biological function, provide the basis for one group of extensively used methods for investigating these receptors. Antibodies directed against these receptors provide another group of reagents with the necessary specificity and sensitivity. For certain purposes, antibodies have clear advantages over the hormones themselves. While hormone binding requires that the native conformation of the receptor be intact, many antibodies will react with the receptors even after they are denatured. Furthermore, receptors for insulin and insulinlike growth factors are large molecules with multiple domains. While insulin and the insulinlike growth factors react with their receptors at a single domain with a specificity that is fixed by nature, it is possible to select different antibodies that react with different sites on the receptors and exhibit different properties. It is therefore at least theoretically possible to use antibodies to map out different domains on the receptor and to learn something about their function. Monoclonal antibodies, because they recognize a single epitope, are ideal for this purpose. However, polyclonal antisera have also been useful. This chapter describes a series of antibodies that react with insulin and insulinlike growth factor-I receptors as well as information acquired by their use.

Steven Jacobs ● The Wellcome Research Laboratories, Research Triangle Park, North Carolina.

I. SOURCES OF ANTIBODY

A. Autoantibodies

Flier *et al.* found that serum from a group of patients with insulin resistance and acanthosis nigricans inhibited insulin-receptor binding; these workers demonstrated that the inhibitors present in these sera were immunoglobulins.[1,2] Subsequently, several similar cases were identified. The properties of antisera from different patients vary (see Table 1), although there are many similarities. Even in an individual patient, the inhibitory antibodies are polyclonal.[2] The IgG class is generally predominant.[2] With most antisera, inhibition of insulin binding appears to be competitive.[3] For the best studied antiserum, B-2, insulin causes reciprocal inhibition of antibody binding.[4] The latter piece of evidence strongly suggests that this antibody binds directly to the insulin receptor and is specific for the insulin receptor. This was confirmed by the demonstration that when cells were labeled by lactoperoxidase iodination or biosynthetically with radioactive precursors and then solubilized, B-2 specifically immunoprecipitated insulin receptors but no other membrane proteins.[5-7] It was recently shown by immunoblotting that antiserum B-2 recognizes both α and β subunits of the receptors.[8]

B. Rabbit Polyclonal Antisera

Two antisera directed against insulin receptors have been produced by immunizing rabbits with insulin receptors purified from rat liver. One of these, A410, was produced against the holoreceptor in its native conformation—at least no purposeful attempt was made to denature it.[9] The other, 505, was prepared against the isolated α subunit, which had been isolated from a band of an SDS gel.[10] Neither antibody inhibits insulin-receptor binding. They do, however, immunoprecipitate solubilized insulin

Table 1. Properties of Some Antireceptor Antibodies

Antibody or antiserum	Class	Target		Subunit	Inhibits hormone binding	Biological activity
		Ins. receptor	IGF-I receptor			
Autoantibodies						
B-2	Polyclonal	Major	Major	α, β	Yes	Agonist
B-10	Polyclonal	Major	Negative	?	Yes	Agonist
Baldwin *et al.*[63]	Polyclonal	Major	Major	?	Yes	Agonist
Other	Polyclonal	Major	—	—	Yes	Agonist
A410	Polyclonal	Major	Major	α, β	No	Agonist
505	Polyclonal	Major	?	α	No	Agonist
αIR-I	IgG1K	Major	Minor	?	No	No
αIR-II	IgG1K	Minor	Major	α	No	?
αIR-III	IgG1K	Minor	Major	α	Yes	?
Roth *et al.*[15]	IgG1K	Major	Minor	α	Yes	Antagonist

receptor labeled with bound [^{125}I]insulin.[9,10] When cells rich in insulin receptors are labeled with ^{125}I by lactoperoxidase or biosynthetically with [^{35}S]methionine and then solubilized, both antibodies specifically immunoprecipitate both the α and β subunits of the insulin receptor but no other proteins,[11] thereby confirming that they interact directly with the insulin receptor and establishing their specificity.

Since the α and β subunits are covalently associated by disulfide bonds, the immunoprecipitation of both subunits in these studies does not necessarily imply that 505 and A410 react with both. One subunit may be precipitated because of its association with the other. In an effort to resolve this ambiguity, immunoprecipitation studies were carried out with ^{125}I-labeled receptor in which the subunits were dissociated by reduction with dithiothreitol and by denaturation with SDS. Under these circumstances, 505 immunoprecipitated only the α subunit[10]; this is not surprising, since it was produced by immunizing with the isolated α subunit, while A410 immunoprecipitated both subunits.[12] The direct interaction of A410 with both subunits, which has also been demonstrated by immunoblotting,[8] does not necessarily imply that α and β have similar immunochemical determinants. Since A410 is polyclonal, a more likely explanation is that it contains several species of antibodies, some reacting with α and some with β.

C. Monoclonal Antibodies

A series of monoclonal antibodies have been produced by using insulin receptor purified from human placenta by insulin Sepharose chromatography as an immunogen.[11,12] Three SJL mice were immunized with this preparation of purified receptor. The mouse that developed the highest serum titer of anti-insulin receptor antibody was sacrificed, and its spleen cells were fused with FO myeloma cells. Because of the extensive structural similarity between insulin and IGF-I receptors, and because IGF-I receptors adsorb to insulin-Sepharose and copurify with insulin receptors,[13] viable clones were screened for the production of antibody against both insulin receptors and IGF-I receptors. The screening assays measured the ability of supernatant media to immunoprecipitate solubilized placenta receptors labeled with either [^{125}I]insulin or [^{125}I]-IGF-I. Six positive clones were identified. Each was positive in both screening assays. One clone eventually died out. Two others stopped producing antibody. The remaining three were serially cloned by limiting dilution five times and grown in large amounts in ascites fluid. These three clones and the antibodies they produce were designated αIR-I, αIR-II, and αIR-III.

On the basis of their ability to immunoprecipitate solubilized receptors labeled with either [^{125}I]insulin or [^{125}I]-IGF-I, αIR-I reacts predominantly with insulin receptors and less potently with IGF-I receptors, while αIR-II and αIR-III react predominantly with IGF-I receptors.[12] αIR-I and αIR-II do not interfere with hormone binding, while αIR-III does. When cells are labeled by lactoperoxidase iodination or biosynthetically with [^{35}S]methionine and then solubilized and immunoprecipitated with these antibodies, αIR-I specifically immunoprecipitates two bands, of 135,000 and 90,000 M_r, corresponding to the α and β subunits of the insulin receptors.[11,12,14] αIR-II and αIR-III also specifically immunoprecipitate two bands.[12,14] These have somewhat different molecular weights than do the corresponding subunits of the insulin receptor: in placenta, 132,000 and 98,000 M_r, and in IM-9 lymphocytes, 136,000 and 98,000 M_r. These are the α and β subunits of the IGF-I receptor.

To determine with which subunit these antibodies react, labeled receptor was reduced with dithiothreitol and denatured with SDS and then reacted with antibody. Under these conditions, αIR-I failed to immunoprecipitate either subunit.[12] Presumably it recognizes an epitope that is destroyed by this treatment. αIR-II and αIR-III immunoprecipitated only the α subunit.[12]

A fourth monoclonal anti-insulin receptor antibody was produced by Roth et al.[15] These workers immunized BALB/c mice with IM-9 cells, a human lymphocytic cell line rich in insulin receptors. Hybrid clones were screened for the production of antibody that inhibited insulin-receptor binding in human placenta membranes. One stable positive clone was identified. This antibody specifically immunoprecipitated only the α and β subunits of the insulin receptor from cells that were biosynthetically labeled with [^{35}S]methionine, demonstrating that it interacts both directly and specifically with the insulin receptor.[15] Exposure of insulin receptor to crude preparations of collagenase extensively degrades the β subunit of the insulin receptor but leaves the α subunit intact. Residual α subunit is still immunoprecipitated by this antibody, indicating that it recognizes a determinant on this subunit.[16]

II. RECEPTOR VALENCE

Although insulin receptor has been purified to homogeneity, it has not been possible to unequivocally determine its valence. This is because during purification, the receptor is easily denatured and loses binding activity. The highest specific activity reported for purified insulin receptor is 28.5 μg of insulin bound per milligram of protein,[17] or about 1.5 molecules of insulin bound per receptor, assuming a 350,000-M_r receptor. The structure of insulin and IGF-I receptors depicted in chapter 1 also suggests that each tetrameric receptor binds more than one hormone molecule. Studies with B-2 antiserum are consistent with this. When solubilized receptors are labeled by binding low concentrations of [^{125}I]insulin (concentrations that will occupy only a small fraction of the available binding sites), B-2 will quantitatively immunoprecipitate the labeled receptors.[18] Clearly, under these circumstances, a molecule of insulin and a molecule of antibody must be bound to the same receptor. However, higher concentrations of insulin completely displace antibody from the receptor[4] (shown by [^{125}I]B-2 binding) and inhibit immunoprecipitation of receptor.[7] This must result from the binding of insulin to a second site on the receptor. Similar results have been obtained with the monoclonal antibody produced by Roth and associates (R. Roth, personal communication). Whether inhibition of binding is due to competition for direct occupancy of the same site, to steric interference, or to an allosteric effect on an adjacent site, but not a distant site, is not entirely clear, although the latter two explanations seem more likely.

III. BIOLOGICAL ACTIVITY OF ANTIRECEPTOR ANTIBODIES

As indicated in Table 1, some but not all antibodies directed against insulin receptors are able to mimic the biological activities of insulin. It is not clear what properties confer agonist activity on the antibodies. Most of the autoimmune sera exhibit agonist activity. These compete with insulin for binding and therefore probably bind at or near the insulin-binding site. However, the monoclonal antibody produced by Roth *et al.*, which lacks agonist activity and in fact is an antagonist, also competes with insulin binding.[15] Furthermore, A410 and 505, which do not inhibit insulin binding, are agonists.[9,10] Clearly, binding to the insulin-binding site is neither necessary nor sufficient for an antibody to be an agonist. Nor can the presence or absence of agonist properties be explained on the basis of which subunit the antibody binds. Both 505 and the antibody produced by Roth *et al.* recognize only the α subunit,[10,16] but 505 is an agonist while the monoclonal antibody is an antagonist.

One property of these antibodies, at least in the case of B-2, that is required for stimulating activity is multivalency.[19] Bivalent F(ab')$_2$ fragments of B-2 retain agonist activity. By contrast, monovalent Fab fragments are devoid of agonist activity, although they still compete with insulin binding.[19] Because of this, they act as antagonists.[19] The agonist activity of Fab fragments can be restored if they are made multivalent by crosslinking them with anti-Fab antibodies.[19] Since αIR-I and the antibody produced by Roth *et al.* are monoclonal and presumably recognize a single epitope on the receptor, it may be difficult for them to form crosslinked aggregates larger than dimers. It is interesting to speculate that the formation of larger aggregates may be required for biological activity and that this may explain why these antibodies are not agonists.

Those antibodies that are agonists mimic all the biological effects of insulin for which they have been tested (Table 2). These include effects of insulin on surface phenomena such as glucose transport[9,10,19–23] and ion flux,[23] both acute effects and long-term effects such as stimulation of lipoprotein lipase,[24] and both metabolic effects as well as effects on differentiation and proliferation.

Evaluation of the ability of these anti-receptor antibodies to mimic the mitogenic effects of insulin is complicated. This is because in some cells, insulin produces its mitogenic effects by interacting with growth factor receptors other than the insulin receptor.[25] It appears that in those cells in which insulin is mitogenic through activation of its own receptor, these antibodies are also mitogenic, while in those cells in which insulin is mitogenic through crossreacting with other growth factor receptors, these antibodies are not mitogenic. For example, in H35 hepatoma cells and F9 embryocarcinoma cells, anti-receptor antibodies simulate thymidine incorporation.[26,69] In these cells, insulin is mitogenic at concentrations as low as 10 ng/ml. By contrast, in human fibroblasts, anti-receptor antibodies do not stimulate thymidine incorporation even though they do stimulate protein synthesis.[25] In these cells, insulin is mitogenic only at concentrations exceeding several micrograms per milliliter, far in excess of that required to saturate insulin receptors. It seems likely that such high concentrations are required because in these cells insulin is producing its mitogenic effects by interacting with some other growth factor receptor. This was very convincingly demon-

Table 2. *Insulinlike Effects of Anti-Insulin Receptor Antibodies*

Rapid membrane effects
    Stimulation of glucose transport and glucose oxidation[9,10,19–23]
    Hyperpolarization of rat skeletal muscle[23]
    Production of insulin mediators[34]
    Stimulation of receptor phosphorylation[35,36]

Rapid intracellular effects
    Inhibition of lipolysis[9,61]
    Stimulation of lipogenesis[64]
    Stimulation of glycogen synthase[22,62,63]
    Stimulation of ribosomal S6 phosphorylation[65]
    Stimulation of amino acid incorporation[61]
    Stimulation of pyruvate dehydrogenase[34]
    Stimulation of H_2O_2 production[66]

Delayed effects
    Induction of tyrosine amino transferase[69]
    Stimulation of mitogenesis[26,69]
    Induction of lipoprotein lipase[24]
    Activation of α-lactalbumin and casein gene expression[67]
    Downregulation of insulin receptors[68]

strated by King *et al.*[25] These investigators showed that in human fibroblasts, the Fab fragment of B-2 immunoglobulin, which blocks insulin binding to its own receptor but has no intrinsic biological activity, blocked the metabolic effects of insulin but did not block its mitogenic effects.

These studies raise two perplexing questions:

1. Why does activation of the insulin receptor stimulate mitogenesis in some cells but not in others, even though they have functional insulin receptors and a mitogenic response to activation of a closely related growth factor receptor?
2. In human fibroblasts, through which receptor does insulin act to stimulate mitogenesis?

Since insulin does not bind to the IGF-II receptor, it cannot be acting through the IGF-II receptor. Since B-2 is almost equally effective in competing with hormone binding to the IGF-I receptor as the insulin receptor (at least in human placenta and IM-9 cells), B-2 should also antagonize activation of the IGF-I receptor. It seems unlikely that insulin-stimulated mitogenesis in fibroblasts is mediated through that receptor either.

Treatment of cells with insulin results in a diverse spectrum of responses. Some of these occur at an intracellular locus (e.g., nucleus, mitochondrion, cytoplasm) and after along delay during which insulin must be present constantly. This has raised the possibilities that different types of insulin receptors may be involved in mediating these effects and that some of these receptors may be intracellular. Since antisera that are agonists appear to mimic all the effects of insulin, these effects must be mediated by the same, or at least immunochemically similar, receptors. (There may be subtle differences between receptors on different cells. See later sections of this chapter.) It is not clear whether anti-receptor antibodies produce all their effects by acting at the cell surface. Several facts are consistent with a possible role for intracellular receptors. Both insulin, recently reviewed by Posner et al.,[27] and anti-receptor antibodies[28] are internalized; intracellular insulin receptors are present on Golgilike membranes[27] and, according to some investigators[29–32] on the nuclear envelope.[32] Although B-2 fails to inhibit insulin binding to nuclear envelope receptors,[32] making it unlikely that these receptors are involved in mediating effects of B-2, both B-2 and A410 react with Golgi receptors (Patel and Posner[33] S. Jacobs and Y.F. Su, unpublished observations). Alternatively, since the insulin receptor is a transmembrane protein, interaction with antibody (or insulin) at the cell surface could generate a signal that is propagated internally. A410 stimulates the release from surface membranes of a low-molecular-weight substance that is thought to mediate many of the effects of insulin.[34] Stimulating antibodies, but not the antibody produced by Roth et al., which is an antagonist, also stimulate insulin-receptor tyrosine-specific protein kinase activity.[35,36] Either effect could result in such transmembrane signaling.

When incubated with intact cells, the monoclonal antibody produced by Roth et al. and the Fab fragment of B-2, are internalized along with the insulin receptor, even though both are antagonists.[37,38] In addition, the monoclonal antibody is able to induce receptor down-regulation.[37] These results show that it is not necessary for an anti-receptor antibody to be stimulating or multivalent in order to be internalized and conversely that internalization of either antibody or receptor is not sufficient to induce the biological effects of insulin.

IV. IMMUNOCHEMICAL CROSSREACTIVITY OF INSULIN AND IGF-I RECEPTORS

In addition to reacting with insulin receptors, sera from several type B patients inhibit IGF-I binding and immunoprecipitate the IGF-I receptor.[39,41] Several of these sera have almost equal titers of anti-insulin receptor and anti-IGF-I receptor activity, although some are relatively selective for insulin receptors, and one, B-10, does not recognize the IGF-I receptor.[40] A410, a rabbit polyclonal antiserum, also immunoprecipitates both receptors.[40,41]

Since these sera are polyclonal, their ability to recognize both insulin and IGF-I receptors does not provide unequivocal evidence for immunochemical crossreactivity; they could contain different species of antibody, some reacting exclusively with insulin receptors and some reacting exclusively with IGF-I receptors. There is circumstantial evidence against this possibility. In one type B patient, titers of antibody against both insulin receptor and IGF-I receptor were found to vary in parallel at different stages of the patient's illness and copurified with each other.[39] In addition, insulin receptor purified from rat liver was used for an immunogen to produce A410. Since rat liver has little if any IGF-I receptor, it is unlikely that the immunized animal would raise antibodies against IGF-I receptors unless they were crossreacting with insulin receptors.

The most convincing evidence that these two receptors have immunochemically similar sites is the ability of several monoclonal antibodies to recognize both receptors. The antibody produced by Roth *et al.* inhibits IGF-I-receptor binding in addition to insulin-receptor binding, although 300-fold higher antibody concentrations are required.[43] αIR-I clearly immunoprecipitates both receptors, although it is approximately 10-fold less potent in immunoprecipitating IGF-I receptors.[14] αIR-II and αIR-III preferentially immunoprecipitate IGF-I receptors, but also appear to immunoprecipitate insulin receptors very weakly.[14]

There does not appear to be immunochemical crossreactivity between IGF-II receptors and insulin or IGF-I receptors. However, this may be because few antisera have been used to evaluate this possibility. Serum B-2 neither inhibits binding to the IGF-II receptor nor immunoprecipitates that receptor.[43] The monoclonal antibody produced by Roth *et al.* does not inhibit binding to the IGF-II receptor.[43]

V. SPECIES SPECIFICITY

αIR-I and the monoclonal antibody produced by Roth *et al.* have strict species specificity.[13,15] Both react with human insulin receptors but not with insulin receptors from other mammalian species against which they have been tested. By contrast, polyclonal antisera have less strict species specificity. Many react almost equally well with insulin receptors from a wide variety of species. However, when examined carefully, many polyclonal antisera do exhibit some degree of species specificity. B-3 crossreacts very poorly with rat liver insulin receptor.[1] B-2 immunoprecipitates both rat and human IGF-I receptors and inhibits IGF-I-receptor binding to the human receptor, but it does not inhibit IGF-I-receptor binding to the rat receptor.[41] The most likely explanation is that B-2 contains several types of antibody. One type reacts with both human and rat IGF-I receptor at a site that does not interfere with IGF-I binding, and one type reacts with human but not rat IGF-I receptor at a site that does.

In view of the ability of αIR-I and the antibody produced by Roth *et al.* to recognize both insulin and IGF-I receptors, their strict species specificity is surprising. Study of insulin and IGF binding to tissues from a number of mammalian and avian species indicates that although insulin and IGF-I receptors probably evolved from a common ancestral protein, they had diverged by the time these two vertebrate families had evolved. One would expect that receptors for insulin in closely related mammalian species (human and rat) would be more similar to each other than to the IGF-I receptors, even in the same species. Certainly they are similar with respect to their relative affinity for different insulinlike peptides. The antibody data, however, indicate that there are epitopes on the human insulin receptor that are more closely related to the human IGF-I receptor than to the rat insulin receptor. This may be an important clue to structural similarities between these two receptors and the mechanisms by which they have evolved.

VI. TISSUE SPECIFICITY

By most criteria, insulin receptors from different tissues appear to be very similar. There is, however, considerable but subtle evidence that insulin receptors may vary from tissue to tissue even in a single species. The clearest example of this is brain insulin receptor, which has 115,000-M_r and 85,000-M_r subunits,[44,45] in contrast to insulin receptors in other tissues, which have 135,000-M_r and 90,000-M_r subunits. The lower molecular weight of the brain receptor appears to be due, at least in part, to less extensive glycosylation.[45] More subtle examples are found in other tissues. Rat adipocyte insulin receptor has a higher affinity for B26 monoiodoinsulin than for A14 monoiodoinsulin.[46] The order of potency is reversed for rat hepatocyte insulin receptor.[46] Treatment with dithiothreitol of placenta insulin receptor decreases its affinity for insulin,[47,48] while treatment of erythrocyte, liver, or adipocyte insulin receptor increases their affinity.[48,50] As described more extensively in other chapters, the insulin receptor is phosphorylated and the extent of phosphorylation is increased by insulin. In most tissues, phosphorylation of solubilized insulin receptor occurs exclusively on tyrosine residues of the β subunit. With solubilized rat liver receptor, but not solubilized rat adipocyte receptor, phosphorylation also occurs on the α subunit.[51] With receptor solubilized from HepG2 cells, a human hepatoma cell line, but not from IM-9 cells or human erythrocytes, phosphorylation also occurs on serine residues.[52]

Possibly because they have been less extensively studied, there is less evidence of tissue-specific differences for IGF-I receptors; however, the α subunit of the human placenta IGF-I receptor has a molecular weight of 132,000 while the α subunit of the IM-9 cell receptor has an apparent molecular weight of 136,000.[12]

There also appear to be tissue-specific immunochemical differences. Again, the most obvious difference is with brain insulin receptor. Two spontaneously occurring autoantibodies that inhibit insulin-receptor binding in a wide variety of tissues fail to inhibit insulin-receptor binding in brain.[44,45] There are more subtle immunochemical differences in other tissues. αIR-I is less effective in immunoprecipitating insulin receptor solubilized from human erythrocytes than from IM-9 cells or human placenta.[11] Several of the autoimmune antisera are more potent in inhibiting insulin binding and IGF-I binding to human placenta as compared with IM = 9 cells.[43] By contrast, the monoclonal antibody produced by Roth et al. has a higher affinity for insulin and IGF-I receptors in IM-9 cells than in human placenta or human adipocytes.[15,43]

There are several possible explanations for these tissue-specific differences. Trivially, some could result from partial denaturation or proteolytic nicking of the receptors during solubilization or purification or during preparation of the tissue. There is no evidence that this does occur, but, it is difficult to rule out. Some could result from differences in the environment of the receptors in different tissues. For example, there may be differences in the phospholipid composition of the membranes of different tissues, or some tissues may contain molecules that interact with these receptors, altering their properties. However, many of the differences described above persist after the receptors have been solubilized and partially purified.[11,45,46,51,52] Some could result from tissue-specific differences in post-translational modification of the receptors. Brain insulin receptor appears to be less extensively glycosylated than does receptor from other tissues.[45] Finally, it is possible that there are multiple, closely related genes coding for these receptors that are differentially expressed in different tissues. Further work is necessary to evaluate these possibilities and to determine their functional significance.

VII. PHORBOL ESTERS AND PHOSPHORYLATION OF RECEPTORS FOR INSULIN AND IGF-I

One of the most important ways in which anti-receptor antibodies have been used has been to identify and quantitate insulin and IGF-I receptors that have been labeled nonspecifically with, for example, ^{32}P-phosphate or labeled biosynthetic precursors. This type of technique has been used extensively to study the biosynthesis, turnover, autophosphorylation, and tyrosine-specific protein kinase activity of these receptors. Since these areas are the subject of other chapters in this volume, they are not discussed here. Antibodies have also been used to study a different type of phosphorylation of insulin and IGF-I receptors, i.e., as occurs in response to phorbol esters.[53]

Protein kinase-C is a serine- and threonine-specific protein kinase, the activity of which is regulated by calcium, diacylglycerol, and phosphatidylserine.[54] In several cells, diacylglycerol is released in response to hormonal stimulation. This activates protein kinase-C, which in turn may mediate many of the actions of the hormone. This scheme has been investigated most extensively in platelets,[54] but it appears to be applicable in a variety of cell types as well. Recently, it has been shown that phorbol esters directly activate protein kinase-C by binding to the diacylglycerol site.[55,56] Phorbol esters are known to produce a variety of cellular effects. Although it certainly has not been proved, it seems likely that all these effects are mediated through activation of protein kinase-C.

Two effects of phorbol esters are to decrease the binding of insulin and IGF-I to their receptors in intact cells.[57-59] Since these effects occur rapidly, within minutes after the addition of phorbol esters, they may result from a direct effect of protein kinase-C on the receptors. The following study was performed to evaluate this possibility.[53] The endogenous adenosine triphosphate (ATP) pool of intact IM-9 cells was labeled with ^{32}P. The cells were then incubated in the presence or absence of phorbol esters. The cells were solubilized, and their receptors for insulin and IGF-I were partially purified on a wheat germ agglutinin column and immunoprecipitated with αIR-I or αIR-III, respectively. The amount of ^{32}P present in the receptors was then quantitated by autoradiography after SDS-polyacrylamide gel electrophoresis of the immunoprecipitates. Phorbol esters increased the extent of phosphorylation of the β subunits of both receptors by approximately fourfold.

Since these studies were perfomed in intact cells, they do not prove conclusively that phosphorylation was directly mediated by protein kinase-C. It has not yet been possible to clearly demonstrate phosphorylation of insulin or IGF-I receptors by purified protein kinase in broken cell preparations. Although possibly the result of technical problems, this failure might suggest a less direct role for protein kinase-C. Phorbol ester-activated protein kinase-C could activate a cascade of kinases or inhibit a phosphatase that indirectly increases the phosphorylation of these receptors. It is also possible that phorbol esters might directly activate some protein kinase other than protein kinase-C, which would result in phosphorylation of these receptors.

The intrinsic protein kinase activity of insulin and IGF-I receptors appears to be tyrosine specific. However, in intact cells, insulin and IGF-I stimulate the phosphorylation of their respective receptors on seryl residues as well as a tyrosyl residues (Kasuga et al.[60], Jacobs, unpublished observation). It is possible that this seryl phosphorylation may be mediated not by the autocatalytic activity of the receptor, but by protein kinase-C. This could occur because hormone binding makes the receptor a better substrate for protein kinase-C or because hormone action increases the intracellular levels of calcium ions or diacyl glycerol. If this is the case then, in intact cells, both phorbol esters and insulin or IGF-I should stimulate the phosphorylation of the same receptor seryl residues. However, there are major differences in phosphopeptide maps of these two receptors when phosphorylated in the presence of hormone versus phorbol ester (Jacobs et al.[53] and Jacobs, unpublished observations).

The functional significance of phorbol ester-mediated phosphorylation of the insulin and IGF-I receptor is not clear. Stimulation of insulin receptor phosphorylation by insulin and phorbol esters is additive.[53] Therefore, phosphorylation by phorbol esters does not appear to either potentiate or inhibit the protein kinase activity of the receptor. Since phorbol esters decrease insulin and IGF-I binding, it is possible that protein kinase-C plays a role in downregulating these receptors under a variety of physiological or pathological conditions.

VIII. REFERENCES

1. Flier, J.S., Kahn, C.R., Roth, J., and Bar, R.S., 1975, Science 190:63–65.
2. Flier, J.S., Kahn, C.R., Jarrett, D.B., and Roth, J., 1976, J. Clin. Invest. 58:1442–2449.
3. Flier, J.S., Kahn, C.R., Jarret, D.B., and Roth, J., 1977, J. Clin. Invest. 60:784–794.
4. Jarrett, D.B., Roth, J., Kahn, C.R., and Flier, J.S., 1976, Proc. Natl. Acad. Sci. USA 73:4115–4119.
5. Kasuga, M., Hedo, J.A., Yamada, K.M., and Kahn, C.R., 1982, J. Biol. Chem. 257:10392–10399.
6. Van Obberghen, E., Kasuga, M., LeCam, A., Hedo, J. A., Itin, A., and Harrison, L.C., 1981, Proc. Natl. Acad. Sci. USA 78:1052–1056.
7. Hedo, J.A., Kasuga, M., Van Obberghen, E., Roth, J., and Kahn, C.R., 1981, Proc. Natl. Acad. Sci. USA 78:4791–4795.
8. Maron, R., Kahn, C.R., Jacobs, S., and Fugita-Yamaguchi, Y., 1984, Diabetes 33:923–928.
9. Jacobs, S., Chang, K.-J., and Cuatrecasas, P., 1978, Science 200:1283–1284.
10. Jacobs, S., Hazum, E., and Cuatrecasas, P., 1980, J. Biol. Chem. 255:6937–6940.
11. Kull, F.C., Jr., Jacobs, S., Su, Y.-F., and Cuatrecasas, P., 1982, Biochem. Biophys. Res. Commun. 106:1019–1026.
12. Kull, F.C., Jr., Jacobs, S., Su, Y.-F., Svoboda, M.E., Van Wyk, J.J., and Cuatrecasas, P., 1983, J. Biol. Chem. 258:6561–6566.
13. Bennett, A., Daly, F.T., and Hintz, R.L., 1981, Diabetes 30(Suppl. 1): 55A.

14. Jacobs, S., Kull, F.C., Jr., and Cuatrecasas, P., 1983, *Proc. Natl. Acad. Sci. USA* **80:**1228–1231.
15. Roth, R.A., Cassell, D.J., Wong, K.Y., Maddux, B.A., and Goldfine, I.D., 1982, *Proc. Natl. Acad. Sci. USA* **79:**7312–7316.
16. Roth, R.A., Mesirow, M.L., and Cassell, D.J., 1983, *J. Biol. Chem.* **258:**14456–14460.
17. Fujita-Yamaguchi, Y., Choi, S., Sakamoto, Y., and Itakura, K., 1983, *J. Biol. Chem.* **258:**5045–5049.
18. Harrison, L.C., Flier, J.S., Roth, J., Karlsson, F.A., and Kahn, C.R., 1979, *J. Clin. Endocrinol. Metab.* **48:**59–65.
19. Kahn, C.R., Baird, K.L., Jarrett, D.B., and Flier, J.S., 1978, *Proc. Natl. Acad. Sci. USA* **75:**4209–4213.
20. Rosen, O.M., Chia, G.H., Fung, C., and Rubin, C.S., 1979, *J. Cell Physiol.* **99:**37–42.
21. Kasuga, M., Akanuma, Y., Tsushima, F., Iwamoto, Y., Kosoka, K., Kibata, M., and Kawanishi, K., 1978, *Diabetes* **27:**938–945.
22. Le Marchand-Brustel, Y., Gorden, P., Flier, J.S., Kahn, C.R., and Freychet, P., 1978, *Diabetologia* **14:**311–317.
23. Zierler, K., and Rogus, E.M., 1981, *Clin. Res.* **29:**A579.
24. Van Obberghen, E., Spooner, P.M., Kahn, C.R., Chernick, S.S., Garrison, M.M., Karlsson, F.A., and Grunfeld, C., 1979, *Nature* **280:**500–502.
25. King, G.L., Kahn, C.R., Rechler, M.M., and Nissley, S.P., 1980, *J. Clin. Invest.* **66:**130–140.
26. Nagarajan, L., and Anderson, W.B., 1982, *Biochem. Biophys. Res. Commun.* **106:**974–980.
27. Posner, B.I., Bergeron, J.J.M., Josefsberg, Z., Khan, M.N., and Khan, R.J., 1981, *Recent Prog. Horm. Res.* **37:**539–82.
28. Carpentier, J.-L., Van Obberghen, E., Gorden, P., and Orci, L., 1981, *Exp. Cell Res.* **134:**81–92.
29. Goldfine, I.D., Jones, A.L., Hradek, G.T., and Wong, K.Y., 1981, *Endocrinology* **108:**1821–1828.
30. Vigneri, R., Goldfine, I.D., Wong, K.Y., Smith, G.J., and Pezzino, V., 1978, *J. Biol. Chem.* **253:**2098–2103.
31. Horvat, A., Li, E., and Katsoyannis, P.G., 1975, *Biochim. Biophys. Acta* **382:**609–620.
32. Goldfine, I.D., Vigneri, R., Cohen, D., Pliam, N.B., and Kahn, C.R., 1977, *Nature* **269:**698–700.
33. Patel, B.A., and Posner, B.I., 1983, *Can. J. Biochem. Cell Biol.* **61:**657–661.
34. Seals, J.R., and Jarett, L., 1980, *Proc. Natl. Acad. Sci. USA* **77:**77–81.
35. Petruzzelli, L.M., Ganguly, S., Smith, C.J., Cobb, M.H., Rubin, C.S., and Rosen, O.M., 1982, *Proc. Natl. Acad. Sci. USA* **79:**6792–6796.
36. Roth, R.A., Cassell, D.J., Maddux, B.A., and Goldfine, I.D., 1983, *Biochem. Biophys. Res. Commun.* **115:**245–252.
37. Roth, R.A., Maddux, B.A., Cassell, D.J., and Goldfine, I.D., 1983, *J. Biol. Chem.* **258:**12094–12097.
38. Kasuga, M., Carpentier, J.-L., Van Obberghen, E., Orci, L., and Gorden, P., 1983, *Biochem. Biophys. Res. Commun.* **114:**230–233.
39. Rosenfeld, R.G., Baldwin, D., Jr., Dollar, L.A., Hintz, R.L., Olefsky, J.M., and Rubenstein, A., 1981, *Diabetes* **30:**979–982.
40. Jonas, H.A., Baxter, R.C., and Harrison, L.C., 1982, *Biochem. Biophys. Res. Commun.* **109:**463–470.
41. Kasuga, M., Sasaki, N., Kahn, C.R., Nissley, S.P., and Rechler, M.M., 1983, *J. Clin. Invest.* **72:**1459–1469.
42. Armstrong, G.D., Hollenberg, M.D., Bhaumick, B., Bala, R.M., and Maturo, J.M. III, 1983, *Can. J. Biochem. Cell Biol.* **61:**650–656.
43. Roth, R.A., Maddux, B., Wong, K.Y., Styne, D.M., Vliet, G.V., Humbel, R.E., and Goldfine, I.D., 1983, *Endocrinology* **112:**1865–1867.
44. Yip, C.C., Moule, M.L., and Yeung, C.W.T., 1980, *Biochem. Biophys. Res. Commun.* **96:**1671–1678.
45. Heidenreich, K.A., Zahniser, N.R., Berhanu, P., Brandenburg, D., and Olefsky, J.M., 1983, *J. Biol. Chem.* **258:**8527–8530.
46. Podlecki, D.A., Frank, B.H., Kao, M., Horikoshi, H., Freidenberg, G., Marshall, S., Ciaraldi, T., and Olefsky, J.M., 1983, *Diabetes* **32:**697–704.
47. Jacobs, S., and Cuatrecasas, P., 1980, *J. Clin. Invest.* **66:**1424–1427.
48. Massague, J., and Czech, M.P., 1982, *J. Biol. Chem.* **257:**6729–6738.
49. Schweitzer, J.B., Smith, R.M., and Jarett, L., 1980, *Proc. Natl. Acad. Sci. USA* **77:**4692–4696.
50. McElduff, A., and Eastman, C.J., 1981, *J. Recept. Res.* **2:**87–95.
51. Zick, Y., Kasuga, M., Kahn, C.R., and Roth, J., 1983, *J. Biol. Chem.* **258:**75–80.

52. Zick, Y., Grunberger, G., Podskalny, J.M., Moncada, V., Taylor, S.I., Gorden, P., and Roth, J., 1983, *Biochem. Biophys. Res. Commun.* **116:**1129–1135.
53. Jacobs, S., Sahyoun, N.E., Saltiel, A.R., and Cuatrecasas, P., 1983, *Proc. Natl. Acad. Sci. USA* **80:**6211–6213.
54. Nishizuka, Y., and Takai, Y., 1981, *Cold Spring Harbor Conf. Cell Prolif.* **8:**237–249.
55. Castagna, M., Takai, Y., Kaibuchi, K., Sano, K., Kikkawa, U., and Nishizuka, Y., 1982, *J. Biol. Chem.* **257:**7847–7851.
56. Niedel, J.E., Kuhn, L.J., and Vandenbark, G.R., 1983, *Proc. Natl. Acad. Sci. USA* **80:**36–40.
57. Grunberger, G., and Gorden, P., 1982, *Am. J. Physiol.* **243:**E319–E324.
58. Thomopoulos, P., Testa, U., Gourdin, M.T., Hervey, C., Titeux, M., and Vainchenker, W., 1982, *Eur. J. Biochem.* **129:**389–393.
59. Rouis, M., Thomopoulos, P., Postel-Vinay, M.-C., Testa, U., Guyda, J.J., and Posner, B.I., 1984, *Mol. PHysiol.* **5:**123–130.
60. Kasuga, M., Zick, Y., Blith, D.L., Karlsson, F.A., Häring, H.U., and Kahn, C.R., 1982, *J. Biol. Chem.* **257:**9891–9894.
61. Kasuga, M., Akanuma, Y., Tsushima, T., Suzuki, K., Kosaka, K., and Kibata, M., 1978, *J. Clin. Endocrinol. Metab.* **47:**66–77.
62. Lawrence, J.C. Jr., Larner, J., Kahn, C.R., and Roth, J., 1978, *Mol. Cell Biochem.* **22:**153–157.
63. Baldwin, D., Jr., Terris, S., and Steiner, D.F., 1980, *J. Biol. Chem.* **255:**4028–4034.
64. Caro, J.F., and Amatruda, J.M., 1981, *Am. J. Physiol.* **240:**E325–E332.
65. Smith, C.J., Rubin, C.S., and Rosen, O.M., 1980, *Proc. Natl. Acad. Sci. USA* **77:**2641–2645.
66. Mukherjee, S.P., Attaway, E.J., and Mukherjee, C., 1982, *Biochem. Int.* **4:**305–314.
67. Nicholas, K.R., and Topper, Y.J., 1983, *Biochem. Biophys. Res. Commun.* **111:**988–993.
68. Caro, J.F., and Amatruda, J.M., 1980, *Science* **210:**1029–1031.
69. Koontz, J.W., 1984, *Mol. Cell. Biochem.* **58:**139–146.

3

The Metabolism of Insulin Receptors: Internalization, Degradation, and Recycling

Kim A. Heidenreich and Jerrold M. Olefsky

I. INTRODUCTION

The concept that receptors function as recognition sites for polypeptide hormones and neurotransmitters on plasma membranes of target cells originated during the early 1900s.[1] The theory that emerged was that binding of hormone to its receptor on the outer face of the membrane is the initial step in hormonal action, which triggers a transmembrane signal such as generation of a second messenger. This theory has now been expanded, since receptors subserve a greater role in hormone function and metabolism. In addition to hormone recognition, receptors mediate the internalization[2-5] and degradation[6] of hormones. This has been documented for many peptide hormone–receptor systems[7] and the process appears to be similar to receptor-mediated endocytosis of other molecules such as low-density lipoproteins,[8] asialoglycoproteins,[9] and α_2-macroglobulins.[10]

More recently, studies have indicated that receptors are themselves internalized within the cell along with hormones,[11-15] and it is possible that the internalized receptors exert biological activity inside the cell. For example, some of the long-term biological effects of insulin may be mediated by the binding of internalized insulin receptors to the nucleus.[16] In receptor systems for growth factors including epidermal growth factor (EGF),[17-20] insulin,[21-26] insulin growth factor I (IGF-I),[27] and platelet-derived growth factor (PDGF),[28-29] the receptor molecules are protein kinases that regulate the autophosphorylation of tyrosine residues within the receptor as well as

Kim. A. Heidenreich and Jerrold M. Olefsky ● Department of Medicine, University of California at San Diego, La Jolla, California.

exogenous substrates. It is possible that autophosphorylation of receptors plays a role in the internalization and downregulation of polypeptide receptors. In addition, occupied receptors may be translated from the surface to an intracellular organelle, where local phosphorylation of proteins could play a role in eliciting various biological actions of insulin. Thus, it is now evident that receptors are multifunctional proteins involved in hormone recognition, transmembrane signaling, phosphorylation of cellular substrates, and possibly transfer of information inside the cell.

In addition to these functional roles, receptors on the plasma membranes of target cells are involved in the regulation of the sensitivity of the cells to hormones. In many cases, the number of hormone receptors is inversely proportional to the level of exposure to hormone. For example, continued exposure of cells to high concentrations of insulin usually leads to a decreased number of surface receptors,[30] which decreases the sensitivity of the cell to the biological effects of insulin.[31] This phenomenon, termed downregulation, has been documented *in vitro* for a variety of hormone–receptor systems,[32] and is also apparent *in vivo*.[33,34] Since the binding of polypeptide hormones to their receptors induces internalization of hormone–receptor complexes, it is evident that the fate of internalized receptors (i.e., sequestration, degradation, or recycling) is important in determining the steady-state level of receptors on the cell surface. Changes in the flux of receptors through these pathways or decreased receptor synthesis probably mediate the downregulation of receptors.

Thus, in view of the importance of the destination of internalized insulin receptors in controlling the steady-state number of receptors on the cell surface and the possible biological activity of internalized receptors, our laboratory[12,15] and others[35] have studied the fate of internalized insulin–receptor complexes in isolated adipocytes. Other related studies have been carried out on isolated hepatocytes,[14] 3T3-L1 adipocytes,[36,37] and chick liver cells.[38,39] Although the biosynthesis of insulin receptors is also an important branch in the metabolism of insulin receptors, this chapter concentrates on the endocytotic-degradative pathway of insulin receptors. We will briefly review the methods that have been used by various laboratories to radiolabel insulin receptors in a variety of different cell types and will describe in more detail the approaches and methods used in our own studies.

II. METHODS FOR LABELING INSULIN RECEPTORS

A. Heavy-Isotope Density Shift

A number of approaches are now available for labeling insulin receptors so that the turnover of receptors can be measured. The heavy isotope density-shift technique has been used to measure rates of synthesis and degradation of insulin receptors in 3T3-L1 adipocytes[36,40,41] and chick liver cells.[38,39] This approach permits the identification of newly synthesized receptors afer shifting cells to medium containing amino acids enriched in nitrogen-15, carbon-13, and deuterium. Since the incorporation of heavy amino acids into receptors increases the density of receptors, the "heavy" receptors can be separated from the "light" receptors by isopycnic density-gradient centrifugation. By following the amounts of heavy and light receptor present at increasing times of exposure of the cells to heavy amino acids, the rates of heavy receptor synthesis and light receptor decay can be calculated. The success of this technique depends on a quantitative binding assay for insulin receptors in the fractions obtained after density-gradient centrifugation. Proteolysis of the receptor must be absent during the receptor isolation procedures and insulin binding must be stable in detergent extracts of the cells. One possible disadvantage in using this technique to measure degradation of insulin receptors is that loss of binding activity may not be synonymous with degradation of receptor proteins.

B. Biosynthetic and Cell-Surface Labeling

With the availability of specific anti-insulin–receptor antibodies, insulin receptors can be identified after external labeling by Na[^{125}I] plus lactoperoxidase[42–44] or endogenous biosynthetic labeling with [^{35}S]methionine or [^{3}H]carbohydrates.[44–48] Cell-surface labeling of membrane proteins involves the iodination of accessible tyrosine and histidine residues in the presence of Na[^{125}I], H_2O_2, and lactoperoxidase. Insulin receptors are distinguished from other radiolabeled proteins by immunoprecipitation with receptor antibodies. This procedure has been used to study the processing of insulin receptors on the cell surface of IM-9 lymphocytes[43,44] and in isolated adipocytes.[49] Biosynthetic labeling of either the carbohydrate or protein moieties of the insulin receptor is carried out by incubating cells with radioactive monosaccharides or amino acids. After incorporation of the radioactive precursors into cellular proteins, insulin receptors are identified by immunoprecipitation. Biosynthetic labeling combined with pulse-chase procedures has provided a means for identifying insulin receptor precursors and quantitating precursor–product relationships in the biosynthetic pathway of insulin receptors.[46–48] Because of the high background of other labeled proteins, external biosynthetic labeling techniques require partial purification of insulin receptors (usually chromatography on lectin affinity columns) prior to immunoprecipitation and electrophoresis. With this approach, the detection of insulin receptors is limited to receptors that are glycosylated (absorb to lectin affinity columns) and immunoreactive. This is a potential problem, since at least one glycosylated precursor of the insulin receptor has been identified[50] and it is possible that the glycosylation properties and immunoreactivity of insulin receptors change during cellular processing.

C. Photoaffinity Labeling

Photoaffinity labeling is another approach used to label insulin receptors.[51–55] With this technique, a photoreactive derivative of insulin is covalently attached to the binding site of the insulin receptor as the result of a photochemical reaction. The insulin derivative contains a functional group such as an arylazide moiety that, upon irradiation, yields a highly reactive intermediate (in the case of arylazide derivatives, a nitrene free radical) capable of inserting into C–H or C–C bonds within the immediate vicinity of the binding site. The use of photoreactive analogs to study insulin receptors was first introduced by Yip et al., who modified the NH$_2$-terminal of the B chain of

insulin by the addition of a phenylazide moiety.[51] Other photoreactive insulin derivatives have been synthesized and characterized by Brandenburg et al.[53,55] The photoreactive derivatives of insulin are receptor agonists and are highly selective in labeling
insulin receptors. The particular insulin analog used in our studies is iodinated (2-
nitro, 4-azido-phenylacetyl-des-Phe^{B1}-insulin [^{125}I]-NAPA-DP-insulin). This derivative, prepared by Peter Thamm, Derek Saunders, and Dietrich Brandenburg, exhibits
80–90% of the biological activity of insulin.[56] Another attractive feature of this
molecule is that it can be activated at a wavelength of light (360 nm), which does not
damage cellular proteins. Thus, photoaffinity labeling offers advantages over other
affinity crosslinking techniques for functional studies of viable cells because covalent
attachment of insulin to its receptor occurs specifically at the ligand binding site without
crosslinking other cellular proteins and without loss of cell viability. Since labeling
occurs by covalent binding of insulin to the receptor, this technique permits selective
evaluation of the metabolism of occupied receptors. A potential drawback of the
photoaffinity labeling technique is that covalent attachment of insulin to its receptor
prevents dissociation of insulin from the receptor following internalization, and it is
theoretically possible that covalently occupied receptors are metabolized differently
than native receptors. However, thus far, all the data suggest that photoaffinity-labeled
receptors are metabolized in the same manner as nonlabeled insulin receptors.

III. EXPERIMENTAL PROCEDURES

Figure 1 is a diagrammatic representation of the general procedure used in our
studies to follow the biological fate of insulin receptors after the initial binding step.
Insulin receptors on the plasma membrane of isolated adipocytes are photoaffinity
labeled at 16°C. The cells are then washed to remove noncovalently bound hormone
and further incubatd at 16°C (a temperature at which internalization of insulin–receptor
complexes is negligible) or warmed to 37°C to allow for internalization. At various
times, one-half the cells in each set are trypsinized to remove cell-surface receptors.
After washing, the cells are then solubilized and analyzed by sodium dodecylsulfate
polyacrylamide gel electrophoresis (SDS-PAGE). The labeled insulin receptors are
visualized by autoradiography or are quantitated by slicing the gels and counting in a
gamma counter. The specific methods used in this scheme are given below and have
been previously reported.[15]

A. Binding and Photolysis of [^{125}I]-NAPA-DP-Insulin

Isolated adipocytes (5 × 10^5 cells/ml) are incubated with [^{125}I]-NAPA-DP-insulin
(20 ng/ml) in minimal essential medium containing 20 mM HEPES (pH 7.5) and 1%
BSA with or without 20 μg/ml unlabeled insulin. Incubations are carried out in plastic
vials in the dark at 16°C for 30 min. At this time, the binding of [^{125}I]-NAPA-DP-
insulin to adipocytes reaches equilibrium. Various additions such as chloroquine or
cycloheximide are included in the incubations as indicated in the figure legends. At
the end of the incubation period at 16°C, the cells are transferred to 60 × 15 mm

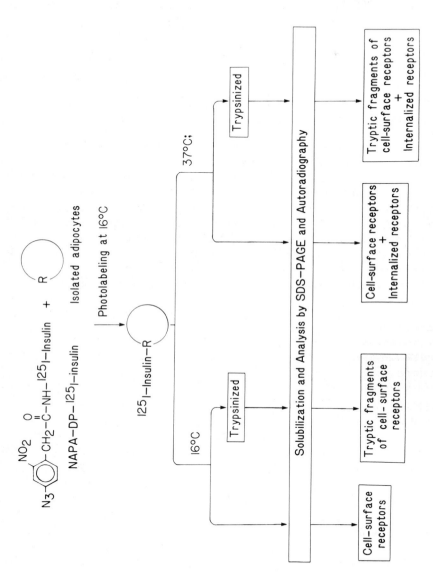

Figure 1. General experimental scheme used to follow the fate of photoaffinity-labeled insulin receptors.

plastic petri dishes and placed on ice 10 cm under a long-wave UV lamp (366 nm) for 3 min. After photolysis, the cells are centrifuged at 400 rpm for 2 min. After removal of the infranatant containing unbound hormone, the cells are resuspended in minimal essential medium containing 10 mM HEPES (pH 7.5) and 1% BSA and further incubated at either 16° or 37°C for various times. The cells are then solubilized in 3% SDS containing 1 mM N-ethylmaleimide (NEM) and 2 mM phenylmethylsulfonylfluoride (PMSF).

B. Degradation and Internalization

Degradation of insulin receptors is assessed by incubating photoaffinity-labeled cells at 37°C. At various times, the cells are solubilized and electrophoresis is performed. Insulin receptors are quantitated by counting the radioactivity in the insulin-receptor band (430,000 daltons).

At 16°C photoaffinity-labeled insulin receptors remain on the cell surface. Following trypsinization, all the labeled insulin receptors are converted into smaller peptide fragments. At 37°C, however, a portion of the labeled receptors are internalized and become insensitive to added trypsin.[12,14,15] Trypsin sensitivity is therefore used to distinguish between insulin receptors on the cell surface and those inside the cell. With this approach, a decrease in the amount of insulin receptors which are susceptible to tryptic digestion represents a decrease in the number of receptors on the cell surface, and this loss of receptors from the cell surface is a measure of insulin-receptor internalization.

C. Polyacrylamide Gel Electrophoresis and Autoradiography

The solubilized samples are centrifuged at 1400 × g for 25 min in the cold. The fat is discarded, and the solubilized material is boiled for 5 min in Laemmli's sample buffer with or without dithiothreitol (DTT) before application to the gels. SDS polyacrylamide gel electrophoresis of the NAPA-DP-insulin-labeled proteins is performed as described by Laemmli[57] with some modifications. One-dimensional slab gel electrophoresis is carried out using a 5–15% linear gradient of acrylamide as the resolving gel and a 3% stacker gel. The bisacrylamide : acrylamide ratio in the resolving gel is 1 : 100 and in the stacking gel, 3 : 100. Electrophoresis is performed at a constant current of 30 mA for approximately 5 hr. Protein molecular-weight standards used for calibration are fibronectin (440,000), thyroglobulin (330,000), ferritin (200,000 and 18,500), myosin (200,000), β-galactosidase (116,500), phosphorylase b (94,000), BSA (67,000), catalase (60,000), ovalbumin (43,000), lactate dehydrogenase (36,000), carbonic anhydrase (30,000), trypsin inhibitor (20,100), and α-lactalbumin (14,400).

After staining with Coomassie brilliant blue and destaining, the gels are either sliced and counted in a gamma counter or dried and subjected to autoradiography at − 70°C using Kodak X-Omat film.

D. Chromatography on Sephadex G-50 Columns

Cells are solubilized in 1% Triton X-100 containing bacitracin (2 mg/ml), 2 mM PMSF, and 1 mM NEM. The solubilized material (2 ml) is applied to Sephadex G-50 columns (50 × 1 cm) and eluted with 1 M acetic acid containing 0.1% BSA into 0.8-ml fractions.

IV. RESULTS AND DISCUSSION

Binding of [^{125}I]-NAPA-DP-insulin (40 ng/ml) to adipocytes reached equilibrium within 30 min at 16°C. All the bound [^{125}I]-NAPA-DP-insulin dissociated within 90 min at 37°C in the absence of photolysis. In contrast, after UV irradiation, 20% of the specifically bound insulin remained associated with the cells, indicating that this fraction of the bound ligand was covalently linked to insulin receptors after photolysis. The proteins covalently labeled with [^{125}I]-NAPA-DP-insulin were identified by electrophoresis and autoradiography (Fig. 2). When electrophoresis was performed under nonreducing conditions, essentially one protein band was labeled (Fig. 2A, lane 1). The labeling of this band was specific in that the labeling was abolished in the presence of excess unlabeled porcine insulin (Fig. 2A, lane 2). The photosensitivity of this labeling was demonstrated by the fact that omission of photolysis resulted in the absence of labeling (data not shown). The apparent molecular weight of this protein band was approximately 430,000. When photolabeled adipocytes were treated with dithiothreitol prior to electrophoresis, three bands were specifically labeled: a predominant binding subunit with an apparent molecular weight of 125,000 and less intensely labeled bands with apparent molecular weights of 115,000 and 90,000 (Fig. 2A, lane 3). The labeling of all three bands was absent in the presence of excess porcine insulin (Fig. 2A, lane 4) and in the absence of photolysis. The molecular weights of the receptor proteins were determined by running the samples on a 5–15% porous acrylamide gel along with 13 proteins of known molecular weights. A semilog plot of the molecular weights versus electrophoretic mobilities of the standard proteins is shown in Fig. 2B. The molecular weights of the radiolabeled receptor proteins were derived from this standard curve. The molecular-weight determination of the nonreduced receptor is significantly higher than that reported previously by our laboratory[12] and others[58] and is most likely due to the porous gradient gel system used and the wider range of molecular standards used for calibration. It should be emphasized that the molecular-weight determinations for the insulin receptor are estimates, since glycosylated proteins have anomalously slow mobilities in SDS acrylamide gels.[59]

The results shown in Fig. 2A and were obtained by solubilizing adipocytes, after photoaffinity labeling, in 3% SDS containing 1 mM NEM and 2 mM PMSF for 20 min at room temperature. NEM and PMSF were included in the solubilizing buffer, since it has been suggested that disulfide bond reduction[60] and proteolysis[61] are responsible for generating multiple forms of the insulin receptor. If NEM was omitted

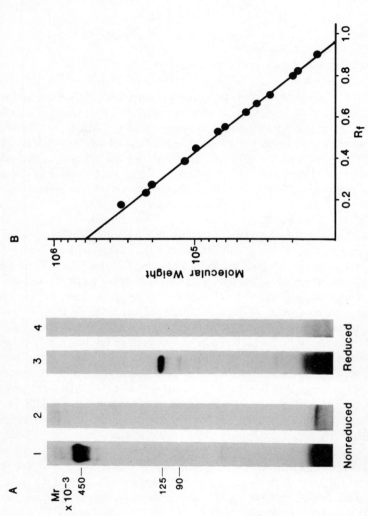

Figure 2. Identification and molecular-weight determination of photoaffinity-labeled insulin receptors on rat adipocytes. (A) Autora-diogram showing proteins on rat adipocytes covalently labeled with [^{125}I]-NAPA-DP-insulin. Isolated adipocytes were incubated with [^{125}I]-1-NAPA-DP-insulin (40 ng/ml) at 16°C for 30 min in the absence (lanes 1 and 3) or in the presence (lanes 2 and 4) of excess unlabeled insulin. After photolysis, adipocytes were solubilized and subjected to electrophoresis on a 5–15% porous acrylamide gel, as described in the text. The apparent molecular weights of the specifically labeled bands are indicated at the left of the figure. (B) Semilog plot of molecular weight versus electrophoretic mobility of standard proteins subjected to electrophoresis as described above. The 13 proteins used for calibration are given in the text.

from the solubilizing buffer, three forms of the receptor were observed when electrophoresis was carried out under nonreducing conditions, even in the presence of a protease inhibitor. The apparent molecular weights of these proteins were 430,000, 350,000, and 270,000 (Fig. 3, lane 4). The amount of each of these different receptor species appeared to be somewhat dependent on the nature of the detergent used initially to solubilize the receptors from the adipocytes after photoaffinity labeling. When samples containing multiple forms of the nonreduced receptor were treated with 50 mM dithiothreitol prior to electrophoresis, the pattern of labeled proteins was the same as that shown in Fig. 2A. Thus, the heterogeneity of receptor forms was only apparent under nonreducing conditions. These data confirm the finding of Yip and Moule that NEM markedly affects the nonreduced receptor by converting the two lower-molecular-weight species into the highest-molecular-weight form of the receptor.[60] On the basis of the observed effects of sulfhydryl alkylating reagents, Yip has postulated that in adipocytes the insulin receptor exists as three redox forms and that their interconversion results from oxidation and reduction reactions catalyzed by a factor within the plasma membrane which is coupled to the redox state of the cell. Using cell-surface labeling techniques combined with immunoprecipitation, other investigators have identified multiple nonreduced forms of the insulin receptor on IM-9 lymphocytes[62] and Reuber

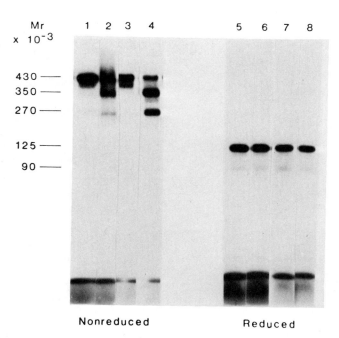

Figure 3. Multiple forms of the nonreduced insulin receptor. Binding of [^{125}I]-NAPA-DP-insulin to rat adipocytes and photolysis were carried out at 16°C. Adipocytes were solubilized in 3% sodium dodecyl sulfate in the presence (lanes 1 and 5) or absence (lanes 2 and 6) of 1 mM N-ethylmaleimide (NEM), or in 1% Triton X-100 in the presence (lanes 3 and 7) or absence (lanes 4 and 8) of 1 mM NEM. The apparent molecular weights of the specifically labeled bands are indicated by the arrows.

H-35 rat hepatoma cells,[63] which range in molecular weight from 520,000 to 190,000. The physiological role of these forms in the binding and action of insulin is under current investigation. Recent studies have indicated that changes in the proportion of the different forms of the insulin receptor are associated with changes in the overall affinity of the cell for insulin. In addition, these forms appear to have different functional roles. For example, downregulation of insulin receptors in Reuber H-35 rat hepatoma cells produces a preferential loss of a high-molecular-weight form of the insulin receptor.[63] Since the following experiments were carried out in the presence of NEM, where only one form of the receptor is apparent, our studies were not able to detect whether the various forms were internalized and degraded differently.

Having demonstrated that insulin receptors could be specifically photoaffinity labeled at 16°C, the next step was to warm the cells to 37°C to determine whether the labeled receptors were internalized and processed. Results from a typical experiment are displayed in the autoradiogram of Fig. 4. At 16°C, the specifically labeled insulin receptor with an apparent molecular weight of 430,000 (Fig. 4, lane 1) is readily seen. If cells were treated with trypsin prior to solubilization, all the labeled insulin receptors were converted into smaller tryptic fragments and (Fig. 4, lane 2). The predominant

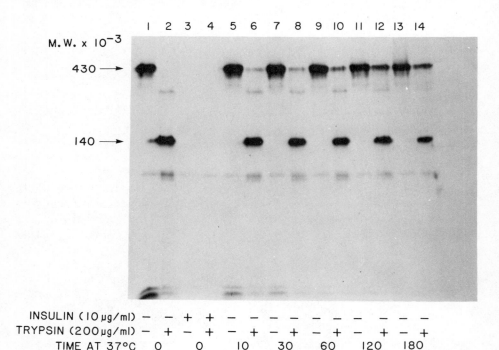

Figure 4. Autoradiogram of surface and internalized insulin receptors following incubation of adipocytes at 37°C. Adipocytes were photoaffinity labeled at 16°C. After incubation at 37°C, one-half the cells were solubilized immediately and the remainder were treated with trypsin (200 μg/ml, 37°C, 10 min) before solubilization. The solubilized samples were subjected to electrophoresis on 5–15% acrylamide gels.

tryptic fragment had an apparent molecular weight of 140,000. This finding indicated that all the photoaffinity-labeled receptors were on the cell surface at 16°C. When cells were photoaffinity labeled at 16°C and then incubated at 37°C, the amount of radiolabeled insulin receptors decreased as a function of time (Fig. 4, lanes 5, 7, 9, 11, 13). To determine the distribution of receptors between the cell surface and the intracellular compartment as a function of time at 37°C, a separate aliquot of cells was exposed to trypsin prior to solubilization at each interval. These experiments demonstrated that an increasing proportion of the labeled insulin receptors became insensitive to enzymatic degradation by trypsin (Fig. 4, lanes 6, 8, 10, 12, 14), indicating internalization of photolabeled insulin receptors. Thus, the autoradiogram presented in Fig. 4 showed (1) a decrease in the total amount of radioactivity in the cellular insulin receptor band, indicating degradation of hormone–receptor complexes, and (2) an increase in the trypsin-insensitive (intracellular) pool of insulin receptors. Similar experiments using the same photoreactive analog of insulin were performed on isolated rat hepatocytes by Fehlman et al.[14] After photoaffinity labeling at 16°C, these investigators noted two reduced forms of the insulin receptor, of 130,000 and 125,000 M_r. These bands disappeared after exposure of the cells to trypsin, indicating that the labeled receptors were at the cell surface. In agreement with our studies on isolated adipocytes, photoaffinity-labeled hepatocytes incubated at 37°C showed a marked decrease in the trypsin sensitivity of the labeled receptor subunits by 30–60 min, reflecting internalization of the hormone–receptor complexes. In contrast to our studies, however, no detectable loss was observed in the total number of receptors associated with the cells. Interestingly, during the subsequent 4-hr of incubation at 37°C, the receptors became trypsin sensitive again, suggesting that they recycled back to the plasma membrane. The results obtained with photoaffinity-labeling experiments on isolated hepatocytes are analogous to the findings of Krupp and Lane; these workers showed that down-regulation of insulin receptors in chick liver cells involved translocation of cell-surface receptors to an intracellular compartment with no change in the rate of receptor degradation measured by heavy-isotope density-shift procedures.[38,39]

Our own studies on adipocytes held the possibility that covalently bound [^{125}I]-NAPA-DP-insulin could be cleaved from the receptor following internalization. This would leave the receptor unlabeled but possibly undegraded. If this occurred, the decrease in total radioactivity in the receptor bands would represent cleavage of the radiolabeled insulin analog for the receptor rather than receptor degradation. Experiments were performed to test this possibility directly. Control studies were first carried out on cells not subjected to photolysis. In these experiments, adipocytes were incubated with 10 ng/ml [^{125}I]-NAPA-DP-insulin at 16°C for 30 min. The cells were then warmed to 37°C for 1 hr to allow for internalization of hormone–receptor complexes. The cells were then incubated in a barbitol–acetic acid buffer (pH 3.5) for 6 min to dissociate cell-surface radioactivity.[64] With this approach, any radioactivity remaining associated with the cells would represent nonextractable (or intracellular) radiolabeled material.[64] After extraction, the cells were solubilized in 1% Triton X-100, 2 mM PMSF, 1 mM NEM, and the extracts were subjected to chromatography on Sephadex G-50 columns. Figure 5A shows the elution profile of cell-associated radioactivity.

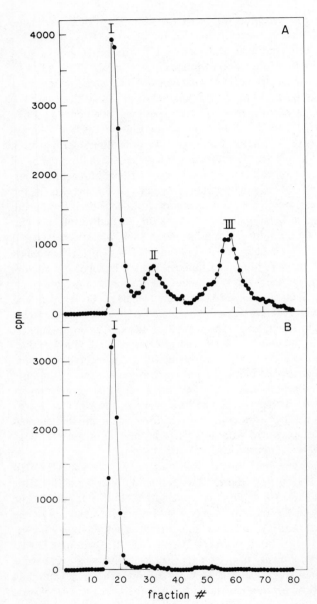

Figure 5. Profile of cell-associated radioactivity after chromatography on Sephadex G-50 columns. (A) Control experiments in which adipocytes were incubated with 10 ng/ml [^{125}I]-NAPA-DP-insulin at 16°C for 30 min. The cells were further incubated at 37°C for 1 hr to allow for internalization. After acid extraction to remove surface radioactivity, the cells were solubilized and subjected to chromatography on G-50 Sephadex. (B) Similar experiments carried out on adipocytes photoaffinity labeled with [^{125}I]-NAPA-DP-insulin.

Three peaks were detected: Peak I eluted with the void volume, peak II represents intact NAPA-DP-insulin, since it migrated with the insulin standard, and peak III represents small degradation products of NAPA-DP-insulin. Recovery of the radioactivity applied to the column was greater than 90%. Of the eluted counts, 42% were recovered in peak I, 18% in peak II, and 37% in peak III. As previously described for iodinated native insulin,[64] Fig. 5A indicates that [^{125}I]-NAPA-DP-insulin bound to insulin receptors on adipocytes is internalized within the cell. Following internalization, the insulin analog dissociates from the receptor. A portion of the [^{125}I]-NAPA-DP-insulin is recovered intact inside the cell (peak III), and some is degraded into low-molecular-weight products (peak III). The precise nature of the material in the void volume is unknown, although it may represent [^{125}I]-NAPA-DP-insulin–receptor complexes that have not dissociated from the receptor. Parallel experiments were performed, except that the cells were subjected to photolysis after the binding of [^{125}I]-NAPA-DP-insulin had reached equilibrium at 16°C. Thus, cells were incubated at 16°C with 10 ng/ml [^{125}I]-NAPA-DP-insulin, exposed to UV light, and washed to remove noncovalently bound hormone. After further incubation at 37°C for 1 hr, the cells were solubilized and the extracts applied to Sephadex G-50 columns. The elution profile derived from the photoaffinity-labeled cells is shown in Fig. 2B. In contrast to the results obtained in control cells, only a single peak that eluted with the void volume (peak I) was apparent. Electrophoresis of this peak demonstrated that it consisted of intact radiolabeled insulin receptors (data not shown). No free [^{125}I]-NAPA-DP-insulin was detected within the cell, indicating that the photoprobe remains covalently attached to the insulin receptor after internalization. Photoaffinity labeling surface receptors with [^{125}I]-NAPA-DP-insulin therefore appears to be a valid method for following the subsequent metabolism of these receptors.

Quantitative measurements of the time course of internalization and degradation of insulin receptors involved photoaffinity labeling adipocytes at 16°C and then incubating the cells at 37°C as described in Fig. 4. At the appropriate intervals, the cells were solubilized and subjected to electrophoresis. Rather than subject the gels to autoradiography, the gels were cut into 2-mm slices and counted in a gamma counter. Typical profiles of the radioactivity associated with cellular proteins at the end of the 16°C incubation and after an additional incubation at 37°C for 120 min are represented in Fig. 6. When electrophoresis was performed under nonreducing conditions, most of the radioactivity was in a peak corresponding to the 430,000–450,000-M_r band seen by autoradiography in Fig. 2. Radioactivity at this peak was abolished when cells were incubated at 16°C in the presence of excess unlabeled insulin. When adipocytes were photoaffinity labeled and then incubated at 37°C for 120 min, the radioactivity at this peak decreased by approximately 50%. In addition to the 430,000-M_r peak, a small amount of radioactivity migrated beyond the tracker dye. This peak decreased when cells were incubated a 37°C for 120 min and may represent noncovalently bound NAPA-DP-insulin associated with the cells when they were solubilized. Degradation was assessed by counting the radioactivity in the insulin-receptor peak (430,000 M_r) as a function of time at 37°C. Susceptibility to tryptic digestion was used to distinguish between insulin receptors on the cell surface and those inside the cell. Accordingly, a decrease in the number of receptors susceptible to trypsin represents the loss of receptors from the cell surface, i.e., internalization. Results of these experiments are shown in Fig. 7. The amount of labeled insulin receptors recovered from the cell (filled circles) decreased rapidly during the first 60 min at 37°C. After this time, there was little net loss of radiolabeled insulin receptors from the cell. A first-order rate plot demonstrated a clearly biphasic degradation curve. An estimated half-life of the initial

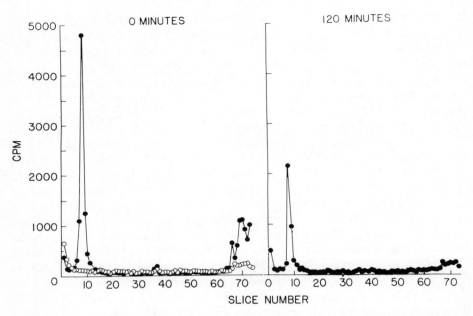

Figure 6. Profile of radioactivity in slices of polyacrylamide gels containing proteins labeled with [^{125}I]-NAPA-DP-insulin. Adipocytes were incubated with [^{125}I]-NAPA-DP-insulin at 16°C for 30 min and then subjected to photolysis. After washing, cells were either solubilized immediately or incubated at 37°C for 120 min prior to solubilization. The tracker dye migrated in slice 60.

phase was 1.5 hr. Insulin receptors (open circles) were internalized at a significantly faster rate than the loss of receptors from the cell, resulting in an accumulation of receptors inside the cell (Fig. 7, inset). The proportion of intracellular insulin receptors reached an apparent steady state after 30 min, representing about 20% of the labeled receptors originally on the cell surface. The faster rate of internalization relative to the net loss of receptors from the cell and the steady-state distribution of receptors between the intracellular and surface pools indicates that internalization is not the rate-limiting step in the overall loss of insulin receptors from the cell. In similar studies carried out on 3T3-L1 adipocytes, photoaffinity-labeled insulin receptors were degraded with a half-life of approximately 7 hr.[37] This rate is considerably slower than that observed in isolated adipocytes and the kinetics are first order as opposed to curvilinear. This suggests a difference in insulin-receptor metabolism by these two cell types. Photoaffinity-labeled receptors on 3T3-L1 adipocytes were degraded at the same rate as noncovalent insulin–receptor complexes, indicating that the inability of insulin to dissociate from the receptor had no effect on the metabolism of the insulin receptor.[36,37]

Receptor occupancy appears to be an important factor in modulating the rate and perhaps the mechanism of receptor degradation as well. When cells are incubated in the absence of hormone, insulin receptors have a half-life of approximately 10 hr. This rate of degradation has been measured in 3T3-L1 preadipocytes[36,41] and chick liver cells,[38,39]

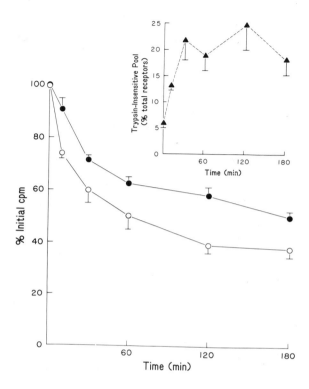

Figure 7. Internalization and degradation of insulin receptors at 37°C. Experiments were carried out as described in Fig. 4. Instead of being subjected to autoradiography, the gels were sliced and counted in a gamma counter. Degradation (●) was assessed by quantitating the amount of radioactivity in the IR band as a function of time at 37°C. Susceptibility to tryptic digestion was used to distinguish between insulin receptors on the cell surface and those inside the cell. A decrease in the number of receptors susceptible to trypsin represents loss of receptors from the cell surface (○). Inset shows the accumulation of insulin receptors inside the cell.

using heavy-iosotope density-shift procedures and in IM-9 lymphocytes using biosynthetic-labeling techniques.[43] In contrast to the basal turnover rate of insulin receptors, when receptors are occupied with insulin, they appear to undergo endocytosis and a portion of them are degraded via a lysosomal pathway. Data from our studies indicate that this pathway is rapid ($t_{1/2}$, 1–2 hr) and sensitive to cycloheximide and cloroquine. Results from other laboratories also support the concept that occupancy regulates degradation of receptors. For example, Kasuga *et al.* reported that the half-life of insulin receptors decreased from 9–12 hr to 3 hr when cultured cells were exposed to insulin.[43] In addition, insulin enhanced the degradation rate of its receptors in 3T3-L1 adiopocytes as assessed by heavy-isotope density-shift procedures.[36]

The effects of cloroquine (0.2 mM) on the degradation, internalization, and intracellular accumulation of photoaffinity-labeled insulin receptors are shown in Fig. 8. Chloroquine markedly inhibited the net loss of insulin receptors from the cell (Fig. 8A) but had no effect on the internalization of surface receptors (Fig. 8B). As a result, the intracellular pool of insulin receptors increased by more than twofold compared with control cells (Fig. 8C). The effect of chloroquine on receptor degradation raises the possibility that lysosomes are involved in the degradation of internalized receptors. On the other hand, chloroquine can also interfere with the processing of internalized cell-surface components by inhibiting fusion of endocytotic vesicles with other intracellular membrane structures.[65] Regardless of the precise mechanisms of this agent,

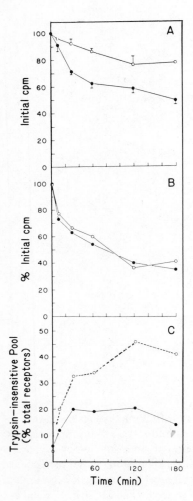

Figure 8. Effect of cycloheximide on the degradation (A), internalization (B), and accumulation (C) of insulin receptors with the cell. Experiments were performed as described in Fig. 7 in the absence (●) and presence (○) of 3.5 μM cycloheximide.

the results clearly indicate that insulin receptors are degraded by a chloroquine-sensitive pathway.

As shown in Fig. 7, after 1 hr at 37°C, the disappearance of insulin receptors from the cell surface slowed markedly with minimal overall loss of insulin receptors from the cell. The apparent plateau in the time course of internalization and overall degradation could be due to cessation of these processes over time. Alternatively, this plateau (approximately 30–60 min at 37°C) could reflect a new dynamic steady state in which the process of receptor internalization is nearly balanced by receptor recycling with only minimal net degradation of the accumulated intracellular receptors. To distinguish between these possibilities, the photolabeled cells were treated with chloroquine at 60 min when net loss of receptors was minimal. One might predict that if both internalization and intracellular degradation of insulin receptors had ceased at 60 min, then chloroquine should have little or no effect when added at this time. However, if the apparent plateau reflects a new equilibrium reached between the processes of internalization and recycling, then chloroquine should cause a perturbation in the plateau. The experiments were carried out by photoaffinity-labeling surface receptors and allowing internalization and degradation to proceed for 60 min. At this time, 0.2 mM chloroquine was added, and the cells were incubated for an additional 2 hr at 37°C. The results of this investigation are shown in Fig. 9. As can be seen, chloroquine had no significant effect on the overall loss of the total cellular receptors. However, a marked change was observed in the distribution of cellular receptors between the intracellular and surface pools. Thus, the addition of chloroquine resulted in a marked increase in loss of receptors from the cell surface, accompanied by a corresponding twofold increase in the intracellular pool of surface-derived insulin receptors. This redistribution was readily appreciated 1 hr after the addition of chloroquine, and a new apparent steady state was reached. These results strongly suggest that the plateau of net receptor loss from the cell surface observed in control cells represents a new steady state between loss of receptors from the cell surface (i.e., internalization) and

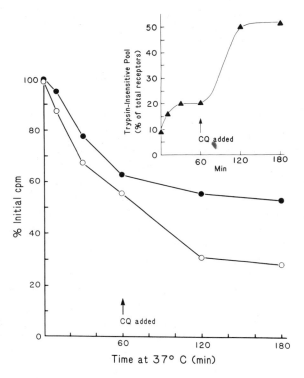

Figure 9. Effect of the addition of chloroquine after incubating photoaffinity-labeled cells at 37°C for 1 hr. Experiments were performed as described in Fig. 7, except that chloroquine (CQ), 0.2 mM, was added at the 1-hr time point.

recycling of receptors back to the plasma membrane. We interpret these data in the following manner. Once insulin binds to receptors, the hormone–receptor complexes are internalized and degraded. Since internalization is more rapid than the intracellular degradation of insulin receptors, a pool of surface-derived receptors accumulates inside the cell. The receptors in this pool are able to recycle back to the plasma membrane. After 60 min at 37°C, recycling from this pool is essentially equal to the rate of internalization, resulting in little net loss of receptors from the cell surface. At this point, degradation of receptors from the intracellular pool is minimal, so that only minimal net loss of receptors from the cells occurs. When chloroquine is added, a redistribution of insulin receptors is seen; there is a marked decrease in surface receptors with a concomitant increase in the intracellular pool resulting in no net loss of receptors from the cell. This could theoretically be due to an increase in internalization or an inhibition of recycling. Since chloroquine did not increase internalization (Fig. 8), the most likely action is an interference with recyling. This possibility is consistent with a chloroquine-induced inhibition of recycling reported by Marshall et al.[66] and with the postulated mechanisms of action for chloroquine proposed by McKanna et al.[65] Thus, these experiments directly indicate that photoaffinity labeled insulin receptors on adipocytes recycle back to the plasma membrane following internalization. As already mentioned, it has been previously reported that insulin receptors on hepatocytes recycle back to the plasma membrane; however, the time course for recycling was much slower than that observed in the present studies.[14]

V. SUMMARY

Regardless of the methods used, all the data from studies of the metabolism of insulin receptors are consistent with the general scheme portrayed in Fig. 10. After the initial binding event, insulin-receptor complexes are internalized via endocytotic mechanisms. The internalization of insulin has been demonstrated in morphological studies using fluorescent[67] and autoradiographic techniques[68] and in biochemical studies[69,70] using radiolabeled insulin. More recently, internalization of insulin receptors has been measured indirectly in isolated adipocytes by measuring the loss of trypsin-sensitive binding activity in solubilized receptor assays in response to insulin.[11] In vivo studies have also been reported using rat liver, in which the distribution of insulin-binding sites among subcellular fractions after exposure to insulin has been measured.[13] Photoaffinity-labeling studies have also provided direct evidence that occupied insulin receptors are internalized within a variety of cell types by either measuring the loss of trypsin sensitivity[12,14,15] or by autoradiography.[14]

Once inside the cell, there are at least three possible fates for the internalized insulin receptor. Receptors can be recycled back to the cell surface,[14,66] degraded (possibly through a lysosomal mechanism)[15,36] or sequestered undegraded within the cell for a period of time. Although the initial internalization event is clearly accelerated when receptors are occupied with insulin, the balance of recycling, degradation, and

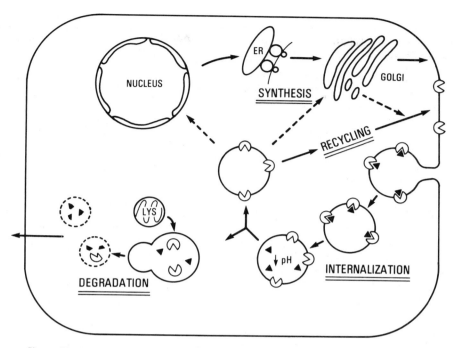

Figure 10. Schematic representation of the fate of occupied insulin receptors in target cells.

intracellular sequestration appears to differ among cell types. The relative magnitudes of recycling, degradation, and sequestration of internalized receptors modulates the final concentration of cellular receptors and the proportion of these receptors found on a cell surface or the cell interior. It seems likely that these events determine the acute effect of insulin to downregulate its receptor. Longer-term regulation of receptor number may involve alterations in the rates of biosynthesis and processing of newly synthesized receptors. While the internalized insulin receptor seems to have multiple fates after the initial endocytotic event, this does not appear to be the case for the ligand. Thus, the current evidence indicates that after endocytosis, the pH of the endocytotic vesicle decreases, permitting rapid dissociation of insulin from its receptor. Once this occurs, the insulin molecule is targeted for ultimate degradation and this is at least partially mediated through lysosomes.[39]

Although techniques are now available to distinguish between surface and intracellular receptors, and the rates of flux of receptors through the various pathways can be assessed, little is known about the subcellular localization of internalized receptors or their potential physiologic function. Recent studies by Wang *et al.* have indicated that following internalization a portion of the receptors lost from the cell surface could be recovered in a Golgi-enriched membrane fraction.[35] The exact identity of the intracellular vesicles containing these receptors remains to be clarified. It seems possible that internalized insulin receptors may serve as a transducing mechanism conveying information from the surface to the cell interior and that internalized receptors exert biological effects at their subcellular sites of localization. The recent demonstration that the subunit of the insulin receptor contains tyrosine-specific protein kinase activity[21-26] is consistent with this view; it is possible that internalized receptors are specifically directed toward intracellular sites, where they mediate phosphorylation reactions eliciting some biologic effects of insulin.

VI. REFERENCES

1. Langley, J.N., 1905, *J. Physiol. (Lond.)* **33**:374–413.
2. Carpenter, G., and Cohen, S., 1976, *J. Biol. Chem.* **71**:159–171.
3. Ascoli, M., and Puett, D., 1977, *FEBS Lett.* **75**:77–82.
4. Terris, S., Hofmann, C., and Steiner, D.F., 1979, *Can. J. Biochem.* **57**:459–468.
5. Djiane, J., Clauser, H., and Kelly, R.A., 1979, *Biochem. Biophys. Res. Commun.* **90**:1371–1378.
6. Terris, S., and Steiner, D.F., 1975, *J. Biol. Chem.* **250**:8389–8398.
7. King, A.C., and Cuatrecasas, P., 1981, *N. Engl. J. Med.* **305**:77–88.
8. Goldstein, J.C., Anderson, R.G.W., and Brown, M.S., 1979, *Nature* **297**:679–685.
9. Wall, D.A., Wilson, G., and Hubbard, A.L., 1980, *Cell* **21**:79–93.
10. Willingham, M.C., Maxfield, F.R., and Pastan, I.H., 1979, *J. Cell. Biol.* **82**:614–625.
11. Green, A., and Olefsky, J.M., 1982, *Science* **200**:1283–1284.
12. Berhanu, P., Olefsky, J.M., Tsai, P., Thamm, P., Saunders, D., and Brandenburg, D., 1982, *Proc. Natl. Acad. Sci. USA* **79**:4069–4073.
13. Desbuquois, B., Lopez, S., and Burlet, H., 1982, *J. Biol. Chem.* **257**:10852–10860.
14. Fehlmann, M., Carpentier, J.-L., Le Cam, A., Thamm, P., Saunders, P., Brandenburg, D., Orci, L., and Freychet, P., 1982, *J. Cell. Biol.* **93**:82–87.
15. Heidenreich, K.A., Berhanu, P., Brandenburg, D., and Olefsky, J.M., 1983, *Diabetes* **32**:1001–1009.
16. Podlecki, D.A., Kao, M., Tsai, P., Brandenburg, D., Lasher, R.S., and Olefsky, J.M., *J. Biochem.* (submitted).
17. Carpenter, G., King, L., Jr., and Cohen, S., 1979, *J. Biol. Chem.* **254**:4884–4891.
18. Cohen, S., Ushiro, H., Stoscheck, C., and Chinkers, M., 1982, *J. Biol. Chem.* **257**:1523–1531.
19. Rubin, R.A., O'Keefe, E.J., and Earp, H.S., 1982, *Proc. Natl. Acad. Sci. USA* **79**:776–780.
20. Rubin, R.A. and Earp, H.S., 1983, *Science* **219**:60–63.
21. Kasuga, M., Karlsson, F.A., and Kahn, C.R., 1982, *Science* **215**:185–187.
22. Zick, Y., Kasuga, M., Kahn, C.R., and Roth, J., 1983, *J. Biol. Chem.* **258**:75–80.
23. Petruzelli, L.M., Ganguly, S., Smith, C.J., Cobb, M.H., Rubin, C.S., and Rosen, O.M., 1982, *Proc. Natl. Acad. Sci. USA* **79**:6792–6796.
24. Shia, M.A., and Pilch, P.F., 1983, *Biochemistry* **22**:717–721.
25. Auruch, J., Nemenoff, R.A., Blackshear, P.J., Pierce, M.W., and Osathanondh, R., 1982, *J. Biol. Chem.* **257**:15162–15166.
26. Van Obberghen, E., Rossi, B., Kowalski, A., Gazzano, H., and Ponzio, G., 1983, *Proc. Natl. Acad. Sci. USA* **80**:945–949.
27. Jacobs, S., Kull, F.C., Jr., Earp, H.S., Svoboda, M.E., Van Wyk, J.J. and Cuatracasas, P., 1983, *J. Biol. Chem.* **258**:9581–9584.
28. Ek, C., and Heldin, C.H., 1982, *J. Biol. Chem.* **257**:10486–10492.
29. Nishimura, J., Huang, J.S., and Deuel, T.F., 1982, *Proc. Natl. Acad. Sci. USA* **79**:4303–4307.
30. Gavin, J.R. III, Roth, J., Neville, D.M., Jr., DeMeyts, P., and Buell, N.D., 1974, *Proc. Natl. Acad. Sci. USA* **71**:84–88.
31. Marshall, S., and Olefsky, J.M., 1980, *J. Clin. Invest.* **66**:763–772.
32. Catt, K.J., Harwood, J.P., Aquilera, G., and Dufau, M.L., 1979, *Nature* **280**:109–116.
33. Kahn, C.R., Neville, D.M., Jr., and Roth, J., 1973, *J. Biol. Chem.* **248**:244–250.

34. Olefsky, J.M., 1976, *J. Clin. Invest.* **57**:1165–1172.
35. Wang, C., Sonne, O., Hedo, J.A., Cushman, S.W., and Simpson, I.A., 1983, *J. Biol. Chem.* **258**:5129–5134.
36. Ronnett, G.V., Knutson, V.P., and Lane, M.D., 1982, *J. Biol. Chem.* **257**:4285–4291.
37. Reed, B.C., 1983, *J. Biol. Chem.* **258**:4424–4433.
38. Krupp, M., and Lane, M.D., 1981, *J. Biol. Chem.* **256**:1689–1694.
39. Krupp, M.N., and Lane, M.D., 1982, *J. Biol. Chem.* **257**:1372–1377.
40. Reed, B.C., and Lane, M.D., 1980, *Proc. Natl. Acad. Sci. USA* **77**:285–289.
41. Reed, B.C., Ronnett, G.V., Clements, P.R., and Lane, M.D., 1981, *J. Biol. Chem.* **256**:3917–3925.
42. Lang, U., Kahn, C.R., and Harrison, L.C., 1980, *Biochemistry* **19**:64–70.
43. Kasuga, M., Kahn, C.R., Hedo, J.A., Van Obberghen, E., and Yamada, K.M., 1981, *Proc. Natl. Acad. Sci. USA* **78**:6917–6921.
44. Harrison, L.C., Itin, A., Kasuga, M., and Van Obberghen, E., 1982, *Diabetologia* **22**:233–238.
45. Van Obberghen, E., Kasuga, M., Le Cam, A., Hedo, J.A., Itin, A., and Harrison, L.C., 1981, *Proc. Natl. Acad. Sci. USA* **78**:1052–1056.
46. Deutsch, P.J., Wan, C.F., Rosen, O.M., and Rubin, C.S., 1983, *Proc. Natl. Acad. Sci. USA* **80**:133–136.
47. Jacobs, S., Kull, F.C., Jr., and Cuatracasa, P., 1983, *Proc. Natl. Acad. Sci. USA* **80**:1228–1231.
48. Hedo, J.A., Kahn, C.R., Hayashi, M., and Yamada, K.M., 1983, *J. Biol. Chem.* **258**:10020–10026.
49. Hedo, J.A., Cushman, S.W., and Simpson, I.A., 1982, *Diabetes* **31**(Suppl. 2):2A.
50. Lane, M.D., Ronnett, G.V., Knutson, V.P., and Simpson, T.L., 1983, In: *Second International Symposium on Insulin Receptors,* p. 36.
51. Yip, C.C., Yeung, C.W.T., and Moule, M.L., 1978, *J. Biol. Chem.* **253**:1743–1745.
52. Jacobs, S., Hazum, E., Schechter, Y., and Cuatracasas, P., 1979, *Proc. Natl.. Acad. Sci. USA* **76**:4918–4921.
53. Wisher, M.H., Baron, M.D., Jones, R.H., Sonksen, P.H., Saunders, D.J., Thamm, P., and Brandenburg, D., 1980, *Biochem. Biophys. Res. Commun.* **92**:492–498.
54. Yip, C.C., Yeung, C.W.Y., and Moule, M.L., 1980, *Biochemistry* **19**:70–76.
55. Thamm, P., Saunders, D., and Brandenburg, D., 1980, *Insulin: Chemistry, Structure and Function of Insulin and Related Hormones,* p. 309, Walter de Gruyter, Berlin.
56. Brandenburg, D., Diaconescu, C., Saunders, D., and Thamm, P., 1980, *Nature* **286**:821–822.
57. Laemmli, U.K., 1970, *Nature* **277**:680–685.
58. Massague, J., Pilch, P.F., and Czech, M.P., 1980, *Proc. Natl. Acad. Sci. USA* **77**:7137–7141.
59. Bretcher, M.S., 1971, *Nature New Biol.* **231**:229–232.
60. Yip, C.C., and Moule, M.L., 1982, *Fed. Proc.* **41**:1341.
61. Massague, J., Pilch, P.F., and Czech, M.P., 1981, *J. Biol. Chem.* **256**:3182–3190.
62. Kasuga, M., Hedo, J.A., Yamada, K.M., and Kahn, C.R., 1982, *J. Biol. Chem.* **257**:10392–10399.
63. Crettaz, M., and Kahn, C.R., 1983, In: *Second International Symposium on Insulin Receptors,* p. 11.
64. Olefsky, J.M., and Kao, M., 1982, *J. Biol. Chem.* **257**:8667–8673.
65. McKanna, J.A., Haigler, H.I., and Cohen, S., 1979, *Proc. Natl. Acad. Sci. USA* **76**:5689–5693.
66. Marshall, S., Green, A., and Olefsky, J.M., 1981, *J. Biol. Chem.* **256**:11464–11470.
67. Schlessinger, J., Schechter, Y., Willingham, M.C., and Pastan, I., 1978, *Proc. Natl. Acad. Sci. USA* **75**:2659–2663.
68. Groden, P., Carpentier, J.-L., Freychet, P., LeCam, A., and Orci, L., 1978, *Science* **200**:782–786.
69. Marshall, S., and Olefsky, J.M., 1979, *J. Biol. Chem.* **254**:10153–10160.
70. Suzuki, K., and Kono, T., 1979, *J. Biol. Chem.* **254**:9786–9794.

4

The Insulin Receptor Protein Kinase

C. Ronald Kahn, Morris F. White, Florin Grigorescu,
Sumiko Takayama, Hans U. Häring, and Marco Crettaz

I. INTRODUCTION

Phosphorylation reactions catalyzed by protein kinases rapidly and reversibly modify the activity of cellular enzymes. Considerable evidence indicates that phosphorylation and dephosphorylation of proteins play an important role in the action of insulin and other hormones.[1–4] Receptors for several growth factors and neurotransmitters[5–14] are also phosphorylated, and in some cases this phosphorylation is increased after binding of the hormone. In addition, protein kinases are associated with certain transforming retroviruses that have profound effects on cellular growth and metabolism.[14–20] These findings suggest that phosphorylation may be an early step in the transmembrane signal process. In 1981, we began to study the possibility that the insulin receptor undergoes a phosphorylation reaction.[21] In this chapter, we review briefly the structure of the insulin receptor kinase, describe the kinetic characteristics of the receptor kinase measured *in vitro*, compare and contrast the phosphorylation of the receptor in broken and intact cells, and finally describe the phosphorylation of the receptor in certain insulin-resistant states.

II. STRUCTURE OF THE INSULIN RECEPTOR KINASE

Using both biosynthetic and surface labeling techniques followed by immunoprecipitation with anti-insulin receptor serum and polyacrylamide gel electrophoresis, we have characterized the structure of the insulin receptor in cultured human lym-

C. Ronald Kahn, Morris F. White, Florin Grigorescu, Sumiko Takayama, Hans U. Häring, and Marco Crettaz ● Research Division, Joslin Diabetes Center and Department of Medicine, Brigham and Women's Hospital, Harvard Medical School, Boston, Massachusetts.

phocytes and rat hepatoma cells.[22] In both cell lines, the receptor consists of two distinct subunits: the α subunit ($M_r \sim 135,000$) and the β subunit ($M_r \sim 95,000$). Under nonreducing conditions, these subunits are disulfide linked to give several distinct receptor forms. These species represent the free α and β subunits, α–β heterodimers, α–α homodimers, and two higher oligomeric forms composed of α–β dimers with estimated molecular weights of 350,000 and 520,000. The model of the insulin receptor, β–α — α–β, shown in Fig. 1, represents the most commonly depicted structure. Biochemical characterization of the purified insulin receptor from human placenta indicates that the predominant form of the receptor is composed of this stoichiometry.[23] It is likely, however, that many other forms occur in the membrane, some of which may be physiologically interconvertible, depending on the redox state of the cell.[24]

Both subunits of the insulin receptor are glycoproteins and both are exposed on the external surface of cells.[25] On the basis of affinity-labeling studies, the α subunit appears to contain the insulin binding domain.[26] The α subunit does not appear to be a transmembrane protein, since it is not surface-labeled in inverted membrane vesicles; the β subunit appears to be a transmembrane protein.[27] These structural characteristics suggest that the two subunits have different functions: the α subunit recognizes the hormonal signal at the extracellular surface, whereas the β subunit is the effector and transmits this signal to the inner face of the plasma membrane and the cytosol.

III. PHOSPHORYLATION REACTIONS OF THE SOLUBILIZED INSULIN RECEPTOR KINASE

A. Insulin-Binding Domain and Tyrosine Kinase: Expressions of the Same Molecular Aggregate

Our characterization of the insulin receptor kinase has been accomplished primarily with insulin receptors purified partially from a rat hepatoma cell line called Fao[28] and with a highly purified preparation of insulin receptors from human placenta kindly provided by Fujita-Yamaguchi.[29] The Fao cell line, derived from the Reuber H-35 (H-4) hepatoma, is well differentiated with respect to hepatic function and contains a large number of insulin receptors.[30–32] Solubilized insulin receptors obtained from these cells can be purified partially by affinity chromatography on immobilized wheat germ agglutinin.[28] This procedure results in a 20-fold purification and nearly 100% yield of the insulin receptor with respect to the Triton X-100 extract.[33]

Phosphorylation of the insulin receptor was detected following incubation of the wheat germ-purified protein with [γ-^{32}P]-ATP, Mn^{2+}, and insulin. The receptor was identified by immunoprecipitation of the reaction mixture with anti-insulin receptor serum and separation of the proteins by sodium dodecylsulfate polyacrylamide gel electrophoresis SDS-PAGE. A representative autoradiogram (Fig. 2) shows that insulin stimulates the phosphorylation of the β subunit of the receptor about 10-fold.[34] This

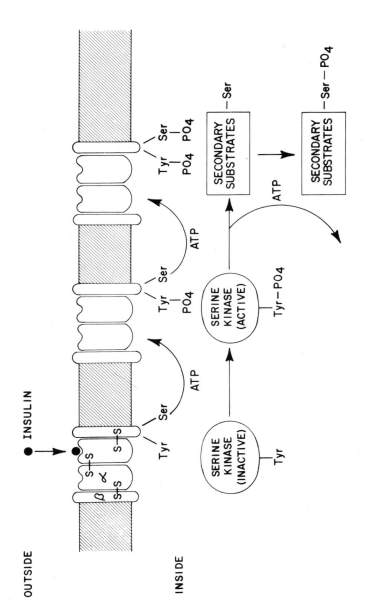

Figure 1. Schematic model of the insulin receptor kinase and possible relations with other cellular kinases.

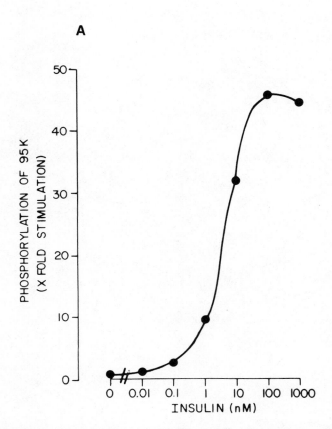

Figure 2. Phosphorylation of the insulin receptor from hepatoma cells (Fao) purified by chromatography on immobilized wheat germ agglutinin agarose. (A) The receptor was incubated in the presence of various insulin concentrations for 1 hr before the phosphorylation reaction. (B) Autoradiogram of a representative sodium dodecylsulfate polyacrylamide gel (reducing conditions).

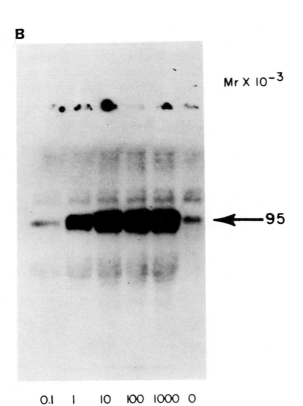

Figure 2. (continued)

autophosphorylation reaction has been confirmed using receptors isolated from a wide variety of cell types, including cultured human lymphocytes,[21] freshly isolated rat hepatocytes,[35] adipocytes,[36] 3T3-L1 adipocytes,[37] human placental membranes,[38,39] human erythrocytes,[40] and cultured mouse melanoma cells.[41]

Although all the experiments reported in the literature are consistent with the notion that the kinase activity is indeed present in the receptor, it was important to show that the insulin receptor purified to near homogeneity still retained protein kinase activity. The insulin receptor has been purified from human placental membranes by sequential affinity chromatography on immobilized wheat germ agglutinin agarose and immobilized insulin-succinyldiaminodipropyl-Sepharose by Fujita-Yamaguchi et al.[29] The insulin-binding activity of this preparation is ~4200 pmole/mg of protein, which represents a 2500-fold purification with respect to the solubilized placental membranes. This purified insulin receptor complex has a specific insulin binding activity of 1.5–1.8 moles of insulin per mole of receptor. This calculation is consistent with the structural model (Fig. 1), which suggests two insulin-binding sites per receptor. By SDS-PAGE, the highly purified receptor is shown to consist of two major proteins: the α subunit and β subunit (Fig. 3). In addition, a peptide of 52,000 M_r is detected in this preparation and is believed to be a degradation product of the β subunit.[23] All three peptides are immunoprecipitated by anti-insulin receptor serum.[42]

When the highly purified insulin receptor from human placenta is incubated in solution with Mn^{2+} and [γ-^{32}P]-ATP, only a single phosphoprotein is detected (Fig. 3). This phosphoprotein has a molecular weight of 95,000 and can be identified as the β subunit of the receptor. As with the partially purified receptor, this phosphoprotein is immunoprecipitated by anti-insulin receptor serum.[42] Increasing insulin concentration stimulates [^{32}P]phosphate incorporation into the β subunit. The highly purified insulin receptor from human placenta immunoprecipitated by anti-insulin receptor serum retains the insulin-stimulated protein kinase activity.[42] These data strongly suggest that the kinase activity is intrinsic to the insulin receptor. This notion is supported by two additional findings: (1) the protein kinase activity exists in the immunoprecipitate of a complex formed by insulin, insulin receptor, and insulin antibodies,[42a] (also M. Crettaz and C.R. Kahn, unpublished results), and (2) insulin receptors immobilized on insulin Sepharose display kinase activity (K. Yu and M. Czech, personal communication). Furthermore, the existence of an ATP-binding site in the β subunit has been confirmed recently by several investigators using ATP affinity-labeling methods.[43–45]

B. Kinetic Characterization of Insulin Receptor Autophosphorylation

Using the solubilized insulin receptor at various stages of purification, we and others have begun to characterize the insulin receptor kinase with respect to its autophosphorylation reaction. The soluble insulin receptor accepts phosphate from [γ-^{32}P]adenosine triphosphate (ATP) but not [γ-^{32}P]guanosine triphosphate (GTP), suggesting that ATP is likely to be the physiological substrate.[46] In vitro, half-maximal stimulation of the receptor kinase occurs at ~5 nM insulin, with a slight inhibition detected above 100 nM insulin.[47,48] The ability of insulin analogs to stimulate receptor

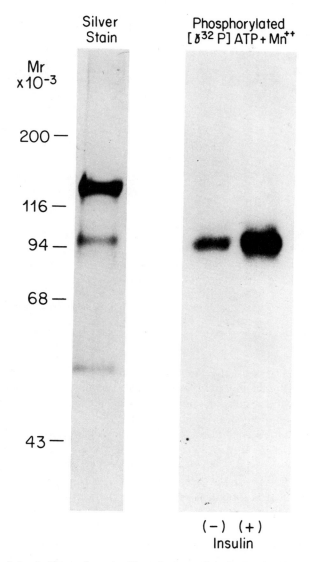

Figure 3. Sodium dodecylsulfate polyacrylamide gel electrophoresis in the presence of 100 mM dithiothreitol of the highly purified insulin receptor from human placental membranes. (Left) Insulin receptor subunits are detected in the polyacrylamide gel by silver staining. (Right) Autoradiograph of the phosporylated β subunit in the presence (+) and absence (−) of insulin.

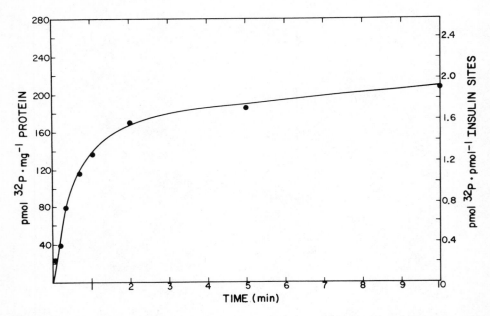

Figure 4. Time course of autophosphorylation of the partially purified insulin receptor from hepatoma cells (Fao) incubated at 22°C with insulin (10^{-7} M), ATP (50 μM), and Mn^{2+} (5 mM). The ordinate on the right shows the incorporation of phosphate per insulin binding site determined from a Scatchard plot.

autophosphorylation corresponds to their known affinity for the insulin-binding site.[40,46,47] Insulinlike growth factors I and II are only 5% as potent as insulin. Analogs that bind to the receptor but that do not exhibit negative cooperativity have been studied as well and appear to be full agonists for this effect.[40,46]

Under optimized conditions, the kinetics of autophosphorylation in the presence of 10^{-7} M insulin are very rapid (Fig. 4). Half-maximal autophosphorylation occurs within about 30 sec, and maximal autophosphorylation is reached by 2 min.[48] This reaction is consistent with the rapid biological responses that typically follow insulin binding to intact cells. Maximal autophosphorylation varies between preparations of the insulin receptor for unknown reasons; in the most active preparations, however, ~200 pmoles of phosphate is incorporated per milligram of protein (Fig. 4). This stoichiometry corresponds to about two molecules of phosphate per insulin binding site.[48] When the concentration of the receptor is decreased but the concentration of all other components in the reaction mixture is unchanged (i.e., by increasing the reaction volume), no significant effect on the initial rate of autophosphorylation can be detected. This observation suggests that the autophosphorylation reaction occurs by an intra- rather than intermolecular mechanism.[48] Other protein kinases have also been shown to undergo autophosphorylation by an intramolecular reaction.[4,49,50] Thus, autophosphorylation of the insulin receptor could provide a very rapid intracellular signal in response to extracellular insulin binding.

The activity of the insulin receptor is regulated by three factors: insulin, a divalent cation, and ATP. Insulin stimulates autophosphorylation by increasing the V_{max} nearly

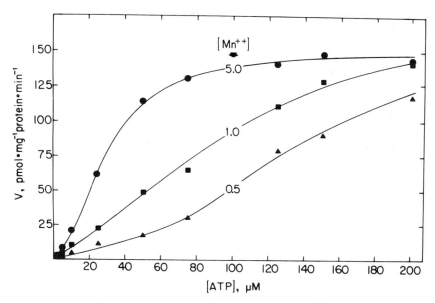

Figure 5. Kinetic curves of autophosphorylation of the partially purified insulin receptor from hepatoma cells (Fao). The initial rates of autophosphorylation were estimated during 30-sec incubations at 22°C in the presence of insulin (10^{-7} M) and the indicated concentrations of Mn^{2+} and ATP. (From White *et al.*[48])

20-fold with little or no change in the K_m for ATP.[48] The kinase activity of the insulin receptor is also regulated by the concentration of divalent cations. Metal ions activate many enzymatic reactions by combining with either the enzyme, the substrate, or both.[51] The latter possibility apepars to be true for the insulin receptor kinase.[48] Within a concentration interval for ATP of 5–200 μM, the most potent divalent cation for regulation of the insulin receptor appears to be Mn^{2+}.[48] In contrast to insulin, Mn^{2+} stimulates the autophosphorylation of insulin receptor kinase by decreasing the K_m for ATP (Fig. 5).[48] Furthermore, Co^{2+} is intermediate in its effect, whereas Mg^{2+} stimulates autophosphorylation slightly, and Ca^{2+} and Zn^{2+} have no effects.[46]

The concentration of Mn^{2+} observed to stimulate autophosphorylation of the insulin receptor is considerably in excess of that necessary to form the presumed substrate, $Mn-ATP^{2-}$. For example, with a Mn^{2+} concentration of 0.5 mM, an ATP concentration of 5–200 μM, and a dissociation constant for $MnATP^{2-}$ of 10 μM,[52] approximately 97% of the ATP is in the form of the complex, $MnATP^{2-}$. The increase in autophosphorylation observed by increasing the concentration of Mn^{2+} from 0.5 to 5 mM occurs without a significant increase in the fraction of $MnATP^{2-}$ (Fig. 5). Therefore, activation of the insulin receptor kinase presumably requires binding of free Mn^{2+} to a specific site on the kinase, in addition to the effect of Mn^{2+} to chelate ATP. The lack of any significant effect of Mg^{2+} also argues against the possibility that the stimulation of the kinase by Mn^{2+} occurs exclusively through chelation of ATP, since both cations form similar complexes with ATP.[51,52] Our results are analogous to the selective stimulation by divalent cations on the hepatic adenylate cyclase system.[53,54] The sigmoidal relationship between the velocity of autophosphorylation versus ATP concentration shown in Fig. 5 may reflect a complex kinetic mechanism involving Mn^{2+}, allosteric effects, or activation by autophosphorylation. However, since Mn^{2+} can stimulate autophosphorylation through several mechanisms, it is not surprising that a variety of complex kinetic curves are observed.[52,55]

Together with insulin and divalent cations, ATP also appears to play a regulatory role in the insulin receptor kinase activity. Thus, as the ATP concentration increases, the K_m for Mn^{2+} approaches zero (Fig. 6). This result suggests that autophosphorylation occurs by an equilibrium-ordered mechanism.[55,56] Another characteristic of this kinetic mechanism is that the V_{max} for the reaction is independent of the concentration of Mn^{2+}, which seems to be the case for the receptor kinase (Fig. 5). According to this model, the insulin receptor kinase will not bind a substrate in the absence of a metal activator. Presumably, the binding of Mn^{2+} to the receptor observed in our *in vitro* studies promotes the binding of MnATP. Thus, the high level of ATP found in the intact cell[4] could drive the autophosphorylation reaction to completion even in the presence of a very low level of metal cation, as long as enough metal cations are present to form the active metal–ATP complex.[55] Furthermore, it is possible that

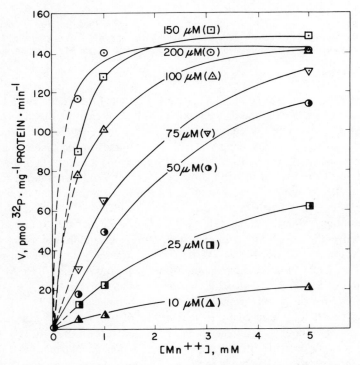

Figure 6. Initial velocities of autophosphorylation of the insulin receptor as a function of the total Mn^{2+} concentration (abscissa) at the various ATP levels shown. (From White *et al.*[48])

other divalent cations not found to stimulate phosphorylation when the ATP level is low, such as Mg^{2+}, could fulfill the requirement of the receptor kinase for a divalent metal ion in the intact cell because of the high cellular concentration of ATP.

Although the *in vivo* roles of metal cations and ATP are difficult to characterize, some observations suggest that the same type of coordinate regulation may occur. Thus, in situations in which cellular ATP is depleted, a decrease in insulin receptor affinity and an uncoupling between insulin binding and action have been detected.[57-60] Insulin action at the cellular level has also been reported to be decreased in a patient with marked Mn^{2+} deficiency and was reversed by Mn^{2+} therapy (A.H. Rubinstein, personal communication).

C. Phosphoamino Acid and Phosphopeptide Analysis of the Insulin Receptor

Most protein kinases catalyze transfer of the phosphate from ATP to the hydroxyl group on serine or threonine residues of proteins.[1-4] Recently, a new group of protein kinases were identified that catalyze the phosphorylation of tyrosine residues of proteins.[4,20] These reactions are rare, so that the amount of phosphotyrosine in cells is about 1 : 3000 of the combined amounts of phosphoserine and phosphothreonine.[15] Tyrosine-specific protein kinase activity is closely associated with oncogene products of some retroviruses[20] and with the receptors for epidermal growth factor[5-10] and platelet-derived growth factor.[11,12] Since insulin has a growth-promoting effect, we were prompted to investigate the phosphoamino acid content of the insulin receptor. When the phosphorylated β subunit was eluted from polyacrylamide gels and hydrolyzed partially in 6N HCl, only phosphotyrosine was observed (Fig. 7).[33,47] In some experiments with the partially purified receptor, phosphoserine has also been detected,[61] but phosphoserine has never been observed with either the affinity-purified receptor or the receptor purified by immunoprecipitation.[42] Therefore, the insulin receptor itself appears to be a tyrosine-specific protein kinase. Whether our partially purified preparations of receptor are contaminated with a serine kinase or whether the specificity can be altered under various experimental conditions remains to be determined.

At least five sites of tyrosine autophosphorylation can be identified in the β subunit of the insulin receptor by reverse-phase high-performance liquid chromatography (RP-HPLC) after trypsin digestion (Fig. 8). Insulin stimulates the phosphorylation of all five sites, but with different time courses.[48] The autophosphorylation of two of these sites, pp4 and pp5, is complete after only 20-sec incubation with ATP and Mn^{2+} at 22°C, whereas pp1 and pp2 are phosphorylated at slower rates; pp1 shows a definite lag (Fig. 9). Rosen et al. have presented data suggesting that autophosphorylation increases the activity of insulin receptor kinase.[62] Possibly, this activation occurs by autophosphorylation of the rapid sites, subsequently affecting that of the slower sites as well as phosphate transfer to substrates.

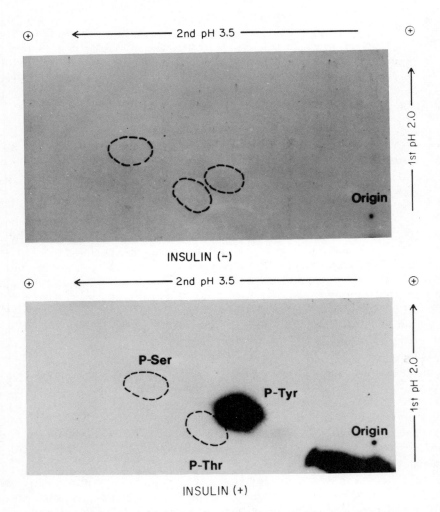

Figure 7. Identification of the phosphoamino acids hydrolyzed from the β subunit of the insulin receptor partially purified from hepatoma cells. Phosphorylation was accomplished in a solution containing receptor, ATP, and Mn^{2+} in the presence ($+$) or absence ($-$) of insulin (10^{-7} M). (From Kasuga et al.[70])

D. Insulin-Receptor-Catalyzed Phosphorylation of Protein Substrates and the Relationship to pp60[src]

The solubilized insulin receptor catalyzes phosphate transfer from ATP to several exogenous substrates: histone 2B,[63,64] casein,[42a] a synthetic peptide that resembles the site of autophosphorylation in pp60[src],[63,65] and synthetic polymers composed of glutamic acid and tyrosine (Y. Zick, personal communication). In considering possible exogenous substrates for the receptor kinase, we have focused our attention on comparing the specificity of this kinase with that of the gene product of the Rous sarcoma virus, pp60[src]. Two substrates were used in these experiments: antibodies to pp60[src] and synthetic peptides that resemble the site of autophosphorylation in pp60[src].[63]

pp60[src] was previously shown to autophosphorylate on tyrosine residues and to phosphorylate the heavy chains of IgG molecules present in antisera to pp60[src].[15,17] Anti-pp60[src] antibodies are also phosphorylated by the epidermal growth factor (EGF) receptor kinase,[66,67] but not by platelet-derived growth factor (PDGF)–receptor kinase (B. Ek and C.-H. Heldin, personal communication). When the purified insulin receptor

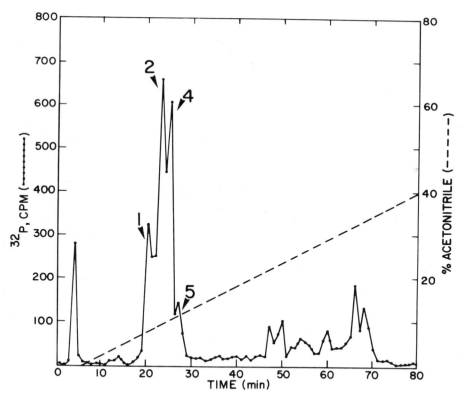

Figure 8. High-performance liquid chromatographic separation of the phosphopeptides obtained by trypsin digestion of the β subunit of the partially purified insulin receptor from hepatoma cells. (Data adopted from Kasuga *et al.*[28] and White *et al.*[48])

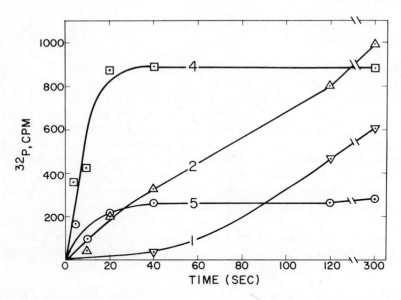

Figure 9. Time course of autophosphorylation of several tryptic phosphopeptides obtained from the β subunit of the insulin receptor. The radioactivity associated with the phosphopeptides identified in Fig. 8, pp1 (▽), pp2 (△), pp4 (□), pp5 (⊙) was determined at the indicated time intervals. (From White *et al.*[98])

was incubated with antiserum to pp60[src], a pronounced insulin-stimulated phosphorylation of the IgG heavy chains occurred (Fig. 10).[63] This phosphorylation reaction was observed with three different antisera to pp60[src]. In contrast, phosphorylation was not detected with four different antisera containing antibodies to the insulin receptor or with three different nonimmune sera. Addition of immobilized protein A to the reaction mixture resulted in precipitation of the phosphorylated IgG molecules, but not of the receptor. These results suggest that the antibody to pp60[src] interacts with the insulin receptor at the catalytic site, but that this interaction is not sufficient for immunoprecipitation.

A synthetic peptide resembling the site of autophosphorylation of pp60[src] is a substrate for the insulin receptor kinase (Fig. 11). Although this reaction can be used to monitor insulin-stimulated kinase activity, the insulin receptor weakly phosphorylates this peptide.[63] The K_m for reaction of this peptide with the insulin receptor is not altered by insulin and has a value of ~1.5 mM, corresponding to the value determined for its reaction with the epidermal growth factor receptor[9] and with the platelet-derived growth factor receptor.[10] Insulin increases the V_{max} ~4.5-fold (Fig. 11); however, the approximate turnover number for phosphorylation of the pp60[src]-derived peptide in the presence of insulin in our assay system is ~4/min at 22°C, a value that is more than 100-fold lower than that calculated for the epidermal growth factor receptor.[9] These differences may reflect different specificities of these kinases or different *in vitro* requirements for optimal activity. The insulin receptor kinase will also phosphorylate other small tyrosine-containing peptides, such as angiotensin.[65] However, none of these reaction can be considered to be of physiological relevance.

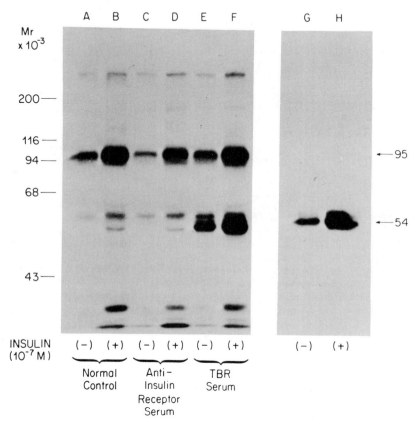

Figure 10. Phosphorylation by the highly purified insulin receptor of antisera from rabbits bearing tumors induced by Rous sarcoma virus (TBR sera). The insulin receptor was incubated with 3 μ normal rabbit serum (lanes A and B), rabbit serum containing antibodies to the insulin receptor (lanes C and D), or TBR serum (lanes E and F) in the absence (lanes A, C, E, and G) or presence of 0.1 μM insulin (lanes B, D, F, and H). The experiment shown in lanes G and H parallels exactly the conditions in lanes E and F, respectively, except that this autoradiogram shows the phosphoprotein that binds specifically to staphylococcal protein A.

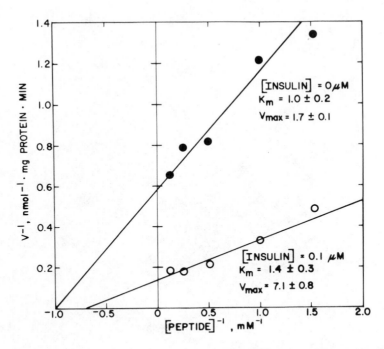

Figure 11. Lineweaver–Burk plot showing insulin-stimulated phosphorylation of the tyrosine residue in a synthetic peptide that resembles the amino acid sequence at the site of tyrosine autophosphorylation in pp60^{src}. The initial velocity of phosphorylation was estimated during 3-min incubation at 22°C of the highly purified insulin receptor from human placental membranes with 25 μM ATP, 10 mM Mg^{2+}, 2 mM Mn^{2+}, and 0.1 μM insulin or with no insulin. The kinetic parameters, K_m (mM) and V_{max} (nmoles/mg protein per min), were determined by nonlinear least-square analysis. (From Kasuga *et al.*[63])

IV. INSULIN RECEPTOR PHOSPHORYLATION IN INTACT CELLS

A. Identification of the Phosphorylated Insulin Receptor with Intact Cells

In an attempt to identify the role of the insulin receptor kinase in biological systems, receptor phosphorylation has been studied with intact cells.[21,35,47,70] Hepatoma cells (Fao) were labeled metabolically with [35S]methionine or [32P]orthophosphate and then incubated with or without insulin. The labeled insulin receptors were solubilized and purified by chromatography on immobilized wheat germ agglutinin followed by immunoprecipitation and SDS-PAGE. An autoradiogram from this experiment is shown in Fig. 12. [35S]methionine-labeled cells yield two peptides of 135,000 and 95,000 M_r, which correspond to the α and β subunits of the insulin receptor, respectively. However, only the 95,000-M_r band is specifically labeled when the cells are incubated with [32P]orthophosphate. Treatment of the cells with 10^{-7} M insulin at 37°C for 15 min after labeling, but before isolation of the insulin receptor, results in a threefold increase in the incorporation of [32P]phosphate into the β subunit with no

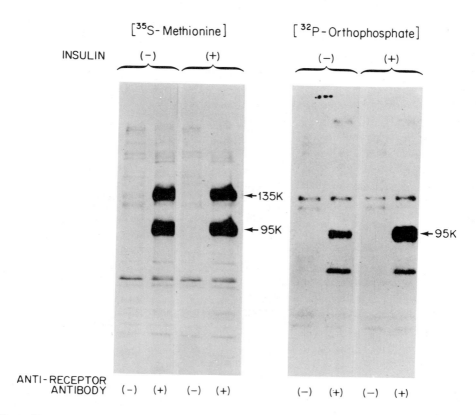

Figure 12. Effect of insulin (10^{-7} M) on the labeling of the 135,000-M_r (α) and 95,000-M_r (β) subunits of the insulin receptor from hepatoma cells (Fao) incubated with [^{35}S]methionine or [^{32}P]phosphate. The labeled receptors were immunoprecipitated with normal human serum ($-$) or anti-insulin receptor serum ($+$).

change in the labeling of the α and β subunits by [^{35}S]methionine (Fig. 12). This phosphorylation reaction occurs in a number of intact cell types, including cultured human lymphocytes and hepatoma cells,[21] freshly isolated hepatocytes,[35] isolated rat adipocytes,[36] cultured mononuclear cells,[68] and mouse melanoma cells.[41]

In the intact cell, receptor phosphorylation is temperature dependent. At 15°C, stimulation was less than at 37°C.[47] This relationship is different from that with insulin binding, which is higher at 15°C than at 37°C. By contrast, most metabolic insulin effects exhibit a decrease at 15°C as compared with 37°C.[60] Internalization of the receptor is almost undetectable at this temperature.[69]

The 95,000-M_r protein labeled by [^{32}P]orthophosphate in Fao cells is identical to the β subunit of the insulin receptor by a number of criteria.[70] The phosphoprotein migrates in the same position by SDS-PAGE as the β subunit of the receptor under both nonreducing and reducing conditions. This phosphoprotein can also be immunoprecipitated by several anti-insulin receptor antibodies according to their titers for precipitation of the insulin receptor. Furthermore, the presence of excess insulin during

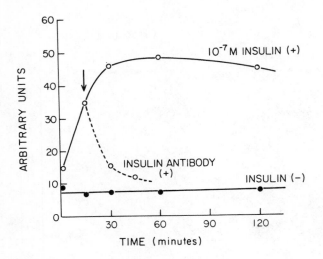

Figure 13. Time course of phosphorylation of the insulin receptor from intact hepatoma cells labeled metabolically for 2 hr with [³²P]orthophosphate and then treated with insulin and insulin antibody as indicated.

immunoprecipitation significantly decreases the recovery of the 95,000-M_r subunit, consistent with our previous observation that high concentrations of insulin block receptor immunoprecipitation. Although it is well known that the structures of the receptors for insulin and IGF-I are similar,[71] this phosphoprotein cannot be the β subunit of the IGF-I receptor because the Fao cell has few IGF-I receptors.[70] Multiplication stimulating activity and epidermal growth factor have no effect on the phosphorylation of the insulin receptor; proinsulin weakly stimulates phosphorylation, a finding consistent with its known decreased potency.[47] We have concluded that insulin stimulates the phosphorylation of the β subunit of its own receptor in intact cells.

B. Comparison of Insulin-Receptor Phosphorylation in Fao Cells and Cell-Free Systems

Although insulin-receptor phosphorylation has been observed in both intact cells and cell-free systems, some differences exist between the two systems with respect to the kinetics of phosphorylation–dephosphorylation and the content of phosphoamino acids in the β subunit of the insulin receptor.[70] In the intact cell, there is a twofold increase in phosphorylation within 1 min after insulin addition to labeled cells, followed by a gradual increase that reaches a maximum after 15–20 min.[47] Since insulin binding is instantaneous at this temperature and insulin concentration, the insulin-stimulated phosphorylation should reflect the kinetics of phosphorylation rather than the kinetics of insulin binding. Phosphorylation of the partially purified receptor from Fao cells reaches a maximum after a 2-min incubation with [γ-³²P]ATP (Fig. 4). Possibly the initial phosphorylation of the receptor in intact cells represents tyrosine autophosphorylation, and the slower phosphorylation occurs by an interaction of the receptor with another serine or threonine kinase, activated by insulin binding. The latter phase would not be expected to occur in the partially purified system because the kinases are not retained on immobilized wheat germ agglutinin.

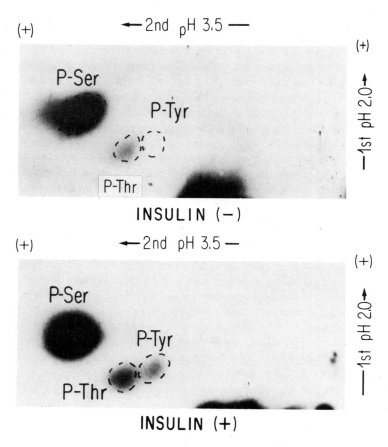

Figure 14. Identification of the phosphoamino acids hydrolyzed from the β subunit of the insulin receptor purified from intact hepatoma cells (Fao) labeled metabolically with [³²P]phosphate and incubated in the absence or presence of insulin.

The reversibility of insulin-receptor phosphorylation was studied using anti-insulin antiserum to remove insulin from the receptor and unlabeled ATP or phosphate to chase or dilute the [^{32}P]phosphate from the receptor. In the first case, hepatoma cells were labeled with [^{32}P]orthophosphate and then incubated with 10^{-7} M insulin for various time intervals between 0 and 120 min. In parallel cultures of labeled cells, insulin (10^{-7} M) was added for corresponding intervals, but a guinea pig anti-insulin antiserum (1 : 400 dilution) was added after 15 min. The results in Fig. 13 show that the antiserum causes dephosphorylation of the β subunit to the basal level of phosphorylation. In a similar experiment,[47] both the dissociation of insulin from the receptor and the dephosphorylation of the insulin receptor were measured after addition of anti-insulin serum. The rates of both processes are identical with a half-time of ~8 min.[47] These results suggest that a bound insulin molecule is important for stimulation of phosphorylation and that cellular phosphatases can act quickly to dephosphorylate the β subunit of the insulin receptor when insulin is removed.

Turnover of the phosphate in the insulin-stimulated receptor was demonstrated in the hepatoma cell by means of a pulse-chase experiment in which unlabeled phosphate was used to dilute the [^{32}P]phosphate added to the cells.[47] Under these conditions, the loss of the radioactive label from the β subunit occurred with a half-time of ~20 min. The slower rate of dephosphorylation in the pulse-chase experiment relative to the antibody-induced dephosphorylation described in the previous paragraph is probably attributable to the time required for dilution of the ^{32}P-labeled ATP pool.

In isotope-dilution experiments with the partially purified receptor from Fao cells, we have obtained evidence suggesting that phosphatase copurifies with the receptor on immobilized wheat germ agglutinin.[47] Addition of unlabeled ATP to the partially purified receptor phosphorylated to steady state with [γ-^{32}P]-ATP in the presence of insulin results in a loss of the radioactive label from the β subunit. Further purification of the receptor by immunoprecipitation presumably separates the receptor from the phosphatase because addition of unlabeled ATP to the purified phosphorylated receptor fails to cause the loss of [^{32}P]phosphate from the β subunit. Although these experiments do not prove the existence of a tyrosine phosphatase, they do suggest that glycoproteins other than the insulin receptor are required to catalyze dephosphorylation of the insulin receptor. Further work will be necessary to characterize these proteins to provide additional insight into the molecular basis of insulin action.

An important difference between insulin-receptor phosphorylation in intact and broken cell experiments lies in the identity of the phosphorylated sites. As mentioned above for the cell-free system, phosphorylation occurs almost exclusively on tyrosine residues. In the intact cell in the basal state, the primary sites of phosphorylation are serine residues and, to a lesser extent, threonine residues (Fig. 14). After insulin stimulation, an increase in both the phosphoserine and phosphothreonine and the appearance of phosphotyrosine are observed. In our early experiments, the amount of phosphotyrosine relative to phosphoserine was low; however, we have recently observed equal amounts of phosphotyrosine and phosphoserine, using additional phosphatase inhibitors during the receptor isolation.[92]

Together, the data suggest that the insulin receptor kinase is stimulated in vivo by insulin to phosphorylate itself at tyrosine residues in the β subunit and by phos-

phorylation or some other mechanism simultaneously alters other endogenous substrates, with the eventual activation of serine and threonine kinases. These kinases
play a role in insulin action intracellularly and also phosphorylate the insulin receptor
at serine and threonine residues. A provisional scheme of reactions is shown in Fig.
1. The endogenous substrates and kinases involved remain to be identified; thus far,
attempts to find cellular substrates for the insulin receptor have been unsuccessful.

It is possible that there is no natural substrate for the receptor kinase except for
the receptor itself. Autophosphorylation could be a sufficient intracellular molecular
signal to (1) initiate the physiological responses of insulin binding by altering some
other property of the receptor (i.e., an ion channel or other enzymatic activity), (2)
alter noncovalent interactions between the receptor and other cellular proteins, or (3)
act to produce site-directed proteolysis. The importance of these mechanisms will be
revealed in future studies.

C. Correlation between Receptor Autophosphorylation and Other Biological Responses

A comparison of the dose–response curve of insulin receptor autophosphorylation
and the dose–response curve of other insulin effects in cultured hepatoma cells is
shown in Fig. 15. Half-maximal stimulation by insulin of glycogen synthase, amino
acid influx, and tyrosine aminotransferase occurs at markedly lower insulin concentrations than does half-maximal stimulation of receptor phosphorylation. The
dose–response curve for the induction of tyrosine aminotransferase is representative

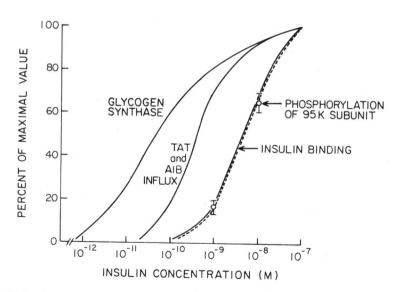

Figure 15. Insulin dose–response curves for stimulation of glycogen synthase, tyrosine aminotransferase
(TAT), aminoisobutyric acid (AIB) influx, insulin binding, and receptor phosphorylation in hepatoma cells
(Fao).

of other metabolic insulin effects in intact cells, which reach half-maximal stimulation at $2-5 \times 10^{-10}$ M insulin, depending on the effect and cell type studied.[72,73] The dose–response curve of receptor phosphorylation is less sensitive and closely follows the dose–response curve of receptor occupancy. Receptor phosphorylation is detectable at 10^{-9} M insulin, a high physiological concentration at which about 5% of the receptors are occupied; half-maximal insulin binding and receptor phosphorylation occur at about $3-5 \times 10^{-9}$ M insulin. Therefore, the dose–response curves of autophosphorylation and insulin binding are related identical to the bioeffects.

This relationship between autophosphorylation and insulin binding might be expected if phosphorylation of the receptor occurs by an intramolecular reaction[48] and is the first molecular event to occur after insulin binding. Since the maximal metabolic effect of insulin in some target tissues, particularly fat cells, is reached when only 2–5% of the total receptors of a cell are occupied,[74] these observations strongly suggest the existence of a signal amplification system occurring through reactions initiated by autophosphorylation. In the partially purified insulin receptor, in which basal autophosphorylation is negligible, some activation of the receptor kinase can be detected at 10^{-10} M insulin, a low physiological concentration. It is likely that insulin activation occurs at this level *in vivo;* however, it is not easily detected due to the relatively high background of basal serine phosphorylation of the 95,000-M_r subunit.

V. INSULIN-RECEPTOR PHOSPHORYLATION IN INSULIN-RESISTANT CELLS

A. Insulin-Receptor Phosphorylation in Mutant Melanoma Cells

Pawelek and co-workers have observed that insulin inhibits the growth of Cloudman S91 melanoma cells in culture.[75] Several variants from this cell line showing altered insulin response have been selected.[76,77] We have characterized insulin binding and receptor phosphorylation in the wild-type Cloudman melanoma (*1A*) and in two of its variants: variant (*111*) is insulin resistant, i.e., insulin neither stimulates nor inhibits growth, and variant (*46*) requires insulin for growth.

Iodine-125-insulin binding to intact cells is similar for the wild-type *1A* and insulin-stimulated variant *46;* however, the insulin-resistant variant *111* shows approximately a 30% decrease in insulin binding.[41] This difference results from a decrease of receptor affinity with no apparent difference in receptor number. A similar pattern of insulin binding is observed when the insulin receptor from each melanoma cell line is partially purified on immobilized wheat germ agglutinin-agarose.

Phosphorylation was studied by incubation of the partially purified receptor with insulin and [γ-^{32}P]-ATP and the receptor was identified by immunoprecipitation and SDS-PAGE. Insulin stimulates with similar kinetics the autophosphorylation of the β subunit of the receptor from all three cell types. However, the amount of phosphate incorporated into the β subunit in the insulin-resistant cell line *111* is decreased approximately 50% from that observed with the two other cell lines.[41] This difference is reflected throughout the entire insulin dose–response curve (10^{-9}–10^{-6} M). Similar results were obtained when phosphorylation was evaluated in the intact cell. Phosphopeptides obtained by trypsin digestion of the β subunit and separation of the phosphopeptides by RP-HPLC indicate at least three sites of autophosphorylation in receptor from the wild-type (*1A*) and variant *46;* only two major sites of phosphorylation are detected in variant *111*.[41] These data suggest that the insulin-resistant variant melanoma *111* possesses a specific defect in the insulin receptor that alters both the affinity of the receptor for insulin and its autophosphorylation properties. They also suggest a possible role of receptor phosphorylation in both the binding and the signaling function of the insulin receptor.

B. Mutation of the Insulin Receptor in Human Erythrocytes

Human erythrocytes are an extremely useful and easily accessible cellular model for biochemical studies. Red cells contain several protein kinases that act on serine and threonine residues. These enzymes have been well characterized in terms of dependence on cyclic adenosine monophosphate cAMP, calcium, or calmodulin,[78–82] and in several cases the cellular protein acceptors have been identified.[83] Recently, a tyrosine-specific protein kinase has been suggested to be associated with the red cell membrane, since the anion transport protein of erythrocytes undergoes phosphorylation at tyrosine.[84]

Human erythrocytes also contain specific insulin receptors[85] that have binding characteristics similar to those found in other classic insulin target cells.[85–87] The erythrocyte insulin receptor has not been well characterized primarily due to the relatively small number of receptors on these cells.[88,89] We have shown that the insulin receptors can be quantitatively extracted from human erythrocytes and that these receptors retain normal binding properties and undergo an insulin-stimulated tyrosine autophosphorylation.[40]

Insulin stimulates four- to fivefold the incorporation of [^{32}P]phosphate from [γ-^{32}P]-ATP into a protein of 95,000-M_r that corresponds to the β subunit of the insulin receptor.[40] Phosphoamino acid analysis demonstrates phosphorylation of the tyrosyl residues exclusively. The dose–response curve for insulin stimulation is sigmoidal; some effect of insulin is observed at 1 ng/ml, but maximal effect is observed at 10 μg/ml. Bovine desalanine–desasparagine insulin, a noncooperative analog of insulin, fully stimulates phosphate incorporation. However, the dose–response curve was found to shift to the right and to become steeper, consistent with the decreased affinity of this analog for the insulin receptor.[90] When insulin binding was performed under the same conditions as the phosphorylation assay, half-maximal stimulation of phosphate incorporation occurred with 6–29% of the fractional occupancy of the receptor.[40] We have used this system to characterize the binding and kinase properties of the insulin receptor in patients diagnosed as having the syndrome of insulin resistance and acanthosis nigricans type A.

The syndrome of insulin resistance and acanthosis nigricans type A is characterized by severe insulin resistance due to a primary defect in insulin action at the cellular level. Some patients are found to have a severe decrease in insulin binding, whereas others show normal insulin binding.[73] In an investigation into the pathophysiology of this syndrome, we have characterized both the binding and kinase activity of the insulin receptor in three patients with this syndrome.[91] Insulin binding was measured on circulating monocytes and erythrocytes, and both binding and kinase activity were measured using the receptor extracted from erythrocytes. Two of the patients showed a marked decrease in binding in both cell types and in the soluble receptor. In one patient, the impaired binding was due to a decrease in receptor number, whereas in the second, it was due to a decrease in receptor affinity. In both patients, maximal phosphorylation was reduced by 80–90%, and in the patient with altered affinity, there was also a rightward shift of the insulin dose–response curve. The third patient had normal insulin binding in both intact red cells and monocytes, and also in the soluble receptor. Interestingly, in this latter patient, a 60% decrease in receptor autophosphorylation was also observed. Thus, some patients with the type A syndrome of insulin resistance show a defect in insulin binding that is retained in the solubilized receptor. In patients with both normal and impaired binding, there is a decrease in insulin-receptor phosphorylation. These data suggest a previously undetected cause of insulin resistance, namely, a defect in the receptor kinase or in the receptor coupling to activation of the kinase.

VI. CONCLUSIONS

During the past few years, we and others have begun to characterize the kinase activity of the insulin receptor in an effort to determine its role in the action of insulin. The insulin receptor appears to contain two functional domains: one for insulin binding (the α subunit) and one for kinase activity (the β subunit). The protein kinase is intrinsic to the insulin receptor and is present in all cells that contain insulin receptors. ATP is the phosphate donor, and a divalent metal ion is required for activity. The receptor acts as a catalyst for autophosphorylation and for phosphorylation of other substrates. Most interestingly, the purified insulin receptor is a tyrosine-specific protein kinase, a property shared with the receptors for epidermal and platelet-derived growth factors and the gene products of certain transforming retroviruses. However, in the intact cell, insulin stimulates phosphorylation of the receptor at tyrosine, threonine, and serine residues. Taken together, these and other data suggest the model of insulin action depicted in Fig. 1. According to this model, insulin binds to the α subunit at the external surface of cells and activates the tyrosine kinase of the β subunit. The insulin receptor then undergoes autophosphorylation at tyrosine residues. This may "activate" the receptor by providing an intracellular modification in response to extracellular insulin binding. The insulin receptor may also phosphorylate other cellular proteins at tyrosine residues. These events then result in an activation of intracellular serine kinases, in turn mediating insulin action through the phosphorylation of other cellular proteins and phosphorylating the receptor itself, perhaps turning off or modifying the insulin signal.

ACKNOWLEDGMENTS. This work was supported in part by grants AM 31036 and AM 29770 to C.R.K. and by a fellowship (grant AM 0716301) to M.F.W. from the Institute of Health and Human Development, National Institutes of Health, United States Public Health Service; by a grant to F.G. from the Délégation Générale de la Recherche Scientifique et Technique (France); by a fellowship to H.U.H. from the Deutsche Forschungs-Gemeinschaft; and by a fellowship (grant 83.830.0.80) to M.C. from the Fonds National Suisse de la Recherche Scientifique.

VII. REFERENCES

1. Rubin, C.S., and Rosen, O.M., 1975, *Annu. Rev. Biochem.* **44**:831–887.
2. Krebs, E.G., and Beavo, J.A., 1979, *Annu. Rev. Biochem.* **48**:923–959.
3. Houslay, M.D., 1981, *Biosci. Rep.* **1**:19–34.
4. Flockhart, D.A., and Corbin, J.D., 1982, *Crit. Rev. Biochem.* **12**:133–186.
5. Cohen, S., Ushiro, H., Stoscheck, C., and Chinkers, M., 1982, *J. Biol. Chem.* **257**:1523–1531.
6. Ushiro, H., and Cohen, S., 1980, *J. Biol. Chem.* **255**:8363–8365.
7. Casnellie, J.E., Harrison, M.L., Pike, L.J., Hellstrom, K.E., and Krebs, E.G., 1982, *Proc. Natl. Acad. Sci. USA* **79**:282–286.
8. Carpenter, G., King, L., and Cohen, S., 1979, *J. Biol. Chem.* **254**:4884–4891.
9. Pike, L.J., Gallis, B., Casnellie, J.E., Bornstein, P., and Krebs, E.G., 1982, *Proc. Natl. Acad. Sci. USA* **79**:1443–1447.
10. Pike, L.J., Bowen-Pope, D.F., Ross, R., and Krebs, E.G., 1983, *J. Biol. Chem.* **258**:9383–9390.
11. Glenn, K., Bowen-Pope, D.W., and Ross, R., 1982, *J. Biol. Chem.* **257**:5172–5176.
12. Ek, B., and Heldin, C.-H., 1982, *J. Biol. Chem.* **257**:10486–10492.
13. Gordon, A.S., Davis, C.G., Milfay, D., and Diamond, I., 1977, *Nature* **267**:539–540.
14. Downward, J., Yarden, Y., Mayes, E., Scarce, G., Totty, N., Stockwell, P., Ullrich, A., Schlessinger, J., and Waterfield, M.D., 1984, *Nature* **307**:521–527.
15. Hunter, T., and Sefton, B.M., 1980, *Proc. Natl. Acad. Sci. USA* **77**:1311–1315.
16. Witte, O.N., Dasgupta, A., and Baltimore, D., 1980, *Nature* **283**:826–831.
17. Collet, M.S., Purchio, A.F., and Erikson, R.L., 1980, *Nature* **285**:167–169.
18. Kawai, S., Ypshida, M., Segawa, K., Sugiyama, H., Ishizaki, R., and Toyoshima, K., 1980, *Proc. Natl. Acad. Sci. USA* **77**:6199–6202.
19. Feldman, R., Hanafusa, T., and Hanafusa, H., 1980, *Cell* **522**:757–765.
20. Bishop, M.J., and Varmus, H., 1982, in *RNA Tumor Viruses* (Weiss, R., Teich, N., Varmus, H., and Coffin, J., eds.), p. 999, Cold Spring Harbor Laboratory, Cold Spring Harbor, N.Y.
21. Kasuga, M., Karlsson, F.A., and Kahn, C.R., 1982, *Science* **215**:185–187.
22. Kasuga, M., Hedo, J.A., Yamada, K.M., and Kahn, C.R., 1982, *J. Biol. Chem.* **257**:10392–10399.
23. Fujita-Yamaguchi, Y., 1984, *J. Biol. Chem.* **259**:1206–1211.
24. Yip, C.C., and Moule, M.L., 1983, *Diabetes* **32**:760–767.
25. Hedo, J.A., Kasuga, M., Van Obberghen, E., Roth, J., and Kahn, C.R., 1981, *Proc. Natl. Acad. Sci. USA* **78**:4791–4795.
26. Yip, C.C., Yeung, C.W.T., and Moule, M.L., 1978, *J. Biol. Chem.* **253**:1743–1745.
27. Hedo, J.A., Cushman, S.W., and Simpson, I.A., 1982, *Diabetes* **31**(Suppl. 2):2A.
28. Kasuga, M., White, M.F., and Kahn, C.R., 1984, *Methods in Enzymol.* (in press).
29. Fujita-Yamaguchi, Y., Choi, S., Sakamoto, Y., and Itakura, K., 1983, *J. Biol. Chem.* **258**:5045–5049.
30. Deschatrette, J., Moore, E.E., Dubois, M., Cassio, D., and Weiss, M.C., 1979, *Somat. Cell. Genet.* **5**:697–718.
31. Crettaz, M., and Kahn, C.R., 1983, *Endocrinology* **113**:1201–1209.
32. Coon, H.E., and Weiss, M.C., 1969, *Proc. Natl. Acad. Sci. USA* **62**:852–859.
33. Hedo, J.A., Harrison, L.C., and Roth, J., 1981, *Biochemistry* **20**:3385–3393.
34. Kasuga, M., Zick, Y., Blithe, D.L., Crettaz, M., and Kahn, C.R., 1982, *Nature* **298**:667–669.
35. Van Obberghen, E., and Kowalski, A., 1982, *FEBS Lett.* **143**:179–182.
36. Häring, H.U., Kasuga, M., and Kahn, C.R., 1982, *Biochem. Biophys. Res. Commun.* **108**:1538–1545.
37. Petruzzelli, L.M., Ganguly, S., Smith, C.R., Cobb, M.H., Rubin, C.S., and Rosen, O., 1982, *Proc. Natl. Acad. Sci. USA* **79**:6792–6796.

38. Machicao, F., Urumow, T., and Wieland, O.H., 1982, *FEBS Lett.* **149**:96–100.
39. Avruch, J., Nemenoff, R.A., Blackshear, P.J., Pierce, M.W., and Osathanondh, R., 1982, *J. Biol. Chem.* **257**:15162–15166.
40. Grigorescu, F., White, M.F., and Kahn, C.R., 1983, *J. Biol. Chem.* **258**:13708–13716.
41. Häring, H.U., White, M.F., Kahn, C.R., Kasuga, M., Lauris, V., Fleischmann, R., Murray, M., and Pawelek, J., 1983, *J. Cell. Biol.* (in press).
42. Kasuga, M., Fujita-Yamaguchi, Y., Blithe, D.L., and Kahn, C.R., 1983, *Proc. Natl. Acad. Sci. USA* **80**:2137–2441.
42a. Zick, Y., Rees-Jones, R.W., Grunberger, G., Taylor, S.I., Moncada, V., Gorden, P., and Roth, J., 1983, *Eur. J. Biochem.* **137**:631–637.
43. Roth, R.A., and Cassell, D.J., 1983, *Science* **219**:299–301.
44. Van Obberghen, E., Rossi, B., Kowalski, A., Gazzano, H., and Ponzio, G., 1983, *Proc. Natl. Acad. Sci. USA* **80**:945–949.
45. Shia, M.A., and Pilch, P.F., 1983, *Biochemistry* **22**:717–721.
46. Zick, Y., Kasuga, M., Kahn, C.R., and Roth, J., 1982, *J. Biol. Chem.* **258**:75–80.
47. Häring, H.U., Kasuga, M., White, M.F., Crettaz, M., and Kahn, C.R., 1984, *Biochemistry* **23**:3298–3306.
48. White, M.F., Häring, H.U., Kasuga, M., and Kahn, C.R., 1983, *J. Biol. Chem.* **259**:255–264.
49. Rangel-Aldao, R., and Rosen, O.M., 1976, *J. Biol. Chem.* **251**:7526–7529.
50. Erlichman, J., Rangel-Aldao, R., and Rosen, O.M., 1983, *Methods Enzymol.* **99**:176–186.
51. Cleland, W.W., 1979, *Methods Enzymol.* **63**:501–513.
52. O'Sullivan, W.J., and Smithers, G.W., 1979, *Methods Enzymol.* **63**:294–336.
53. Londos, C., and Preston, M. Sue, 1977, *J. Biol. Chem.* **252**:5951–5956.
54. Londos, C., and Preston, M. Sue, 1977, *J. Biol. Chem.* **252**:5957–5961.
55. Segel, I.H., 1975, *Enzyme Kinetics*, pp. 227–272, Wiley, New York.
56. Morrison, J.F., 1979, *Methods Enzymol.* **63**:257–294.
57. Chandamouli, V., Milligan, M., and Cater, J.R., 1977, *Biochemistry* **16**:1151–1158.
58. Kono, T., Robinson, F.W., Sarver, J.A., Vega, F.W., and Pointer, R.H., 1977, *J. Biol. Chem.* **252**:2226–2233.
59. Siegel, J., and Olefsky, J.M., 1980, *Biochemistry* **19**:2183–2190.
60. Häring, H.U., Bierman, E., and Kemmler, W., 1981, *Am. J. Physiol.* **240**:E556–E565.
61. Zick, Y., Grunberger, G., Podskalny, J.M., Moncada, V., Taylor, S.I., Gorden, P., and Roth, J., 1984, *Biochem. Biophys. Res. Commun.* (in press).
62. Rosen, O.M., Herrera, R., Olowe, Y., Petruzelli, L.M., and Cobb, M.H., 1983, *Proc. Natl. Acad. Sci. USA* **80**:3237–3240.
63. Kasuga, M., Fujita-Yamaguchi, Y., Blithe, D.L., White, M.F., and Kahn, C.R., 1983, *J. Biol. Chem.* **258**:10973–10979.
64. Zick, Y., Wittaker, J., and Roth, J., 1983, *J. Biol. Chem.* **258**:3431–3434.
65. Stadtmauer, L.A., and Rosen, O.M., 1983, *J. Biol. Chem.* **258**:6682–6685.
66. Chinkers, N., and Cohen, S., 1981, *Nature* **290**:526–529.
67. Kudlow, J.E., Buss, J.E., and Gill, G.N., 1981, *Nature* **290**:519–521.
68. Grunberger, G., Zick, Y., Roth, J., and Gorden, P., 1983, *Biochem. Biophys. Res. Commun.* **115**:560–566.
69. Gorden, P., Carpentier, J.L., Freychet, P., and Orci, L., 1980, *Diabetologia* **18**:263–274.
70. Kasuga, M., Zick, Y., Blithe, D.L., Karlsson, F.A., Häring, H.U., and Kahn, C.R., 1982, *J. Biol. Chem.* **257**:9891–9894.
71. Kasuga, M., Van Obberghen, E., Nissley, P., and Rechler, M.M., 1981, *J. Biol. Chem.* **256**:5305–5308.
72. Kahn, C.R., 1975, in *Methods in Membrane Biology* (Korn, E.D., ed.), Vol. 3, pp. 81–146, Plenum, New York.
73. Kahn, C.R., Baird, K.L., Flier, J.S., Grunfeld, C., Harmon, J.T., Harrison, L.C., Karlsson, F.A., Kasuga, M., King, G.L., Lang, U.C., Podskalny, J.M., and Van Obberghen, E., 1981, *Recent Prog. Horm. Res.* **37**:477–538.
74. Kono, T., and Barham, F.W., 1971, *J. Biol. Chem.* **246**:6210–6216.
75. Pawelek, J., Sansome, M., Koch, N., Christie, G., Halaban, R., Hendee, J., Lerner, A.B., and Varga, J., 1975, *Proc. Natl. Acad. Sci. USA* **72**:951–955.
76. Pawelek, J., Murray, M., and Fleischmann, R., 1982, *Cold Spring Harbor Conf. Cell Prolif.* **9**:911.

77. Kahn, R., Murray, M., and Pawelek, J., 1980, *J. Cell. Physiol.* **103:**109–119.
78. Avruch, J., and Fairbanks, G., 1974, *Biochemistry* **13:**5507–5514.
79. Fairbanks, G., and Avruch, J., 1974, *Biochemistry* **13:**5514–5521.
80. Rubin, C.S., Erlichman, J., and Rosen, O.M., 1972, *J. Biol. Chem.* **247:**6135–6139.
81. Roses, A.D., and Appel, S.H., 1973, *J. Biol. Chem.* **248:**1408–1411.
82. Nelson, M.J., Daleke, D.L., and Heuslis, W.H., 1982, *Biochim. Biophys. Acta* **686:**182–188.
83. Waxman, L., 1979, *Arch. Biochem. Biophys.* **195:**300–314.
84. Dekowski, S.A., Rybicki, A., and Drickamer, K., 1983, *J. Biol. Chem.* **258:**2750–2753.
85. Tsibris, J.C.M., Raynor, L.O., Buhi, W.C., Buggie, J., and Spellacy, W.N., 1980, *J. Clin. Endocrinol. Metab.* **51:**711–717.
86. Wachslicht-Rodbard, H., Gross, H.A., Rodbard, D., Ebert, M.H., and Roth, J., 1979, *N. Engl. J. Med.* **300:**882–887.
87. Pederson, O., Beck-Nielson, M., and Heding, L., 1980, *N. Engl. J. Med.* **302:**886–892.
88. Im, J.H., Meezan, E., Rackley, C.E., and Kim, H.D., 1983, *J. Biol. Chem.* **258:**5021–5026.
89. Hara, H., Hidaka, H., Kosmakos, F.C., Mott, D.M., Wasquez, B., Howard, B-Y., and Bennett, P.H., 1981, *J. Clin. Endocrinol. Metab.* **52:**17–22.
90. DeMeyts, P., 1980, in *Hormone and Cell Regulation* (J. Dumont and J. Nunet, eds.), Vol. 4, Elsevier Biomedical Press, Amsterdam, pp. 107–121.
91. Grigorescu, F., Flier, J.S., and Kahn, C.R., 1984, *Clin. Res.* **32:**519A.
92. Takayama, S., White, M.F., Lauris, V., and Kahn, C.R., 1984, *Proc. Natl. Acad. Sci. USA*, in press.

5

What Do Receptor Kinases Do?

Paul F. Pilch, Michael A. Shia, and Joshua B. Rubin

There is now compelling evidence that the β subunit of the insulin receptor is an insulin-regulated, tyrosine-specific protein kinase. Some of this evidence is detailed in the previous chapters of this volume (Chapter 4). Data from our own laboratory on this point include the following observations: (1) Affinity-purified insulin receptor is capable of ligand-dependent autophosphorylation as well as kinase activity toward exogenous substrates.[1] (2) An immunoprecipitate of partially purified insulin receptor retains kinase activity.[2] (3) The β subunit of the receptor has an adenosine triphosphate (ATP) binding site.[2] (4) Treatment of purified receptor with elastase abolishes kinase activity concomitant with cleavage of the β subunit. Elastase action has no effect on insulin binding to the α subunit nor does it alter the structure of the α subunit.[1] Similar studies from several laboratories have given essentially identical results on one or more aspects of the above considerations.[3-6] It thus seems highly likely that tyrosine-specific protein kinase activity is an intrinsic property of the insulin receptor and is not due to the action of an unseen but tightly receptor-associated protein.

No physiological function has yet been identified that is regulated as a result of this insulin-dependent tyrosine phosphorylating activity. Identifying such a function is an obvious important goal of future research in this area, and there are several biological processes whose activity might be changed as a consequence of receptor kinase activity. The action of insulin on cells is pleiotropic and is thus potentially separable in terms of signals such as tyrosine phosphorylation. Insulin can mediate rapid changes in fuel metabolism (e.g., glucose transport and its cellular utilization) that can be detected within a matter of minutes after exposure of cells to hormone.[7] On a time scale requiring several hours of insulin exposure, insulin regulates its own cell-surface receptor number[8] as well as many other events requiring protein synthesis.[9] Finally, insulin is mitogenic for certain cell lines,[10] a process that also

Paul F. Pilch, Michael A. Shia, and Joshua B. Rubin • Department of Biochemistry, Boston University School of Medicine, Boston, Massachusetts.

A   **B**

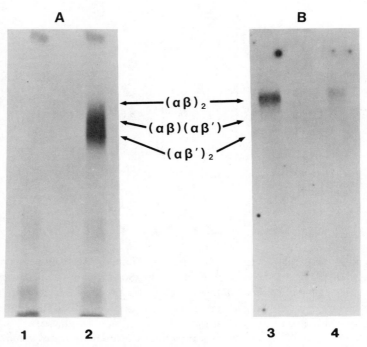

$(\alpha\beta)_2$

$(\alpha\beta)(\alpha\beta')$

$(\alpha\beta')_2$

1   2   3   4

Figure 1. Comparison of affinity crosslinked and phosphorylated insulin receptor. Affinity-purified insulin receptor was prepared from human placenta and pure receptor was affinity crosslinked with [^{125}I]insulin or stimulated to autophosphorylate by insulin, according to the procedures described by Shia *et al.*[1] (see their figure 5). Both reactions contained 1μg/100 μl of affinity-purified receptor and were analyzed by electrophoresis on 5% polyacrylamide gel (acrylamide–bisacrylamide, 30 : 0.3), followed by autoradiography. Depicted are the autoradiographs of the stained dried gels. Lanes 1 and 2 are the results of affinity crosslinking reactions performed in the presence and absence of excess unlabeled insulin. Two bands (320,000 M_r and 290,000 M_r) are specifically labeled by [^{125}I]insulin. These bands correspond to receptor forms $(\alpha\beta)$ (α) (β_1) and $(\alpha\beta_1)_2$, respectively. The results of affinity crosslinking indicate that there is a conspicuous shortage of the 350,000-M_r $(\alpha\beta)_2$ holoreceptor in our preparation. However, the products of autophosphorylation done in the presence and absence of insulin, as shown in lanes 3 and 4, demonstrate that this receptor form is in fact present, and moreover, that it is the only phosphorylated form of the receptor.

requires prolonged exposure of cells to hormone. The primary signal(s) resulting from insulin–receptor interaction is unknown; it is tempting to speculate that the intrinsic tyrosine-specific kinase activity of the receptor may represent such a signal that may in turn regulate some or all of the above biological processes. In this chapter, we speculate on the possible physiological role of the insulin receptor kinase activity within the context of both our data and the above discussion as well as what is known in general about tyrosine-specific protein kinases.

Stimulation of receptor autophosphorylation or histone phosphorylation in response to insulin can be observed within minutes at room temperature[2,11] and is therefore sufficiently fast to account for the action of insulin on glucose transport, which is also rapid.[7] It is conceivable that receptor kinase activity could activate glucose transport directly in appropriate target cells or could initiate a cascade of events toward this same end. As yet, there are no direct data on this point, but indirect evidence obtained in part in our laboratory suggests that receptor-associated kinase activity may not mediate activation of fat cell glucose transport. We have observed that when rapidly autophosphorylated receptor (≤ 8 min ATP exposure) is analyzed by sodium dodecylsulfate polyacrylamide gel electrophoresis (SDS-PAGE) and autoradiography in the absence of reductant, the only form of the receptor seen was that of native $\alpha_2\beta_2$ tetramer[1,2] (Fig. 1). In contrast, SDS-PAGE analysis of the same preparation of pure affinity-labeled receptor shows the typical pattern of heterogeneous receptor forms that results from proteolysis during tissue preparation[1,2] (Fig. 1). As shown in Fig. 1, the preparation of the receptor can be quite depleted in tetrameric holoreceptor, yet it is only this intact tetramer that is capable of autophosphorylation. These results tend to indicate that intact β subunit is required for kinase activity, since receptor heterogeneity results from proteolytic cleavage of this subunit.[12] Moreover, we have shown that elastase treatment of pure receptor results in loss of kinase activity concomitant with β-subunit proteolysis.[1] Previously it had been shown that treatment of intact rat fat cells with elastase resulted in β-subunit cleavage under conditions where there is no change in the ability of insulin to enhance the rate of glucose oxidation.[13] Thus, studies of purified receptor in vitro and studies of affinity-labeled receptor in intact cells give qualitatively similar results. Elastase treatment leads to specific cleavage of the β subunit in both cases. In vitro, kinase activity is abolished, whereas in vivo, there is no apparent effect on the ability of insulin to activate glucose transport. A tentative conclusion that can be drawn from these considerations is that receptor autophosphorylation and regulation of hexose transport by insulin are unrelated phenomena. A similar conclusion was recently reached by Simpson and Hedo.[14] These workers demonstrated the ability of anti-insulin receptor antibody to stimulate glucose transport and to phosphorylate certain microsomal membrane proteins in a dose-dependent fashion. These same doses of antiserum were ineffective in eliciting receptor autophosphorylation.[14]

Our results and those cited above[14] do not completely rule out the possiblity that receptor-mediated kinase activity is still present under the experimental conditions described, but autophosphorylation is lost. To test this point directly, it will be necessary to demonstrate that abolishing the kinase activity of the receptor in fat cells toward a putative, physiologically relevant, endogenous substrate other than the receptor itself results in no alteration of insulin-dependent glucose utilization in this same cell preparation. At present, this may not be possible because of several unknown factors concerning the structure and function of the insulin receptor and the lack of a physiologically relevent endogenous substrate for the receptor kinase. There may in fact be no such substrate other than the receptor itself, although this appears unlikely (see below).

Tyrosine-specific protein kinases have only recently been discovered. They appear to act rarely as compared with kinases that phosphorylate serine and threonine,[15] hence the lack of a generally accepted body of data describing functional cellular targets for these enzymes. The first tyrosine-specific protein kinase discovered is a gene product of the Rous sarcoma virus (RSV) called pp60[v-src], which catalyzes an autophosphorylation of a 60,000-dalton protein, the src kinase.[16] The tyrosine kinase activity of this protein appears to be essential for cellular transformation by Rous sarcoma virus,[17] but autophosphorylation is not.[18] Several other retrovirus-associated tyrosine kinases have been described that appear to be necessary for oncogenesis that autophosphorylate as well.[19] In the case of the most heavily studied kinase of RSV, phosphorylation of endogenous substrates has been observed, but it has not been possible to correlate phosphorylations of these proteins with transformation activity. This is probably due to the very low degree of tyrosine phosphorylation of the presumed substrates seen in these experiments.[20] That is, the tyrosine phosphorylation of vinculin and glycolytic enzymes due to RSV transformation of cells is increased by 1% and 5%, respectively, as compared with uninfected cells.[20] Another potentially important substrate for the pp60^{src} is a 36,000-dalton protein (p36) of unknown function that is plasma membrane-associated.[21] A protein of similar size is also phosphorylated due to the epidermal growth factor (EGF) receptor kinase [22] and may therefore be a common substrate for tyrosine-specific kinases. However, it is not clear that the same protein is being studied in the literature cited, and there have been no reports of p36 phosphorylation due to the insulin receptor kinase.

The discovery by Cohen and colleagues that the receptor for the mitogenic peptide EGF is a tyrosine-directed protein kinase[23] and the subsequent discoveries that this activity is common to receptors for insulin,[24–27] insulinlike growth factor 1 (IGF-1),[28,29] and platelet-derived growth factor (PDGF)[30,31] have been integrated with the data described above for the retroviral kinases in developing the notion that tyrosine-specific protein kinase activity is intimately coupled with the control of cellular growth. The very recent finding that the EGF receptor has a protein sequence homologous to the oncogene *erbB*[32] has further fueled this speculation. According to this hypothesis, a constitutive and substantial increase in tyrosine kinase activity due to *onc* gene activation would lead to cellular transformation, whereas the controlled exposure of a cell to a mitogenic ligand would activate a receptor kinase and lead to controlled cellular division. As of now, this hypothesis lacks experimental verification, but it remains intriguing nevertheless. A possible mechanism that could link *onc* gene kinases and receptor kinases would be a common substrate such as the previously noted p36 protein. It should also be noted that receptors for the mitogens nerve growth factor (NGF) and IGF-2 have recently been purified[33–35] but have not been described as tyrosine kinases.

Another common feature of several mitogen receptors worth considering in attempts to assign potential physiological significance to their tyrosine kinase activity is the ability of these receptors to regulate their cell-surface number in response to ligand. Cells exposed to ligand for a prolonged period (~16 hr) show a substantial loss in cell-surface receptor number in the process of downregulation, which was first demonstrated for the insulin receptor.[8] Receptors for EGF,[36] PDGF,[37] and IGF-1[38] also undergo downregulation in proportion to the concentration of ligand used in the experiment. It follows from the dose dependency of this phenomenon that cells have a mechanism for recognizing ligand-occupied receptors. One such way could be receptor autophosphorylation. We have demonstrated that autophosphorylation of the insulin receptor can proceed via an intramolecular mechanism.[1] Thus, insulin binding will stimulate phosphorylation of ligand-occupied receptors, permitting cells to distinguish this receptor population from other insulin receptors. One possible consequence of this discriminatory ability of cells could be receptor downregulation. Therefore, it will be highly desirable to find conditions that alter receptor kinase activity in a specific fashion so that the effects of such alterations can be correlated with possible changes in the rate or magnitude of downregulation.

I. REFERENCES

1. Shia, M.A., Rubin, J.B., and Pilch, P.F., 1983, *J. Biol. Chem.* **258:**14,450–14,455.
2. Shia, M.A., and Pilch, P.F., 1983, *Biochemistry* **22:**717–721.
3. Roth, R.A., and Cassell, D.J., 1983, *Science* **219:**299–301.
4. Van Obberghen, E., Rossi, B., Kowalski, A., Gazzano, H., and Ponzio. G., 1983, *Proc. Natl. Acad. Sci. USA* **80:**945–949.
5. Kasuga, M., Fujita-Yamaguchi, Y., Blithe, D.L., and Kahn, C.R., 1983, *Proc. Natl. Acad. Sci. USA* **80:**2137–2141.
6. Roth, R.A., Mesirow, M.L., and Cassell, D.J., 1983, *J. Biol. Chem.* **258:**14,456–14,460.
7. Karnielli, E., Zarnowski, M.J., Hissin, P.J., Simpson, I.A., Salans, L.B., and Cushman, S.W., 1981, *J. Biol. Chem.* **256:**4772–4777.
8. Gavin, J.R., Roth, J., Neville, D.M., Jr., DeMeyts, P., and Buell, D.N., 1974, *Proc. Natl. Acad. Sci USA* **71:**84–88.
9. Jefferson, L.S., Flairn, K.E., and Peavy, D.E., 1981, *Handbook of Diabetes Mellitus, Vol. 4 pp. 133–177, Garland STPM Press, New York.*
10. Koontz, J.W., and Iwahaski, M., 1981, *Science* **211:**946–949.
11. Rosen, O.M., Herrera, R., Olowe, Y., Petruzelli, L.M., and Cobb, M.H., 1983 *Proc. Natl. Acad. Sci. USA* **80:**3237–3240.
12. Massague, J., Pilch, P.F., and Czech, M.P., 1981, *J. Biol. Chem.* **256:**3182–3190.
13. Pilch, P.F., Axelrod, J.D., and Czech, M.P., 1981, in *Current Views on Insulin Receptors* (Andreani, D., DePirro, R., Lauro, R., Olefsky, J., and Roth, J. eds) pp. 255–260, Academic Press, London.
14. Simpson, I.A., and Hedo, J.A., 1984, *Science* **223:**1301–1304.
15. Hunter, T.B., Sefton, B.M., and Cooper, J.A., 1981, *Cold Spring Harbor Conf. Cell Prolif.* **8:**1189–1202.
16. Collet, M.S., and Erikson, R.L., 1978 *Proc. Natl. Acad. Sci. USA* **75:**2021–2024.
17. Sefton, B.M., Hunter, T. Beemon, K., and Eckhart, W., 1980, *Cell* **20:**807–816.
18. Snyder, M.A., Bishop, J.M., Colby, W.W., and Levinson, A.D., 1983, *Cell* **32:**891–901.
19. Land, H., Parada, L.F., and Weinberg, R.A., 1983 *Science* **222:**771–778.
20. Cooper, J.A. Reiss, N.A., Schwartz, R.J., and Hunter, T., 1983, *Nature* **302:**218–223.
21. Courtneidge, S., Ralston, R., Alitalo, K., and Bishop, J.M., 1983, *Mol. Cell Biol.* **3:**340–350.
22. Fava, R., and Cohen, S., 1984, *J. Biol. Chem.* **259:**2636–2645.
23. Cohen, S., Carpenter, G., and King, L., 1980 *J. Biol. Chem.* **255:**4834–4842.
24. Kasuga, M., Karlsson, F.A., and Kahn, C.R., 1982, *Science* **215:**185–187.
25. Zick, Y., Kasuga, M., Kahn, C.R., and Roth, J., 1983, *J. Biol. Chem.* **258:**75–80.
26. Petruzzelli, L.M., Sabyasachi, G., Smith, C.J., Cobb, M.H., Rubin, C.S., and Rosen, O.M., 1982, *Proc. Natl. Acad. Sci. USA* **79:**6792–6796.
27. Avruch, J., Nemenoff, R.A., Blackshear, P.J., Pierce, M.W., and Osathanondh, R., 1982, *J. Biol. Chem.* **257:**15,162–15,166.
28. Rubin, J., Shia, M.A., and Pilch, P.F., 1983, *Nature* **305:**438–440.
29. Jacobs, S., Kull, F.C., Jr., Earp, H.S., Svoboda, M.E., Van Wyk, J.J., and Cuatrecasas, P., 1983, *J. Biol. Chem.* **258:**9581–9584.
30. Ek, B., Westermark, B., Wasteson, A., and Heldin, C.H., 1982, *Nature* **295:**419–420.
31. Nishimura, J., Huang, J.S., and Deuel, T., 1982, *Proc. Natl. Acad. Sci. USA* **79:**4303–4307.
32. Downward, J., Yarden, Y., Mayes, E., Scrace, G., Totty, N., Stockwell, P., Ullrich, A., Schlessinger, J., and Waterfield, M.D., 1984, *Nature* **307:**521–527.
33. Oppenheimer, C.L., and Czech, M.P., 1983, *J. Biol. Chem.* **258:**8539–8542.

34. August, G.P., Nissley, S.P., Kasuga, M., Lee, L., Greenstein, L, and Rechler, M.M., 1983, *J. Biol. Chem.* **258:**9033–9036.
35. Puma, P., Buxser, S.E., Watson, L., Kelleher, D.J., and Johnson, G.L., 1983, *J. Biol. Chem.* **258:**3370–3375.
36. Aharonov, A., Pruss, R.M., and Herschman, H.R., 1978, *J. Biol. Chem.* **253:**3970–3977.
37. Heldin, C.-H., Wasterson, A., and Westemark, B., 1982, *J. Biol. Chem.* **257:**4216–4221.
38. Rosenfeld, R.G., Hintz, R.L., and Dollar, L.A., 1982, *Diabetes* **31:**375–381.

II

*Intracellular Signaling Mechanisms of
Insulin Action*

6

The Role of Calcium in the Transduction of Insulin Action

Jay M. McDonald and Harrihar A. Pershadsingh

I. INTRODUCTION

The year 1983 has particular significance for progenitors of the role of calcium in neurohumoral action and cell regulation, since it was precisely 100 years earlier that Ringer's classic paper, describing the vital role of calcium in the generation of mechanical events in the heart, was published.[1] Since then, calcium has been shown to be an essential regulator of a diverse constellation of fundamental intracellular processes.[2–5] A fairly large body of evidence has accumulated over the past 20 years that implies an important role of calcium in the molecular mechanism of action of insulin. Suffice it to say that a reexamination of the role of calcium in the action of insulin is thereby both timely and fitting, especially in view of recent developments in this area. These are the issues that are dealt with in this brief review of some new and interesting findings that may once again reemphasize the importance of calcium and permit its reconsideration as an essential factor in the mechanism of insulin action.

The mechanism by which insulin produces its diverse metabolic responses in target tissues remains an enigma despite intensive research efforts. These metabolic responses may be divided into two general categories: (1) plasma membrane events, e.g., transport of organic solutes and ions, and (2) intracellular events regulated by intermediary metabolism, e.g., increased glycogen synthesis, inhibition of lipolysis, increased protein synthesis, and increased glucose oxidation. It is generally accepted that the initial event in the mechanism of insulin action is the binding of insulin to specific receptors at the plasma membrane, but the plasma membrane triggering event(s) that occur immediately distal to this interaction are still poorly understood. Historically,

Jay M. McDonald and Harrihar A. Pershadsingh ● Division of Laboratory Medicine, Departments of Pathology and Medicine, Washington University School of Medicine, St. Louis, Missouri.

the search for this signal or messenger system has included consideration of glucose, cyclic adenosine monophosphate (cAMP), cyclic guanosine monophosphate (cGMP), membrane hyperpolarization, potassium, magnesium, calcium, H_2O_2, and, more recently, phospholipids and protein mediators.[6,7] Modification of cation homeostasis appears to be central to the transduction process, and calcium has emerged as a viable candidate at least for this purpose. Calcium was originally proposed as the mediator of insulin action by Clausen et al.[8] and Kissebah et al.[9] Yet, evidence to support this hypothesis has been largely indirect. Recent observations, however, have provided more direct evidence for insulin-dependent modification of intracellular calcium homeostasis via subcellular organelles, particularly the plasma membrane.

The informational role of calcium is made possible by its asymmetric distribution across the plasma membrane, endoplasmic reticulum, and mitochondrial inner membrane. Extracellular calcium is $>10^{-3}$ M, whereas cytoplasmic calcium ($[Ca^{2+}]_{cyt}$) is estimated to be $<10^{-7}$ M.[10] Calcium enters the cell by flowing down an exceedingly steep electrochemical gradient (calcium equilibrium potential $\simeq -160$ mV); however, $[Ca^{2+}]_{cyt}$ is maintained in the submicromolar range by active processes as shown schematically in Fig. 1. Calcium is extruded from the cell by an active ATP-dependent calcium pump.[11] The existence of an Na^+/Ca^{2+} exchange mechanism for calcium extrusion may be limited to excitable tissues[12] and has not yet been directly demonstrated in many of the target cells for insulin. Intracellularly, the calcium concentration in the region of the endoplasmic reticulum is controlled by an ATP-dependent calcium pump analogous to but distinctly different from the one present in the plasma membrane.[13,14] Cytoplasmic calcium is electrophoretically driven into the mitochondria, where the energy source is derived from the proton-motive force generated by substrate metabolism via the respiratory chain.[15]

Less appreciated is the fact that the $[Ca^{2+}]_{cyt}$ is not uniform. Both experimental data[16] and theoretical analysis[17] support the concept that changes in $[Ca^{2+}]_{cyt}$ are heterogeneously distributed, being restricted to small domains near the sites of origin

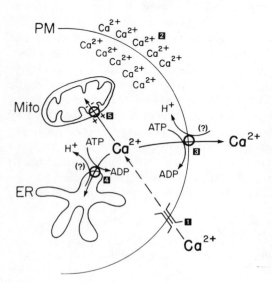

Figure 1. Schematic depicting generally accepted mechanisms of subcellular calcium homeostasis in the adipocyte. Passive entry of extracellular calcium (designated 1) and calcium binding to the plasma membrane via adsorption–desorption homeostasis (designated 2) are shown. Other processes shown are mediated via energy-dependent mechanisms. The (Ca^{2+}, Mg^{2+})ATPase–calcium pumps of the plasma membrane and endoplasmic reticulum are designated 3 and 4, respectively. The electrophoretic calcium uniporter in the mitochondrial inner membrane is designated 5. The possible involvement of Ca^{2+}–H^+ countertransport exchange mechanisms with activity of the plasma membrane and endoplasmic reticulum calcium pumps are still uncertain.

of the change. Thus, discrete calcium gradients may be generated within the cytoplasm, resulting in constantly fluctuating messenger domains. This is due to the rapidity with which calcium distribution can be regulated by active systems in mitochondria, endoplasmic reticulum, and plasma membrane (Fig. 1), in conjunction with the calcium-buffering capacity of the cytosolic proteins, which exceeds the diffusion rate of calcium within the cytosol.

The purpose of this brief review is to highlight rather than extensively review the evidence that calcium is involved in insulin action; those aspects that have already been adequately reported are not dealt with here. Emphasis is placed on recent work and on studies in which subcellular calcium homeostasis has been investigated. Finally, a model for the role of calcium in the early events in the mechanism of insulin action is presented.

II. EVIDENCE RELATING CALCIUM TO INSULIN-SENSITIVE CELLULAR ACTIVITIES

A. Plasma Membrane Events

Calcium appears to play an essential role in the regulation of a variety of plasma membrane-related phenomena, many of which are also regulated by insulin. The role of calcium in the regulation of glucose transport has been the subject of intense investigation. A body of largely indirect evidence has accumulated demonstrating that calcium serves as an activator of glucose transport, an effect similar to that of insulin in target tissues such as muscle and adipocytes. There are, however, data that oppose this hypothesis, and the issue remains controversial. Since this has been the subject of recent reviews,[18,19] it is not dealt with further here.

It is generally accepted that the primary event responsible for initiation of the insulin-dependent cascade of metabolic events is binding of the hormone to its specific receptor.[20] The specific binding of insulin to its high-affinity receptor has been shown to be enhanced with concentrations of calcium similar to that of plasma. Desai et al.[21] showed a 50% decrease in the number of high-affinity sites in the absence of calcium and an increased dissociation rate of insulin from its high-affinity receptor. These workers concluded that extracellular calcium increased the number of high-affinity receptors in the adipocyte by altering the ratio of high- to low-affinity receptors. Others have recently confirmed the observation that calcium increases insulin binding to the receptor.[22] Furthermore, by using Tb^{3+} as a competitive probe, they have shown that the insulin receptor possesses a high-affinity calcium-binding site[22a] that may be responsible for a conformational change, resulting in a consequent increase in receptor affinity.[22] However, the importance of this observation in the mechanism of insulin action is unclear. It seems unlikely that calcium-dependent increase in receptor affinity has substantial intracellular metabolic consequences because (1) there are many more receptors present ("spare receptors") than are required for insulin to initiate its metabolic signal, and (2) the extracellular calcium concentration normally remains within a restricted range with only subtle fluctuations. On the other hand, the finding that the insulin receptor binds calcium with high affinity supports the concept that the receptor, calmodulin, and the high-affinity (Ca^{2+},Mg^{2+})ATPase may have a fundamental structural and/or functional relationship, since these proteins bind calcium with high affinity (see Section VI).

The phosphorylation of the 95,000-M_r subunit of the insulin receptor by insulin is apparently dependent on physiological extracellular concentrations of calcium and magnesium.[23] The effect was shown to require energy, since ^{32}P incorporation could be prevented with 2,4-dinitrophenol. Interestingly, the calmodulin antagonist, trifluoperazine, also substantially inhibited ^{32}P incorporation, suggesting that the calcium-dependent component of receptor phosphorylation may be modulated by calmodulin.[23] Calmodulin has been identified as an ubiquitous calcium-dependent regulator of many metabolic activities in eukaryotic cells.[5] Again, the meaning of these observations remains unknown, but they do suggest that calcium is important in this phosphorylation process and that it may be related to the recent observation that the insulin receptor binds calmodulin (see Section IV.D).

Finally, membrane calcium may be involved in the regulation of other effects of insulin on plasma membrane events such as potassium and/or sodium fluxes and membrane hyperpolarization. Evidence for this is based largely on studies derived from a variety of tissues in which the effects of insulin were not studied. Their role in insulin action remains speculative (see Section VI).

B. Intracellular Events

In contrast to the relatively consistent demonstration of the role for calcium in some of the insulin-mediated plasma membrane events, evidence supporting a role for calcium in the effects of insulin on intracellular metabolism is less convincing. Clearly, as discussed in Section VI, the effects of insulin on intracellular enzymes and enzyme systems cannot be explained by a uniform change in cytoplasmic calcium as has been proposed for the α_1-receptor-related hormones such as vasopressin and angiotensin II.[24] Only some of the enzymes regulated by insulin are regulated in the same manner by changes in calcium ion concentration. Nevertheless, because of the restriction and compartmentalization of both target enzymes and calcium and the existence of putative cytoplasmic calcium messenger domains (see Section I) to specific regions within the cell, there need not be a consistent pattern for calcium to function as the end-point mediator of some of the target effects of insulin. Since this topic has been the subject of several recent reviews,[7,19,25] it is only emphasized here with significant examples.

An important example of an enzyme complex that is regulated by both calcium and insulin is mitochondrial pyruvate dehydrogenase, which plays a pivotal role in intermediary metabolism. Pyruvate dehydrogenase phosphate phosphatase is activated by insulin and micromolar concentrations of calcium, respectively.[26] Therefore, the regulation of pyruvate dehydrogenase through regulation of mitochondrial calcium homeostasis by insulin (see Section V.B) remains a distinct possibility that may or may not be dependent on changes in cytoplasmic calcium.

There are several intracellular enzymes whose sensitivity to insulin has been proposed to be dependent on calcium, but there are an increasing number of reports to the contrary. For instance, Kissebah et al.[27] found a dependence of insulin-sensitive lipolysis on cellular calcium. However, Fain and Butcher[28] found no such calcium dependence for the antilipolytic effect of insulin in the adipocyte. Also, based on their observed inhibition of lipolysis using calcium chelating agents, Siddle and Hales[29] concluded that the physiological effects of insulin did not require changes in calcium fluxes or participation of intracellular calcium as a mediator. Similarly, although extracellular calcium has been shown to be required for insulin-induced conversion of glycogen synthase from the inactive D form to the active I form,[30] Lawrence and Larner demonstrated that calcium had no effect on insulin-sensitive glycogen synthase activity in adipocytes.[31]

III. EFFECTS OF INSULIN ON CELLULAR CALCIUM FLUXES

Reports on the effect of insulin on calcium fluxes are inconsistent. Cellular calcium flux studies present numerous technical and interpretive problems[32] that may account for much of the conflicting data regarding the effects of insulin. These data have been reviewed in general by Czech,[7] and in particular, pertaining to the adipocyte, by McDonald et al.,[19] and therefore are not considered in detail here. The effects of insulin on cellular calcium fluxes have been studied primarily in skeletal muscle, myoblasts, and adipocytes. As an example of the conflicting data obtained with these studies, Kissebah and co-workers[33,34] reported that insulin inhibited calcium efflux, whereas Clausen and Martin[35] reported that insulin stimulated calcium efflux. Both studies were performed with rat adipocytes, and the experimental designs appear to be similar. However, a knowledge of the strict time frame of these responses is essential in order to adequately assess the implications of these effects, an issue that was not addressed.

A new probe, orthovanadate ($+5$ oxidation state), has recently been found to have insulinlike effects on certain aspects of cellular metabolism and to have major effects on net cellular calcium fluxes and active calcium transport by subcellular organelles. Vanadate (0.1–5 mM), when added to intact cells, has been reported to (1) stimulate glucose oxidation[36] and glucose transport[37] in adipocytes, (2) stimulate glucose transport in adipose tissue and skeletal muscle,[38] (3) stimulate $^{45}Ca^{2+}$ efflux in adipose tissue and skeletal muscle,[38] (4) hyperpolarize muscle,[39] and (5) mimic the effect of insulin on glycogen synthase in adipocytes.[40] Vanadate, like insulin, has been reported to stimulate the phosphorylation of the solubilized β subunit of the insulin receptor.[40] Interestingly, vanadate inhibits the (Ca^{2+},Mg^{2+})ATPase of muscle sarcolemma[41] and, like insulin, inhibits the (Ca^{2+},Mg^{2+})ATPase of the adipocyte plasma membrane.[42] Vanadate also inhibits ATP-dependent calcium transport in the same preparation.[42] Finally, vanadate alters the activity of NADH oxidation,[43] a system that has been reported to be altered by insulin.[44] The correlation between the effects of vanadate on calcium homeostasis and the insulinlike effects of this ion on cellular metabolism is quite striking. Nevertheless, caution is warranted in interpreting these correlations, as it is with interpreting all studies using so-called calcium probes (e.g., local anesthetics, A23187, lanthanum). None of the probes is truly calcium specific, so that data obtained using these compounds merely imply involvement of calcium through differing and incompletely understood mechanisms. Also, the interpretation that the source of calcium seemingly mobilized by vanadate is apparently the endoplasmic reticulum system[38] may serve as an additional explanation for its insulinlike effects, since vanadate also inhibits the (Ca^{2+},Mg^{2+})ATPase/transport system in the adipocyte endoplasmic reticulum.[42]

IV. EFFECTS OF INSULIN AND RELATED FACTORS ON CALCIUM HOMEOSTASIS AT THE PLASMA MEMBRANE

The largest and most detailed accumulation of evidence concerning the effects of insulin on subcellular calcium homeostasis is derived from studies using adipocyte plasma membranes.

A. Insulin-Dependent Effects on Calcium Binding

Treatment of adipocytes with physiological concentrations of insulin results in an increase in calcium binding to adipocyte plasma membranes by increasing the maximum binding capacities of both the high- and low-affinity sites without altering their affinities for calcium (Table 1). The ability of insulin treatment of cells to increase calcium binding to subsequently isolated plasma membranes was also demonstrated in skeletal muscle[45] and independently confirmed in adipocytes.[46] Sandra and Fyler[46] further studied the relationship between the calcium binding to adipocyte plasma membranes, phosphatidylserine, and glucose transport. On the basis of their observations, these workers conclude that the effect of insulin on calcium binding to adipocyte plasma membranes was related to the phospholipid moiety, which may be functioning as an integral component for the modulation of glucose transport by insulin. On the other hand, insulin has been reported to decrease calcium binding to plasma membranes isolated from liver,[47] to an adipocyte ghost preparation,[34] and to human placental membranes.[48] However, the first study used pharmacologically high concentrations of insulin, the second contained nuclei and subcellular organelles, and the third preparation was ill defined and of questionable purity. Therefore, although the latter two studies used physiological concentrations of insulin, it is difficult to interpret these findings as they relate specifically to the plasma membrane.

B. Inhibition of the (Ca^{2+},Mg^{2+})ATPase/Calcium Transport System by Insulin

The adipocyte plasma membrane contains a calmodulin-sensitive high-affinity (Ca^{2+},Mg^{2+})ATPase[14,49] that drives the active transport of calcium.[13] This system, like that which has been well characterized in the human erythrocyte plasma membrane, appears to be the primary mechanism for the extrusion of calcium from the cell to the extracellular space.[11] This system also shares many features with a variety of plasma membrane pumps in other eukaryotic cells.[50] The differing characteristics between plasma membrane and endoplasmic reticulum (Ca^{2+},Mg^{2+})ATPase and calcium transport systems in the adipocyte have recently been reviewed.[14]

Direct addition of insulin to isolated adipocyte plasma membranes had an inhibitory effect on the high affinity (Ca^{2+},Mg^{2+})ATPase (Table 2). This inhibitory effect of insulin has been confirmed by others,[51] and is of particular significance, since the enzyme represents a primary active transport system. This effect on the

Table 1. Enhancement of Calcium-Binding Capacity of
Adipocyte Plasma Membranes by Insulin[a]

Binding site	Maximum binding capacity nmoles calcium/mg protein		% increase after insulin treatment
	Control	Insulin	
High-affinity site	1.26 ± 0.15	1.61 ± 0.16	27.8 ± 3.0
Low-affinity site	13.0 ± 1.1	17.1 ± 0.7	31.5 ± 8.6

[a] Data obtained with plasma membranes isolated from insulin-treated and control cells and reproduced from Table 2, McDonald et al.[73]

(Ca^{2+},Mg^{2+})ATPase also could be produced by treatment of intact cells with insulin $(10^{-9}$ M) prior to isolation of the plasma membranes.[52]

An insulin-sensitive $110,000-M_r$ calcium sensitive phosphoprotein has been identified in the adipocyte plasma membrane.[53] The $K_{0.5}$ for phosphoprotein formation was determined to be 0.07 μM free calcium, and several other characteristics strongly suggest that this molecular entity represents a phosphorylated intermediate of the high-affinity (Ca^{2+},Mg^{2+})ATPase.[14] The phosphorylation of this calcium-dependent $110,000-M_r$ protein was inhibited in plasma membranes from insulin-treated $(10^{-9}$ M) cells as compared with those from untreated cells.[53] The 30–40% inhibition of phosphorylation induced by insulin is similar in magnitude (39%) to the effect of insulin on the (Ca^{2+},Mg^{2+})ATPase,[54] but the similarity may be fortuitous; thus, pursuit of the relationship should prove interesting.

The kinetic parameters obtained for the high-affinity (Ca^{2+},Mg^{2+})ATPase as a function of free calcium revealed that under initial rate conditions addition of 10^{-9} M insulin resulted in a 39% inhibition of the (Ca^{2+},Mg^{2+})ATPase.[54] The inhibitory effect of insulin occurred primarily through an increase in the Hill coefficient and, to a smaller extent, by a decrease in reaction velocity (Table 2). In other words, insulin

Table 2. Inhibition of the (Ca^{2+},Mg^{2+})ATPase of Adipocyte Plasma Membranes
by Insulin[a]

1. Direct addition of physiological concentrations of insulin $(10^{-10}-10^{-9}$ M) to plasma membranes inhibits activity in a concentration-dependent manner.
2. The maximal inhibition ranged from 50% at low calcium $(3.5 \times 10^{-8}$ M) to 13.3% at higher calcium $(1.6 \times 10^{-7}$ M).
3. The insulin inhibition resulted in a 21% decrease in the V_{max} and an increase in the Hill coefficient from 1.36 ± 0.08 to 1.73 ± 0.06.
4. Inactive insulin analog (desoctapeptide) had no effect.
5. Effect was rapid (1–2 min, the shortest time period measured).
6. Inhibition reproduced by treating the intact cell with insulin before isolating and assaying the plasma membrane.
7. Similar inhibition produced by treating the intact cell with the insulin-mimetic agent, concanavalin A.

[a] Data obtained from Pershadsingh and McDonald.[52,54,88]

inhibited the (Ca^{2+},Mg^{2+})ATPase over a narrow threshold via an increase in calcium substrate cooperativity. Interaction of calcium and ATP with the enzyme in the erythrocytes has been shown to cause the system to oscillate between active and inactive states that may underlie a calcium-dependent mechanism for inhibition of the (Ca^{2+},Mg^{2+})ATPase by insulin.[55]

No direct effect of insulin has been seen on ATP-dependent calcium transport under conditions thus far tested. This, however, is not unexpected since transport is assayed under isotonic conditions (as opposed to the hypotonic (Ca^{2+},Mg^{2+})ATPase conditions), wherein ATP-dependent calcium transport is seen only in vesicles oriented inside-out with their insulin receptors facing the interior. Obviously, externally added insulin would be inaccessible to the receptor site if one assumes that the hormone cannot cross the vesicular membrane barrier in an isotonic medium. Nevertheless, it seems reasonable to assume that the (Ca^{2+},Mg^{2+})ATPase is the essential driving machinery for calcium extrusion from the cell; the effect of insulin *in vivo* would therefore be to retard this process.

One can estimate the effect of the inhibition of (Ca^{2+},Mg^{2+})ATPase by insulin on calcium concentration in the cytosol. Using calculated and measured estimates of physical and metabolic characteristics of the adipocyte (see ref 54, for a complete discussion), we were able to show that, within 1 sec after the onset of inhibition by insulin, there would be at least a 10-fold increase in $[Ca^{2+}]_{cyt}$ (from ~0.06 to ~0.60 μM) of the entire cytoplasmic calcium pool, assuming homogeneous distribution throughout the intracellular H_2O space. Since the cytoplasmic calcium pool is heterogeneously distributed, the above calculations could greatly underestimate the effect that insulin would have on the calcium concentration in localized plasma membrane domains. Although this value can only be a very rough approximation, it is clear that relatively small changes in the activity of the (Ca^{2+},Mg^{2+})ATPase would result in major intracellular changes in cytoplasmic calcium.

C. Stimulation of the (Ca^{2+},Mg^{2+})ATPase/Calcium Transport by Endogenous Factors

Both the (Ca^{2+},Mg^{2+})ATPase[14] and its associated high-affinity active calcium transport system[13] are stimulated by calmodulin and by a low-molecular-weight heat-labile, acid-stable, water-soluble factor obtained from fat cell plasma membranes which is considered to constitute one of the putative "second messengers" generated by insulin.[56] This supernatant factor has been shown to mimic the effects of insulin on glycogen synthetase,[57] pyruvate dehydrogenase,[58] and phosphodiesterase.[59] Recently, the calcium concentration dependency on the ability of calmodulin and the supernatant factor to activate calcium transport by the plasma membranes was compared; both agents were shown to activate calcium transport over the identical superimposable narrow calcium threshold.[54] Although the active component(s) of this supernatant factor obtained from adipocyte plasma membranes[56,58,59] is not likely to be a calcium-dependent regulatory protein, it may contain phospholipids, phosphatidic acid and/or proteases[60] that could activate the plasma membrane (Ca^{2+},Mg^{2+})ATPase.[61] The mechanistic interpretation of these data is difficult at this point; however, it may represent a negative feedback signal that functions to terminate the calcium signal. Nevertheless, the significance of these observations must await identification of the active materials in the factor that are responsible for its insulin-mimetic properties.

D. Effect of Insulin on Calmodulin Binding

This laboratory previously described two classes of specific calcium-dependent calmodulin binding sites in the adipocyte plasma membrane.[62] It was recently determined that insulin increases the binding of calmodulin to the high-affinity site in a calcium-dependent manner.[63] The general characteristics of this effect are summarized in Table 3.

These observations may be of fundamental significance, since calmodulin is an ubiquitous calcium-dependent regulatory ligand found in all eukaryotic cells investigated so far[64] and that directly modulates key protein kinases, phosphoprotein phosphatases, and other target proteins, including the plasma membrane (Ca^{2+},Mg^{2+}) ATPase.[65] Furthermore, preliminary observations made in this laboratory indicate that the insulin receptor possesses a calmodulin-binding domain.[66] This observation may explain the calcium dependence of insulin receptor phosphorylation (Section II.A), and thus represents an initial crucial step in understanding the molecular interaction between the insulin receptor, the (Ca^{2+},Mg^{2+})ATPase and possibly other plasma membrane target proteins (kinases, proteases) that may be essential to the transduction of the insulin signal.

V. EFFECTS OF INSULIN ON CALCIUM HOMEOSTASIS OF OTHER ORGANELLES

A. Insulin-Stimulated Calcium Transport by the Endoplasmic Reticulum

Treatment of adipocytes with physiological concentrations of insulin prior to fractionation results in a significant modification of ATP-dependent calcium uptake by the endoplasmic reticulum[67] (Table 4). Also in experiments using skinned muscle fibers, insulin has been shown to stimulate calcium uptake into sarcoplasmic reticulum.[68] However, insulin was shown either to have no effect[69] or to decrease[70]

Table 3. Effect of Insulin on Calcium-Dependent Calmodulin Binding to Adipocyte Plasma Membranes[a,b]

Direct addition of physiological concentrations of insulin (10^{-10}–10^{-9} M) increases calmodulin binding by 75 ± 17% in a concentration-dependent manner.
Inactive insulin analogue (desoctapeptide insulin) has no effect.
Insulin increases B_{max} of high-affinity site without affecting the $K_{0.5}$.
Effects observed in <5 min.
Effect of insulin greatest above the $K_{0.5}$ for calcium (~2.0 μM).

[a] B_{max}, maximum binding constant; K_a, apparent half-maximal saturation constant for calmodulin binding; $K_{0.5}$, apparent half-maximal saturation constant.
[b] Data obtained from Goewert et al.[63]

Table 4. Stimulation of ATP-Dependent Calcium Uptake by Adipocyte Endoplasmic
Reticulum by Insulin[a]

Insulin (10^{-9}) treatment of cells result in a 20% increase in the steady-state filling capacity of the endoplasmic reticulum.
This increase resulted from a 28% increase in the apparent V_{max} of calcium uptake without alteration in the K_a.
The effect was dependent on insulin concentration and was saturable.
The relatively inactive insulin analog, desoctapeptide insulin, had no effect.
Insulin treatment had no effect on calcium efflux from the endoplasmic reticulum initiated by the chelator EGTA or the calcium ionophore A23187.

[a] Data obtained from McDonald et al.[67]

calcium uptake into endoplasmic reticulum vesicles from liver. Apart from the difference in tissues employed, the significance of the contrasting effects of insulin is unknown.

The apparent paradox that the effects of insulin would be expected to (1) decrease $[Ca^{2+}]_{cyt}$ through stimulation of the endoplasmic reticulum calcium uptake pump, and (2) increase $[Ca^{2+}]_{cyt}$ through blockade of the plasma membrane calcium extrusion pump, may be explained by the observation that the stimulatory effect of insulin on the endoplasmic reticulum calcium pump could be reproduced simply by increasing the extracellular calcium concentration.[71] Such a maneuver would increase the calcium concentration gradient across the plasma membrane, and consequently the passive influx of calcium thereby increasing $[Ca^{2+}]_{cyt}$, an effect similar to that arising from an increase in $[Ca^{2+}]_{cyt}$ through inhibition of the plasma membrane pump by insulin. The increased $[Ca^{2+}]_{cyt}$ in turn serves to accelerate active calcium sequestration by the endoplasmic reticulum through substrate activation, resulting in an increase in the V_{max},[71] similar to the stimulatory effect of insulin on this calcium transport system.[67] Although the significance of the effect of insulin on adipocyte endoplasmic reticulum calcium transport is presently unknown, it may be related to the regulation of some intracellular hormone-regulated, calcium-sensitive enzymatic processes (see Section II.B).

B. Effect of Insulin on Mitochondrial Calcium Homeostasis

Although mitochondrial calcium homeostasis is known to be intimately linked to energy metabolism, the nature of this relationship is unclear.[72] Pyruvate dehydrogenase phosphatase is a calcium-activated enzyme that plays a pivotal role in glucose oxidation; insulin is known to increase this phosphatase activity.[25] McDonald et al.[74] reported that insulin treatment of adipocytes increased a labile (exchangeable) pool and decreased a stable (nonexchangeable) pool of calcium in subsequently isolated mitochondria. Increased cytoplasmic calcium via inhibition of calcium extrusion at the plasma membrane could also result in increased calcium cycling[15] across the mitochondrial inner membrane, resulting in an increase in the labile pool. One may speculate that an increased exchangeable labile calcium pool may signify increased accessibility of calcium ions to pyruvate dehydrogenase calcium binding sites resulting in activation of the phosphatase when insulin is present. On the other hand, Severson et al.[75] were unable to demonstrate increased incorporation of extracellular $^{45}Ca^{2+}$ into insulin-treated adipocyte mitochondria. It is important to point out that both studies described above suffer from the possible artifacts induced by subcellular fractionation and may not reflect in vivo changes in mitochondrial calcium.

More recently, Davis et al.[76] reported that physiological concentrations of insulin cause a 19 mV depolarization of the mitochondrial membrane, which they propose to be secondary to an increase in cytoplasmic calcium. The mitochondrial membrane depolarization could directly result from stimulation of the electrophoretic calcium uptake uniporter. Subsequently, respiration is stimulated as the mitochondrial energetic state is ultimately reestablished. These studies involve both intact adipocytes and isolated mitochondria and thus circumvent some of the problems associated with the previous studies analyzing mitochondrial calcium.

VI. SPECULATIONS ON CALCIUM AS A TRANSDUCER OF INSULIN ACTION

Although accumulated evidence indicates that calcium participates in the cellular mechanism of insulin action, it cannot be considered as a second messenger or mediator exclusively responsible for the diverse effects of insulin on intracellular metabolic activities. Rather, it is more likely that the complex response of the target cell to insulin (and this may differ from one tissue to another) is the result of a highly integrated metabolic cascade which, although initiated at the plasma membrane, involves the interaction of a variety of substances, perhaps including those previously proposed as being involved in insulin action.[7]

The most compelling evidence for the role of calcium in the mechanism of insulin action is obtained from the effects of insulin on (1) adipocyte plasma membrane calcium binding, (2) (Ca^{2+},Mg^{2+})ATPase, and (3) calmodulin binding. Furthermore, preliminary data indicating that an integral component of the insulin receptor is a calmodulin-binding protein[66] suggest a model for the early effects of insulin action in which the insulin receptor, the (Ca^{2+},Mg^{2+})ATPase and calmodulin interact in order to generate at least one fundamental early signal. The effects of insulin on the plasma membrane (Ca^{2+},Mg^{2+})ATPase and calcium binding may also provide the mechanism whereby insulin regulates some of the plasma membrane effects of the hormone in the immediate vicinity of the plasma membrane.

Therefore, we propose the following model for the initial molecular events effecting insulin-induced activation of the plasma membrane (Fig. 2). The signal is initiated by the interaction of insulin primarily with the α subunit of the insulin receptor. Another part of the receptor (e.g., the β subunit) is the calmodulin-binding domain, which is endowed with an increased affinity for calmodulin immediately subsequent to insulin binding. This activated state of the insulin–receptor complex then causes a competitive dissociation of calmodulin from the structurally related (Ca^{2+},Mg^{2+})ATPase/ calcium-transport complex. This latter event, which represents the conversion of the highly active holoenzyme to the less active apoenzyme, results in decreased calcium extrusion and an increase in calcium concentration in related microdomains of the

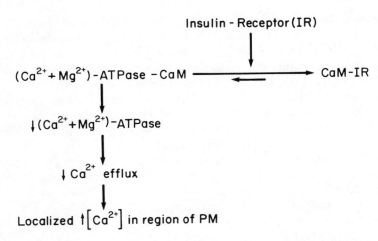

Figure 2. Hypothetical model for the initial molecular events immediately distal to the interaction of insulin with its high-affinity receptor resulting in activation of the plasma membrane (PM). CaM represents calmodulin.

plasma membrane. These events, in conjunction with the generalized surface-activating properties of insulin, result in an increase in the specific binding of calcium to the plasma membrane that could directly regulate a number of insulin-dependent phenomena, such as the stimulation of glucose transport and modification of other ion fluxes.

Calcium may regulate plasma membrane permeability to potassium and/or sodium[77] by direct regulation of specialized pores or by modulation of their conductive pathways via calcium-dependent changes in membrane fluidity.[78] These changes may underlie the known effect of insulin to hyperpolarize target cells (see ref. 79 for review). Hyperpolarization of the cell (negative on the cytoplasmic side) may explain the fact that insulin stimulates primarily sodium-dependent amino acid transport,[80] which is sensitive to changes in the sodium electrochemical gradient.[79,81] Interestingly, Zierler and Rogus[82] demonstrated that only a 1.5 mV hyperpolarization (compared with the 5–10 mV induced by insulin) enhanced glucose transport by as much as 40%. These investigators speculated that an electrically induced change in conformation of the glucose carrier facilitated accelerated transit and consequently increased net glucose uptake by the cell. Clearly, these findings warrant further investigation.

The observation that insulin-induced activation of target cells is accompanied by an increase in intracellular pH[79] is compatible with our proposal. Increased $[Ca^{2+}]_{cyt}$ has been shown to be accompanied by a decrease in cytoplasmic pH.[83] Indeed, ATP-dependent transport of calcium by erythrocyte plasma membranes has been shown to be accompanied by an exchange of proton(s).[84] Therefore, inhibition of the calcium pump could lead to a coupled increase in cytosolic pH (decreased hydrogen ion concentration). In related studies it has been shown that calcium transport in plasma membranes isolated from the livers of streptozotocin-induced diabetic rats is increased twofold[85] and that the intracellular pH in tissues from diabetic animals is decreased.[86] These observations are consistent with the existence of a pump-mediated Ca^{2+}/H^+

exchange mechanism in which the increase in calcium transport in the diabetic state results in an increase in cytoplasmic hydrogen ion concentration (decreased pH).

The evidence related above supports the conclusion that calcium is of fundamental importance in transmission of the insulin-induced metabolic signal. The precise mechanism and sequence of events involved in transduction and amplification of the signal require an understanding of the structural and functional relationships among the insulin receptor, the (Ca^{2+},Mg^{2+})ATPase, and calmodulin.

VII. REFERENCES

1. Ringer, S., 1883, *J. Physiol. (Lond.)* **4**:29–42.
2. Anghileri, L.F., and Tuffet-Anghileri, A.M. (eds.), 1982, *The Role of Calcium in Biological Systems,* Vols. I–III, CRC Press, Boca Raton, Florida.
3. Triggle, D.J., 1980, in *Membrane Structure and Functions* (E.E. Bittar, ed.), Vol. 3, pp. 1–58, Wiley, New York.
4. Harold, F.M., 1982, *Curr. Top. Membr. Transport.* **16**:485–516.
5. Siegel, F.L., Carafoli, E., Kretsinger, R.H., MacLennan, D.H., and Wasserman, R.H. (eds.), 1980, *Calcium Binding Proteins: Structure and Function,* Elsevier–North-Holland, New York.
6. Czech, M.P., 1977, *Annu. Rev. Biochem.* **46**:359–384.
7. Czech, M.P., 1981, in: *Handbook of Diabetes* (M. Brownlee, ed.), Vol. 2, pp. 117–149, Garland STPM Press, New York.
8. Clausen, T., Elbrink, J., and Martin, B.R., 1974, *Acta Endocrinol. (Copenh.) Suppl.* **191**:137–143.
9. Kissebah, H.A., Tulloch, B.R., Hope-Gill, H., Clarke, P.V., Vydelingum, N., and Fraser, T.R., 1975, *Lancet* **1(7899)**:144–147.
10. Borle, A.B., 1980, *Rev. Physiol.* **90**:13–153.
11. Schatzmann, H.J., 1982, in: *Membrane Transport of Calcium* (E. Carafoli, ed.), pp. 49–108, Academic Press, London.
12. Blaustein, M.P., 1974, *Rev. Physiol. Biochem. Pharmacol.* **70**:33–84.
13. Pershadsingh, H.A., Landt, M., and McDonald, J.M., 1980, *J. Biol. Chem.* **225**:8983–8986.
14. McDonald, J.M., Chan, K-M., Goewert, R.R., Mooney, R.A., and Pershadsingh, H.A., 1982, *Ann. NY Acad. Sci.* **402**:381–401.
15. Carafoli, E., 1979, *FEBS Lett.* **104**:1–5.
16. Rose, B., and Lowenstein, W.R., 1975, *Science* **190**:1204–1206.
17. Matthews, E.K., 1980, in *Secretory Mechanisms* (C.R. Hopkins and C.J. Duncan, eds.), pp. 225–249, Cambridge University Press, London.
18. Clausen, T., 1980, *Cell Calcium* **1**:311–325.
19. McDonald, J.M., Graves, C.B., and Christensen, L., 1984, in: *Calcium and Cell Function* (W.Y. Cheung, ed.), Vol. 5, Academic Press, San Diego, in press.
20. Roth, J., Kahn, C.R., Lesniak, M.A., Gorden, P., Demeyts, P., Megyesi, K. , Neville, D.M., Jr., Gavin, J.E. III, Sol, A.H., Freychet, P., Goldfine, I.D., Bar, R.S., and Archer, J.A., 1975, *Recent Prog. Horm. Res.* **31**:95–139.
21. Desai, K.S., Sinman, B., Steiner, G., and Hollenberg, C.H., 1978, *Can. J. Biochem.* **56**:843–848.
22. Williams, P.F., and Turtle, J.R., 1981, *Proc. Aust. Biochem. Soc.* **14**:39.
22a. Williams, P.F., 1984, *Diabetes* (Suppl. 1) **33**:10A.
23. Plehwe, W.E., Williams, P.F., Caterson, I.D., Harrison, L.C., and Turtle, J.R., 1983, *Biochem. J.* **214**:361–366.
24. Exton, J.H., 1981, *Mol. Cell. Endocrinol.* **23**:233–264.
25. Denton, R.M., Brownsey, R.W., and Belsham, G.J., 1981, *Diabetalogia* **21**:347–362.
26. Martin, B.R., and Denton, R.M., 1970, *Biochem. J.* **117**:861–877.
27. Vydelingum, N., Kissebah, A.H., and Wynn, V., 1978, *Horm. Metab. Res.* **10**:38–46.
28. Fain, J.N., and Butcher, F.R., 1976, *J. Cyclic. Nucleotide Res.* **2**:71–78.

29. Siddle, K., and Hales, C.N., 1980, *Horm. Metab. Res.* **12:**509–515.
30. Hope-Gill, H.R., 1976, *Horm. Metab. Res.* **8:**321–322.
31. Lawrence, J.C., Jr., and Larner, J., 1978, *J. Biol. Chem.* **253:**2104–2113.
32. Borle, A.B., 1981, *Cell Calcium* **2:**187–196.
33. Hope-Gill, H., Kissebah, A., Tulloch, B., Clarke, P., Vydelingum, N., and Fraser, T.R., 1975, *Horm. Metab. Res.* **7:**195–196.
34. Kissebah, A.H., Clarke, P., Vydelingum, N., Hope-Gill, H., Tulloch, B., and Fraser, T.R., 1975, *Eur. J. Clin.* **5:**339–349.
35. Clausen, T., and Martin, B.R., 1977, *Biochem. J.* **164:**251–255.
36. Shechter, Y., and Karlish, S.J.D., 1980, *Nature* **284:**556–558.
37. Dubyak, G.R., and Kleinzeller, A., 1980, *J. Biol. Chem.* **255:**5306–5312.
38. Clausen, T., Anderson, T.L., Sturup-Johansen, M., and Petkova, O., 1981, *Biochim. Biophys. Acta* **646:**261–267.
39. Zemkova, H., Teisenger, J., and Vyskocil, F., 1982, *Biochim. Biophys. Acta* **720:**405–410.
40. Tamura, S., Brown, T.A., Dubler, R.E., and Larner, J., 1983, *Biochem. Biophys. Res. Commun.* **113:**80–86.
41. Caroni, P., and Carafoli, E., 1981, *J. Biol. Chem.* **256:**3263–3270.
42. Delfert, D.M., Pershadsingh, H.A., and McDonald, J.M. (manuscript in preparation).
43. Ramasarma, T., MacKellar, W.C., and Crane, F.L., 1981, *Biochim. Biophys. Acta* **646:**88–98.
44. Low, H., Crane, F.L., Grebing, C., Tolly, M., and Holl, K., 1978, *FEBS Lett.* **91:**166–168.
45. Cheng, L.C., Rogus, E.M., and Zierler, K., 1978, *Biochim. Biophys. Acta* **513:**141–155.
46. Sandra, A., and Fyler, D.J., 1982, *Endocr. Res. Commun.* **9:**107–120.
47. Shlatz, L., and Marinetti, G.V., 1972, *Science* **176:**175–177.
48. Williams, P.F., and Turtle, J.R., 1981, *Biochim. Biophys. Acta* **676:**113–117.
49. Pershadsingh, H.A., and McDonald, J.M., 1980, *J. Biol. Chem.* **255:**4087–4093.
50. Barritt, G.J., 1981, in: *The Role of Calcium in Biological Systems*, Vol. II (L.J. Anghileri and A.M. Tuffet-Anghileri, eds.), pp. 17–30, CRC Press, Boca Raton, Florida.
51. Schoenle, E., and Froesch, E.R., 1981, *FEBS Lett.* **97:**283–287.
52. Pershadsingh, H.A., and McDonald, J.M., 1981, *Biochem. Int.* **2:**243–248.
53. Chan, K-M., and McDonald, J.M., 1982, *J. Biol. Chem.* **257:**7443–7448.
54. Pershadsingh, H.A., and McDonald, J.M., 1983, *Cell Calcium* **5:**111–130.
55. Pedmonte, C.H., and Balegno, H.F., 1982, *Molec. Cell. Biochem.* **47:**31–34.
56. McDonald, J.M., Pershadsingh, H.A., Kiechle, F., and Jarett, L., 1981, *Biochem. Biophys. Res. Commun.* **100:**857–864.
57. Larner, J., Galasko, G., Cheng, K., Depaoli-Roach, A.A., Huang, L., Daggy, P., and Kellog, J., 1979, *Science* **206:**1408–1410.
58. Seals, J.R., and Jarett, L., 1980, *Proc. Natl. Acad. Sci. USA* **77:**77–81.
59. Kiechle, F.L., and Jarett, L., 1981, *FEBS Lett.* **133:**279–282.
60. Kiechle, F.L., Strauss, J.F., Tanaka, T., and Jarett, L., 1982, *Fed. Proc.* **41:**1087 (abst.).
61. Niggli, V., Adunyah, E.S., and Carafoli, E., 1981, *J. Biol. Chem.* **256:**8588–8592.
62. Goewert, R.R., Landt, M., and McDonald, J.M., 1982, *Biochemistry* **221:**6310–6315.
63. Goewert, R.R., Klaven, N.B., and McDonald, J.M., 1983, *J. Biol. Chem.* **258:**9995–9999.
64. Kretsinger, R.H., 1979, *Adv. Cyclic Nucleotide Res.* **11:**2–26.
65. Cheung, W.Y., 1979, *Science* **207:**19–27.
66. McDonald, J.M., Goewert, R.R., and Graves, B., 1984, *Diabetes* (Suppl. 1)**33:**10A.
67. McDonald, J.M., Bruns, D.E., and Jarett, L., 1978, *J. Biol. Chem.* **253:**3504–3508.
68. Brautigan, D.L., Kerrick, W.G., and Fischer, E.H., 1980, *Proc. Natl. Acad. Sci. USA* **77:**936–939.
69. Taylor, W.M., Bygrave, F.L., Blackmore, P.F., and Exton, T.H., 1979, *FEBS Lett.* **104:**31–34.
70. Waltenbaugh, A.M.A., and Friedmann, N., 1978, *Biochem. Biophys. Res. Commun. Vol.* **82:**603–608.
71. Black, B.L., McDonald, J.M., and Jarett, L., 1981, *J. Biol. Chem.* **256:**322–329.
72. Saris, N.-E., and Akerman, K.E.O., 1980, *Curr. Top. Bioener.* **10:**103–179.
73. McDonald, J.M., Bruns, D.E., and Jarett, L., 1976, *Proc. Natl. Acad. Sci. USA* **73:**1542–1546.
74. McDonald, J.M., Bruns, D.E., and Jarett, L., 1976, *Biochem. Biophys. Res. Commun.* **71:**114–121.
75. Severson, D.L., Denton, R.M., Bridges, B.J., and Randle, P.J., 1976, *Biochem. J.* **154:**209–223.

76. Davis, R.J., Brand, M.D., and Martin, B.R., 1981, *Biochem. J.* **196**:133–147.
77. Putney, J.W., 1982, in *The Role of Calcium in Biological Systems,* Vol. II (L.J. Anghileri and A.M. Tuffet-Anghileri, eds.), pp. 47–66, CRC Press, Boca Raton, Florida.
78. Storch, J., Schachter, D., Inoue, M., and Wolkoff, A.W., 1983, *Biochim. Biophys. Acta* **727**:209–212.
79. Moore, R.D., 1983, *Biochim. Biophys. Acta* **737**:1–49.
80. Guidotti, G., Borghetti, A.F., and Gazzola, G.C., 1978, *Biochim. Biophys. Acta* **515**:329–366.
81. Johnstone, R.M., 1979, *Can. J. Physiol. Pharmacol.* **57**:1–15.
82. Zierler, K., and Rogus, E.M., 1981, *Fed. Proc.* **40**:121–124.
83. Gerstenblith, G., Hoerter, J.A., Jacobus, W.E., Lakatta, E.D., Miceli, M.V., and Renlund, D.G., 1983, *J. Physiol. (Lond.)* **334**:105P–106P.
84. Niggli, V., Sigel, E., and Carafoli, E., 1982, *J. Biol. Chem.* **257**:2350–2356.
85. Chan, K.M., and Junger, K., 1983, *Diabetes* (Suppl. 1) **32**:127A.
86. Brunder, D.G., Oleynek, J.J., and Moore, R.D., 1983, *J. Gen. Phys.* **82**:15A.
87. Pershadsingh, H.A., and McDonald, J.M., 1979, *Nature* **281**:495–497.

7

Actions of Insulin on Electrical Potential Differences across Cell Membranes

Kenneth Zierler

I. INTRODUCTION: HYPOTHESIS THAT MEMBRANE FUNCTIONS DEPEND ON MEMBRANE POTENTIAL

Electrical force is one of the basic forces of nature. It can be much stronger than gravity, as is its complementary force, magnetism. Magnetic and electrical forces act over distances; that is, there are fields of force. As in the schoolchild's experiment in which iron filings sprinkled on a sheet of paper were made to show lines of magnetic force fields in response to a horseshoe magnet held beneath the paper, so electrical fields can be demonstrated to orient charged particles and dipoles within their territory.

External or plasma membranes of cells of multicellular animals are modified lipid phases separating two aqueous phases: one intracellular, the other extracellular. It is a fact that there is an electrical potential difference between these two aqueous phases. Because this electrical potential difference, or membrane potential, as it is loosely called, was studied so much more extensively in excitable cells (i.e., nerve, skeletal muscle, myocardium), it was thought that the steady-state electrical potential difference, or resting membrane potential, of excitable cells serves as a baseline upon which action potentials could be inscribed. But we have learned that there is a steady-state electrical potential difference between intra- and extracellular aqueous phases of all mammalian and amphibian cells examined so far and that in all these nucleated mammalian cells the membrane potential is of the same order of magnitude, between about -40 and -80 mV, inside negative with respect to outside, whether or not the cell is classically excitable.

Kenneth Zierler ● Department of Physiology, The Johns Hopkins University School of Medicine, Baltimore, Maryland.

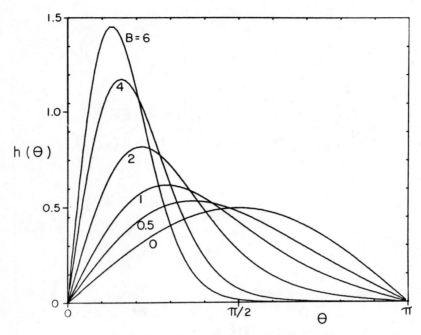

Figure 1. Families of probability density functions $h(\theta)$ that a dipole angle will lie between θ and $(\theta + d\theta)$. $h(\theta)$ is a function of the product of dipole movement μ and electrical field strength E. B is the ratio of this product to thermal energy, $B = \mu\mathscr{E}/kT$. As B increases, the peak of the density function shifts toward zero θ, and the function becomes more skewed.

A membrane potential of -40 to -80 mV is impressed across a plasma membrane of only about 70-Å thickness, i.e., the electrical potential difference is of the order of 100,000 V/cm length. The change in voltage per unit length dV/dx is the measure of strength of the electrical field.

We can consider a cell membrane, for purposes of electrical analysis, to be a pair of parallel plates separating two aqueous conducting solutions, with some dielectric material between the plates. A field strength of 10^5 V/cm is so enormously powerful that one wonders why a breakdown of the membrane dielectric does not occur—indeed an unsolved question. A sense of the size of membrane electrical field strength may be gained by the reminder that gel electrophoresis is carried out at a field strength of only about 200–600 V/cm. We see at once that naturally occurring membrane potentials are far more powerful orienting forces than those we use in the laboratory. Remember, this membrane electrical potential difference exists only across the cell membrane, from the interface between the external aqueous phase and the outer surface of the membrane to the interface between the internal aqueous phase and the inner surface of the membrane. Every charged particle and every dipole between these two interfaces, in the substance of the membrane, is stressed to orient itself in response to the trans-membrane electrical field strength.

The work done in rotating a dipole through an angle $d\theta$, where θ is the angle between the dipole and the direction of the electrical field (taken as perpendicular to the plane of the membrane), is a function of the product of dipole moment μ and electrical field strength $\mathscr{E} = d\psi/dx$. Every CONH group of a protein is a dipole; proteins have large dipole moments, from ~200 to 1200 debye ($200 - 1200 \times 10^{18}$ esu-cm). The likelihood of orientation of dipoles, as opposed to random distribution of dipole angles, is a function of the ratio of the orienting energy $\mu\mathscr{E}$ to thermal energy kT, where k is Boltzmann's constant and T is absolute temperature. For a given ratio $B = \mu\mathscr{E}/KT$ there is a specifiable probability that the dipole angle will lie between some stated θ and $\theta + d\theta$. The probability density function $h(\theta)$ undergoes characteristic changes as B increases (Fig. 1), over a range of values of B found in mammalian cells [See Debye[1] for a more complete treatment of $h(\theta)$.] As B increases (i.e., as dipole moment increases for a given electrical field strength or as electrical field strength increases at a given dipole moment) the peak of $h(\theta)$ increases in height and shifts toward a smaller dipole angle, $h(\theta)$ becomes more skewed, with narrowing of the range of values of θ at half-maximum $h(\theta)$. There is increased clustering of dipole angles about the mode, but there is also increased separation between mode and median $h(\theta)$.

Note in Fig. 1 that for a given small dipole angle θ, the probability can increase dramatically as B increases. For example, the fraction of dipoles lying between angle 10° and 11° of the direction of the electrical field increases from 0.0016 when there is no electrical field, to 0.0146 when $B = 5$, to 0.0173 when $B = 6$. Even when there is no electrical field across a membrane, as is the case for some preparations of plasma membrane vesicles, some dipole angles of some membrane proteins will probably be small, but the number of protein molecules so oriented will be few compared with the number oriented in a field of 10^5 V/cm. Therefore, if function (e.g., specific D-glucose transport) requires that dipole angles be at some small specific value, that function may be realized in the absence of an electrical field, but quantitatively it will be a poor caricature of the function in the presence of the normal field across the intact cell membrane *in situ*, to the extent that quantitative expression of that function depends on the number of protein molecules in the required configuration.

Whatever forces tend to lock proteins in position in lipid bilayer cell membranes serve to tether the protein as its dipoles are prodded to orient in the electrical field. The protein as a whole cannot rotate, but dipole angles must tend to move toward directions commanded by the field. As a result, strains may occur on the spine of the protein. With sufficient strain, the protein may flip from one stable configuration to another stable configuration, possibly creating new dipole–dipole interactions to stabilize the second configuration.

A familiar example of this sort of altered configuration causing altered membrane function may be a voltage-dependent ion channel,[2] such as the tetrodotoxin-sensitive Na^+ channel of excitable cells, in which a 2–4-mV depolarization causes an e-fold increase in Na^+ conductance,[3] i.e., a change in electrical field strength by less than 5% more than doubles the rate at which Na^+ moves, in the net, through those channels in response to a given chemical potential gradient for Na^+.

A cell must spend a substantial part of its resting energy to maintain the electrical potential difference across its surface membrane. Why is this so? We conjecture that the cell must maintain configurations of membrane proteins that impart functions necessary to the cell. A change in membrane potential, then, is a force that acts over the entire cell surface, capable of stressing membrane proteins toward new configurations with which a different level of functional activity may be associated. Specificity of response to a given change in membrane potential might depend on dipole moments of the proteins or on the existence of critical angles for certain of the dipoles in a particular protein with a flip to a new stable configuration which may be associated.

II. INSULIN-INDUCED HYPERPOLARIZATION OF ITS MAJOR TARGET TISSUES, EXCEPT LIVER

A. Cross-Striated Muscle

In 1957 I discovered that insulin hyperpolarizes rat extensor digitorum longus muscle.[4] We have repeatedly confirmed insulin-induced hyperpolarization (IIH) in extensor digitorum longus[5-7] and in rat caudofemoralis muscle.[8,9] IIH has been reported in rat soleus muscle (Flatman and Clausen[10]), in rat diaphragm (Otsuka and Ohtsuki,[11] Bolte and Lüderitz,[12] Takamori et al.[13]), in mouse diaphragm (Zemkova et al.[14]), and in frog sartorius muscle (De Mello,[15] Moore and Rabovsky[16]). There is one report of failure to find IIH in rat soleus muscle,[17] but it should be dismissed on the basis of technical flaws in the experiment and of failure to find any biological effect of the insulin under conditions of the experiment. Our unpublished experience confirms that insulin hyperpolarizes rat soleus muscle and rat diaphragm.

All measurements of membrane potential ψ_m reported above were made by impaling muscle fibers with small glass capillaries containing 2–3 M KCl or K acetate, leading to Ag–AgCl electrodes. The potential difference between this intracellular probe and a reference electrode in the bathing solution is taken as ψ_m. In all the experiments cited above, the muscle was excised and insulin effect was determined in response to the addition of insulin to the bathing solution.

Under these conditions, in rat caudofemoralis muscle, for example, supramaximal concentrations of insulin hyperpolarize by -8 to -10 mV.[9] Approximately half-maximal hyperpolarization, -4 to -4.5 mV, occurs at 100 μU/ml insulin.[9] (By convention, the interior of the muscle fiber at rest under normal conditions is negative with respect to the bathing solution. ψ_m of rat caudofemoralis is about -78 mV. Supramaximal concentrations of insulin increase it to -86 to -88 mV, i.e., hyperpolarization is an increase in absolute magnitude of ψ_m, but does make ψ_m more negative.)

There is a report of IIH in muscles of rat hindlimbs (not otherwise specified) in response to intraperitoneal injection of a large dose of insulin.[18] Such an experiment is difficult to interpret because systemic effects of insulin, such as hypokalemia, make interpretation ambiguous.

Otsuka and Ohtsuki[11] reported, in diaphragms from K^+-deficient rats, that IIH decreased, disappeared, and was replaced by insulin-induced depolarization as external K^+ concentration $[K^+]_o$ was increased. We[19] have found what may be the same phenomenon in rat caudofemoralis muscle, from rats on a normal diet. IIH decreased and disappeared, although we cannot be sure that it was converted to true depolarization as $[K^+]_o$ was increased. However, we cannot ascribe the effect to increased $[K^+]_o$. It occurred when we raised $[K^+]_o$ by equimolar substitution of KCl for NaCl, which is the way in which Otsuka and Ohtsuki elevated $[K^+]_o$, but it did not occur when $[K^+]_o$ was elevated such that it maintained constant the product of $[K^+]_o$ and $[Cl^-]_o$ by substitution of $[K^+]_o$ for $[Na^+]_o$. Equimolar substitution of KCl for NaCl leads to increased $[K^+]_o[Cl^-]_o$ product with water shift from interstitial fluid into cells, and the cells swell.

Two points arise from our experiments with increased $[K^+]_o$. First, insulin-stimulated glucose uptake in rat caudofemoralis correlated with IIH, as IIH was altered or not altered by increased $[K^+]_o$ by the two methods of raising $[K^+]_o$. Second, with constant product increase in $[K^+]_o$, ψ_m in the absence of insulin decreased approximately as a function of log $[K^+]_o$, but IIH was nearly the same at all ψ_m, and so was insulin-stimulated glucose uptake. Basal glucose uptake (not in the presence of insulin) actually increased with increased $[K^+]_o$, in the constant $[K^+]_o[Cl^-]_o$ product experiments. If insulin-stimulated glucose uptake is a function of IIH, it does not seem to depend on the absolute level of ψ_m produced by insulin. The experiments are difficult to interpret unequivocally because a change in $[K^+]_o$ does more than simply change ψ_m. We have yet to learn all the things high $[K^+]_o$ may do in cell membranes and cell functions. In rat adipocytes, permeability of cell membrane to K^+ is increased at high $[K^+]_o$[20]; we do not know the effect of such a permeability change on membrane functions, such as glucose transport.

B. Adipose Tissue

In 1962 Beigelman and Hollander[21] first reported IIH of cells of rat epididymal fat pads, detected with microelectrodes. It is technically difficult to impale adipose cells so as to measure ψ_m, because the aqueous phase of mature adipocytes is only about 4% of cell volume. Those first records were quite noisy, reflecting the technical difficulty; there was also some uncertainty as to whether the probe electrode was recording from adipocytes or from fibroblasts, which make up a large minority of the bulk of fat pads. Accordingly, indirect methods for estimating ψ_m were tried.

These indirect methods, useful for suspensions of free floating cells or for very small or fragile cells that defy measurement by intracellular electrode techniques, depend on the use of a substance having some property sensitive to an electrical field or to an electrical potential difference between the two aqueous phases. The two most widely used types are (1) ions whose equilibrium distribution between internal and external aqueous phases depends solely on the electrical potential difference (i.e., there are no pumps for the substance) and (2) fluorescent dyes whose signal is voltage dependent either because they fall in the same class as ions in group (1) or because fluorescence quantum yield is, in ways that are not well understood, dependent on either electrical fields or membrane surface charges. There are pitfalls and traps with all techniques for estimating ψ_m (e.g., see for critiques, Armstrong and Garcia-Diaz[22] and Leader[23]), more so with indirect methods.

Using one or another of these indirect methods, Petrozzo and Zierler,[24] Cheng et al.,[25] and Davis et al.[26] demonstrated IIH in suspensions of adipose cells from rat epididymal fat pads. The latter two studies were more complete. Davis and co-workers calculated that the basal ψ_m of adipocyte plasma membrane was -75 mV, somewhat greater polarization than estimated previously by others, and that a supramaximal concentration hyperpolarized by -9 mV, the same amount we observed in rat skeletal muscle and diaphragm.

There is a report of failure to detect IIH in cells of rat epididymal fat pads,[27] using intracellular electrodes, from the same laboratory that reported failure to find IIH in muscle (see Stark and O'Doherty,[17] referred to earlier). The same criticism is raised about this report; it is technically flawed, and it presents no evidence that the insulin it used displayed biological activity under conditions of the experiment.

Davis and co-workers[26] also estimated fat cell mitochondrial membrane potentials, by methods described by Davis and Martin in Chapter 8 of this volume, found that a supramaximal concentration of insulin depolarized mitochondrial membrane by 19 mV.

C. Myocardium

Insulin has been reported to hyperpolarize dog heart muscle,[28] cultured chick myocardial cells,[29] and canine false tendons and papillary muscles from kittens and rats.[30] Although there is a negative report,[31] that study did not demonstrate that the insulin used was biologically active under the conditions of the experiment.

In addition to these studies, there are a number of reports of insulin effects on other electrophysiological aspects of myocardium, such as height and configuration of action potentials from normal and experimentally infarcted tissue. These studies were stimulated by a series of papers by Sodi-Pallares et al., beginning in 1962,[32] in which intravenous infusion of a solution containing glucose–insulin–potassium was claimed to have desirable electrocardiographic and clinical effects. The complicated experimental conditions make these reports difficult to interpret unambiguously in terms of underlying mechanisms, nor are observations consistent. (See Kinoshita and Yatani[33] and Anderson et al.[34] for two recent articles that refer to the earlier literature, illustrating the diversity of results.)

D. Liver

Insulin does not hyperpolarize hepatocytes. Hepatocytes respond metabolically to insulin in ways not observed in skeletal muscle, heart, and fat cells. In particular, there is no insulin-stimulated D-glucose uptake in hepatocytes. Friedmann et al.[35] observed no effect of insulin alone on membrane potential in perfused rat liver, but found that insulin prevented hyperpolarization that glucagon would have produced in the absence of insulin. Wondergem,[36] in cultured hepatocytes, found a slow depolarization in response to insulin.

E. Other Cells

Insulin has been reported to hyperpolarize toad bladder and toad colon mucosa,[37] rat gastric mucosa,[38] rabbit ciliary epithelium,[39] and cultured toad kidney cells.[40] Closely related are studies of short-circuit current (active Na^+ transport) in frog and toad skin and toad bladder. (See Fidelman et al.[40] and references cited therein.) Not all of these tissues are usually thought of as insulin sensitive, and in most cases large concentrations of insulin were used. However, cultured toad kidney cells responded to as little as 10 μU/ml. These cells also showed other typical metabolic responses to insulin. Insulin proved effective only when added to the basal surface, not to the apical surface, suggesting that there are insulin receptors on the basal surface, not on the apical surface, and that insulin does not penetrate from basal to apical surface.

Williams et al. reported that insulin had no effect on ψ_m of pancreatic acinar cells[41] measured in fragments of pancreas from mice made diabetic by streptozotocin. Insulin was effective, however, when large concentrations were used to stimulate uptake of glucose analogs by suspensions of acinar cells dissociated from pancreases.

III. THE IMMEDIATE MECHANISM OF INSULIN-INDUCED HYPERPOLARIZATION

Bioelectricity is static electricity; bioelectrical potentials occur because charges have been separated. With respect to electrical potential differences across biomembranes, between bulk aqueous phases on the two sides, the observed potential is a weighted sum of contributions due to diffusion (in response to chemical potential differences, such as differences between K^+ concentrations) and to electrogenic pumping (i.e., to an energy-coupled net transmembrane transport excess of a cation or anion). That is,

$$\psi_m = T_D\psi_D + T_P\psi_P$$

where ψ_D is the diffusive component of the potential, ψ_P is the pump component, $T_D = g_D/(g_D + g_P)$, and $T_P = g_P/(g_D + g_P)$, the fractions of electrical conductance contributed by current due to the diffusive and the pump component, respectively.

The diffusive component is given by the classic Goldman–Hodgkin–Katz equation[42,43]:

$$\psi_D = -\frac{RT}{F} \ln \frac{P_K[K^+]_i + P_{Na}[Na^+]_i + P_{Cl}[Cl^-]_o}{P_K[K^+]_o + P_{Na}[Na^+]_o + P_{Cl}[Cl^-]_i}$$

where P values are permeability coefficients (dimensions of velocity, and are diffusivities through the membrane per unit membrane thickness), R is the gas constant, T temperature, F the Faraday constant ($\sim 10^5$ coulombs/equivalent), and ln the natural logarithm.

In skeletal muscle it is generally held that there is no Cl^- pump and that the equilibrium distribution of Cl$^-$ is determined entirely by ψ_m. Therefore, to the extent that ψ_m and ψ_D are the same (and they are in resting mammalian skeletal muscle), the equation can be reduced to

$$\psi_m \cong \psi_D = -\frac{RT}{F} \ln \frac{[K^+]_i + (P_{Na}/P_K)[Na^+]_i}{[K^+]_o + (P_{Na}/P_K)[Na^+]_o}$$

The pump component is attributed entirely to a ouabain-sensitive Na$^+$,K$^+$-activated ATPase that can pump out more Na$^+$ than it pumps in K$^+$. In resting rat caudofemoralis muscle, ψ_P accounts for less than -1 mV out of the observed -78 mV.

Hyperpolarization must be due to either increased ψ_P or increased ψ_D, or both. There is disagreement among various reports as to whether IIH is due to increased ψ_P or ψ_D. The literature is reviewed extensively by Moore[44] and to some extent by Zierler.[45] A problem is that in studies in which it was deduced that IIH was due to stimulation of ψ_P, interpretation hinges on results of experiments with ouabain given either for too long or in too high a concentration, or both, to assure that the effect was only on ψ_P and not also on ψ_D. Eventually ψ_D is effected by the ouabain-induced changes in ion concentration, and there are probably increases in P_{Na} as well, given a large enough ouabain concentration. Our own experience[9] is that, in rat caudofemoralis, ouabain, in a concentration and during a time at which it completely blocks isoproterenol-induced hyperpolarization (known to be due to activation of the electrogenic Na$^+$/K$^+$ exchange pump), had no effect on insulin-induced hyperpolarization.

Thus, in rat caudofemoralis muscle, IIH is probably due to increased absolute magnitude of ψ_D. We find, as do others, that ψ_m observed in response to insulin never exceeds the K$^+$ equilibrium potential, which is the potential predicted by the modified ψ_D equation when there is no permeability to Na$^+$; i.e., $P_{Na} = 0$.

The question then arises as to whether ψ_D increases in absolute magnitude in response to shifts in ion concentration or in response to altered P_{Na} or P_K.

In normal rat caudofemoralis muscle, P_{Na} is no more than 8% of P_K and may be only about 5%. Therefore, shifts in Na$^+$ alone have little effect on ψ_m. Shifts in K$^+$ occur slowly, too slowly to account for the observed rapidity of IIH. (The normal time constant for K$^+$ efflux from rat muscle at room temperature is about 5 hr.[46,47]) Insulin hyperpolarizes rat caudofemoralis in less than 1 sec.[8] Therefore, it is likely that in rat skeletal muscle insulin hyperpolarizes by decreasing the ratio P_{Na}/P_K, as we concluded elsewhere[48] on the basis of calculation of P_{Na}/P_K from the modified ψ_D equation with measured values of ψ_m and all ion concentrations. Insulin decreased P_{Na}/P_K from 0.064 to 0.040 in muscles from normal rats, and from 0.053 to 0.027 in hypophysectomized rats. We calculated P_K on the basis of studies of efflux of ^{42}K from rat skeletal muscle and found it reduced from 1.46×10^{-7} cm/sec without insulin to 0.87 with insulin.[48] We concluded that since P_{Na}/P_K was reduced and P_K was reduced, P_{Na} must have been reduced to an even greater degree than was P_K.

Not everyone agrees with this. Lantz et al.,[29] based on current clamp studies of cultured chick heart cells, interpret their data as showing that insulin increases P_K. However, De Mello[15] reported, in agreement with our conclusion from flux studies, that insulin decreased K$^+$ conductance across frog sartorius muscle fibers.

The subject may be resolved by experiments in progress with voltage clamping, myoball clamping, and patch clamping of muscle.

IV. IS INSULIN-INDUCED HYPERPOLARIZATION AN EVENT IN TRANSDUCTION OF THE INSULIN SIGNAL?

A. Net Potassium Shift from Extracellular to Intracellular Space

There is a high likelihood that at least some of the classic target cell responses to insulin are attributable to insulin-induced hyperpolarization. One of the most prominent clinical responses to insulin is decreased serum K^+ concentration. This hypokalemia is mainly the result of net K^+ transfer from plasma and interstitial tissue into skeletal muscle. Studies of blood flow and arteriovenous concentration differences in the forearm of human subjects, in response to constant infusion of insulin into the brachial artery, showed that less insulin was required to demonstrate insulin-induced net K^+ uptake by muscle than was required to demonstrate stimulated glucose uptake.[49] In a series of experiments on excised rat muscle, in which external K^+ concentration was held constant, insulin increased K^+ content of muscle, but this increase was not detectable until 1–2 hr had elapsed, and then it was too small to account for the observed hyperpolarization.[5] That is, if the observed hyperpolarizations were due only to an increased ratio of $[K^+]_i/[K^+]_o$, the required increase in ratio was several times greater than observed in 1–2 hr. This means that, although the increased ratio of $[K^+]_i/[K^+]_o$ could not be the cause of the observed hyperpolarization, the observed insulin-induced hyperpolarization was energetically adequate to explain the observed shift in K^+ simply on the basis of the Goldman–Hodgkin–Katz equation. The inference is rather good that insulin-induced hyperpolarization is the cause of observed net shifts of K^+ from extracellular to intracellular space.

B. Insulin-Stimulated D-Glucose Uptake

We undertook a series of experiments designed to investigate a relationship between hyperpolarization and increased D-glucose uptake. The experiments were based on three questions, the answers to which would test the hypothesis that IIH stimulates D-glucose uptake (see below). If these questions are answered in the negative, the hypothesis is unlikely to be correct. If the answers are affirmative, evidence for the hypothesis is necessary but not sufficient. It is, then, a matter of seeing whether enough of such circumstantial evidence is forthcoming to give confidence in the merit of the hypothesis.

1. Does IIH precede stimulated glucose uptake?

Obviously, if insulin-stimulated glucose uptake precedes IIH, then IIH cannot be the cause of glucose uptake. Our experiment consisted of delivering a concentrated insulin solution, by pressure injection, to the surface of caudofemoralis muscle, excised from the rat, over a distance of tens of micrometers. Hyperpolarization never occurred with injections of control solutions, but did occur after insulin injection within 1 sec in about two-thirds of the trials.[8] In some cases, the response occurred in <500 msec.

Because there was still a diffusion delay over a distance of tens of micrometers, insulin probably hyperpolarizes in somewhat ≪500 msec—at least two orders of magnitude faster than measurable stimulation of glucose uptake.[50] Therefore, insulin-stimulated glucose uptake is not likely to be the cause of hyperpolarization, but hyperpolarization might be a cause of glucose uptake. Furthermore, we had shown in our earliest observations that insulin hyperpolarized rat muscle to the same extent whether or not glucose was present in the bathing solution.

2. Can insulin be bypassed?Can we hyperpolarize by electrical means in the absense of insulin and bring about increased specific D-glucose uptake?

We devised a method, a triple-sucrose gap, by which we could impose a small hyperpolarization on a 4-mm segment of whole caudofemoralis muscle, by an average of only about -2 mV.[51] The principle of the method (Fig. 2) is that, by replacing the extracellular electrolyte fluid with nonconducting sucrose, current can pass between the two electrodes only by penetrating muscle fiber membranes in the compartment under one electrode, passing longitudinally through myoplasm, and finally coursing transversely through cell-surface membranes to the outside in the compartment under the other electrode. Two current paths are provided in order to reduce effects of electrotonic decay on the distribution of hyperpolarization over the length of the anodal segment. The space constant for electrotonic decay in rat caudofemoralis muscle is ~1.6 mm, so that the 4-mm segment is ~2.5 space constants in length. The longitudinal resistance of muscle segments in a sucrose gap, in series with neighboring longitudinal and membrane resistance networks, serves as a voltage divider, attenuating applied voltage by a factor of the order of 10 at the point of maximum hyperpolarization. Maximum hyperpolarization occurs at an interface between a sucrose gap and Krebs–Ringer solution in the anodal compartment, from which it would decay to 8% of maximum at the other end of the segment, were the second conducting pathway not provided. The combination of voltage-dividing and electrotonic decay placed prac-

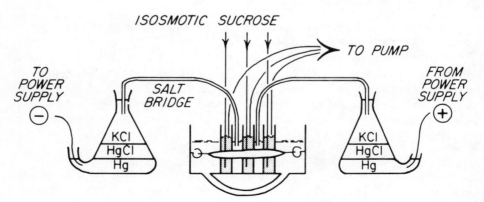

Figure 2. Muscle chamber and electrodes from triple sucrose-gap experiments. See text. (From Zierler and Rogus.[51])

tical limits on the amount of average hyperpolarization over the length of the fiber. Note also that while the anodal segment is hyperpolarized, the cathodal segment is depolarized, quantitatively as a mirror image.

With this technique we were able to hyperpolarize by only about −2 mV averaged over the length of the anodal segment. This small hyperpolarization, only about one-fourth or less of maximum produced by insulin, increased intracellular radiolabeled 2-deoxy-D-glucose (2-DG) content by 40% compared with paired depolarized controls (Table 1). In unpaired experiments in which incubation in sucrose gaps without application of voltage was compared with depolarized segments, there was no effect of depolarization on the 2-DG content of muscle; that is, electrically produced hyperpolarization did increase muscle 2-DG uptake. Furthermore, the effect was stereospecific. Radiolabeled L-glucose uptake was the same in paired hyper- and depolarized segments. The results do not contradict the hypothesis.

3. *If we can find a way to prevent insulin from hyperpolarizing without preventing insulin binding to its receptor, can stimulation of glucose uptake be prevented as well?*

We sought to challenge the hypothesis by seeing whether depolarizations produced by increased $[K^+]_o$ caused a dissociation of the effects of insulin on membrane potential and on glucose uptake. (For a discussion of these experiments, see Section II.A.) We demonstrated that when $[K^+]_o$ was increased so as to maintain constant the product of $[K^+]_o[Cl^-]_o$, despite depolarization proportional to $\log[K^+]_o$ in absence of insulin, there was little blunting of either insulin-induced hyperpolarization or insulin-stimulated glucose uptake. However, when $[K^+]_o$ was increased by equimolar substitution

Table 1. Effects of Electrically Produced Hyperpolarization on 2-Deoxyglucose Uptake by Rat Caudofemoralis Muscle

Treatment	$(2\text{-DG})_i$	$(\text{L-GLU})_i$
None	1.79 ± 0.13 (14)	—
Insulin (100 mU/ml)	3.10 ± 0.43 (9)	—
Depolarizede	1.84 ± 0.12 (10)	0.090 ± 0.014
Hyperpolarized	2.42 ± 0.20 (10)	0.082 ± 0.026
Paired ratio		
Insulin/No treatment	2.01 ± 0.28 (9)	—
Hyperpolarized/Depolarized	1.39 ± 0.12 (10)	—
Unpaired ratio		
No treatment/Depolarized	0.97 ± 0.18 (14/10)	—

Data are from experiments with triple sucrose-gap apparatus. $(2\text{-DG})_i$ and $(\text{L-GLU})_i$ are, respectively, the apparent intracellular space occupied by 2-deoxy-D-glucose and L-glucose, per unit extracellular (sucrose) space. Values are means ± SEM, with unit numbers of experiments in parentheses. Insulin used in supramaximal concentration; Hyper- and depolarizations by ~2 mV. Paired ratios are from experiments on the same muscle. Data are from Zierler and Rogus[51], and from unpublished results of those experiments.

of KCl for NaCl, there was a progressive decrease in, and eventually absence of, insulin-induced hyperpolarization; there was also a decrease in, but not obliteration of, insulin-stimulated glucose uptake. A highly significant nonlinear correlation was found between the two effects of insulin (Fig. 3).

Nevertheless, the hypothesis that IIH is an obligatory event in the transduction chain upstream from stimulated glucose uptake cannot permit even a small amount of stimulated glucose uptake in the absence of hyperpolarization. Either (1) the hypothesis is wrong, (2) it is incomplete and there is an alternate pathway by which there is some insulin-stimulated glucose uptake, or (3) the experiment is faulty. We know that at least the last is true. Membrane potentials are measured only in surface fibers, whereas 2-DG uptake is assessed by measuring 2-DG content of the whole muscle. We found, at 38 mEq $[K^+]_o$, with equimolar substitution, a consistent increase in absolute amplitude of membrane potential (increasing polarization) as one impaled more or less radially through the muscle from the surface through a depth of six fibers. This increasing polarization appeared in both controls and insulin-treated muscles. Membrane potentials of the set of fibers sixth from the surface in 38 mEq $[K^+]_o$ were about the same as those of surface fibers exposed to 9.6 mEq $[K^+]_o$. Thus, persistent stimulated 2-DG uptake of whole muscle at 38 mEq $[K^+]_o$ (when there was no hyperpolarization of surface fibers) may be attributed to the likelihood that deeper fibers were exposed only to smaller K^+ concentrations. We do not know why the $[K^+]_o$ of the bathing solution did not equilibrate with that of interstitial solution throughout the whole muscle, but we suspect that failure to equilibrate may have been related to the shift of water from interstitial space into cells, propelled by the increased product of $[K^+]_o[Cl^-]_o$ when $[K^+]_o$ was increased by equimolar substitution of KCl for NaCl.

The experiments with increasing $[K^+]_o$, maintaining constant a product of $[K^+]_o[Cl^-]_o$, are interesting because, unless $[K^+]_o$ itself modifies membrane structure in some fashion different from the effect of transmembrane electrical fields, they

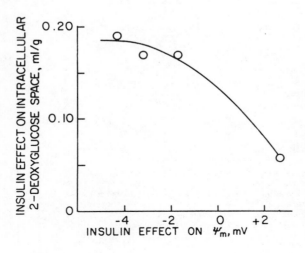

Figure 3. A measure of insulin-stimulated glucose uptake plotted against insulin-induced hyperpolarization when $[K^+]_o$ was increased by equimolar substitution of KCl for NaCl. Each point is a mean of series of experiments at a given $[K^+]_o$, with $[K^+]_o$ doubling from normal, moving from left to right. The relationship is nonlinear and fits the second-order equation $y = 0.134 - 0.0217x - 0.00227x^2$, with correlation coefficient 0.992, where y is stimulated glucose uptake and x is millivolts of hyperpolarization. (Unpublished data from Zierler and Rogus.)

suggest that the absolute level of membrane potential is not a crucial element in the function of membrane D-glucose transportation, but that hyperpolarization from whatever initial level may be the required event. That is, these studies suggest that it may not be the absolute electrical field strength that determines function of membrane protein. However, the effects of $[K^+]_o$ and of the way in which $[K^+]_o$ is changed may be quite complicated and may modify responses to changes in electrical field strength significantly. Therefore, it is premature to speculate about possible basic relationships between field strength and membrane function. More clean-cut experiments are in the works, in which the effects on membrane functions and insulin effects on membrane functions will be analyzed during graded voltage clamping and patch clamping of suitable preparations. In these experiments desired voltages are impressed and held across cell membranes only by electrical means.

C. Transduction Steps Preceding and Following Hyperpolarization

Insulin-induced hyperpolarization cannot be the initial event in response to association between insulin and its receptor. There must be at least two earlier steps (Fig. 4).

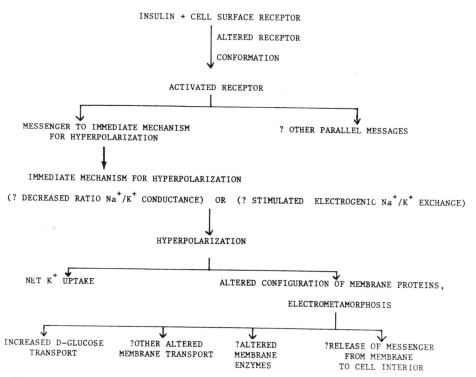

Figure 4. Tentative scheme of major steps in transduction of insulin–receptor signal to classical responses to insulin (by muscle, heart, and adipose cells) via the hyperpolarization path.

The step immediately preceding hyperpolarization must be the immediate mechanism by which insulin hyperpolarizes, whether that be increased electrogenic pumping (which our data do not support), increased P_K (which our data do not support), or decreased P_{Na} with decreased P_{Na}/P_K (which our data do support). The subject needs to be settled by critical experiments.

Whatever the immediate mechanism, something must cause it. We conjecture that the first step in response to association between insulin and its receptor must be a structural change in the receptor that alters some physical property. For example, the receptor in its naked configuration may have certain affinities for, say, Ca^{2+}, which may be altered when insulin is bound. Ca^{2+} dissociated from the receptor may block nearest-neighbor Na^+ and K^+ channels, causing the observed hyperpolarization. We do know that, just as antiserum to insulin receptor can mimic insulin action with respect to stimulated glucose uptake, so insulin receptor antiserum also hyperpolarizes rat caudofemoralis muscle.[52]

In Fig. 4, we move distally from the level at which hyperpolarization occurs. We already have a body of evidence that hyperpolarization is a cause of insulin-stimulated net K^+ uptake and specific D-glucose uptake. We can attribute the net K^+ uptake to thermodynamic necessity; it is not necessary to invoke an intermediate step. Nor may it be for the case of glucose uptake, which might occur simply because hyperpolarization may alter the character of glucose transporters already in the membrane, as suggested in Section I. However, in view of the possibility that glucose transporters may cycle between the membrane and some intracellular loci, as suggested by Suzuki and Kono[53] and by Cushman and Wardzala,[54] it may be that the hyperpolarization, again by its action on configuration of some membrane protein, causes release of a substance from the membrane—be it Ca^{2+} or a peptide, or whatever—that triggers movement of transporters into the membrane.

V. REFERENCES

1. Debye, P., 1979, *Polar Molecules*, The Chemical Catalog Co., New York.
2. Hille, B., 1979, in *Membrane Transport Processes*, Vol. 3 (C.F. Stevens and R.W. Tsien, eds.), pp. 5–16, Raven Press, New York.
3. Hodgkin, A.L., and Huxley, A.F., 1952, *J. Physiol. (Lond.)* **117**:500–544.
4. Zierler, K.L., 1957, *Science* **126**:1067–1068.
5. Zierler, K.L., 1959, *Am. J. Physiol.* **197**:515–523.
6. Zierler, K.L., 1959, *Am. J. Physiol.* **197**:524–526.
7. Hazlewood, C.F., and Zierler, K.L., 1967, *Johns Hopkins Med. J.* **121**:188–193.
8. Zierler, K., and Rogus, E.M., 1981, *Biochim. Biophys. Acta* **640**:687–692.
9. Zierler, K., and Rogus, E.M., 1981, *Am. J. Physiol.* **241**:C145–C149.
10. Flatman, J.A., and Clausen, T., 1979, *Nature* **281**:580–581.
11. Otsuka, M., and Ohtsuki, I., 1965, *Nature* **207**:300–301.
12. Bolte, H.-D., and Luderitz, B., 1968, *Pfluegers Arch.* **301**:354–358.
13. Takamori, M., Ide, Y., and Tsujihata, M., 1981, *J. Neurol. Sci.* **50**:89–94.
14. Zemkova, H., Teisinger, J., and Vyskocil, F., 1982, *Biochim. Biophys. Acta* **720**:405–410.
15. De Mello, W.C., 1967, *Life Sci.* **6**:959–963.
16. Moore, R.D., and Rabovsky, J.L., 1979, *Am. J. Physiol.* **236**:C249–C254.
17. Stark, R.J., and O'Doherty, J., 1982, *Am. J. Physiol.* **242**:E193–E200.

18. Frolkis, V.V., 1980, *Neurosci. Behav. Physiol.* **10**:92–98.
19. Zierler, K., and Rogus, E.M., 1983, *Trans. Assoc. Am. Physicians* **46**:203–208.
20. Horn, L.W., and Zierler, K.L., 1975, *J. Physiol. (Lond.)* **253**:207–222.
21. Beigelman, P.M., and Hollander, P.B., 1962, *Proc. Soc. Exp. Biol. Med.* **110**:590–595.
22. Armstrong, W.McD., and Garcia-Diaz, J.F., 1981, in *Epithelial Ion and Water Transport* (A.D.C. MacKnight and J.P. Leader, eds.), pp. 43–53, Raven Press, New York.
23. Leader, J.P., 1981, *Epithelial Ion and Water Transport* (A.D.C. MacKnight and J.P. Leader, eds.), pp. 55–62, Raven Press, New York.
24. Petrozzo, P., and Zierler, K., 1976, *Fed. Proc.* **35**:602.
25. Cheng, K., Groarke, J., Osotimehin, B., Haspel, H.C., and Sonenberg, M., 1981, *J. Biol. Chem.* **256**:649–655.
26. Davis, R.J., Brand, M.D., and Martin, B.R., 1981, *Biochem. J.* **196**:133–147.
27. Stark, R.J., Read, P.D., and O'Doherty, J., 1980, *Diabetes* **29**:1040–1043.
28. Imanaga, I., 1978, in *Recent Advances in Studies on Cardiac Structure and Metabolism*, (T. Kobayashi, T. Sano, and N.S. Dhalla, eds.), Vol. II, pp. 441–450, University Park Press, Baltimore.
29. Lantz, R.C., Elsas, L.J., and DeHaan, R.L., 1980, *Proc. Natl. Acad. Sci. USA* **77**:3062–3066.
30. LaManna, V., and Ferrier, G.R., 1981, *Am. J. Physiol.* **240**:H636–H644.
31. Malinow, M.R., 1958, *Acta Physiol. Latin. Am.* **8**:125–128.
32. Sodi-Pollares, D., Testelli, M.R., Fishleder, B.L., Bistent, A., Medrano, G.A., Friedland, C., and DeMicheli, A., 1962, *Am. J. Cardiol.* **9**:166–181.
33. Kinoshita, K., and Yatani, A., 1982, *Jpn. Heart J.* **23**:791–804.
34. Anderson, G.J., Swartz, J., Dennis, S.C., and Reiser, J., 1983, *J. Am. Coll. Cardiol.* **1**:1290–1295.
35. Friedmann, N., Somlyo, A.V., and Somlyo, A.P., 1971, *Science* **171**:400–402.
36. Wondergem, R., 1983, *Am. J. Physiol.* **244**:C17–C23.
37. Crabbe, J., 1969, in *Protein and Polypeptide Hormones* (M. Margoulies, ed.), pp. 260–263, Excerpta Medica Foundation, Amsterdam.
38. Rehm, W., Schumann, H., and Heinz, E., 1961, *Fed. Proc.* **20**:193.
39. Miller, J.E., and Constant, M.A., 1960, *Am. J. Ophthalmol.* **50**:855–862.
40. Fidelman, M.L., May, J.M., Biber, T.U., and Watlington, C.D., 1982, *Am. J. Physiol.* **242**:C121–C123.
41. Williams, J.A., Bailey, A.C., Preissler, M., and Goldfine, I.D., 1982, *Diabetes* **31**:674–682.
42. Goldman, D.E., 1943, *J. Gen. Physiol.* **27**:37–60.
43. Hodgkin, A.L., and Katz, B., 1949, *J. Physiol. (Lond.)* **108**:37–77.
44. Moore, R.D., 1983, *Biochim. Biophys. Acta* **737**:1–49.
45. Zierler, K., 1984, in *Insulin, Its Receptor and Diabetes* (M.D. Hollenberg, ed.), Marcel Dekker, New York.
46. Zierler, K.L., Rogus, E., and Hazlewood, C.F., 1966, *J. Gen. Physiol.* **49**:433–456.
47. Zierler, K.L., 1960, *Am. J. Physiol.* **198**:1066–1070.
48. Zierler, K.L., 1972, in *Handbook of Physiology, Endocrinology*. Vol. I (D.F. Steiner and N. Freinkel, eds.), pp. 347–368, American Physiological Society, Washington, D.C.
49. Zierler, K.L., and Rabinowitz, D., 1964, *J. Clin. Invest.* **43**:950–962.
50. Haring, H.U., Kemmler, W., Renner, R., and Hepp, K.D., 1978, *FEBS Lett.* **95**:177–180.
51. Zierler, K., and Rogus, E.M., 1980, *Am. J. Physiol.* **239**:E21–E29.
52. Zierler, K., and Rogus, E.M., 1983, *Am. J. Physiol.* **244**:C58–C60.
53. Suzuki, K., and Kono, T., 1980, *Proc. Natl. Acad. Sci. USA* **77**:2542–2545.
54. Cushman, S.W., and Wardzala, L.J., 1980, *J. Biol. Chem.* **255**:4758–4762.

8

Effect of Insulin on the Membrane Potentials of Rat White Epididymal Fat Cells

Roger J. Davis and B. Richard Martin

I. INTRODUCTION

Insulin produces a number of rapid changes in the membrane transport processes of fat cells. Examples are the stimulation of the transport of glucose,[1] amino acids,[2] potassium,[3,4] calcium,[5] and phosphate.[6] It is possible that the changes in the ionic balance caused by insulin may play a role in mediating or maintaining some of the effects of insulin. An understanding of ionic transport by mechanisms that are electrogenic or electrophoretic requires a knowledge of the electrical potential across the plasma membrane. We have therefore studied the electrical potentials across the plasma membrane as well as the organelle membranes of fat cells stimulated by insulin.

Direct electrophysiological techniques using intracellular microelectrodes have been used to measure the plasma membrane potential ($\Delta\psi p$) of many cell types. White fat cells are not ideally suited to this method, as they have a small intracellular water space and contain large globules of fat that can block the microelectrode tip.[7,8] Ionic leaks at the point of insertion will also limit the reliability of this estimate of $\Delta\psi p$.[9] However, Beigelman and co-workers have used this technique and have observed values of the resting $\Delta\psi p$ from 10–60 mV.[10–16] More recently, Stark et al.[17] reported a mean $\Delta\psi p$ of 46 mV.

In view of the wide variation in results obtained using microelectrodes, cationic probes of membrane potential have been used. The distribution of these probes can be monitored by either changes in fluorescence (rhodamine and carbocyanine dyes[18,19]),

Roger J. Davis ● Department of Biochemistry, University of Massachusetts Medical Center, Worcester, Massachusetts. *B. Richard Martin* ● Department of Biochemistry, University of Cambridge, Cambridge, England.

the accumulation of radioactivity (methyltriphenylphosphonium$^+$; [^{3}H]-TPMP$^{+(20)}$), or nuclear magnetic resonance (NMR) spectroscopy.[21] A problem with these probes is that eukaryotic cells are multicompartmental systems. We have determined that the principal sites of accumulation of [^{3}H]-TPMP$^+$ in fat cells are the mitochondria.[22] Similar observations have been made with [^{3}H]-TPMP$^{+(23-26)}$ and carbocyanine dyes[27,28] in other cell types. It is therefore not possible to estimate the $\Delta\psi p$ from cationic probes because the data obtained represent a complex function of the $\Delta\psi p$ and the membrane potentials of the organelles.

The problems encountered with cationic probes of the $\Delta\psi p$ suggest that anionic probes may yield data that are more easily interpretable because they will not be accumulated by organelles. Bis-oxonol is an anionic dye the fluorescence yield of which is enhanced when cells are polarized.[29] We were unable to obtain a stable fluorescence signal from fat cells incubated with this dye, however, probably because of the difficulty in stirring the fat cells to keep them in suspension without causing cell breakage. An alternative approach that we have used is to investigate the accumulation of [^{14}C]-SCN$^-$.[30]

Using a combination of the approaches with [^{14}C]-SCN$^-$ and [^{3}H]-TPMP$^+$, we have been able to monitor plasma membrane potential and the mitochondrial membrane potential, $\Delta\psi p$ and $\Delta\psi m$, respectively, of intact fat cells simultaneously.[22] This review summarizes the data we have obtained on the effect of insulin on these parameters.

II. CHARACTERIZATION OF THE PLASMA MEMBRANE POTENTIAL

The plasma membrane potential of fat cells has been estimated by two different methods using lipophilic ions. First, the accumulation of [^{3}H]-TPMP$^+$ in the presence of uncouplers of oxidative phosphorylation has been used.[31] The uncouplers were used in order to depolarize the mitochondria so that these organelles do not accumulate the lipophilic ion. This technique has been shown to work well with lymphocytes,[26] but is not very satisfactory when applied to fat cells because of the large decrease in the adenosine triphosphate (ATP) level observed with the incubation conditions used. However, we have estimated the $\Delta\psi p$ to be 60.6 mV by this method.[30,31] A second method is to use lipophilic anions that are excluded by fat cells and are not accumulated by mitochondria. This method has the advantage that the accumulation ratio is principally determined by $\Delta\psi p$. We have studied the accumulation of [^{14}C]-SCN$^-$ by fat cells.[30,32] Substantial binding of [^{14}C]-SCN$^-$ was corrected for by calibrating the observed [^{14}C]-SCN$^-$ accumulation ratios with $\Delta\psi p$. This was achieved by comparing the [^{14}C]-SCN$^-$ and ^{86}Rb$^+$ accumulation by fat cells incubated with the potassium ionophore valinomycin in media containing different K$^+$ concentrations. The $\Delta\psi p$ of control fat cells was estimated to be between 50–65 mV. The estimate of $\Delta\psi p$ of about 60 mV obtained from experiments with lipophilic ions is consistent with the higher values observed by Biegelman and co-workers[10-16] using intracellular microelectrodes.

To further characterize the $\Delta\psi p$, we have investigated the dependence of the $\Delta\psi p$ on the ionic composition of the medium.[30] The accumulation of [^{3}H]-TPMP$^+$ by cells is proportional to $\log(\Delta\psi p)$. It is therefore possible to investigate changes in $\Delta\psi p$

Table 1. Estimates of the Plasma Membrane Potential of Fat Cells

Method	Control $\Delta\psi p$ (mV)	Insulin-induced hyperpolarization (mV)	Reference
1 Microelectrodes	10–60	15–30	10–16
2 Microelectrodes	46	0	17
3 Thiocyanate exclusion	50–65	0	32 (Fig. 1)
4 TPMP$^+$ accumulation^a	58.3	3	48
5 TPMP$^+$ accumulation	60.6	0	32 (Fig. 1)
6 Carbocyanine dye	Uncalibrated	Yes	47

^{a} In this study, the accumulation of the lipophilic ion by mitochondria was not taken into account (see ref. 31 for discussion).

with [^{3}H]-TPMP$^+$ independently of the accumulation of [^{3}H]-TPMP$^+$ by organelles. Using this technique, we have reported the effect of alterations in the concentrations of K$^+$, Na$^+$, and Cl$^-$ concentrations on the $\Delta\psi p$.[30] Application of the Goldman–Hodgkin–Katz equation[33,34] (which relates the diffusion potentials, E, of monovalent ions to $\Delta\psi p$) to these data permits calculation of the relative electrophoretic permeability coefficients, P, of the plasma membrane to these ions. The coefficients were determined to be $P_{K^+} = 1$, $P_{Na^+} = 0.18$, and $P_{Cl^-} = 0.54-0.91$.[30] These data are in contrast to those obtained with electrically excitable tissues, where $P_{K^+} \gg P_{Na^+}$ and $P_{K^+} \approx P_{Cl^-}$. A consequence of this difference is that whereas $\Delta\psi p$ of excitable cells is similar to E_{K^+} and E_{Cl^+}, $\Delta\psi p$ of fat cells is very different from the diffusion potentials of K$^+$, Na$^+$, and Cl$^-$.

In addition to the diffusion potentials of ions, electrogenic processes occurring in the plasma membrane affect $\Delta\psi p$. An example is the sodium pump, which has been demonstrated to be electrogenic in several tissues.[35,36] We have investigated the effect of ouabain, an inhibitor of the sodium pump, on $\Delta\psi p$ of fat cells, using [^{3}H]-TPMP$^+$.[30] Ouabain was observed to depolarize fat cells by 1.3 and 2.2 mV under conditions in which the mitochondria were polarized or depolarized (dinitrophenol), respectively. This depolarization of $\Delta\psi p$ could not be accounted for by changes in the accumulation of K$^+$ or Na$^+$ by the cells and is tentatively proposed to be a component of $\Delta\psi p$ contributed by the sodium pump.[30] The role of other electrogenic processes in the plasma membrane is yet to be determined.

III. CHARACTERIZATION OF THE MITOCHONDRIAL MEMBRANE POTENTIAL

In order to estimate the mitochondrial membrane potential, $\Delta\psi m$, it is assumed that the accumulation of [^{3}H]-TPMP$^+$ by the mitochondria is very much greater than that by other organelles. This assumption has been discussed previously.[22–25] The main justification for the assumption is that $\Delta\psi m$ is large and is logarithmically related to the accumulation of a lipophilic ion. The available evidence[37–41] suggests that organelles other than the mitochondria have a low membrane potential, i.e., <60 mV. In contrast, mitochondria have a membrane potential in vitro of at least 150 mV.[42]

The Nernst equation predicts distribution ratios of monovalent ions at equilibrium with the membrane potential to be 10-fold and 300-fold for a $\Delta\psi$ of 60 and 150 mV, respectively. Thus the mitochondrial accumulation of lipophilic cations is significantly greater than that of other organelles.

Once the mitochondrial volume has been estimated, $\Delta\psi m$ can be calculated.[22] The calculated $\Delta\psi m$ is a function of the estimate made of the $\Delta\psi p$, i.e., 168 mV ([^{3}H]-TPMP$^+$) to 178 mV ([^{14}C]-SCN$^-$). To ensure that the calculated $\Delta\psi m$ does reflect the mitochondrial membrane potential, the effect of ionophores and mitochondrial inhibitors was investigated. The calculated $\Delta\psi m$ was markedly reduced by uncouplers of oxidative phosphorylation annd the potassium ionophore, valinomycin.[22] Oligomycin, an inhibitor of mitochondrial ATP synthetase, had little effect on the calculated $\Delta\psi m$, but inhibitors of electron transport, such as CN$^-$, did reduce the $\Delta\psi m$. Addition of oligomycin to CN$^-$-treated cells caused a further decrease in the calculated $\Delta\psi m$.[32] All these data are consistent with the calculated $\Delta\psi m$ representing a $\Delta\psi$ localized to the mitochondria.

IV. EFFECT OF INSULIN ON THE ACCUMULATION OF [^{3}H]-TPMP$^+$ AND [^{14}C]-SCN$^-$

A. Plasma Membrane Potential

Insulin has been reported to cause a hyperpolarization of $\Delta\psi p$ of white fat cells within a range of 14.7–30.3 mV, using microelectrodes.[10,11,13,14] We therefore expected to observe a hyperpolarization when we monitored $\Delta\psi p$ of fat cells with [^{14}C]-SCN$^-$ and [^{3}H]-TPMP$^+$. Surprisingly, no significant effect of insulin was observed (Fig. 1). The accumulation of [^{14}C]-SCN$^-$ is not consistent with the large hyperpolarization observed with microelectrodes. The difficulty of accurately determining the [^{14}C]-SCN$^-$ accumulation means that the data do not exclude the possibility of a small hyperpolarization. Zierler[43] reported that skeletal muscle is hyperpolarized by about 5 mV; similar observations have been made with several other tissues (see Chapter 7). If fat cells are hyperpolarized by 5 mV this would be observed in experiments using [^{3}H]-TPMP$^+$, which is very sensitive to changes in $\Delta\psi p$. No change in [^{3}H]-TPMP$^+$ accumulation was found, however (Fig. 1). It is possible that the hyperpolarization was not observed because the cells were perturbed by the addition of dinitrophenol (which was added to abolish the mitochondrial accumulation of [^{3}H]-TPMP$^+$). Further experiments were therefore carried out in the absence of dinitrophenol, to determine whether the results were artifactual.

If fat cells are hyperpolarized by insulin, it would be expected that this could occur by three basic mechanisms. First, a change in the electrophoretic permeability of the plasma membrane to ions may occur. We have measured the relative electrophoretic permeabilities of Na$^+$, K$^+$, and Cl$^-$ after insulin treatment. The dependence of the accumulation of [^{3}H]-TPMP$^+$ (in the absence of dinitrophenol) on the concentration of monovalent ions in the medium was fitted to the Goldman–Hodgkin–Katz equation. The relative permeabilities ($P_{K^+} = 1$, $P_{Na^+} = 0.18$, $P_{Cl^-} = 0.54$–0.91) were not significantly different before and after insulin treatment.[44] Second, a hy-

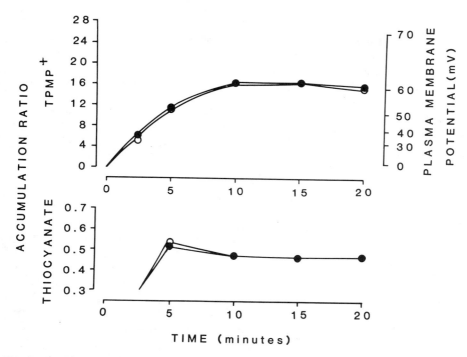

Figure 1. Effect of insulin on the accumulation of [¹⁴C]thiocyanate and [³H]-TPMP⁺ in the presence of dinitrophenol. Fat cells were incubated at 37°C in a bicarbonate-buffered medium containing 120 mM NaCl, 4.8 mM KCl, 1.28 mM $CaCl_2$, 1.12 mM KH_2PO_4, 1.2 mM $MgSO_4$, 25,3 mM $NaHCO_3$, 1% (w/v) bovine serum albumin, 20 mM fructose and gassed with O_2 : CO_2 (19 : 1). Accumulation ratios were measured by the oil-flotation method.[22] (A) Accumulation of 0.3 μM [³H]-TPMP⁺ (0.5 μCi/ml) was determined in media containing 0.5 mM dinitrophenol. Fat cells were shown to be insulin responsive under these conditions by investigating the uptake of ⁸⁶Rb⁺. (B) The accumulation of 70 μM [¹⁴C]-SCN- (2.5 μCi/ml) was measured in media in the absence of dinitrophenol. Control and insulin-treated (1 unit/ml) fat cell suspensions are represented by filled and open circles, respectively. (Data from Davis.[32])

perpolarization may be caused by a change in the diffusion potential of an ion. Insulin has been observed to increase the K⁺ accumulation by fat cells.[3,4,22] However, Zierler[45] reported that in muscle the hyperpolarization precedes the increased accumulation of K⁺. Third, a hyperpolarization may be caused by a change in the rate of electrogenic processes that occur in the plasma membrane. We have observed that inhibition of the sodium pump with ouabain causes a rapid depolarization of the $\Delta\psi p$ by 1.3–2.2 mV, in the presence and absence of dinitrophenol, respectively. There is a small increase in the ouabain-induced depolarization if the fat cells are pretreated with insulin for 30 min.[32] This result is consistent with the suggestion that insulin hyperpolarizes $\Delta\psi p$ by stimulating the electrogenic sodium pump.

In fat cells we have obtained only poor evidence for a hyperpolarization of the $\Delta\psi p$ induced by insulin using lipophilic ions. Stark *et al.*[17] made similar observations using microelectrodes. This is in contrast to the data reported for skeletal muscle (see Chapter 7), but is similar to observations that have been made on two tissues that are very responsive to insulin: brown fat cells[7] and liver.[46] As the increased accumulation of K^+ and the stimulation of the sodium pump are consistent with a hyperpolarization, we cannot exclude the possibility that the $\Delta\psi p$ is increased. However, we would argue that our data indicate a hyperpolarization not larger than 1–3 mV. Zierler has suggested that the hyperpolarization observed in muscle may be an important signal generated after insulin binding. However, we have observed that insulin will stimulate ouabain-sensitive $^{86}Rb^+$ uptake by dinitrophenol-treated fat cells. Under these conditions, no change in $\Delta\psi p$ is detected with $[^3H]$-TPMP$^+$.[32] This experiment suggests that a hyperpolarization of the plasma membrane is not required for action of insulin on fat cells.

B. Mitochondrial Membrane Potential

We have reported that the $[^3H]$-TPMP$^+$ accumulation by isolated fat cells (in the absence of an uncoupler of oxidative phosphorylation) is decreased after the addition of insulin.[22] This result is anomalous, however, because of the reported lack of effect[17] or hyperpolarization[10,11,13,14,47,48] of $\Delta\psi p$. We resolved this problem by suggesting that additional factors may be affecting the accumulation of $[^3H]$-TPMP$^+$ after the addition of insulin. There are many possibilities.

First, $[^3H]$-TPMP$^+$ exhibits complex binding characteristics and insulin may alter this binding so that the apparent $[^3H]$-TPMP$^+$ accumulation changes independently of the $\Delta\psi$. The binding of $[^3H]$-TPMP$^+$ to fat cells was investigated by the addition of the potassium ionophore valinomycin. Pretreatment of cells with insulin had no effect on the accumulation of $[^3H]$-TPMP$^+$ by valinomycin-depolarized cells.[22] Binding of $[^3H]$-TPMP$^+$ by components in the medium can also affect the distribution of the lipophilic cation. Bovine serum albumin (BSA) is known to bind hydrophobic ligands such as free fatty acids[49] and was found to bind $[^3H]$-TPMP$^+$.[31] The antilipolytic effect of insulin will decrease the level of free fatty acids in the medium, this effect may alter the binding of $[^3H]$-TPMP$^+$ to the albumin. However, it was found that although free fatty acids do increase the binding of $[^3H]$-TPMP$^+$ to BSA at high albumin concentrations, little effect of fatty acids or binding of $[^3H]$-TPMP$^+$ was observed at the low concentrations of albumin (1% w/v) used in the experiments with fat cells.[22] A second possibility is that a change in the apparent accumulation of $[^3H]$-TPMP$^+$ could occur independently of $\Delta\psi p$ if there was a change in the accumulation by organelles. This could be achieved by a change in the organelle volume or $\Delta\psi$. We have suggested that the decrease in the accumulation of $[^3H]$-TPMP$^+$ by intact fat cells caused by insulin is a result of a change in the accumulation of the lipophilic cation by the mitochondria.[22] To account for the data, the inner mitochondrial volume would have to decrease to 82% of the original volume or depolarize by 5.4 mV if there is no change in the $\Delta\psi p$ (60 mV). If the $\Delta\psi p$ hyperpolarizes after the addition of insulin by 5 mV, the changes would be 32% (decrease in volume) or 10.4 mV ($\Delta\psi m$).

If the electrical potential is reduced across the mitochondrial inner membrane, what effect would this be expected to have on the fat cells? We have suggested that a decrease in $\Delta\psi m$ could occur if insulin were to alter the mitochondrial transport of ions or synthesis of ATP.[22] There is extensive evidence that insulin does alter the ionic balance of fat cells[50] and that changes in mitochondrial ion transport may occur in a primary or secondary manner in relation to this. Furthermore, it has been observed that the ATP turnover is enhanced by insulin.[51] All these processes are associated with an increase in the oxygen uptake by the mitochondria, and insulin has been reported to stimulate the oxygen uptake by fat cells.[52]

Under some conditions *in vitro*, the O_2 uptake by mitochondria is under thermodynamic control.[53] Thus, addition of ADP or Ca^{2+} to respiring mitochondria causes an increase in O_2 uptake, which is associated with a decrease in the proton electrochemical gradient ($\Delta\bar{\mu}_{H^+}$) across the inner mitochondrial membrane.[53] However, care should be taken in exptrapolating from the *in vitro* data to the *in vivo* setting. It is quite likely that the O_2 uptake by mitochondria is regulated kinetically in intact cells. In this case, changes in O_2 consumption could occur independently of alterations in the $\Delta\bar{\mu}_{H^+}$. A further complication is that the calculated $\Delta\psi m$ represents, at best, the bulk-phase potential difference rather than the potential difference at the surface of the inner mitochondrial membrane, which may be of greater functional significance than the bulk-phase potential.

V. PROSPECTS FOR FUTURE WORK

The outstanding question that needs to be answered concerns the localization of the potential change caused by insulin in intact cells. We have suggested that it is mitochondrial in origin, but have been unable to determine whether this is a true change in the mitochondrial membrane potential or simply a reduction in the mitochondrial matrix space. To do this will require the application of new techniques capable of resolving the membrane potentials of the cellular organelles. Possible methods are the use of NMR probes[21] or the quantitation of the fluorescence of potential-sensitive dyes by fluorescence microscopy.[27,28]

Other physiological changes in $\Delta\psi m$ have been reported, and it will be of interest to compare the changes caused by insulin with those caused by other agents. We have observed that β-adrenergic agonists cause a considerable depolarization of the mitochondria.[31] It was proposed that the free fatty acids released during lipolysis uncouple the mitochondria.[31] Insulin antagonizes this effect probably because of its antilipolytic effect. An α-adrenergic effect on $\Delta\psi m$ of white fat cells was also reported.[54] Levenson *et al.*[55] have investigated the accumulation of cyanine dyes by murine erythroleukemia cells during erythroid differentiation. A large decrease in cell fluorescence was observed at the time of commitment to differentiation, which was interpreted as a decrease in the $\Delta\psi m$. In two other systems, a change in the mitochondrial accumulation of rhodamine 123 was found to be the result of an increased number of mithochondria (blastogenic transformation of lymphocytes[56]) or a decrease in number of functional mitochondria (contact inhibition of fibroblast growth[57]).

The study of changes in $\Delta\psi m$ caused by perturbation of cells with growth factors, hormones, or inducers of differentiation will provide evidence for the role of these organelles in the control of cell growth.

VI. REFERENCES

1. Czech, M.P., 1980, *Diabetes* **29**:399–409.
2. Guidotti, G., Borghetti, A.F., and Gazzola, G.C., 1978, *Biochim. Biophys. Acta* **515**:329–366.
3. Gourley, D.R.H., and Bethea, M.D., 1964, *Proc. Soc. Exp. Biol. Med.* **115**:821–823.
4. Resh, M.D., Nemenoff, R.A., and Guidotti, G., 1980, *J. Biol. Chem.* **255**:10938–10945.
5. Clausen, T., and Martin, B.R., 1977, *Biochem. J.* **164**:251–255.
6. Clausen, T., 1977, *FEBS Meet.* 229–238.
7. Krishna, G., Mosowitz, J., Dempsey, P., and Brodie, B.B., 1970, *Life Sci.* **9(1)**:1353–1361.
8. Cheng, K., Haspel, H.C., Vallano, M.L., Osotimehin, B., and Sonnenberg, M., 1980, *J. Membr. Biol.* **56**:191–201.
9. Lassen, U.V., Nielsen, A.M.T., Pape, L., and Simonsen, L.O., 1971, *J. Membr. Biol.* **6**:269–288.
10. Beigelman, P.M., and Hollander, P.B., 1962, *Proc. Soc. Exp. Biol. Med.* **110**:590–595.
11. Beigelman, P.M., and Hollander, P.B., 1963, *Diabetes* **12**:262–267.
12. Beigelman, P.M., and Hollander, P.B., 1964, *Proc. Soc. Exp. Biol. Med.* **115**:14–16.
13. Beigelman, P.M., and Hollander, P.B., 1964, *Proc. Soc. Exp. Biol. Med.* **116**:31–35.
14. Beigelman, P.M., and Hollander, P.B., 1965, *Acta Endocrinol. (Copenh.)* **50**:648–656.
15. Miller, L.V., Schlosser, G.H., and Beigelman, P.M., 1966, *Biochim. Biophys. Acta* **112**:375–376.
16. Beigelman, P.M., and Shu, M.J., 1972, *Proc. Soc. Exp. Biol. Med.* **141**:618–621.
17. Starke, R.J., Read, P.D., and O'Doherty, 1980, *Diabetes* **29**:1040–1043.
18. Bashford, C.L., and Smith, J.C. 1979, *Methods Enzymol.* **LV**:569–586.
19. Waggoner, A.S., 1979, *Annu. Rev. Biophys. Bioeng.* **8**:47–68.
20. Skulachev, V.P., 1971, *Curr. Top. Bioenerg.* **4**:127–190.
21. Cafiso, D.S., and Hubbell, W.L., 1981, *Annu. Rev. Biophys. Bioeng.* **10**:245–276.
22. Davis, R.J., Brand, M.D., and Martin, B.R., 1981, *Biochem. J.* **196**:133–147.
23. Deutsch, C., Erecinska, M., Werrlein, R., and Silver, I.A., 1979, *Proc. Natl. Acad. Sci. USA* **76**:2175–2179.
24. Hoek, J.B., Nicholls, D.G., and Williamson, J.R., 1980, *J. Biol. Chem.* **255**:1458–1464.
25. Scott, I.D., and Nicholls, D.G., 1980, *Biochem. J.* **186**:21–33.
26. Felber, S.M., and Brand, M.D., 1982, *Biochem. J.* **204**:577–585.
27. Johnson, L.V., Walsh, M.L., and Chen, L.B., 1980, *Proc. Natl. Acad. Sci. USA* **77**:990–994.
28. Johnson, L.V., Walsh, M.L., Bockus, B.J., and Chen, L.B., 1981, *J. Cell Biol.* **88**:526–535.
29. Rink, T.J., Montecucco, C., Hesketh, T.R., and Tsien, R.Y., 1980, *Biochim. Biophys. Acta* **595**:15–30.
30. Davis, R.J., and Martin, B.R. (in preparation).
31. Davis, R.J., and Martin, B.R., 1982, *Biochem. J.* **206**:611–618.
32. Davis, R.J., 1982, Ph.D. Thesis, University of Cambridge, Great Britain.
33. Goldman, D.E., 1943, *J. Gen. Physiol.* **27**:37–60.
34. Hodgkin, A.L., and Katz, B., 1949, *J. Physiol. (Lond.)* **108**:37–77.
35. Thomas, R.C., 1972, *Physiol. Rev.* **52**:563–594.
36. Trachtenberg, M.C., Packey, D.G., and Sweeney, T., 1981, *Curr. Top. Membr. Transp.* **19**:159–218.
37. Feldherr, C., 1973, *Experientia* **29**:546–547.
38. Palmer, L.G., and Civan, M.M., 1975, *Science* **188**:1321–1322.
39. Palmer, L.G., and Civan, M.M., 1977, *J. Membr. Biol.* **33**:41–61.
40. Civan, M.M., 1978, *Am. J. Physiol.* **234(4)**:F261–F269.
41. Reijngoud, D.J., and Tager, J.M., 1977, *Biochim. Biophys. Acta* **472**:419–449.
42. Mitchell, P., and Moyle, J., 1969, *Eur. J. Biochem.* **7**:471–484.
43. Zierler, K.L., 1957, *Science* **126**:1067–1068.
44. Davis, R.J., and Martin, B.R. (in preparation).
45. Zierler, K.L., 1959, *Am. J. Physiol.* **197**:515–523.
46. Friedmann, N., and Dambach, G., 1980, *Biochim. Biophys. Acta* **596**:180–185.
47. Petrozzo, P., and Zierler, K.L., 1976, *Fed. Proc. Biol.* **35**:602.
48. Cheng, K., Groarke, J., Osotimehin, B., Haspel, H.C., and Sonnenberg, M., 1981, *J. Biol. Chem.* **256**:649–655.

49. Goodman, D.S., 1958, *J. Am. Chem. Soc.* **80**:3892–3898.
50. Moore, R.D., 1983, *Biochim. Biophys. Acta* **737**:1–49.
51. Hepp, D., Challoner, D.R., and Williams, R.H., 1968, **243**:4020–4026.
52. Jungas, R.L., and Ball, E.G., 1963, *Biochemistry* **2**:383–388.
53. Nicholls, D.G., 1982, in: *Bioenergetics: An Introduction to the Chemiosmotic Theory,* Academic Press, New York, pp. 22–56.
54. Davis, R.J., and Martin, B.R., 1982, *Biochem. J.* **206**:619–626.
55. Levenson, R., Macara, I.G., Smith, R.L., Cantley, L., and Housman, D., 1982, *Cell* **28**:855–863.
56. Darzynkiewicz, Z., Staiano-Coico, L., and Melamed, M.R., 1981, *Proc. Natl. Acad. Sci. USA* **78**:2383–2387.
57. Goldstein, S., and Korczack, L.B., 1981, *J. Cell Biol.* **91**:392–398.

9

The Case for Intracellular pH in Insulin Action

Richard D. Moore

I. INTRODUCTION

At the outset, it should be emphasized that it is impossible to change only one species of ion inside the cell. Moreover, although one ion may be predominant in a given signal system (e.g., Ca^{2+} in excitation–contraction coupling), it is unlikely that ions are the only signal system, or that only one ion is involved, in the case of insulin.

Changes in pH_i are intimately associated with the Na^+ electrochemical gradient across the plasma membrane and therefore with the (Na^+,K^+) pump. In order to maintain perspective, this chapter covers some aspects of the role of Na^+ in insulin action. Possible involvement with Ca^{2+} is mentioned as well.

In my opinion, the reductionist approach implicit in emphasizing only one factor at a time will inevitably retard development of a correct understanding of any regulatory system, especially in the case of the action of a pleotropic hormone such as insulin.

A. Why Have the Ionic Aspects of Insulin Been Ignored?

A case can be made that it is a historical accident that until recently ions and electrochemical events were essentially ignored in studies of insulin action. Within 1 year after the discovery of insulin in 1922, it had been demonstrated that insulin affected not only organic metabolism, reflected by blood glucose, but also ionic events, as reflected by plasma potassium and phosphate levels. The effect on plasma glucose was immediately, and understandably, explored. This is true not only because the

Richard D. Moore • Biophysics Laboratory, State University of New York, Plattsburgh, New York; and Department of Physiology and Biophysics, College of Medicine, University of Vermont, Burlington, Vermont.

means for measuring plasma and urine glucose levels was widely available, but also because a conceptual framework for thinking about the role of glucose was at hand. Indeed, the revolutionary developments by Krebs, Warburg, and others that pioneered our understanding of carbohydrate metabolism were then at center stage.

By contrast, as late as 1946, in at least one major U.S. medical center, it was considered a "research project" to measure a plasma potassium level. Perhaps even more important, it was not until the late 1970s that a conceptional framework began to crystallize that makes clear the critical role of ionic and electrochemical events in cell regulation.

B. Intracellular pH As a Metabolic Regulator

The idea that intracellular pH may be a regulatory factor has been especially slow to gain acceptance. In contrast to Ca^{2+}, for which the evidence for this ion's key role in excitation–contraction coupling has long been clear and has paved the way for establishing this cation as a regulatory factor in other cell events, the commonly held assumption has been that intracellular pH—pH_i—is invariant, being held constant except in the most extreme perturbations such as disease states or extreme physical fatigue.

This widely held assumption is illustrated by several quotes made over the past few decades. For example, according to Thieden and associates, "It is generally accepted . . . that the hydrogen ion concentration in the cell . . . is nearly constant."[1] On the basis of inspection of pH optima, another investigator observed that many enzymes must function away from their pH optima and would therefore be very sensitive to changes in pH. From this, he concluded: "If the cell is to maintain a reasonable balance between its various metabolic pathways the intracellular pH must be maintained at constant value."[2]

Between 1967 and 1977, I asked perhaps two dozen biochemists and cell biologists whether they believed intracellular pH to be constant or variable. The response was invariably that pH_i must be constant. When asked why the response was that since almost all enzymes are sensitive to changes in pH, pH_i must be held constant to prevent metabolic chaos.

Busa and Nuccitelli[3] pointed out two other factors that prolonged the pH_i constancy hypothesis: (1) a lack of adequate techniques to measure pH_i and (2) the fact that blood pH is normally regulated within about a few hundredths. Manifestations of this pH_i constancy hypothesis that still persist include the infrequency with which complete pH profiles of enzymes are published and the fact that textbook authors often fail to take protons explicitly into account in chemical reactions such as ATP hydrolysis.

Perhaps the first clear suggestion that pH_i is an intracellular regulator of metabolism was made by Trivedi and Danforth in 1966.[4] By 1983, the view that pH_i plays a role in regulation of important cell functions has gained a significant following and is supported by a widening base of data (e.g., see Nuccitelli and Deamer[5] and Busa and Nuccitelli[3]). Changes in pH_i have been implicated as significant for several metabolic transitions, including gamete activation, dormancy–metabolism transitions, the cell cycle (including cell division), stimulus–response coupling in muscle and blood cells, and insulin action.

II. EFFECT OF INSULIN ON pH_i

A. Evidence for the Effect of Insulin on pH_i

The suggestion that pH_i may be the intracellular signal for insulin was first advanced by Manchester[6] in 1970 and was based on the sharp pH profile of some intracellular enzymes, especially phosphofructokinase. A kinetic analysis, based on a physical model of the (Na^+,K^+) pump, led Moore to suggest in 1973[7] that insulin might be increasing the rate of proton extrusion from muscle and suggest that changes in pH_i are part of the signaling system in insulin action. In 1970, Manchester[6] had reported three experiments on rat diaphragm muscle that demonstrated an insulin-induced increase of pH_i of 0.04, 0.07, and 0.15.

B. Effect of Insulin on pH_i in Amphibian Tissue

Moore and co-workers confirmed, in frog skeletal muscle, the prediction that *in vitro* addition of insulin increases pH_i, using both the weak acid [^{14}C]-5,5-dimethyl-oxazolidine-2,4-dione (DMO)[8–12] and the noninvasive technique of [^{31}P]-NMR.[13] Neither growth hormone nor albumin, at the same concentration (2 μM) as insulin, affects pH_i, indicating that the ΔpH_i produced by insulin is not a nonspecific protein effect.[11] In the absence of CO_2/HCO_3^-, the magnitude of the effect of insulin is $+0.16 \pm 0.03$ units when measured with DMO[9] and $+0.16 \pm 0.05$ when measured with [^{31}P]-NMR, using the difference between the resonance peaks of intracellular inorganic phosphate and phosphocreatine.[14] In the presence of CO_2/HCO_3^-, the effect of insulin is about $+0.13 \pm 0.02$ (Fidelman *et al.*).[12]

More recently, Putnam and Roos[15] used pH-sensitive microelectrodes to confirm the elevation of pH_i when frog semitendinosus muscle is exposed to 1 mU insulin/ml in the presence of 0.1% bovine serum albumin. By 50 min after addition of this concentration of insulin, pH_i had risen by about 0.08 ± 0.01, while the plasma membrane potential increased in magnitude by 5.4 ± 1.90 mV. Although in the absence of albumin, insulin (400 mU/ml) did not increase pH_i recovery after 5% CO_2 acidification, when Ringer K^+ was increased from 2.5 to 15 mM, the hormone nearly tripled recovery of pH_i after the acid load. This recovery was inhibited by 1 mM amiloride, leading these investigators to suggest that insulin can also activate Na/H exchange in frog skeletal muscle and thereby stimulate extrusion of acid.

The effect of insulin on pH_i in amphibians is not lmited to skeletal muscle. Morrill and co-workers[16] have used [^{31}P]-NMR to follow the effect of insulin on pH_i in frog prophase-arrested oocytes. In frog Ringer without albumin, control pH_i is 7.38 and 0.1–10 μM insulin elevates pH_i to 7.75–7.8 over a 1–2-hr period. Presoaking the oocytes in Na^+-free Ringer's solution for 30–60 min lowers control pH_i to 7.25 and blocks the effect of insulin on pH_i. Addition of Na^+ restores the effect of insulin to elevate pH_i. One mM amiloride blocks not only the elevation of pH_i by insulin, but the ability of insulin to stimulate cell division as well (G. Morrill, personal communication).

C. Effect of Insulin on pH_i in Mammalian Tissue

There have been two reports, both based on $[^{31}P]$-NMR measurement of pH_i, that insulin does not affect pH_i in mammalian muscle. In rat heart, insulin ($\sim$3 mU/ml, or 20 nM) does not affect pH_i as measured by $[^{31}P]$-NMR using the resonance peak of 2-deoxyglucose 6-phosphate.[17] The possibility has not been ruled out that the presence of the nonphysiological agent 2-deoxyglucose alters the insulin response.[18] Another group has measured pH_i in perfused cat soleus and biceps brachii using the $[^{31}P]$-NMR resonance peak of intracellular inorganic phosphate and 2-deoxyglucose phosphate and reported that in the presence of high (250 mU/ml) levels of insulin, pH_i does not change.

In both studies, which report no effect of insulin on pH_i, the concentrations of insulin were unusually high. It is an old observation that large elevations of the concentration of some hormones reverses the effect. Another possibility is that when the experiments began, sufficient insulin may still have been present on the receptor of the hormone to activate the acid-extrusion mechanism.

On the other hand, Podo et al.[19] have also used $[^{31}P]$-NMR and reported that in the isolated rat diaphragm insulin increases pH_i, as determined by the resonance peak of intracellular inorganic phosphate. The muscles were preincubated in the presence or absence of insulin in a Warburg respirometer containing Ringer's solution with glucose 6-phosphate. pH_i was then measured by next placing the muscles in the NMR spectrometer at 4°C. Those muscles preincubated with insulin had pH_i values averaging 0.15 higher than that of the controls.

Moreover, there is evidence that insulin can elevate pH_i in other mammalian cells. Moolenaar et al.[20] used an internalized fluorescent pH_i indicator to follow pH_i continuously in diploid human fibroblasts. Insulin is known to act synergistically with growth factors such as epidermal growth factor (EGF) in stimulating DNA synthesis. Dialyzed fetal calf serum (FCS) elevates pH_i by $\sim$0.2 pH unit within 15–20 min. Depleted FCS, which lacks mitogenic activity, has a negligible effect on pH_i. The polypeptide growth factor EGF also increases pH_i by $\sim$0.1 pH unit after 20 min. Although insulin (8×10^{-7} M) by itself does not produce a significant elevation of pH_i, in the presence of EGF the hormone does produce a statistically significant elevation of pH_i. When the pH_i elevation due to EGF has reached a steady state, addition of insulin produces a further rise of pH_i by almost 0.1 pH unit. Moolenaar et al.[20] concluded that the effects of both EGF and insulin on pH_i are due to stimulation of Na^+/H^+ exchange, because in these cells pH_i recovery following an acid load is (1) blocked by amiloride, (2) accompanied by an increase in amiloride-sensitive ^{22}Na influx and H^+ efflux, and (3) dependent on the extracellular Na^+ concentration.

Perhaps the most decisive test of the physiological significance of a hormone effect is to demonstrate the predicted consequences of reduced levels of the hormone.

Our laboratory[21] had previously demonstrated in rats that 75 mg/kg streptozotocin (SZ) produces mild diabetes accompanied by an elevation of intracellular Na^+, followed by a significant decrease in intracellular ATP.

If a physiological role of insulin is to elevate pH_i by activation of Na/H exchange, hypoinsulinemia sufficiently moderate as to not decrease blood pH consequent to metabolic acidosis should produce an observable decrease in pH_i. Recently, we used [^{14}C]-DMO to determine *in vivo* pH_i in soleus muscles of rats made diabetic by injection of 65 or 75 mg/kg SZ.[22] SZ-injected rats show classic signs of diabetes, i.e., elevated (greater than twofold) plasma glucose and reduced (by ~50%) immunoreactive insulin levels. However, as reflected by blood pH, none of the diabetic rats had metabolic acidosis. Intracellular Na^+ was elevated as observed previously. pH_i was significantly decreased by 0.07 ± 0.024 (S.E.) and by 0.076 ± 0.031 7 days after doses of 65 and 75 mg/kg SZ, respectively.

Table 1 illustrates the effect of 75 mg streptozotocin/Kg upon pH_i in rat soleus muscle. As expected, intraperitoneal injection of streptozotocin significantly lowered plasma insulin levels and raised blood glucose. However, the degree of diabetes was sufficiently mild that there was no evidence of metabolic acidosis as reflected by the lack of decrease in blood pH. However, pH_i of the soleus muscle was significantly decreased by 0.076 units 7 days after streptozotocin injection, and by 0.152 units 10 days after streptozotocin injection.

Clancy et al.[23] reported similar results in rats made diabetic with alloxan or SZ. Although blood pH was significantly depressed (to 7.07) in rats made diabetic for 2 days, blood pH had returned to normal by 7 and 28 days. In rats made diabetic for 7 days, pH_i was decreased by 0.28 in cardiac muscle, by 0.23 in skeletal muscle, and by 0.16 in liver. After 28 days, pH_i was decreased by about the same amount in both cardiac and in skeletal muscle, but the decrease in liver was now 0.24. Administration of insulin for 4–5 hr restores pH_i of cardiac muscle and skeletal muscle to normal while blood pH remains depressed. In hemidiaphragm preparations from normal and from 2-day diabetic rats, *in vitro* administration of 100 mU insulin/ml increases pH_i by 0.1–0.25 units, and this effect was blocked by amiloride.

Table 1. Effect of Streptozotocin Diabetes in Sprague–Dawley Rats Postinjection of 75 mg Streptozotocin/Kg[a,b]

	Control (sham-injected)	Days postinjection	
		7	10
Plasma glucose (mg/dl)	165 ± 8.6	437.5 ± 30.3[c]	448.7 ± 19.6[c]
Insulin (μU/ml)	58.7 ± 7.6	13.5 ± 5.1[c]	6.0 ± 2.1[c]
pCO_2 (mm Hg)	35.2 ± 1.0	35.6 ± 0.6	37.2 ± 3.6
Blood pH (pH_o)	7.485 ± 0.009	7.472 ± 0.012	7.474 ± 0.031
Intracellular pH (pH_i)	7.026 ± 0.017	6.950 ± 0.030	6.874 ± 0.043[c]

[a] From Brunder, Oleynek, and Moore (1983).
[b] Measurements taken from soleus muscle.
[c] Values significantly different from controls at P less than 0.05.

D. Summary

The evidence seems to indicate beyond a reasonable doubt that insulin increases pH_i in amphibian tissue. In frog skeletal muscle, this effect has been confirmed by three totally different techniques: [^{14}C]-DMO, [^{31}P]-NMR determination of the difference between inorganic phosphate and phosphocreatine peaks, and pH-sensitive microelectrodes. That this effect is not limited to amphibian muscle is indicated by the finding that insulin elevates pH_i in frog oocytes.

In mammalian tissue, however, the reports are inconsistent, and the question as to whether insulin affects pH_i in mammalian tissues is not yet resolved. Nevertheless, in spite of two negative reports of the effect of insulin on muscle, the fact remains that two other laboratories report that *in vitro* addition of insulin to rat diaphragm elevates pH_i as measured by [^{31}P]-NMR in one study and by [^{14}C]-DMO in the other. Moreover, one of these groups, plus our own, have observed the expected decrease in pH_i in nonketoacidotic (as reflected by normal blood pH) diabetic rats.

Possible explanations for the failure of the two groups to observe an elevation of pH_i by insulin include the use of very high concentrations of insulin and intracellular buffering. Intracellular buffering power reported for rat skeletal muscle (~68 mmoles/pH per liter) and rat heart (51 and 77 mmoles/pH per liter) is twice that reported for frog sartorius muscle (35 mmoles/pH per liter)[24] thus an effect of insulin on pH_i in mammals might be significantly slower than in amphibians. Another factor to consider is possible lack of a necessary cofactor. The finding by Moolenaar et al.[20] that insulin does elevate pH_i in human fibroblasts, but only in the presence of EGF, is especially provocative in this regard. It may well be that, at least in mammals, the presence of some other serum factor(s), such as EGF, is required for insulin to elevate pH_i.

III. MECHANISM OF INSULIN EFFECT ON pH_i

A. Regulation of pH_i

It is now generally recognized that the proton is not at equilibrium across the plasma membrane. For example, Kostyuk and Sorokina[25] pointed out that in frog sartorius muscle the equilibrium pH_i (calculated from the measured pH_o of 7.4 and membrane potential of -90 mV) is about 5.9—far lower than the value they observed—7.1—using pH-sensitive microelectrodes. This equilibrium persists in the face of the general tendency of metabolism to produce protons (Roos and Boron).[24] Physicochemical and organellar buffering and biochemical proton sources and sinks can produce short-term effects on the amount of free intracellular protons produced by the acidifying effects of metabolism. However, the central role in maintaining pH_i above its equilibrium level is due to transport systems that remove acid from the cell, i.e., either extrude H^+ or accumulate HCO_3^- or OH^- against their electrochemical gradient or both.

In squid axons, snail neurons, and barnacle muscle, acid extrusion is achieved by coupling the entry of Na^+ and HCO_3^- to the exit of Cl^- and possibly H^+. In crayfish neurons, in amphibian proximal tubular cells, and in amphibian skeletal muscle, Na/H exchange is used to extrude acid. In mammalian muscle, both Na/H exchange and Cl/HCO$^-_3$ exchange operate in parallel (for a complete discussion, see Roos and Boron[24]).

B. Effect of Insulin on Na/H Exchange

With some significant exceptions, most of the data pertaining to the mechanism whereby insulin affects pH$_i$ have come from studies on frog skeletal muscle.

The effect of insulin on pH$_i$ is not blocked by 10^{-3} M ouabain.[11] Considerable evidence now indicates that the change in pH$_i$ by insulin is due to activation of Na/H exchange.

The Na/H exchange mechanism operates by using the energy made available by Na$^+$ moving down its free-energy gradient δG_{Na} to drive the proton up its own free-energy gradient δG_H. Therefore, the average free energy change, $\langle \Delta G \rangle_{Na/H}$, for an Na/H exchange mechanism that couples Na$^+$ influx to proton efflux is the sum of the free energy required to transport n Na$^+$ ions inward and that required to transport m H$^+$ ions outward:

$$\langle \Delta G \rangle_{Na/H} = n\Delta G_{Na} + (-m\Delta G_H)$$

or

$$\langle \Delta G \rangle_{Na/H} = n \left[eV_m + kT \ln \frac{\gamma_{Na_i}[Na^+]_i}{\gamma_{Na_o}[Na^+]_o} \right] - m \left[eV_m + kT \ln \frac{\alpha_{H_i}}{\alpha_{H_o}} \right] \quad (1)$$

where e is the protonic charge, k is Boltzmann's constant, T is absolute temperature, V_m is the membrane potential, γ_{Na_i} (γ_{Na_o}) is the intracellular (extracellular) Na$^+$ activity coefficient, $[Na^+]_i$ ($[Na^+]_o$) is the intracellular (extracellular) Na$^+$ concentration, and α_{H_i} α_{H_o}) is intracellular (extracellular) H$^+$ activity. This mechanism will have sufficient energy available from ΔG_{Na} to transport protons outward when $\langle \Delta G \rangle_{Na/H} < 0$. If $[Na^+]_o$ is decreased sufficiently to null the average free energy, i.e., $\langle \Delta G \rangle_{Na/H} = 0$, this system should not transport protons. If $[Na^+]_o$ is decreased below this null point so that $\langle \Delta G \rangle_{Na/H} > 0$, activation of the system should transport protons inward. Under physiological conditions, $\Delta G_{Na} < 0$, $\Delta G_H < 0$, and $\langle \Delta G \rangle_{Na/H} < 0$.

Therefore, the hypothesis that the increase in pH$_i$ caused by insulin is due to stimulation of Na/H exchange yields the following four predictions:

Prediction #1. The elevation of pH$_i$ should be associated with an increased influx of Na$^+$. In the presence of sufficient ouabain to inhibit the stimulation of the (Na$^+$,K$^+$) pump by insulin[26] this increased Na$^+$ influx would produce an increase in intracellular Na$^+$, which should correlate with the increase in pH$_i$ produced by insulin. Interpretation of ^{22}Na influx is less clear due to the presence of the Na/Na exchange in some cells, especially frog skeletal muscle.

Confirmation. This prediction was confirmed by the observation that in the presence of 10^{-3} M ouabain, the change in pH_i produced by insulin in each frog sartorius muscle is positively correlated ($r = 0.689, p < 0.01$) with net Na^+ influx, as reflected by the elevation in Na_i^+ produced by the hormone in the same muscle.[11]

Consistent with these results is the observation by Clausen and Kohn[27] that in the presence of 10^{-3} M ouabain, insulin produced a significant 30 to 40% increase in $^{22}Na^+$ influx in rat soleus muscle. In the presence of ouabain, insulin also increases $^{22}Na^+$ uptake by hepatocytes.[28] However, in rat adipocytes preincubated 15 min with 10^{-3} M ouabain, insulin (3 nM) does not increase $^{22}Na^+$ uptake (Resh *et al.*).[29]

Prediction #2. In frog sartorius, $[Na^+]_i \sim 7$–8 mM. For $pH_o = 7.4$ and Eq. (5) (assuming $n/m = 1$) indicates that decreasing $[Na^+]_o$ to about 6.8 mM should null the free energy for Na/H exchange ($\langle \Delta G \rangle_{Na/H} = 0$) and at that value of $[Na^+]_o$, insulin should have no effect on pH_i.

Confirmation. In Ringer's solution containing CO_2 and bicarbonate, decreasing $[Na^+]_o$ to this calculated value (a 15-fold reduction) completely blocks ($p > 0.5$) the effect of insulin on pH_i in frog skeletal muscle,[9] thereby confirming this prediction. This prediction is also supported by the finding that depleting frog oocytes of Na^+ by soaking in Na^+-free Ringer's solution (which, by decreasing the Na^+ driving force, should bring $\langle \Delta G \rangle_{Na/H}$ toward zero) blocks the effect of insulin on pH_i.[16]

Prediction #3. Lowering $[Na^+]_o$ still further reverses the sign of $\langle \Delta G \rangle_{Na/H}$ and thus reverses the direction of Na/H exchange. Therefore, removing extracellular Na^+ should convert the action of insulin from an increase to a decrease in pH_i.

Confirmation. When either Mg^{2+} or choline is used to replace the Na^+ in the Ringer's solution (actually about 0.12 mM Na^+,[11] the effect of insulin on pH_i is converted to a statistically significant *decrease* ($p < 0.005$).[11] This decrease in pH_i produced by insulin in Mg^{2+} Ringer's solution has been confirmed using the noninvasive method of [^{31}P]-NMR to measure pH_i.[13]

Of equal importance, the magnitude of ΔG_{Na} is sufficient to move H^+ against its energy gradient, as indicated by the fact that $\langle \Delta G \rangle_{Na/H}$ always has the proper sign for the observed flux of H^+.[11]

It is possible that all the above results could be due to the operation of a Na^+–CO_3^{2-} cotransport system.[30] However, the elevation of pH_i by insulin in HCO_3^- free Ringer's solution rules out this possibility[10,11] and also argues against the possibility that the effect is due to Cl^-/HCO_3^- exchange.

Prediction #4. Because the diuretic drug amiloride (3,5-diamino-6-chloropyrazinoyl guanidine) blocks Na/H exchange,[31] this drug should block all the above effects of insulin, i.e., it should block the effect of insulin on both H^+ efflux and the associated Na^+ influx and on the decrease in pH_i produced by insulin in Na-free Ringer's solution.

Confirmation. In the presence of 5% CO_2/30 mM HCO_3^-, 0.5 mM amiloride does block the elevation by insulin of pH_i in frog sartorius muscle.[10] In an identical Ringer's solution, but lacking amiloride, insulin significantly ($p < 0.001$) increases pH_i by

$0.096 \pm 0.016.$[9] In Ringer's solution lacking CO_2/HCO_3^-, amiloride still blocks the elevation of pH_i by insulin.[11] The increase in Na_i^+ due to exposure of muscles to insulin for 90 min in the presence of both 1 mM ouabain and 0.5 mM amiloride is essentially zero ($p > 0.5$).[11] Finally, in muscles placed in Ringer's solution in which Na^+ is replaced by osmotically equivalent amounts of Mg^{2+}, 0.5 mM amiloride inhibits the decrease in pH_i produced by insulin.[11]

These effects of amiloride are not limited to amphibian skeletal muscle, as the elevation of pH_i in frog oocytes by insulin is also blocked by this agent.[16] Moreover, the elevation of pH_i in rat hemidiaphragms by *in vitro* addition of insulin is also blocked by (5×10^{-4} M) amiloride (Clancy *et al.*).[23]

Of considerable importance is the finding that 1 mM amiloride inhibits the stimulation by insulin of pH_i recovery that occurs in acid-loaded frog semitendinosus muscles in the presence of 15 mM K_o^+.[15] This observation is especially important because it clearly demonstrates that the effect of amiloride is to block the stimulation by insulin of acid transport, as opposed to metabolic- or buffer-induced changes in pH_i.

There is no evidence that the change in pH_i is secondary to metabolic changes. To the contrary, insulin stimulates both glycolysis and ATP hydrolysis by increased activity of the (Na^+-K^+)-pump. In the cell, more than 90% of ATP^{4-} is bound to Mg^{2+}.[14] It is the Mg ATP^{2-} form that is hydrolyzed by such cell processes as the $(Na^+,K^+)ATPase$. Although glycolysis does not produce intracellular H^+ under anaerobic conditions, the accompanying hydrolysis of ATP does according to the reaction (at $pH \geq 8$):

$$Mg\ ATP^{2-} + H_2O \rightarrow Mg\ ADP^{1-} + P_i^{2-} + H^+$$

Even at lower pH values, the combined effect of glycolysis and hydrolysis of the ATP produced is still to produce protons according to the followng reaction[3]:

$$Glucose \rightarrow 2\ lactate^{1-} + 2\ H^+$$

Accordingly, under anaerobic conditions, if insulin does not stimulate acid extrusion, the hormone would be expected to produce a decrease in pH_i due to stimulation of ATP hydrolysis by the (Na^+,K^+) pump. Thus, the insulin-induced increase in pH_i that occurs in the face of increased production of acid under anoxic conditions (Fidelman *et al.*[12]) and especially the evidence of recovery from acid-loaded cells (Putnam and Roos[15]) adds considerable weight to the argument that, at least in frog muscle, this hormone can stimulate acid extrusion by Na/H exchange.

C. Conclusions

It seems reasonably well established that in frog skeletal muscle and oocytes, insulin can stimulate the Na/H exchange system in the plasma membrane and that amiloride blocks the stimulation by insulin of this transport system. The evidence that insulin exerts this effect on pH_i in mammalian cells is not yet as clear, but a growing body of evidence suggests that under physiological conditions insulin regulates pH_i in mammals as well. What little evidence exists suggests this is also due to activation of Na/H exchange.

IV. EFFECTS OF INSULIN-MEDIATED CHANGES IN pH_i ON GLYCOLYSIS

Although the elevation of pH_i may play a role in the action of insulin on several cell functions, including protein synthesis, nucleic acid synthesis, and cell division, experimental results to date indicate that the stimulation of glycolysis by insulin is the most likely possibility.

The hypothesis that the insulin-induced stimulation of glycolysis is due to activation of PFK by an increase in pH_i consequent to stimulation by insulin of Na/H exchange at the plasma membrane leads to the several predictions:

1. Blocking the effect of insulin on pH_i should block the stimulation of glycolysis by this hormone. This prediction was verified by the finding that, in a glucose-free Ringer's solution (30 mM HCO_3^-/5% CO_2), either amiloride or a 15-fold reduction in the concentration of extracellular Na^+, $[Na^+]_o$, blocks the action of insulin on glycolysis as well as on pH_i.[10]

2. The effect of insulin on glycolysis should be a function of $[Na^+]_o$, with a reversal of ΔG_{Na} by decreasing $[Na^+]_o$ resulting in an inhibition of glycolysis by insulin. Variations in $[Na^+]_o$ have been shown to have little effect on V_m in frog sartorius muscle.[32] By holding pH_o constant, and since $[Na^+]_i$ and pH_i have relatively small variations under the conditions of these experiments,[12] it follows from Eq. (1) that changes in $\langle \Delta G \rangle_{Na/H}$ are determined primarily by extracellular Na^+:

$$\langle \Delta G \rangle_{Na/H} = A - B \log [Na^+]_o \qquad (2)$$

where A and B are positive and are approximately constant in the present experimental conditions. Therefore, since the direction of change in pH_i is determined by the sign of $\langle \Delta G \rangle_{Na/H}$,[11] a test of prediction #1 is that the effect of insulin on glycolytic flux should be reversed as $\log [Na^+]_o$, a component of the Na^+ free-energy gradient, is lowered sufficiently to reverse the sign of $\langle \Delta G \rangle_{Na/H}$ in Eq. (2). Although the magnitude of ΔpH_i, and therefore of the change in glycolytic flux, is not expected to be determined by $\langle \Delta G \rangle_{Na/H}$ alone, the free-energy term should be one of the variables determining ΔpH_i. The effect of insulin on glycolytic flux varied approximately linearly with $\log [Na^+]_o$ (Fig. 1). Although not theoretically predicted, the data indicate a reasonably linear correlation (see Fig. 1) of percentage change in glycolytic flux with $\log [Na^+]_o$, with the intercept at $[Na^+]_o = 5.8$ mM representing the value at which $\langle \Delta G \rangle_{Na/H}$ should equal zero in the above hypothesis. Reducing $[Na^+]_o$ from 104 mM to 0.12 mM changes the insulin effect on gly-

Figure 1. Relationship between log $[Na^+]_o$ and glycolytic flux as measured by anaerobic lactate production (% ΔALP) in frog sartorius. (●) mean ± SE for seven to ten determinations. (○) mean ± SE for three or four determinations. A least-squares linear regression was performed on the 60 actual data points ($r = 0.823$, $p < 0.001$; slope > 0, $p < 0.001$). In view of the nonlinearities present in the pH profile of phosphofractokinase (PFK) and that likely exist in the coupling of $\langle \Delta G \rangle_{Na/H}$ to ΔpH_i, the apparent linear correlation between the change in glycolytic flux and log $[Na^+]_o$ might seem surprising. It should be emphasized that the linear correlation does not indicate that the data are actually linear; more likely, the data lie along an approximately linear region of a functional curve. Regardless of initial pH_i, other effectors of PFK, such as AMP, fructose 6-phosphate (F6P), ATP, and ADP must have moved the pH profile (see Fig. 3), so that the intersection of the curve with pH_i lies near the midpoint of the quasilinear region.

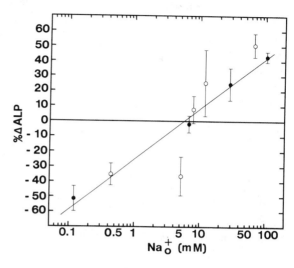

colytic flux from an increase of 42.9 ± 3.5% ($p < 0.001$) to a decrease of 51.5 ± 8.4% ($p < 0.001$).[12]

3. Changing pH_i by a means other than insulin (e.g., by varying CO_2) should mimic the action of insulin on glycolysis. If the central hypothesis that ΔpH_i represents the connection between insulin activation of Na/H exchange at the plasma membrane and the effect of insulin on glycolysis is correct, changes in pH_i produced by any other means should produce the same effect on glycolysis as the ΔpH_i produced by insulin. Figure 2 demonstrates that the relationship between ΔpH_i and percentage change in glycolytic flux was independent of whether the effects are produced by insulin or by varying CO_2, since the means of the experimental points all lie on essentially the same line. The results demonstrate that the change in pH_i produced by insulin was sufficient to account for the entire effect of insulin on glycolysis in the time period (130 min) of these experiments.

4. Stimulating glycolysis either by insulin or by increasing pH_i secondary to a decrease in CO_2 should be associated with activation of PFK. Also, inhibiting glycolysis either by insulin in a Na-free Ringer's solution by increasing CO_2 to decrease pH_i should be associated with a decrease in the activity of this rate-limiting enzyme. For a nonequilibrium process such as that catalyzed by PFK,[33,34] an inverse change in the amount of substrate fructose 6-phosphate (F6P) in these experiments, and the flux through the enzyme, represented by anaerobic lactate production in these experiments, will unambiguously indicate a change in the activity of the enzyme (see discussion in Fidelman et al.[12]).

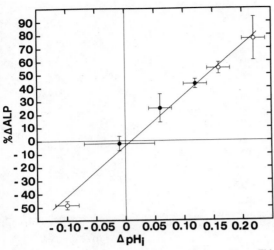

Figure 2. Relationship between changes in intracellular pH (ΔpH_i) and percentage change in glycolytic flux, as measurd by percentage changes in anaerobic lactate production (%ΔALP) in frog sartorius produced either by insulin or by varying level of CO_2. (●) Changes due to insulin at 5% CO_2 in Ringer's solution containing (from left to right) $[Na^+]$ = 6.8 mM, 28.7 mM, and 104 mM Na^+. (○) Changes produced (in the absence of insulin) by CO_2 levels of (from left to right) 10%, 2.7%, and 2%, compared with paired controls at 5% CO_2. To mimic the effect of insulin when the Na^+ gradient is reversed, CO_2 was increased from 5% to 10% in the same Na-free Ringer's solution used in the insulin experiments. Each data point represents the means ± SE. The plotted line is a least-squares fit to the points. (From Fidelman *et al.*[12])

In each individual experiment, stimulation or depression of glycolysis by insulin was associated with, respectively, an activation or a deactivation of PFK. In 21 of 23 individual experiments, using CO_2 to produce an elevation or depression of pH_i of the same magnitude produced by insulin was associated with respectively, an activation or a deactivation of PFK.

A. Summary

These results clearly indicate that the Na^+ concentration, or activity, component of $\langle\Delta G\rangle_{na/H}$ can determine both the magnitude as well as the direction of the acute action of insulin on glycolysis. That merely lowering $[Na^+]_o$ reverses the action of insulin on glycolytic flux indicates that ionic phenomena play the predominant role in mediating the acute action of insulin on glycolysis. The confirmation of this particular prediction provides especially powerful support for the model and strongly implies a direct, functional relationship between the Na^+ concentration gradient and the intracellular signal that mediates the effect of insulin on glycolysis.

The correlation between changes in pH_i, whether induced by insulin or by changes in CO_2 levels, supports the thesis that the immediate signal that mediates this insulin effect is the change in pH_i. The results, obtained in live muscle, of the plot of substrate versus rate indicate that the effect of changes in pH_i is to affect the activity of phosphofructokinase, PFK, the pacemaker or rate-limiting enzyme[35] of glycolysis. This is consistent with the extreme sensitivity[4] to small changes in pH of PFK; phosphofructokinase isolated from frog skeletal muscle can be maximally activated by pH elevations as small as 0.1–0.2 units,[4] i.e., a 20–37% decrease in H^+ activity (Fig. 3).

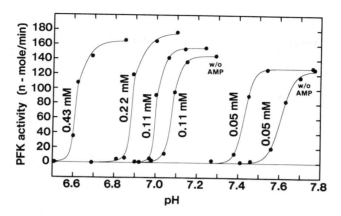

Figure 3. Effect of pH on phosphofructokinase (PFK) activity at various concentrations of fructose 6-phosphate (F6P) in the presence or absence of 0.1 mM 5'-AMP. (Data are from Figs. 1 and 2 in Trivedi and Danforth.[4])

B. Dependence on ΔpH_i Rather Than on pH_i

Regardless of initial pH_i, other effectors of PFK, such as 5–AMP, F6P, ATP, and ADP must have moved the pH profile[4] so that the intersection of the curve with pH_i lies near the midpoint of the linear region. This indicates that more important than initial values of pH_i are changes in pH_i. It must be kept in mind that the glycolytic regulatory system is a complex, dynamic system with multiple feedback loops. Therefore, changes in glycolysis produced by any one effector will eventually be at least partially nullified by resulting changes in levels of other effectors, e.g., ATP and ADP. Accordingly, absolute values of glycolytic flux could only be correlated with pH_i if the dynamics of the non-steady-state changes found in such complex systems were taken into account. It is for this reason that relative changes in glycolytic flux, over the same time interval, were used in these studies for quantitative analysis.

C. Other Mechanisms

It has been reported that over a longer period, insulin increases amounts of PFK in diabetic rats[36] by controlling the level of a peptide-stabilizing factor that mediates the rate of degradation of PFK.[37] However, PFK is not increased significantly until 24 hr after insulin.[37] The results summarized above indicate that the ΔH_i must be the predominant mediator of the acute (130 min) action of insulin on glycolysis. This suggests that the effect of insulin on glycolysis is mediated by two mechanisms which may or may not be interrelated.

Jarett and Seals[38] and Larner *et al.*[39] also reported a peptide generated by insulin that appears to be part of the intracellular signal system whereby this hormone stimulates pyruvate dehydrogenase,[38] activates the phosphoprotein phosphatase that converts glycogen synthase to its active form, and inhibits AMP-dependent protein kinase[39] (see also Chapters 10, 11, and 12 in this volume). Moreover, there is evidence that changes in intracellular Ca^{2+} (see Chapter 6) and the membrane potential V_m (see Chapter 7) may be part of the signal system for insulin. Taken together, these results suggest that different aspects of insulin action may be mediated by different intracellular signals. The possible context of pH_i in this signal network is discussed in the next section.

V. ROLE OF INTRACELLULAR pH IN INSULIN ACTION

Increasingly, evidence supports the hypothesis that the rise in pH_i that results from insulin stimulation of Na/H exchange is part of the intracellular signal for perhaps several effects of this hormone. As outlined above, the increase in pH_i most likely mediates the acute effect of insulin on glycolysis, at least in frog skeletal muscle. Moreover, insulin probably stimulates type A amino acid transport by increased synthesis of new transport sites; there is evidence that this may be mediated at least in part by an increase in pH_i (see review by Moore).[40] A provocative finding is that in rat adipocytes alkaline pH causes a minor but definite enhancement of hexose transport in the absence of insulin.[41]

Elevation of pH_i not only appears to be involved in insulin action, but a growing body of evidence suggests that elevation of pH_i is part of the signal system for several stimuli that trigger anabolic cell functions such as DNA synthesis[41] and/or protein synthesis[42,43] (see review by Busa and Nuccitelli[3]). In the action of mitogens, elevation of pH_i appears to play a permissive role, as other signals are also required.

Specific protein phosphorylation and dephosphorylation reactions are probably involved in amplification of many effects of insulin.[44] Initiation of DNA synthesis by insulin and other growth factors is preceded by phosphorylation of ribosomal S6 proteins and by amiloride-sensitive Na^+-dependent H^+ efflux. Either amiloride or dissipation of H^+ gradients, with dinitrophenol (DNP) or carbonyl cyanide m-chlorophenyl hydrazone (CCCP), blocks both the phosphorylation of ribosomal S6 proteins and the amiloride-sensitive Na^+-dependent H^+ efflux.[45] Moreover, in isolated hepatocytes, within 30–40 min, insulin increases phosphorylation of two protein fractions and decreases phosphorylation of four others, and all these phosphorylation effects are inhibited by amiloride.[46] These studies cannot yet be considered conclusive, since amiloride enters the cell,[44] pH_i was not measured, nor were changes in pH_i by physicochemical methods used to mimic the biochemical effects. Nevertheless, the most likely interpretation is that an elevation of pH_i is part of the signal whereby insulin triggers changes in the phosphorylation state of key cell proteins.

A. Regulation of the Energy State of the Cell

It is widely recognized that insulin causes increased storage of energy, in the form of glycogen and fat, which can later be used for metabolic purposes, including work. Insulin also stimulates the synthesis of macromolecules such as nucleic acids and proteins. Moreover, insulin stimulates production of the energy currency of the cell, ATP, by stimulating both glycolysis and oxidative phosphorylation. While this aspect of insulin action on energy flow in the cell is widely recognized, it may not be so well appreciated that much of the increase in energy stored in the cell due to insulin is not in molecular form, but in biophysical parameters such as electric fields, activity (or concentration) gradients, as well as an actual increase in the average energy available from the hydrolysis of each ATP molecule.

For example, insulin energizes the plasma membrane, e.g., it increases energy gradients across it. The effect of insulin to stimulate the (Na^+/K^+) pump and to activate Na/H exchange results in an increase in several physical energy gradients across the plasma membrane, including the electrical potential across the plasma membrane, V_m, the electrochemical potential difference for Na^+ ($\Delta\mu_{Na}$) and for H^+ ($\Delta\mu_H$). Secondary to these changes or other processes, or both, insulin also increases the energy gradient across the membrane for type A amino acids ($\Delta\mu_{AA}$), and possibly Ca^{2+} ($\Delta\mu_{Ca}$) (see review by Moore[40]).

Moreover, because the proton is part of the chemical reaction for hydrolysis of Mg ATP:

$$Mg\ ATP^{2-} + H_2O \rightarrow Mg\ ADP^{1-} + P_i^{2-} + H^+$$

For this chemical reaction, it is clear that $\Delta G_{Mg\ ATP}$ is a function of the proton activity:

$$\Delta G_{Mg\ ATP} = \Delta G^\circ_{Mg\ ATP} - RT\ \ln\left[\frac{\alpha_{Mg\ ATP}}{\alpha_{Mg\ ADP}\cdot\alpha_{P_i}\cdot\alpha_H}\right] \tag{3}$$

Therefore, the increase in pH_i produced by insulin results in an increase in the free-energy gradient, $\Delta G_{Mg\ ATP}$, available from the hydrolysis of Mg ATP.

Since pH is defined as the negative log of the hydrogen ion activity, α_H:

$$pH \equiv -\log_{10}\alpha_H$$

the free energy available from Mg ATP hydrolysis in the cell may be written explicitly as a function of the intracellular pH:

$$\Delta G_{Mg\ ATP} = \Delta G^\circ_{Mg\ ATP} - RT\ \ln\left[\frac{\alpha_{Mg\ ATP}}{\alpha_{Mg\ ADP}\cdot\alpha_{P_i}}\right] - 2.3026\ RT\cdot pH_i \tag{4}$$

Most textbooks incorporate the term $2.3026\ RT\cdot pH$ into the standard free-energy gradient, $\Delta G^\circ_{Mg\ ATP}$, leading to the statement that this term is pH dependent. However, when the equation is expressed in its more complete form, as above, $\Delta G^\circ_{Mg\ ATP}$ becomes more clearly a true constant. From Eq. (4), it is apparent that at 37°C, an elevation of pH_i typical of insulin, say from 7.15 to 7.3, would result in an additional 0.212 kcal/mole available from ATP hydrolysis. This is only about 2% of the total available but, because of the ability of insulin to decrease intracellular P_i,[17] the increase in energy available from ATP hydrolysis could be still greater. In any case, some processes have energy thresholds and in these, even a 2% increase could be critical.

How can the energy available from—and therefore stored in—an ATP molecule be increased? The answer is clear from inspection of Eq. (3) or (4). The energy available from Mg ATP hydrolysis *per se*, i.e., breaking the terminal phosphate bond, is given by $\Delta G^{\circ}_{Mg\ ATP}$. This term is actually positive because breakng a bond does not release energy, but rather requires energy. If that is the case, where does the energy come from? Inspection of Eq. (3) indicates that it must come from the right-hand term, i.e., the ratio

$$\alpha_{Mg\ ATP}/\alpha_{Mg\ ADP} \cdot \alpha_{P_i} \cdot \alpha_H$$

must be sufficiently >1.0. By factoring out the proton activity, this ratio can be separated into two terms, as in Eq. (4): $\alpha_{Mg\ ATP}/(\alpha_{Mg\ ADP} \cdot \alpha_{P_i})$, the phosphorylation potential, and pH_i. The fact that the intracellular proton activity is low, $\alpha_H \sim 7.9 \times 10^{-8}$, or $pH_i = 7.1$, alone accounts for about -10 kcal/mole at 37°C. Taking Mg $ATP_i = 5.8 \times 10^{-3}$ M, Mg $ADP_i = 3.1 \times 10^{-6}$ M,[14] and assuming P_i to be in the millimolar range, accounts for about another -4.6 kcal/mole, the total available, $\Delta G_{Mg\ ATP}$, being lower in magnitude than -14.6 kcal/mole, since $\Delta G^{\circ}_{Mg\ ATP}$ is positive but sufficiently small that $\Delta G_{Mg\ ATP}$ is negative. In other words, energy is available from the reaction, precisely because of mass action effects; the concentration, or activity, of the products is much less than that of the reactants. Thus, it is primarily the environment of the Mg ATP molecule that determines the energy available from hydrolysis of this molecule. Therefore, since the environment, in particular $\alpha_{Mg\ ADP}$, α_{P_i}, and pH_i, is a function of the whole cell (metabolism and membrane transport processes), $\Delta G_{Mg\ ATP}$ clearly is determined by the functional state of the whole cell.

As discussed above, the effect of insulin on this functional state is to change the intracellular environment (increase pH_i, decrease P_i) in a manner that increases the magnitude of $\Delta G_{Mg\ ATP}$, i.e., to increase the energy available from hydrolysis of Mg ATP. This increased energy is not a violation of the first law of thermodynamics because it takes work to change the environment of the cell; e.g., energy is required to elevate pH_i.

Thus, insulin cannot only vary the rate of energy delivery via ATP synthesis—it must regulate the energy intensity, i.e., ΔG/molecule of ATP hydrolyzed, as well. This provides one more opportunity for insulin to exert a cascade of regulatory influences.

This suggests the concept that insulin increases not only the biochemical, but the biophysical energy state of the whole cell.[40] Zierler and Rogus[47] have proposed a related concept that the increased electrical field across the plasma membrane mediates the action of insulin on glucose permeability. One advantage of these biophysical forms of energy storage is that, once increased, some of them are immediately accessible. An electrochemical gradient across the plasma membrane may be tapped literally within milliseconds, without the delays inherent in turning on metabolic production of available energy, (e.g., the synthesis of ATP) and in waiting for levels of metabolites to reach new values.

This view is consistent with the classic view that insulin is an anabolic (or growth) hormone but enlarges the concept to view the effect of insulin on the cell as a whole biophysical system. The cell becomes somewhat like an atom in an excited state. In contrast to energizing an atom or molecule, it takes several minutes to change the energy state of the whole cell. For example, because of intracellular buffering, as well as the energy required to extrude protons, the effect of insulin on pH_i is relatively slow, taking more than 1 hr to reach maximum. However, once in the higher energy state, the energized cell is poised for energy-demanding chores, such as synthesis of macromolecules, cell division, or the performance of other forms of work. Thus, in this view, by taking the cell into a higher biophysical energy state, the effect of insulin on the ionic insulin transduction system is preparing anabolic processes in the cell to respond to more specialized intracellular signals, other than pH_i, such as Ca^{2+} or messenger peptides.

B. Relationship of ΔpH_i to Other Ionic Effects of Insulin

In the cell, it is impossible to change only one ionic species.[40] Interactions between ionic species occur at the level of transport across the plasma membrane (between the cell and its environment), transport across intracellular membranes (between cytoplasm and the interior of cell organelles), and competition for intracellular binding sites.[40]

1. Interaction between pH_i and Ca_i^{2+}

In addition to the membrane effects of insulin on the high-affinity Ca^{2+}-ATPase (see Chapter 6) and to increased Na : Ca exchange secondary to stimulation of the (Na^+, K^+) pump by insulin, changes in pH_i would be expected to affect intracellular free Ca^{2+}. Intracellular H^+ and Ca^{2+} 'buffer' each other by competition for the same binding sites and by Ca : H exchange across the inner mitochondrial membrane. Calmodulin mediates the Ca^{2+} responses of several important enzymes and is one of the most primitive of all proteins. This Ca^{2+}-binding protein is also very pH sensitive. Busa and Nuccitelli[3] have pointed out that because of the pH sensitivity of the Ca^{2+} dissociation constant of calmodulin, an increase of pH_i by 0.5 unit would have the same effect as a fivefold increase in free Ca_i^{2+} at constant pH_i. They suggest that calmodulin might more accurately be considered a Ca^{2+}/H^+-binding protein, and may serve as a sensor of both $[Ca^{2+}]_i$ as well as pH_i changes.

2. Possible Interactions between pH_i and cAMP

There is some evidence that cAMP elevates pH_i by stimulation of the Cl^-/HCO_3^- acid extrusion mechanism (see review by Busa and Nuccitelli[3]). Conversely, circumstantial evidence suggests that moderate changes in pH_i may regulate

[cAMP]$_i$. The cAMP-mediated stimulation of lipolysis and calorigenesis is markedly depressed by acidosis and increased by alkalosis (see Busa and Nuccitelli[3]).

On the basis of these and other considerations, Busa and Nuccitelli[3] have suggested that "H^+, Ca^{2+}, and cAMP might all function interdependently as 'synarchic messengers'[48] with pH$_i$ providing a metabolic context within which, e.g., a hormonal stimulus might have rather different consequences at two different pH$_i$s . . . ."

C. Feedback Relationships

The interactions between the level of pH$_i$ and other ions affected by insulin suggest the existence of novel ionic feedback mechanisms, both positive and negative, of probable biological significance. Since feedback mechanisms produce nonlinear responses, simple linear cause-and-effect analysis is precluded when considering the action of this hormone. Nevertheless, potentially important feedback loops may be identified.

Of all the membrane transport processes increased by insulin, the one most likely stimulated by an intramembrane-mediated signal, independent of any intracellular signal, is the (Na^+, K^+) pump. The clearest evidence for this possibility is that *in vitro* addition of insulin to purified fragments of plasma membrane from frog skeletal muscle reproducibly increases the activity of the ouabain-inhibitable (Na^+, K^+) ATPase. This occurs at levels of Na^+ and ATP found within the cell.[26] The fact that storing the membrane fragments at liquid nitrogen temperatures but not at $-8°C$ destroys the response to insulin without blocking the (Na^+, K^+)ATPase activity suggests that hydrophobic interactions within the membrane might be involved in transmitting the insulin effect. This is consistent with the report that insulin increases membrane fluidity.[50]

It is not known whether the activation of Na/H exchange by insulin is due to an intramembrane signal. Regardless of the mechanism, activation of Na/H exchange is not secondary to stimulation of the (Na^+, K^+) pump, since ouabain does not block the elevation of pH$_i$ by this hormone.[11]

Skou[49] has shown that (Na^+, K^+)ATPase is quite sensitive to changes in pH$_i$. An elevation of pH increases the affinity of the enzyme system for Na^+. This suggests that in the intact cell, activity of the (Na^+, K^+) pump may be increased by an elevation of pH$_i$. This would provide a positive feedback loop for the ionic effects of insulin.

Although insulin stimulates both transport systems independently, the consequences produced by stimulation of either transport system feeds back on the other. Elevation of pH$_i$ might provide a further stimulation of the (Na^+, K^+) pump. Conversely, to the extent that increased activity of the (Na^+, K^+) pump would lead to a decrease in $[Na^+]_i$, increasing the driving force for Na/H exchange, the effect of insulin on the (Na^+, K^+) pump would provide a boost to Na/H exchange after its activation by insulin. Both interactions would provide positive feedback. Such regenerative feedback loops act like a switch. However, there would be obvious limitations to the extent that activity of these two transport systems could be increased, if nothing else, imposed by energy limitations.

Another positive feedback loop is suggested by the fact that an increase in pH_i increases Na/Ca exchange (see review by Moore[40]).

VI. UNIQUE PROPERTIES OF THE PROTON FOR INTRACELLULAR REGULATION

It seems reasonable to assume that the very first cells to appear on this planet needed to regulate very basic or fundamental functions, for example, the ability to vary the rate of energy flow within the cell and the related problem of taking the cell from a state of relative dormancy to one of activity. To do this, something in those first cells would be required to act as an intracellular signal, and those first cells would have to have used whatever was at hand. It would be much later that the development of the more complex functions required for differentiation would appear and thus require the need for intracellular signals of increased specificity, and therefore more complex structure—molecular second messengers. As Busa and Nuccitelli[3] have pointed out, "It seems safe to assume that, just as it does today, from the very beginnings of life on earth biochemistry has involved weakly ionized compounds and has relied heavily on acid-catalysis; regulation of pH_i therefore provided a powerful means of regulating the metabolism of early cells *without requiring the evolution of special receptor molecules*" (italics added). From these considerations, one may begin to infer that the first primitive metabolic functions were controlled in part by protons.

For a substance to be an intracellular signal, three characteristics are necessary: (1) it must bind tightly to organic macromolecules in order to stay bound and to have sufficient energy of binding to induce conformational changes, (2) it must have specificity for certain sites on these macromolecules, and (3) its levels within the cell must be regulatable.

A. The Proton As an Intracellular Signal in the First Cells

As is true today, of those substances dissolved in the primitive ocean, most were ions, and in all likelihood the same ions as in today's seas. Of these, Ca^{2+} and H^+ (1) have the highest field strength and therefore the highest energies of binding to macromolecules, (2) have specificities of binding that vary with the structure of the macromolecule, and (3) could have been regulated provided appropriate transport systems were present in the primitive membrane.

It is not without significance that the intracellular thermodynamic activities of the proton and of Ca^{2+} ($\sim 10^{-7}$ M for both) are not only of the same order of magnitude, but are both much less than that of other cations, including not only Na^+ and K^+, but also Mg^{2+}.[14] The fact that both H^+ and Ca^{2+} bind tightly to macromolecules would require that their thermodynamic activity be kept low if changes in such activity were to play a role in regulation of the conformational state, and therefore function, of macromolecules and macromolecular complexes. The fact that the specificity of these two cations overlaps paves the way for their interaction, and therefore for the immediate development of some complexity of regulation in the early cells.

B. Unique Physical Properties of the Proton

The proton has certain features unique among all ions. The fact that in an aqueous environment its mobility exceeds that of any other ion by an order of magnitude makes it ideal not only for rapid signaling, but also for producing a more pervasive, environmental, change in the cell interior than possible with molecular signals.[51] Moreover, by nature of its small (unhydrated) size, the proton may participate in quantum mechanical tunneling whereas the probablility for this with other ions or with molecules is vanishingly small.

C. The Proton and Energy-Transducing Systems

The proton plays a key role in cellular energy transduction. Not only is it part of the reaction, and therefore a determinate of the energy released from ATP hydrolysis (see above), it plays a central role in ATP synthesis by chemiosmotic mechanisms. By using the same currency for both energy transduction and information transduction, the earliest cells would have immediately provided opportunities for close coordination of these two critical processes. As Busa and Nuccitelli[3] have pointed out, the proton is especially suited to both sense and coordinate energy-processing systems. In their words, "the apparent potential of pH_i to communicate information regarding cellular *energy balance* to enzymes and structures which may share no other common effector further emphasizes the *integrative* character of pH_i—precisely the characteristic required of a central effector of metabolism" (italics added). It is not surprising that the proton, or pH_i, can both sense and regulate the energy state of the cell since the proton is part of that energy state.

D. The Proton As a Metabolic Regulator

In their 1966 paper, Trivedi and Danforth[4] had remarked that "it would be surprising if physiological regulation of enzyme activity by pH were not a widespread and perhaps primitive method of control." Insulin is a primitive hormone, found widely throughout the animal kingdom. It would therefore be expected to affect very primitive, functions, such as energy flow, essential to survival of the cell. Such primitive functions would likewise be expected to use a primitive intracellular signal system.

A corollary of the thesis that pH_i is a normal regulatory factor in the cell is that there must be a pattern to the pH profiles of enzymes. Otherwise, changes in pH_i would produce metabolic chaos. During the course of evolution, enzymes with inappropriate responses to pH changes would of necessity have been selected against. But pH profiles *per se* are not sufficient to identify possible control points for ΔpH_i. Because of the complex nature of allosteric enzymes, pH may not necessarily affect the activity of the enzyme directly (i.e., its pH profile may be flat), but rather may affect the interaction of another allosteric effector with that enzyme. Such examples are known.[51]

In view of the fundamental differences in physical behavior of protons and discrete molecules, it would perhaps be wise to confine the term "second messenger" to molecules. This would emphasize the profound differences between these two types of signal systems. Molecular signals represent discrete signals analogous to letters, whereas the proton is a less discrete factor, producing a more pervasive, or environmental— and therefore general—effect.

VII. MODEL OF IONIC PART OF MECHANISM OF INSULIN ACTION

In view of factors discussed above, it appears that the proton, and more specifically changes in pH_i, represent the intracellular signal for insulin activation of glycolysis and probably part of the signal for the hormone's effect on several cell processes.

On the other hand, there is reason to believe that Ca^{2+} also plays a role in mediating insulin action. The inhibition of the high-affinity Ca^{2+}-ATPase by physiological concentrations of insulin in itself strongly suggests an important role for Ca^{2+}. Moreover, both the increase in $\Delta\mu_{Na}$ and the increase in pH_i would be expected to increase Na/Ca exchange (see review by Moore[40]). Finally, H^+ and Ca^{2+} may be expected to compete for binding to macromolecules and for transport sites on intracellular organelles.

The model proposed here is a holistic, as opposed to purely molecular, system driven by interaction of thermodynamic gradients. In this model, the entire cell is the transducing system for insulin with the main energy and information transduction processes occurring at the plasma membrane (see Fig. 4). The primary energizer of this system is the (Na^+, K^+) pump. Stimulation of this primary active transport system by insulin results in maximization of both the electrical and chemical (or activity) component of $\Delta\mu_{Na}$, while activation of the Na/H exchange system serves primarily as an information transduction process. The pH_i-regulating aspects of the insulin transduction system (e.g., activation of the Na/H exchange) are almost certainly involved not only in insulin action, but in that of other hormones including mitogens.[3,52] This may be the first example of a model of hormone action that successfully predicts a means to reverse the direction of the hormone effect.[12]

An analogy to this model, in which the intact cell is required for transduction of a primary (extracellular) message into an intracellular signal, would be the operation of a pipe organ. To play a tune, the system must be energized by turning on the bellows of the pipe organ, or by stimulating the (Na^+, K^+) pump. The identity of each tune is then determined by which stops are pushed, or by which Na-coupled transport systems are activated. In the cell model presented here, Na/H exchange is also involved in energization of intracellular processes including levels of ATP and energy available from the hydrolysis of ATP.[42] Although these processes probably operate in the action of any hormone or mitogen that requires the cell to enter an energized state, activation of other specific transport systems provides the opportunity

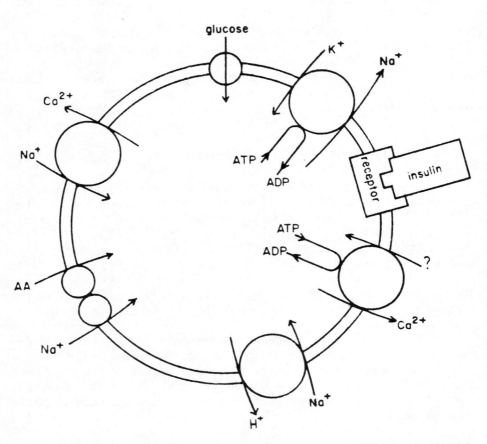

Figure 4. Transport systems affected by insulin include stimulation of active Na^+, K^+ transport (the (Na^+, K^+) pump) and inhibition of the high-affinity Ca^{2+} pump, both of which are primary active transport systems obtaining their energy from the hydrolysis of ATP. Insulin also activates the Na/H exchange system. Of these effects, the stimulation of the (Na^+, K^+) pump may be mediated by membrane phase changes. The increased glucose permeability is probably due to translocation of transport sites into the plasma membrane. Insulin also stimulates Na^+-coupled amino acid transport (cotransport), where most if not all of the energy for concentrating these amino acids in the cells is obtained from $\Delta\mu_{Na}$. This is probably due to an increased number of these transport sites, resulting from stimulation of protein synthesis. It is also likely that the increase in $\Delta\mu_{Na}$, secondary to insulin stimulation of the (Na^+, K^+) pump may explain part of the effect of this hormone on type A amino acid transport. It would also be expected that the increase in $\Delta\mu_{Na}$ might increase Na/Ca exchange, especially since the stoichiometry of this process is 4 Na^+/1 Ca^{2+},[61] making the process very sensitive to changes in membrane potential such as that produced by insulin. The coordinated stimulation of the (Na^+, K^+) pump and Na/H exchange has been called part of the ionic insulin transduction system. In this model, a change in intracellular pH represents a generalized intracellular signal for insulin action (see text). Other factors, including Ca^{2+} and a molecular factor (probably a peptide) are probably involved as well in the signal system for insulin.

for individualization of the message. In the case of insulin, the other transport systems very likely include the high-affinity Ca^{2+}-ATPase, Na/Ca exchange, and possibly Na/Mg exchange. Moreover, the increase in $\Delta\mu_{Na}$ must play some role in stimulation of amino acid uptake by this hormone, since uptake of type A amino acids is directly coupled to $\Delta\mu_{Na}$ and uptake of other amino acids is coupled to the distribution of type A amino acids.

This model does not imply that other mechanisms such as a peptide mediator or direct intracellular action[53] do not play a role in insulin action (see Section IV.C). However, the ionic aspects of insulin action have the advantage of accounting for previously unexplained observations and have the unique virtue of being detectable in humans by noninvasive means, making them operationally testable in the clinical setting (see Section VIII).

If we were to assume that H^+ is perhaps the first intracellular signal, we might expect that during evolution molecular second messengers would have been added in order to enable insulin to assume the ability to influence more specific regulatory processes as they evolved.

VIII. CLINICAL IMPLICATIONS

A. Implications for Diabetes Mellitus

The model for mediation of the ionic effects of insulin—the insulin transduction system—predicts that in hypoinsulinemic states such as diabetes, not only will Na^+_i be elevated, but that pH_i will be decreased with a consequent decrease in production of ATP. Since the (Na^+,K^+) pump contributes to the plasma membrane potential, V_m, the model also implies that the increase in Na^+_i would be associated with a decrease in the magnitude of V_m.

The predicted increase in Na^+_i has been confirmed by the finding that in rats made hypoinsulinemic by SZ diabetes[20,54] or by fasting,[20] the level of Na^+_i in the soleus muscles of these animals increases by about 30%. In the diabetic rats, the elevation of Na^+_i correlated at least as well with the diabetic state as did the elevation of plasma glucose.[55] In the hypoinsulinemic rats, either starved or diabetic, ATP_i decreased after Na_i increased.[20] The predicted decrease in the magnitude of V_m has been confirmed by Grossie[55] (see Section 2.3 and Table I) in skeletal muscle of rats made diabetic with alloxan. Preliminary evidence[22] supports the predicted decrease in pH_i in nonacidotic SZ diabetic rats.

All four of these parameters—ATP_i, pH_i, Na^+_i, and V_m—play important roles in cell regulation. Both ATP_i and pH_i play roles not only in energy metabolism, but as allosteric effectors of several enzymes. The increase in Na^+_i, by decreasing the Na^+ driving force, $\Delta\mu$, would tend to decrease both Na/H and Na/Ca exchange. Since, unlike Na/H, Na/Ca exchange is not electrically neutral, this Ca^{2+} extruding system is very sensitive to V_m as well as to elevation of Na^+_i. Therefore, it is almost certain that Na/Ca exchange would be decreased in the hypoinsulinemic states described above. Moreover, the decrease in V_m would be expected to increase Ca^{2+} leaking into cells through voltage-sensitive Ca^{2+} channels. Elevated levels of intracellular Na^+ of arteriolar smooth muscle cells have been implicated in essential hypertension, since a decrease in Na/Ca exchange would be expected to elevate intracellular Ca^{2+} and thereby increase the contractile tone of these muscle cells (see Fregly and Kare[56]). Thus, the increase in Na^+_i and the decrease in V_m observed in hypoinsulinemia could provide an explanation for the observation that diabetics have a higher incidence of hypertension than do nondiabetics.

Since an elevation of Na^+_i will decrease the force driving Na^+ into the cell, application of the Hodgkin–Huxley equations leads to the explicit quantitative prediction that with elevation of Na^+_i, such as reported in diabetes by Moore and co-workers,[21,54] the conduction velocity for both nerve and muscle would be decreased.[57] Elevation of Na^+_i in peripheral nerve or diabetic animals has been observed using voltage-clamp techniques.[58] On the basis of their report that (Na^+, K^+) ATPase activity is decreased in sciatic nerves of rats made diabetic with streptozotocin, Das and co-workers[59] have already suggested this explanation for the observation that in diabetic patients, peripheral nerve conduction velocities are decreased.

B. Implications of Insulin Effect on Intracellular pH for Therapy of Diabetes

It has long been known that untreated juvenile diabetics become acidotic. This acidosis, reflected by a decrease in blood pH, is attributed to increased production of inorganic acids, i.e., ketoacids. However, while this undoubtedly is true, the finding that insulin increases intracellular pH by activation of an Na/H exchange system in the plasma membrane predicted that in hypoinsulinemic states, such as diabetes, there should be an initial acidosis; this is attributable to decreased activity of this biophysical transport mechanism, which may occur before metabolic acidosis becomes established.

Even in the state of mild diabetes, where metabolic acidosis is not in evidence, pH_i in rat tissues may be decreased by at least 0.15 units. Moreover, in rats made diabetic under identical conditions, intracellular ATP is decreased by about 24%.[21] This preliminary verification, plus confirmation that Na^+_i is increased, suggests a fundamental reconsideration of the pathophysiology of diabetes. Moreover, the fact that this increase in intracellular Na^+ could be correlated with the diabetic state as well as could plasma glucose[54] suggests that parameters within the cell may reflect the pathophysiological state of a diabetic at least as well as do extracellular levels of glucose. It is not an unreasonable hypothesis that if intracellular parameters, such as levels of Na^+_i, ATP_i, and pH (pH_i), can be normalized in the diabetic, the pathophysiology of this disease might be decreased.

Until recently, this concept would have been of academic interest. However, $[^{31}P]$-NMR has been used for noninvasive evaluation of the effect of insulin on pH_i in frog skeletal muscle.[13] It is now possible[60] to follow pH_i and ATP_i noninvasively in patients, using wide-bore FT-NMR. This line of reasoning suggests the urgency of using FT-NMR to measure pH_i and other intracellular parameters to establish whether they provide a more rational, and more successful, basis for determining the dosage of insulin in diabetes therapy than does the present exclusive reliance on plasma glucose. This approach would use cellular parameters to determine insulin dosage; the diet would then be adjusted accordingly in order to normalize plasma glucose.

IX. REFERENCES

1. Thieden, H.I.D., Henson, S.E., and Dich, J., 1961, *Acta Chem. Scand.* **26:**2771–2776.
2. Davies, D.D., 1973, in Society of Experimental Biology. *Symposia Vol. 27: Rate Control of Biological Processes* (D.D. Davies, ed.), pp. 513–529, Cambridge University Press, Cambridge.
3. Busa, W.B., and Nuccitelli, R., 1984, *Am J. Physiol.* **246:**R409–R438.
4. Trivedi, B., and Danforth, W.H., 1966, *J. Biol. Chem.* **241:**4110–4114.
5. Nuccitelli, R., and Heiple, J.M., 1982, in *Intracellular pH: Its Measurement, Regulation and Utilization in Cellular Functions* (R. Nuccitelli and D.W. Deamer, eds.), pp. 161–169, Alan R. Liss, New York.
6. Manchester, K.L., 1970, *Hormones* **1:**342–351.
7. Moore, R.D., 1973, *J. Physiol. (Lond.)* **232:**23–45.
8. Moore, R.D., 1977, *Biophys. J.* **17:**259a.
9. Moore, R.D., 1977, *Biochem. Biophys. Res. Commun.* **91:**900–904.
10. Moore, R.D., Fidelman, M.L., and Seeholzer, S.H., 1979, *Biochem. Biophys. Res. Commun.* **91:**905–910.
11. Moore, R.D., 1981, *Biophys. J.* **33:**203–210.
12. Fidelman, M.L., Seeholzer, S.H., Walsh, K.B., and Moore, R.D., 1982, *Am. J. Physiol.* **242:**c87–c93.
13. Moore, R.D., and Gupta, R.K., 1981, *Int. J. Quantum Chem. Quantum Biol. Symp.* **7:**83–92.
14. Gupta, R.K., and Moore, R.D., 1980, *J. Biol. Chem.* **255:**3987–3993.
15. Putnam, R.W., and Roos, A., 1983, *The Physiologist,* **26:**70.
16. Morrill, G., Kostellow, A., Weinstein, S.P., and Gupta, R.J., 1983, *Fed. Proc.* **42:**1791.
17. Bailey, I.A., Radda, G.K., Seymour, A.L., and Williams, S.R., 1983, *Biochem. Biophys. Res. Commun.* **720:**17–27.
18. Meyer, R.A., Kushmerick, M.J., Dillon, P.F., and Brown, T.R., 1983, *Fed. Proc.* **42:**1248.
19. Podo, R., Carpinelli, G., and D'Agnolo, G., 1982, *Tenth International Conference on Magnetic Resonance in Biological Systems,* Stanford, Aug. 29–Sept. 3, 1982, p. 14.
20. Moolenaar, W.H., Tsien, R.Y., van der Saag, P.T., and de Laat, S.W., 1983, *Nature* **304:**645–648.
21. Moore, R.D., Munford, J.W., and Pillsworth, J.R., 1983, *J. Physiol. (Lond.)* **338:**277–294.
22. Brunder, D.G., Oleynek, J.J., and Moore, R.D., 1983, *J. Gen. Physiol.* **82:**15a.
23. Clancy, R.L., Gonzalez, N.C., Shaban, M., and Cassmeyer, V., 1983, *Cell Physiol. II:* A-107.
24. Roos, A., and Boron, W.F., 1981, *Physiol. Rev.* **61:**296–433.
25. Kostyuk, P.G., and Sorokina, Z.A., 1961, in *Membrane Transport Metabolism* (A. Kleinzeller and A. Kotyk, eds.), pp. 193–203, Academic Press, New York.
26. Gavryck, W.A., Moore, R.D., and Thompson, R.C., 1975, *J. Physiol. (Lond.)* **252:**43–58.
27. Clausen, T., and Kohn, P.G., 1977, *J. Physiol. (Lond.)* **265:**19–42.
28. Fehlmann, M., and Freychet, P., 1981, *J. Biol. Chem.* **256:**7449–7453.
29. Resh, M.D., Nemenoff, R.A., and Guidotti, G., 1980, *J. Biol. Chem.* **255:**10938–10945.
30. Funder, J., Tosteson, D.C., and Wieth, J.O., 1978, *J. Gen. Physiol.* **71:**721–746.
31. Benos, D.J., 1982, *Am. J. Physiol.* **242:**C131–C145.
32. Mullins, L.J., and Noda, K., 1963, *J. Gen. Physiol.* **47:**117–132.
33. Rolleston, F.S., and Newsholme, E.A., 1967, *Biochem. J.* **104:**524–533.
34. Williamson, J.R., 1970, *Adv. Exp. Med. Biol.* **6:**117–136.
35. Karpatkin, S., Helmreich, E., and Cori, C.F., 1964, *J. Biol. Chem.* **239:**3139–3145.
36. Dunaway, G.A., and Weber, G., 1974, *Arch. Biochem. Biophys.* **162:**629–637.
37. Dunaway, G.A., Leung, G.L.-Y., Thrasher, J.R., and Cooper, M.D., 1978, *J. Biol. Chem.* **253:**7460–7463.
38. Jarett, L., and Seals, J.R., 1979, *Science* **206:**1407–1408.
39. Larner, J., Galasko, G., Cheng, K., DePaoli-Roach, A.A., Huang, L., Daggy, P., and Kellogg, J., 1979, *Science* **206:**1408–1410.
40. Moore, R.D., 1983, *Biochim. Biophys. Acta* **737:**1–49.
41. Sonne, O., Gliemann, J., and Linde, S., 1981, *J. Biol. Chem.* **256:**6250–6254.

42. Winkler, M.M., and Steinhardt, R.A., 1981, *Dev. Biol.* **84**:*432–439*.
43. Winkler, M.M., 1982, in *Intracellular pH: Its Measurement, Regulation, and Utilization in Cellular Functions* (R. Nuccitelli and D.W. Deamer, eds.), pp. 325–340, Alan R. Liss, New York.
44. Straus, D.S., 1981, *Life Sci.* **29**:2131–2139.
45. Pouyssegur, J., Chambard, J.C., and Paris, S., 1982, in *Symposium on Ions, Cell Proliferation and Cancer* (A.L. Boynton, W.L. McKeehan, and J.F. Whitfield, eds.), pp. 205–218, Academic Press, New York.
46. Le Cam, A., Auberger, P., and Sampson, M., 1982, *Biochem. Biophys. Res. Commun.* **106**:1062–1070.
47. Zierler, K.L., and Rogus, E.M., 1980, *Am. J. Physiol.* **239**:E21.
48. Rasmussen, H., 1981, *Calcium and cAMP as Synarchic Messengers,* Wiley, New York.
49. Skou, J.C., 1982, *Ann. NY Acad. Sci.* **II**:169–184.
50. Bryszewska, M., and Leyka, W., 1983, *Diabetologia* **24**:311–313.
51. Moore, R.D., Fidelman, M.L., Hansen, J.C., and Otis, J.N., 1982, in *Intracellular pH: Its Measurement, Regulation, and Utilization in Cellular Functions* (R. Nuccitelli and D.W. Deamer, eds.), pp. 385–416, Alan R. Liss, New York.
52. Moore, R.D., 1981, *Int. J. Quantum Chem. Quantum Biol. Symp.* **8**:365–371.
53. Goldfine, I.D., 1981, *Biochim. Biophys. Acta* **650**:53–67.
54. Moore, R.D., Munford, J.W., and Popolizio, M., 1979, *FEBS Lett.* **106**:375–378.
55. Grossie, J., 1982, *Diabetes* **31**:194–202.
56. Fregly, M.J., and Kare, M.R., (eds.), 1982, *The Role of Salt in Cardiovascular Hypertension,* Academic Press, New York.
57. Hodgkin, A.L., and Huxley, A.F., 1952, *J. Physiol. (Lond.)* **117**:500–544.
58. Brismar, T., and Anders, A.F.S., 1981, *Acta Physiol. Scand.* **113**:499–506.
59. Das, P.K., Bray, G.M., Aguayo, A.J., and Rasminsky, M., 1976, *Exp. Neurol.* **53**:285–288.
60. Havrankova, J., Roth, J., and Brownstein, M., 1978, *Nature* **272**:827–829.
61. Mullins, L.J., 1977, *J. Gen. Physiol.* **70**:681–695.

10

Multiple Intracellular Peptide Mediators of Insulin Action

K. Cheng, M. Thompson, C. Schwartz, C. Malchoff, S. Tamura, J. Craig, E. Locher, and J. Larner

I. INTRODUCTION

This chapter discusses a cluster of insulin mediator peptides that duplicate *in vitro* the following insulin actions in the cell: (1) inhibit cyclic adenosine monophosphate (cAMP)-dependent protein kinase, (2) stimulate pyruvate dehydrogenase (PDH) phosphatase, (3) inhibit hormone-stimulated adenylate cyclase, and (4) stimulate glycogen synthase phosphoprotein phosphatase. Also discussed is a post-insulin receptor defect in human leprechaun fibroblast cells that sheds light on the mechanism of action of insulin. Finally, the possible relationship between insulin receptor phosphorylation and mediator formation is considered. It is clear that several mediators of varying molecular size is formed from proteins in the cell membrane following insulin binding to the receptor and that these mediators carry out some of the actions of insulin constituting the pleiotypic program. A model of mediator formation similar to the pro-opiomelanocortin model of peptide hormone formation is proposed. A recent review of this topic has appeared.[1]

II. cAMP-Dependent Protein Kinase Inhibition

This was the first mediator to be described.[2] It has been purified 1000–2000-fold from rat and rabbit muscle by anion-exchange high-performance liquid chroma-

K. Cheng, M. Thompson, C. Schwartz, S. Tamura, and J. Larner • Department of Pharmacology, University of Virginia School of Medicine, Charlottesville, Virginia. J. Craig and E. Locher • Department of Internal Medicine, University of Virginia School of Medicine, Charlottesville, Virginia. C. Malchoff • Departments of Pharmacology and Internal Medicine, University of Virginia School of Medicine, Charlottesville, Virginia.

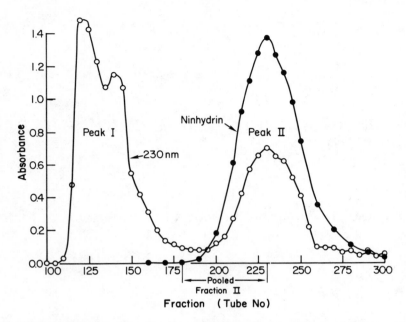

Figure 1. Sephadex G-25 chromatogram of insulin-treated rat muscle extract. Dry muscle extract powder from 80 g muscle was dissolved in 10–15 ml 50 mM formic acid containing 0.1 mM EDTA, 0.1 mM cysteine, and 10 mM NaCl and applied onto a Sephadex G-25 column (5.0 × 85 cm). The column was eluted with the same buffer with a flow rate of 1 sec/drop and 95 drops/fraction.

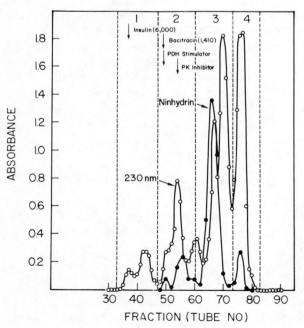

Figure 2. Sephadex G-15 chromatogram of pooled fraction II of Sephadex G-25. Lyophilized fraction II from Sephadex G-25 was redissolved in 1 ml 50 mM formic acid and applied onto a Sephadex G-15 column (2.5 × 90 cm). The column was eluted with 50 mM formic acid with a flow rate of 5 sec/drop and 95 drops/fraction.

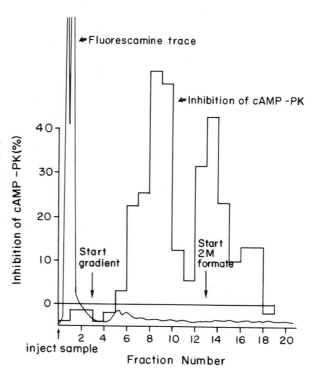

Figure 3. DEAE-high performance liquid chromatography of Sephadex G-15 fraction 2. Sephadex G-15 fraction 2 sample was applied to a DEAE-column (13 × 0.5 cm, particle size 10 μm) in 5 mM pyridine/acetic acid, pH 6. The column was eluted at a flow rate of 0.5 ml/min with 5 mM pyridine/acetic acid, pH 6 for 30 min, then a linear gradient from 5 mM to 2 M pyridine/acetic acid, pH 6 over 100 min, followed by 2 M acetic acid for 40 min. The fractions were continuously monitored for fluorescamine-reactive material; samples were tested for inhibitory effect on cyclic adenosine monophosphate-dependent protein kinase (cAMP-PK).

tography (HPLC) followng Sephadex chromatography. The mediator has been identified and followed by its presence in greater amount or activity in tissues exposed to insulin compared with controls. This mediator is present in peak II of the Sephadex G-25 column (Fig. 1) and in peak II of the subsequent Sephadex G-15 column, where it is positioned slightly behind (i.e., it has a lower molecular weight) the pyruvate dehydrogenase stimulatory mediator (Fig. 2). The highly purified mediator which contains peptide has low reactivity with fluorescamine after purification on HPLC (Fig. 3). On acid hydrolysis, the following provisional amino acid composition is obtained: Asp or Asa (1), Ser (1), Glu or Gln (3), Gly (5), Ala (1), Lys (0.5), Leu (0.5). It is a hydrophilic acidic peptide with an isoelectric point of about 2–3. The highly purified mediator displays no activity on either mitochondrial pyruvate dehydrogenase phosphatase or glycogen synthase phosphatase.

III. PDH MEDIATOR

This mediator, which acts on mitochondrial pyruvate dehydrogenase phosphatase is isolated using the same steps outlined for the cAMP-dependent protein kinase mediator (see Figs. 1 and 2). As shown in Fig. 4, the only fraction that stiumulates PDH in a dose-dependent manner and shows the control insulin difference is fraction 2 from the Sephadex G-15 column. Fraction 1 for the insulin has a small stimulating

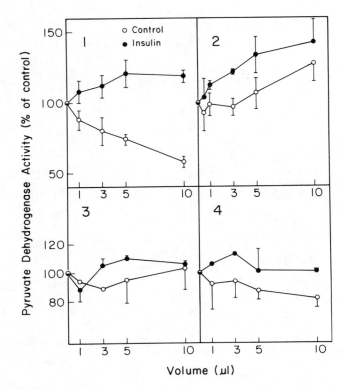

Figure 4. Effect of Sephadex G-15 fractions from both control (○) and insulin-treated (●) muscle on adipocyte mitochondrial pyruvate dehydrogenase. Data represent means ± SE from four experiments performed on different days. Where no error is shown, the standard error is less than the size of the symbol.

action, while the control is inhibitory and possibly contains the inhibitory mediator described by Saltiel and Cuatrecasas.[3] However, for the HPLC step, a strong (quaternary ammonium) anion-exchange resin is used instead of the weaker amino anion-exchange column. As shown in Fig. 5, the stimulatory mediator is adsorbed on the column and is thus separated from the bulk of the fluorescamine reactivity. It is then eluted with increased salt concentration. As shown previously with the cAMP-dependent protein kinase inhibitory mediator, the PDH mediator has very low reactivity with fluorescamine after purification on HPLC. On acid hydrolysis, the following provisional amino acid composition is obtained: Asp or Asn (1), Ser (1), Glu or Gln (2), Gly (2), Phe (2), Lys (4), Leu (2). Like the previous mediator, it is hydrophilic, acidic, and contains peptide, with an isoelectric point of about 4–5. On treatment of the column with dilute (50 mM) formic acid, a PDH inhibitory mediator is eluted, first described by Saltiel *et al.*[3] Since the PDH phosphatase stimulatory mediator has no activity on cAMP-dependent protein kinase and a different amino acid composition, this suggests two distinct molecules.

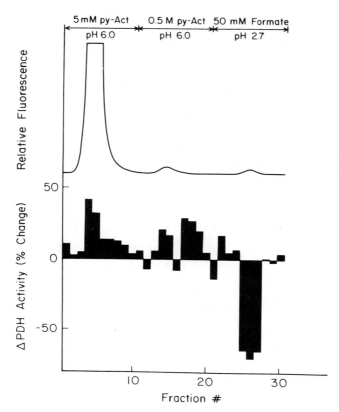

Figure 5. Chromatography of pyruvate dehydrogenase phosphatase mediators from a high-performance liquid chromatography (HPLC) anion-exchange column (quaternary ammonium). Lyophilized pyruvate dehydrogenase phosphatase mediator from Sephadex G-15 column was redissolved in 15 ml of 5 mM pyridine-acetic acid (pH 6.0) and applied onto a 4.6 × 160 mm anion-exchange column with a flow rate of 0.5 ml/min and collected 3 ml per fraction. The column was then eluted in a stepwise manner with 0.5 M pyridine–acetic acid (pH 6.0), followed by 50 mM formic acid (pH 2.7). The fractions were continuously monitored for fluorescamine-reactive material and tested for their effects on pyruvate dehydrogenase phosphatase.

IV. ADENYLATE CYCLASE INHIBITORY MEDIATOR

When heat-denatured charcoal-treated rat muscle or liver extracts are chromatographed over Sephadex G-25 columns and scanned for this mediator, it is observed in a higher-molecular-weight region than the preceding mediators, i.e., 2000–5000-M_r range (Fig. 6a,b). This mediator acts to inhibit hormone-stimulated or GTP γS-stimulated cyclase in liver, fat cell, and mouse S-49 lymphoma membranes. Experiments are in progress to purify this mediator further and study its mechanism of action. The mediator inhibits both GTP and glucagon or GTP γS-stimulated cyclase. Inter-

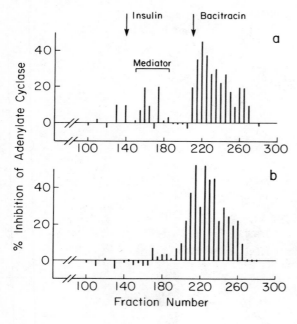

Figure 6. The crude extracts from the livers of insulin-treated (a) and control (b) rats were eluted from a Sephadex G-25 (5 × 90 cm) column in 5-ml fractions. The indicated fractions were assayed for their ability to inhibit the guanosine triphosphate-isoproterenol-stimulated adenylate cyclase of S-49 mouse lymphoma cell plasma membranes.

ingly, the maximum inhibition of the lymphoma cell cyclase is about 40% (Fig. 7). We suspect that mediator acts at either the level of the G/F subunit or the cyclase itself.

V. GLYCOGEN SYNTHASE PHOSPHATASE MEDIATOR

This mediator has been identified in the Sephadex G-25 chromatogram by the enhanced activity observed in extracts of muscle pretreated with insulin as compared

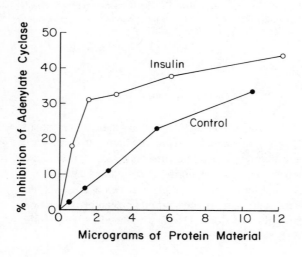

Figure 7. Mediator fractions from the Sephadex G-25 column were combined, desalted, and assayed for their ability to inhibit the guanosine triphosphate-isoproterenol-stimulated adenylate cyclase of S-49 mouse lymphoma cell plasma membranes. (○) mediator from insulin-treated animals; (●) mediator from control animals.

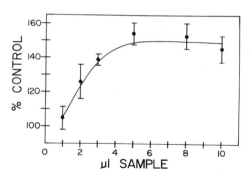

Figure 8. Dose-response curve of insulin mediator on type 1 protein phosphatase. Increasing volumes of pooled, lyophilized saline, and insulin-treated Sephadex G-25 chromatographed extracts were assayed with type 1 protein phosphatase. Each point is the mean of five determinations with the indicated standard deviations.

with control (Fig. 8). As with the adenylate cyclase mediator, the glycogen synthase phosphatase mediator appears to have a high-molecular-weight range (2000–5000 M_r), as compared with the first three mediators described. It is now assayed by measuring $^{32}P_i$ release from labeled phosphorylase *a*, using a highly purified preparation of muscle phosphoprotein phosphatase 1. A dose–response relationship between amount of mediator added and stimulation of activity is observed, thereby validating the assay. Again further purification is under way.

VI. GENERAL PROPERTIES

All putative insulin mediators are relatively low-molecular-weight heat-stable, non-charcoal adsorbable peptides or peptidelike compounds (conjugated peptides). All are more acid stable and alkali labile, and some are degraded by proteases such as trypsin or by carbohydrases,[1] suggesting peptide and perhaps glycopeptide structures.

There is now evidence for the presence of multiple mediators. Since the amino acid composition of two (see present data) appear to be unrelated to each other, and since there is no biological cross-reactivity, a mechanism of formation analogous to that of the splitting of pro-opiomelanocortin to a number of hormones seems more likely than a mechanism in which one mediator is converted to a second (i.e., angiotensin I to angiotensin II). In this model, the higher-molecular-weight mediators, when assayed biologically, would be expected to exhibit multiple mediator activities, while the low-molecular-weight mediators, when tested biologically, would display one or a smaller number of mediator activities (Fig. 12). The advantage of multiple mediators in explaining the pleiotypic pattern of insulin action, which is complex in time and subcellular location (i.e., in multiple intracellular sites), is obvious.

VII. PROTEOLYSIS IN LEPRECHAUN FIBROBLASTS: A GENETIC APPROACH TO THE MECHANISM OF INSULIN ACTION

Skin fibroblasts from a patient with the familial congenital condition of leprechaunism and severe insulin resistance are unresponsive to insulin in terms of glycogen synthase activatability (Fig. 9). The cells also have diminished insulin-binding capacity,

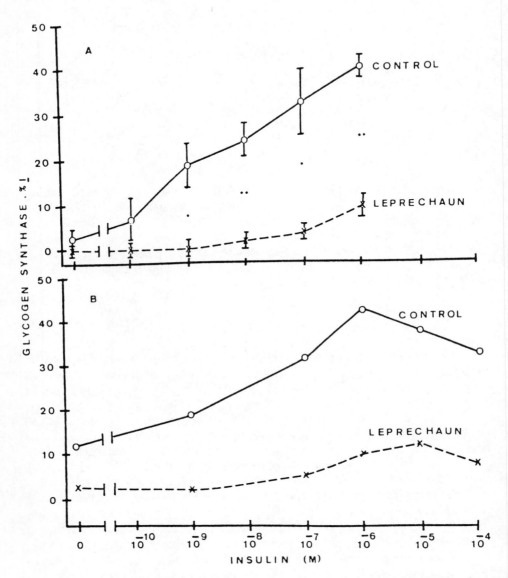

Figure 9. Effect of insulin concentration on activity of glycogen synthase in cultured skin fibroblasts from a patient with leprechaunism and from a control subject. Monolayers were incubated for 24 hr in medium that was free of serum and glucose and then exposed to the indicated concentrations of insulin for 2 hr. (A) Data expressed as the mean ± SEM of results from three experiments in which fibroblasts from the two subjects were studied in parallel. (*) the differences between leprechaunism and control values were significant at a $p < 0.01$ level for insulin concentrations of 10^{-9} and 10^{-7} M. (**) Differences significant at a $p < 0.005$ level for insulin concentrations of 10^{-8} and 10^{-6} M. (B) Effect of higher concentrations of insulin. Fibroblasts from both subjects were studied in the same experiment. Each point represents the mean of the results from duplicate culture dishes.

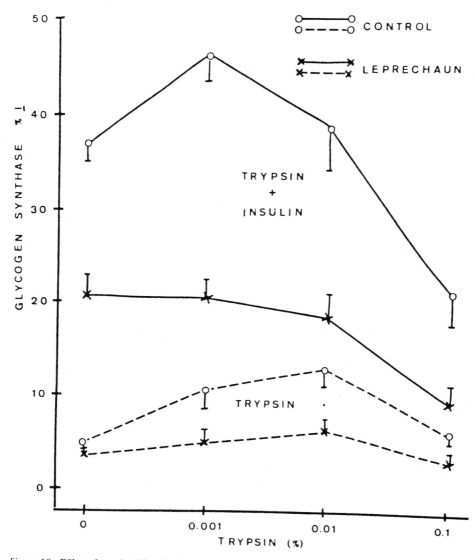

Figure 10. Effect of trypsin with and without insulin on glycogen synthase activity in cultured skin fibroblasts from a patient with leprechaunism (X) and from a control patient (O). Monolayers of the cells were exposed to indicated concentrations of trypsin in Dulbecco's phosphate-buffered saline (DPBS) for one min. After removal of trypsin-containing medium and rinsing of monolayers with soybean trypsin inhibitor at twice the concentration of the preceding trypsin, cells were incubated in DPBS supplemented with glutamine (2 mM), Eagle's basal medium vitamins, Eagle's basal medium amino acids, and bovine serum albumin (1 mg/ml) without glucose or calf serum and with (——) or without (– – – –) insulin (10^{-6} M) for 2 hr at 37°C in air. Cell extracts were then prepared and assayed for glycogen synthase activity. Data are presented as the mean ± SEM for results from four experiments in which cells from the two subjects were studied in parallel; vertical lines represent SEM. Differences between control and leprechaun fibroblasts were significant ($p < 0.05$), with trypsin alone at a concentration of 0.01% and with insulin plus trypsin at concentrations of 0, 0.001%, and 0.01%.

but this is not severe enough to explain the post-receptor defect. Thus, the ratio of diminished insulin binding to decreased activatability of glycogen synthase is markedly reduced in the leprechaun cells as compared with the control cells. An analogous situation holds for another insulin-like agent, multiplication stimulating activity (MSA) (data not shown). Since trypsin was shown previously to activate glycogen synthase in adipocytes and to produce mediator proteolytically, control and leprechaun fibroblasts were challenged with trypsin, and their ability to respond by glycogen synthase activation was studied. As shown in Fig. 10, leprechaun fibroblasts were unresponsive to trypsin as well as to insulin or MSA. This finding suggests that mediator formation via proteolysis is depressed in the leprechaun cells. Thus, the substrate from which mediator is produced may be unresponsive to trypsin. It does not rule out an abnormality in the endogenous protease as well. The substrate may be unavailable for chemical reasons (i.e., altered primary structure) or for steric reasons (i.e., altered other components in the membrane, making it unavailable for cleavage). These data again strongly suggest an important role for proteolysis in the mechanism of insulin action.

VIII. RECEPTOR PHOSPHORYLATION AND MEDIATOR FORMATION

Recent reports have demonstrated that the addition of insulin to cells, to detergent extracts of membranes, and to purified receptors enhances tyrosine phosphorylation on the β subunit of the insulin receptor. In experiments with intact cells, serine and threonine phosphorylation on the β subunit is also increased. What is the possible relationship of this hormonally enhanced receptor phosphorylation and mediator formation?

Several possibilities can be considered. First, the two processes may be unrelated to each other, i.e., phosphorylation of the receptor may be a signal for one function, such as receptor internalization, hence separate from mediator formation. Second, the two processes may be related to each other in one of several possible ways. Phosphorylation may control receptor aggregation or clustering, a process that may be essential to receptor activation. An analogous control is known to occur with certain enzymes the activity and aggregation state of which are well known to be regulated by covalent phosphorylation. For example, dephosphorylated glycogen synthase (active or I form) is known to aggregate much more readily than the phosphorylated form of the enzyme (D, or inactive)[4] Phosphorylation may control proteolytic mediator formation or the inverse may be true, i.e., proteolysis may control phosphorylation. Several examples of proteolytically activated protein kinases have been described. Probably the best characterized is the proteolytically activated C kinase described by Nishizuka and co-workers.[5] Other examples include two kinases activated by limited proteolysis as described by Traugh and co-workers.[6] Recently, one of these has been shown to be activated by insulin and responsible for ribosomal protein S6 phosphorylation. Adenosine triphosphate (ATP)-dependent proteolysis has also been recognized for a number of years. The mechanism of ligation of a peptide in an isomide linkage as a signal for proteolysis was recently discovered.[7] Thus, there are several ways in which proteolysis may be closely intertwined with phosphorylation.

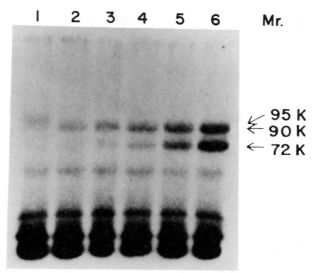

Figure 11. Time course of trypsin-dependent phosphorylation of insulin receptor. The solubilized insulin receptor fraction was incubation with trypsin (0.01%) for 0 min (lane 1), 3 min (lane 2), 6 min (lane 3), 10 min (lane 4), 20 min (lane 5), or 30 min (lane 6) at 4°C and then incubated with [γ-^{32}P]-ATP and MnCl$_2$. Phosphorylated insulin receptors were immunoprecipitated with anti-insulin receptor antibodies and electrophoresed on 6% SDS-polyacrylamide gel (from ref. 8, with permission).

In some very recent experiments, S. Tamura showed that trypsin will also enhance tyrosine phosphorylation on the insulin receptor in partially purified and in highly purified receptor preparations. As shown in Fig. 11, trypsin in low concentrations enhances phosphorylation in 90,000- and 72,000-M_r bands on tryosine, which are immunoprecipitated by anti-insulin receptor antibody. These results suggest that the

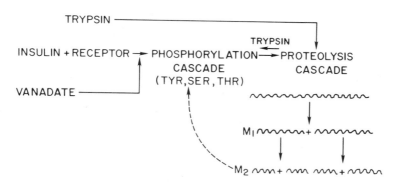

Figure 12. Model of an interrelated phosphorylation and proteolysis initiated by insulin binding to its receptor. The phosphorylation cascade precedes and triggers the proteolysis cascade, creatng the formation of two larger mediators (M1) and 4 smaller mediators (M2). The action of two insulinlike agents, trypsin and vanadate, is also shown. Trypsin directly activates the proteolysis cascade, which then activates phosphorylation as well as mediator formation. Vanadate activates phosphorylation directly.

stimulation by trypsin of tyrosine phosphorylation may be related to the insulin-like metabolic actions of trypsin.

A model for these events is shown in Fig. 12. It suggests that insulin binding to the receptor initiates a phosphorylation cascade, which then triggers a proteolytic cascade, producing multiple mediators from a single substrate in a manner similar to the splitting of pro-opiomelanocortin into a number of hormones. The results with trypsin suggest that the proteolysis events may control or regulate the phosphorylation of the receptor as well.

IX. SUMMARY

This chapter details the present status of the purification and characterization of four mediators of insulin action in cytoplasm, mitochondria, and cell membrane. Post-receptor insulin resistance in fibroblasts from a patient with leprechaunism as well as the possible interrelationship of receptor phosphorylation to proteolytic mediator formation are documented and discussed. We conclude that both receptor phosphorylation and limited proteolysis are involved in the formation of multiple peptide-containing mediators.

X. REFERENCES

1. Larner, J., *J. Cyclic Nucleotide Res.*, 1982, **8**:289–296.
2. Larner, J., Huang, L.C., Brooker, G., Murad, F., and Miller, T.B., 1974, *Fed. Proc.* **33**:61.
3. Saltiel, A., Siegel, M.I., Jacobs, S., and Cuatrecasas, P., 1982, *Proc. Natl. Acad. Sci. USA* **79**:3513–3517.
4. Smith, C.H., and Larner, J., 1972, *Biochem. Biophys. Acta* **264**:224–228.
5. Takai, Y., Kishimoto, A., Iwasa, Y., Kawahara, Y., Mori, T., and Nishizuka, Y., 1979, *J. Biol. Chem.* **254**:3692–3695.
6. Perisic, O., and Traugh, J.A., 1983, *J. Biol. Chem.* **258**:9589–9592.
7. Hershko, A., and Ciechanover, A., 1982, *Annu. Rev. Biochem.* **51**:335–364.
8. Craig, J.W., Larner, J., Locker, E.F., and Elders, M.J., 1984, *Mol. and Cell. Biochem.*, in press.

11

Intracellular Mediators of Insulin Action

Leonard Jarett, Frederick L. Kiechle, S. Lance Macaulay, Janice C. Parker, and Kathleen L. Kelly

I. INTRODUCTION

Insulin is the key anabolic hormone regulating the metabolism of carbohydrates, lipids, and proteins, especially in muscle, liver, and adipose tissue. This regulation is brought about by changes in membrane events that occur within seconds, in various intracellular processes that occur within minutes, and in nuclear functions involved in growth that occur within hours. The initial event responsible for these metabolic alterations results from the interaction of insulin with its receptor on the plasma membrane of responsive cells. Yet, the molecular events that occur subsequent to the hormone—receptor interaction and which lead to the metabolic alterations are unknown. A number of compounds, including cyclic nucleotides, calcium, a fragment of insulin, and hydrogen peroxide, have been suggested to be the second messenger or intracellular mediator of insulin action. None of these candidates has been demonstrated to mimic all the metabolic actions of insulin and to work *in vivo* as well as *in vitro*. Therefore, they are unlikely to be insulin mediators. This subject has been reviewed extensively in recent years.[1–8]

   The purpose of this chapter is to review the substantial body of data that supports the concept that the interaction of insulin with its receptor, generates from the membrane a low-molecular-weight molecule which acts as a second messenger or mediator for many of the actions of insulin. This molecule has been shown to be generated in an insulin-sensitive manner in subcellular systems, intact cells, and whole animals. It affects a variety of enzyme systems and metabolic pathways, both in subcellular

Leonard Jarett, S. Lance Macaulay, Janice C. Parker, and Kathleen L. Kelly ● Department of Pathology and Laboratory Medicine, University of Pennsylvania School of Medicine, Philadelphia, Pennsylvania. *Frederick L. Kiechle* ● Department of Clinical Pathology, William Beaumont Hospital, Royal Oak, Michigan.

systems and in intact cell systems, in a manner similar to insulin. This mediator may be part of a new family of mediator substances involved in the action of other hormones and cellular regulatory substances. Unfortunately, the chemical nature and molecular structure of this mediator have not been elucidated. This step is mandatory to establish the role played by these proposed mediator substances in the signaling process and to determine the precise mechanism by which they control enzymatic pathways. A number of reviews on various aspects of this subject have appeared during the past several years.[9-18]

II. SUBCELLULAR SYSTEMS

A. Adipocyte System

The approach taken by this laboratory to the study of insulin action has been based on the concept that the interaction of insulin with its receptor in the plasma membrane elicits the generation of a signal or mediator molecule that causes alterations in various metabolic pathways. Therefore, direct effects of insulin should be observed experimentally in subcellular preparations of plasma membranes, if there is preservation of the components and organization of the membrane involved in the series of reactions necessary for insulin to generate its intracellular signal. Such a plasma membrane preparation has been developed in this laboratory.[19] Jarett and Smith[20] first used this membrane preparation to show that concanavalin A and physiological concentrations of insulin, within seconds of addition, increased the hydrolysis of adenosine triphosphate (ATP), apparently by activation of a Mg^{2+} ATPase. Since alterations in phosphorylation are important and early events in the regulatory mechanisms involved in metabolic processes[21] and insulin is known to alter phosphorylation of adipocyte proteins,[7] the effect of insulin on phosphorylation of this subcellular membrane system from adipocytes was studied.

In the first report of a series of studies, Seals et al.[22] used [γ-^{32}P]-ATP to label the plasma membrane preparation in the presence and absence of insulin. Insulin addition to the assay mixture containing plasma membranes produced a concentration-dependent and specific inhibition of phosphorylation throughout the incubation from 30 sec to 10 min as compared with the control. This direct effect of insulin on the autophosphorylation of plasma membranes was further characterized by electrophoretic analysis of the phosphorylated membrane.[23,24] Under the conditions used in the first study, two major and three minor bands of phosphorylation were observed, with the major bands being phosphoester phosphoproteins with molecular weights of 42,000 and 120,000. Insulin rapidly decreased the phosphorylation of 120,000 and 42,000-dalton phosphoproteins, which accounted for the decrease in phosphorylation found in the earlier study. The 120,000-M_r phosphoprotein was determined to be of plasma membrane origin, and its identity is yet to be determined. The insulin-sensitive phosphoprotein of 42,000 M_r was shown to be the α subunit of pyruvate dehydrogenase (PDH) contained in the mitochondria, which contaminated the plasma membrane preparation. Insulin had no effect on the phosphorylation of the α subunit of PDH when the hormone was added directly to purified mitochondrial preparation. However, the insulin effect was restored when adipocyte plasma membranes were added to the adipocyte mitochondria. These studies suggested that insulin stimulated the generation of a signal compound from the plasma membranes that acted on the α subunit of PDH in the mitochondria, causing dephosphorylation. These *in vitro* effects of insulin on the phosphorylation of the α subunit are similar to the *in vivo* effects of insulin on this enzyme system as described by Denton and Hughes,[25] suggesting that the material generated from the plasma membrane could be a putative mediator of insulin action.

Insulin has been shown to activate PDH *in vivo* by causing dephosphorylation of the α subunit of the enzyme, subsequent to the activation of the phosphatase component of that enzyme complex.[25] Data from the subcellular system studies suggest that the effect of insulin on the phosphorylated state of the α subunit of PDH would result in activation of the enzyme system. This was demonstrated by Seals and Jarett.[26] Insulin stimulated PDH activity in a concentration-dependent manner in the subcellular mixture of plasma membranes and mitochondria from rat adipocytes, reaching a maximum stimulation at 50–100 μU/ml. In addition, it was shown that concanavalin A (Con A) and anti-insulin receptor antibody, both insulin-mimetic agents, rapidly stimulated PDH activity when added to such a subcellular mixture. None of these ligands affected PDH activity when added to mitochondria alone, demonstrating that generation of the material that caused dephosphorylation of the α subunit and activation of the enzyme was dependent on ligand–plasma membrane interaction. These studies not only further substantiated that the material generated from the plasma membrane could act as an insulin mediator, but also suggested that the material was not a piece of the insulin molecule.

Concurrent with these studies on subcellular systems, Larner et al.[27] reported isolation of a low-molecular-weight (1000–1500), acid-stable mediator generated by insulin in rabbit skeletal muscle. This group showed that the mediator stimulated glycogen synthase by activating glycogen synthase phosphoprotein phosphatase and inhibiting cyclic adenosine 5′-phosphate (cAMP)-dependent protein kinase. Jarett and Seals[28] showed that this material isolated from insulin-treated muscle stimulated PDH in adipocyte mitochondria more than did material from control muscle. This activation of PDH, which was analogous to the stimulation seen when insulin was added to a plasma membrane mitochondrial mixture, suggested that the mediator from adipocyte plasma membrane was similar, if not identical, to the muscle mediator.

The mediator generated from adipocyte plasma membranes was further characterized by two laboratories. In both cases, the mediator was obtained in a supernatant fraction after centrifugation of the incubated plasma membranes. The mediator activity in the supernatant was detected on the basis of its ability to stimulate PDH in isolated mitochondria. Seals and Czech[29] demonstrated that insulin increased the amount of mediator compared with control in the supernatant of adipocyte plasma membranes prepared in phosphate buffer. Kiechle et al.[30] showed that under certain conditions the mediator was released spontaneously from the plasma membrane without insulin treatment of the membrane. Insulin stimulation of mediator release was observed with plasma membranes prepared in phosphate buffer, but not with plasma membranes prepared in Tris buffers. The mediator could be depleted from the plasma membranes following repeated centrifugation and washes with phosphate buffer. The mediator present in the supernatant was stable at pH 7.0 and produced linear concentration-response curves. In another study, Seals and Czech[31] showed that maximal supernatant activity from insulin treatment was obtained with 100 μU/ml insulin after 1-min treatment of the membranes.

B. Other Insulin-Sensitive Systems

Table 1 summarizes the various plasma membrane preparations used to prepare the mediator by insulin treatment. The adipocyte membrane system has been discussed. Jarett and Kiechle[9] found that liver plasma membranes would spontaneously release a low-molecular-weight mediator that stimulated PDH in adipocyte and liver mitochondria. The adipocyte membrane mediator was also effective on liver mitochondrial PDH. Saltiel et al.[32] reported that insulin treatment of liver plasma membranes increased the amount of mediator generated, assessed on the basis of stimulation of PDH in adipocyte or liver mitochondria. Recently, Sakamoto et al.[33] demonstrated that insulin treatment of human placental plasma membranes produced a low-molecular-weight factor capable of activating PDH in adipocyte and liver mitochondria. The mediators from the plasma membrane preparations from adipocytes and liver have been shown to alter glycogen synthase, low K_m cAMP phosphodiesterase, adenylate cyclase, and acetyl CoA carboxylase activity. Details of these studies are presented in Section A.

A series of reports have illustrated that alteration of the physiological state of the animal, from which plasma membranes are subsequently isolated, can modify the production of mediator from the membranes by insulin. Begum et al.[34-36] have shown diet to affect the insulin-dependent generation of the mediator from adipocyte and liver plasma membranes. High-carbohydrate diet had no effect on insulin or Con A generation of the mediator from either membrane source, whereas a high-fat diet prevented insulin and Con A stimulation of mediator production by both membrane sources. It

Table 1. Isolated Plasma Membrane Preparations Reported to Produce Insulin Mediator

Membrane source	Condition of animal	Ligand causing increased mediator
Adipocyte (rat)	Normal	Insulin, anti-insulin receptor antibody, Concanavalin A
	High-carbohydrate diet	Insulin
	High-fat diet	Not responsive to ligands
Liver (rat)	Normal	Insulin, anti-insulin receptor antibody
	High-carbohydrate diet	Insulin
	High-fat diet	Not responsive to ligands
	Diabetes	Not responsive to ligands
	Diabetes treated	Insulin
	Fasting	Not responsive to ligands
	Fasting–Refed	Insulin
Placenta (human)	Normal	Insulin

is not clear whether this effect of high-fat diet was the result of decreased insulin binding to the membranes or of a true defect in the ability of the membranes to produce the mediator. Amatruda et al.[37] have shown that hepatocyte plasma membranes from diabetic or fasted animals failed to release the mediator of PDH activity after insulin stimulation. Treatment of the diabetic rat or refeeding the fasted animal restored the ability of insulin to stimulate mediator release. The binding of insulin to hepatocyte membranes was normal or increased in these altered states. These data support the concept that alterations at the plasma membrane can be responsible for or accompany the insulin resistance of liver in fasting or diabetic rats.

III. WHOLE CELL AND TISSUE STUDIES

To further substantiate the proposal that this low-molecular-weight material generated from the plasma membrane in the presence of insulin represents the mediator of insulin action, additional criteria must be met. First, the mediator should have an ubiquitous cellular or tissue distribution and the amount or activity of the material must be altered by insulin in a manner consistent with the known effects of insulin on that cell or tissue type. Second, the mediator should alter the activity of a variety of insulin-sensitive enzymes or systems in a manner analogous to insulin in intact cells. This section reviews evidence for this proposal in the various cell types studied.

Table 2 lists the various cells or tissues from which the mediator has been extracted and the manner in which insulin affects the amount or activity of the mediator. Larner et al.[27] were the first to extract the mediator from intact muscle tissue and to show that insulin treatment of rabbits increased the amount of mediator that could be extracted. They injected rabbits with insulin or saline, freeze-clamped skeletal muscle, and extracted the tissue in a boiling acid buffer, containing EDTA and a reducing agent. The extract was chromatographed on Sephadex G-25 and eluted at a molecular weight of 1000–1500. The mediator was identified by its ability to inhibit cAMP-dependent protein kinase and to stimulate glycogen synthase phosphoprotein phosphatase. Jarett and Seals[28] showed that this same material stimulates PDH activity in adipocyte mitochondria. Recent studies in our laboratory (S.L. Macaulay and L.

Table 2. Intact Cell Systems or Tissues in Which Insulin Altered the Amount or Activity of Mediator

Source	Mediator response to insulin
Adipocytes (rat)	Increase
Muscle (rabbit and rat)	Increase
Liver (rat)	Increase
H₄ hepatoma (rat)	Increase
IM-9 (human)	Decrease

Jarett, unpublished observations) have shown that rats treated with insulin have increased amounts of mediator in both skeletal muscle and liver as compared with control rats. After acid extraction and ultrafiltration, a 1000–2000-dalton mediator was identified by its ability to stimulate PDH and low K_m cAMP phosphodiesterase and to inhibit adenylate cyclase. Previously, Kiechle et al.[38] had shown that treatment of rat adipocytes with insulin increases the amount or activity of an acid-stable, low-molecular-weight material. This mediator was found to stimulate PDH and low K_m cAMP phosphodiesterase activity. The addition of insulin to insulin-sensitive H_4-II-E-C3' hepatoma cells produced a concentration-dependent increase in the amount of acid extractable low-molecular-weight mediator as determined by effects on PDH and low K_m cAMP phosphodiesterase activity.[39]

The ability of insulin to generate the mediator in IM-9 cultured human lymphocytes was investigated.[40] Despite extensive studies on the binding of insulin to its receptor on IM-9 lymphocytes, there have been no reports of a biological response of this cell to insulin. However, a low-molecular-weight material was extracted from the IM-9 lymphocytes that stimulated PDH activity. Yet, in contrast to the other cells or tissues studied, insulin treatment of IM-9 lymphocytes significantly reduced the amount or activity of PDH-stimulatory material present.

Thus, these studies showed that an insulin-sensitive, low-molecular-weight material could be found in intact cells or tissue similar to the mediator substance obtained from plasma membrane preparations both in size and ability to affect insulin-sensitive enzymes.

IV. INSULIN-SENSITIVE ENZYMES

The ability of insulin-generated material from different sources to modulate several enzyme systems in a manner analogous to insulin treatment of whole cells has been examined (Table 3). The activity of these enzymes in intact cells may be altered by insulin or by the mediator of insulin action through dephosphorylation (glycogen synthase, PDH,($Ca^{2+}-Mg^{2+}$-ATPase), phosphorylation (acetyl CoA carboxylase), or some other mechanism (low K_m cAMP phosphodiesterase and adenylate cyclase). Larner et al.[27] demonstrated that the mediator from skeletal muscle inhibits the activity of cAMP-dependent protein kinase and stimulates the activity of phosphoprotein phosphatase when glycogen synthase is used as a substrate. These changes in enzyme activity would result in dephosphorylation and activation of glycogen synthase as observed following insulin treatment of intact cells. The mediator was not observed to alter the activity of several cAMP-independent protein kinases tested.

In this laboratory and others,[29,32–37] PDH has been used as the major biological assay for the mediator. PDH is a multienzyme complex the activity of which is regulated by the degree of phosphorylation of the α subunit of pyruvate decarboxylase.[41] The phosphorylated form of the enzyme is inactive, while the dephosphorylated form is active.[41,42] The activity of the enzyme complex is controlled by a cAMP-independent protein kinase and phosphoprotein phosphatase.[25] The mediator of insulin action could alter PDH activity through changes in phosphorylation of the α subunit by increasing

Table 3. Enzyme Systems Tested for Response to the
Insulin-Sensitive Mediator

Enzyme	Alteration in activity by mediator
cAMP-dependent protein kinase	Decrease
Glycogen synthase phosphoprotein Phosphatase	Increase
cAMP-independent protein kinase	No effect
Pyruvate dehydrogenase	Increase
cAMP-independent kinase	No effect
Phosphatase	Increase
Low K_m cAMP phosphodiesterase	Increase
High K_m cAMP phosphodiesterase	No effect
$(Ca^{2+} + Mg^{2+})$ATPase	Increase
Acetyl CoA carboxylase	Increase
Adenylate cyclase	Decrease

phosphatase activity, by decreasing kinase activity, or by both. A series of studies were performed with the adipocyte subcellular system[43] and with the partially purified mediator from adipocytes,[38] hepatoma cells,[39] adipocyte plasma membranes,[30] and liver plasma membranes.[9] In each case, the increase in PDH activity was attributable to activation of the PDH phosphatase and not to any alteration of the PDH kinase.[13] Recently, Newman et al.[44] have shown that the insulin-generated mediator from adipocyte plasma membranes activates the PDH phosphatase directly. The mechanism by which the mediator activates the phosphatase is unknown.

All the above studies have been performed on nonintact mitochondria, raising questions of the physiological significance of the findings. Parker and Jarett[45] have completed a set of studies using intact isolated liver mitochondria. Insulin mediator from liver membranes, hepatoma cells, liver, and muscle stimulated PDH activity in these mitochondria up to 10-fold. This evidence supports the proposal that the mediator does function in intact cells. Furthermore, Jarett et al.[46] have shown that addition to adipocytes of mediator from skeletal muscle of control and insulin-treated rats resulted in stimulation of PDH activity of the intact cells. The mediator from insulin-treated rats increased PDH activity more than did mediator from control rats. The mediator effect was additive to the effect of maximal concentrations of insulin on intact cells. The time for peak activation was similar to insulin, but whereas insulin had a continuous effect on PDH activity, the mediator effect peaked at 10 min and declined thereafter. The mediator effect was concentration dependent, and maximal stimulation was five times basal. Thus, the intact mitochondrial studies and whole cell studies are consistent with the proposal that the mediator effect on PDH is of physiological relevance.

The earlier studies[26,28,29] and many subsequent studies[32,34–36] that used stimulation of PDH as the measure of mediator activity reported stimulations of up to 30%. More recent studies report greater stimulations. There were several reasons for the comparatively small stimulations reported in earlier studies. First, no attempts were made to prevent dephosphorylation of the enzyme complex during isolation, so that the control enzyme was already in a dephosphorylated, highly active state, with little activity left to stimulate. Second, when experiments were performed with ATP in the preincubations and assay reactions, a competition was set up between the increased dephosphorylation and stimulation of enzyme activity caused by the mediator and rephosphorylation and inactivation of the enzyme. The result was that the mediator established a new equilibrium between phosphorylation and dephosphorylation with a modest increase in activity. Kiechle et al.[30,38] addressed this problem as follows. The isolated mitochondria were incubated with ATP, in order to phosphorylate and inactivate PDH. The mitochondria were washed, stored frozen, and subsequently used for assay in the absence of ATP. Thus, the PDH complex starts in a form that is only 10–15% active, allowing for marked stimulation. Subsequent studies by our laboratory have found stimulations from 2- to 10-fold, depending on the amount of mediator tested. In many cases, the mediator would evoke maximum activity.

Another enzyme system that has been tested is the insulin-sensitive high-affinity Ca^{2+}-stimulated Mg^{2+}-dependent ATPase (Ca^{2+}–Mg^{2+})-ATPase found in the plasma membrane of rat adipocytes. This system may represent an enzymatic basis for a calmodulin-sensitive plasma membrane Ca^{2+} transport system.[54,55] Treatment of adipocytes or the adipocyte plasma membranes with physiological concentrations of insulin, up to 100 μU/ml, decreased enzyme activity and decreased the phosphorylation of a phosphoprotein with a molecular weight of 110,000. This phosphoprotein was identified as (Ca^{2+}–Mg^{2+})-ATPase.[54,56,57] The mediator from adipocyte plasma membranes stimulated the activity of (Ca^{2+}–Mg^{2+})-ATPase fourfold and more than doubled Ca^{2+} transport.[51] The mediator altered both enzyme and transport systems opposite to the effect observed with physiological concentrations of insulin. The explanation for this phenomenon is unknown.

Acetyl CoA carboxylase appears to be activated by insulin via phosphorylation by a cAMP-independent protein kinase.[52,53] This kinase phosphorylates sites distinct from those phosphorylated by a cAMP-dependent protein kinase activated by glucagon or epinephrine, leading to an inhibition of enzyme activity. Saltiel et al.[54] have demonstrated that the mediator generated by insulin from liver plasma membranes activates acetyl CoA carboxylase.

Insulin treatment of hepatocytes[55] or adipocytes[56,57] increased the activity of the low K_m cAMP phosphodiesterase present in membranes of the microsomal fraction. This enzyme is also activated by insulin in the plasma membranes of adipocytes.[58] The enzyme system, as it exists in the microsomal membrane fraction of the adipocyte, was stimulated by the addition of the mediator from several sources in the absence of ATP.[39,59] The mediator increased the V_{max} of the enzyme and had no effect on the K_m. The low K_m cAMP phosphodiesterase in liver plasma membranes has been reported to be activated by phosphorylation.[60] Our laboratory has been unable to demonstrate control of the adipocyte microsomal or plasma membrane enzyme by phosphorylation. We conclude that low K_m cAMP phosphodiesterase from adipocytes is regulated by a mechanism other than phosphorylation. The mediator from various sources had no effect on the high K_m cAMP phosphodiesterase in the adipocytes microsomes, as would be expected, since this enzyme is not modulated by insulin in intact cells.[55,56]

Insulin has been reported to inhibit both the basal and hormonally stimulated adenylate cyclase in hepatocytes and adipocytes.[61,62] Saltiel et al.[54,63] have shown that the ethanol extractable material from mediator released from liver plasma membranes will inhibit both the basal and hormonally stimulated enzyme. The ethanol residue that activates PDH and acetyl CoA carboxylase had no effect on adenylate cyclase activity. Work in this laboratory has shown that the insulin-sensitive mediator from rat skeletal muscle and liver inhibited adenylate cyclase. However, ethanol fractionation revealed the inhibition to be in the ethanol-insoluble precipitate that contained the PDH activator and low K_m cyclic AMP phosphodiesterase stimulator. In contrast, insulin mediator from cultured hepatoma cells, which stimulated PDH and inhibited adenylate cyclase, was found only in the ethanol-soluble fraction (S.L. Macaulay, J.O. Macaulay, J.C. Parker, and L. Jarett, unpublished observations). The mechanism controlling the activity of adenylate cyclase does not appear to involve phosphorylation.

A number of recent studies have shown that the insulin mediator can work on isolated intact cells (Table 4). Work by Zhang et al.[64] shows that the insulin mediator probably regulates key enzymes involved in lipogenesis, antilipolysis, and lowering of cAMP of intact adipocytes. These investigators added insulin-generated mediator from adipocyte plasma membrane to adipocytes and found elevated cAMP levels to be lowered, stimulated lipolysis to be suppressed, and lipogenesis to be stimulated. Caro et al.[65] reported that insulin-generated mediator from liver particulate fraction stimulated lipid synthesis and caused downregulation of insulin receptors in primary cultures of isolated rat hepatocytes. Jarett et al.[46] showed that addition of insulin-generated mediator from rat skeletal muscle to intact adipocytes stimulates PDH activity. The use of intact cells provides an interesting alternative method for studying insulin-sensitive metabolic pathways that are not amenable to in vitro methods.

Initially, it was proposed that the primary function of the mediator of insulin action was to modify the state of phosphorylation, hence the activity of various regulatory enzymes. The mediator appeared to alter phosphorylation by modulating the activity of phosphatase and kinase enzymes. However, it was evident that some insulin-sensitive enzymes are not regulated by a phosphorylation–dephosphorylation mechanism. Therefore, the mediator(s) may alter enzyme activity by fulfilling a cofactor function for specific phosphatases, kinases, or other coenzymes, or by direct allosteric effects.

Table 4. Response of Intact Cells to Direct Addition of Hormone Mediators

Hormone-generated mediator	Cells	Response
Insulin	Adipocyte	Increase lipogenesis
		Decrease lipolysis
		Decrease cAMP levels
		Increased PDH
	Cultured hepatocyte	Increase lipogenesis
		Downregulation of insulin receptors

V. PHYSICAL AND CHEMICAL CHARACTERISTICS OF THE MEDIATOR

The previous discussion has described the putative second messenger of insulin action as a single entity. However, there is direct and indirect evidence for the existence of more than one mediator. Present evidence suggests that these molecules may represent a family of related compounds.

Indirect evidence for the existence of more than one mediator can be found in the biphasic or bell-shaped response detected in some test systems to increasing concentrations of insulin, insulin-mimetic agents (Table 5), or insulin mediators. Turakulov et al.[66,67] reported that insulin treatment of rats resulted in an increase in a low-molecular-weight material in the cytosol of liver, insulin-dependent cytoplasmic regulator, which produced a biphasic response in calcium uptake by mitochondria.[66] A biphasic response of glycogen synthase phosphatase activity was reported by Larner et al.[27] with an increasing concentration of a low-molecular-weight gel chromatography fraction from rabbit skeletal muscle. Seals and Jarett[26] reported that the direct addition of insulin, Con A, or anti-insulin receptor antiserum to the adipocyte subcellular system composed of mitochondria and plasma membranes also activated PDH in a biphasic manner. Seals and Czech[31] showed that the supernatant obtained from adipocyte plasma membranes incubated in increasing concentrations of insulin produced biphasic stimulation of PDH when added to isolated adipocyte mitochondria. Saltiel et al.[32,54] demonstrated the same phenomenon on the activity of PDH and acetyl CoA carboxylase using the supernatant from the incubation of a crude preparation of liver plasma membranes with increasing concentrations of insulin.

Direct evidence for the existence of more than one mediator has been reported by two laboratories. Using high-voltage electrophoresis, Chen et al.[68] separated the low-molecular-weight mediator from the skeletal muscle of insulin-treated rabbits into two fractions. One fraction stimulated the activity of glycogen synthase phosphoprotein phosphatase and inhibited the activity of cAMP-dependent protein kinase. The other fraction had the opposite effect, i.e., it inhibited glycogen synthatase phosphoprotein

Table 5. Sources on Which Increasing Concentrations of Insulin or Other Ligands Produce a Biphasic Response in Various Test Systems

Source	Test system	Ligand
Liver	Calcium uptake in liver mitochondria	Insulin
Skeletal muscle	Glycogen synthase phosphatase	Insulin
Adipocyte subcellular system	Pyruvate dehydrogenase	Insulin, concanavalin A, anti-insulin receptor antibody
Adipocyte plasma membranes	Pyruvate dehydrogenase	Insulin
Liver plasma membranes	Pyruvate dehydrogenase Acetyl CoA carboxylase	Insulin

phosphatase activity and stimulated cAMP-dependent protein kinase. Although these effects were concentration dependent, the biphasic response was not observed. Saltiel et al.[54,63] showed ethanol precipitation of mediator generated from liver plasma membranes to separate the material into (1) an ethanol-insoluble residue that contains a substance that activates PDH and acetyl CoA carboxylase and (2) an ethanol-soluble material that inhibits PDH, acetyl CoA carboxylase, and basal and hormonally stimulated adenylate cyclase. Both the stimulatory and the inhibitory activities demonstrate a monophasic dose–response curve, suggesting that the relative ratio of the two counteracting substances may be responsible for the biphasic response observed in crude extracts. These two substances must have similar molecular weights, since both the inhibitor and stimulator of PDH coelute on a high-performance liquid chromatographic (HPLC) molecular sieve column. Ethanol precipitation studies in Jarett's laboratory on insulin-generated mediator from IM-9 lymphocytes, H_4-II-E-C3'-hepatoma cells, rat skeletal muscle, and rat liver have not separated PDH stimulatory activity from adenylate cyclase inhibitor or low K_m PDE stimulator. The effective material from the former two sources was found in the ethanol supernatant, while from the latter two sources it was found in the precipitate.

In view of these findings, it is possible that insulin treatment of IM-9 lymphocytes only stimulates the generation of inhibitor of PDH.[40] These cells are the only system studied to date in which insulin treatment has reduced the amount and/or activity of mediator as compared with control. By contrast, in H_4-II-E-C3'-hepatoma cells,[39] insulin did not produce a biphasic curve but simply increased mediator activity to a maximum, even at high insulin concentrations, suggesting that only an activating material was generated. These two cell lines may provide sufficient quantities of both the inhibitor and stimulator to permit further chemical characterization.

The insulin mediators have yet to be purified and chemically identified. The mediators from various tissues have several common characteristics, including low molecular weight (1000–3000), acid stability, thermostability, and water solubility. Since the mediators are derived from plasma membranes, they must be generated from membrane components or their derivatives, such as proteins, glycoproteins, glycolipids, phospholipids, proteolipids, fatty acids, prostaglandins, and leukotrienes.

Early studies suggested that the mediators might be peptides.[14,27,30,38] This observation was based on the fact that the mediator copurified with a peak absorbance at 230 nm as well as ninhydrin positivity. No correlation was found between the amount or activity of the mediator and the absorbance, probably because the material was impure. Some investigators have been able to destroy mediator activity with protease (trypsin) digestion,[29,33] and others have seen no effect or partial inactivation following protease digestion.[11,15] The protease experiments must be cautiously interpreted, since large quantities of enzyme were used, and trace contaminants of other enzymes could have accounted for the inactivation. Several investigators report that the addition of inhibitors of arginine-specific proteases to plasma membranes prevents the generation of the mediator by insulin. Seals and Czech[29] have proposed that after insulin binds to its receptor, a membrane-bound protease is activated that cleaves the mediator at an arginine residue. Muchmore et al.[69] have shown that the effects of most proteolytic inhibitors on the effects of insulin on adipocytes may be secondary to decreased cellular ATP. Originally, we were unable to confirm that peptide bonds were necessary for biological activity of the mediator, since proteolytic enzymes, dansylation, or treatment with fluorescamine failed to inactivate the mediator. More recent experiments using the mediator from the skeletal muscle of insulin-treated rats have demonstrated that trypsin has little if any effect on mediator activity, whereas chymotrypsin has a greater effect and pepsin, the greatest effect on activity, but complete inactivation was not obtained (Macaulay and Jarett, unpublished observations). These data indicated that while the mediator may contain amino acids, the biological activity was not totally dependent on their presence. Partial inactivation was only observed when high concentrations of enzymes were used (mg/ml). When smaller amounts of enzyme (μg/ml) were used, no inactivation was found. Several investigators have reported the presence of amino acids in partially purified preparations of the mediator. Gainutdinov et al.[70] reported an amino acid composition for the insulin-dependent cytoplasmic regulator consisting of 42 amino acids: aspartic acid 3, threonine 1, serine 7, glutamic acid 6, proline 4, glycine 6, alanine 3, valine 1, isoleucine 1, leucine 1, tyrosine 1, phenylalanine 1, histidine 1, lysine 5, and arginine 1. They also reported that the compound contained sugar residues and organic phosphorus. Larner et al.[17] reported a provisional amino acid composition of Phe, Lys, Glu, Asp, Gly, Leu, Ser for the mediator extracted from insulin-treated skeletal muscle. Seals[71] also reported the presence of a total of 32 amino acids in mediator preparations from insulin-treated and control adipocyte plasma membranes. Sixteen different amino acids were identified. The number of amino acids identified by Seals and Gainutdinov and co-workers would yield a peptide much larger than the mediator found by Larner's or Jarett's groups. Whether these amino acids represent essential structure components of the mediators of insulin action and what additional molecular constituents are associated with the mediator composition is not yet known. In the meantime, caution must be exercised in interpreting these data. No proof has been provided that the specimens analyzed were pure, therefore, the amino acids may be contaminants.

We could not substantiate that common sugars or amino sugars were part of the mediator structure, based on the inability of the material to bind to a variety of immobilized lectins. However, Begum et al.[36] have shown that both neuraminidase and β-D-galactosidase treatment inactivates the mediator generated by liver plasma membranes. These observations have yet to be confirmed.

The above data led us to consider other membrane components, such as phospholipids and their derivatives, as potential mediators. Farese et al.[72] showed that insulin treatment of adipocytes increased the content of several phospholipids. Walaas et al.[73] reported that insulin increased the phosphorylation of a muscle sarcolemmal proteolipid with a molecular weight of 3600. Wasner[74,75] found that insulin increased the amount of a low-molecular-weight compound called cAMP-antagonist, which may contain PGE_1, phosphate, and inositol as structural components. This compound inhibits adenylate cyclase and protein kinases and activates phosphoprotein phosphatases and PDH. A preliminary report from Begum et al.[76] demonstrates that indomethacin, an inhibitor of prostaglandin synthesis, prevented mediator generation from liver plasma membrane by insulin. This inhibitory effect of indomethacin was reversed by the addition of prostaglandin E_2. It has been suggested that mediator is generated following activation of a membrane-bound phospholipase A_1 or A_2.[77,78]

We have investigated the role of phospholipids and related compounds in insulin action. The addition of aqueous dispersions of these phospholipids to assays for PDH and low K_m cAMP phosphodiesterase demonstrated specific effects. Phosphatidylserine and phosphatidylinositol 4-phosphate were found, respectively, to stimulate and inhibit PDH activity specifically, while other phospholipids had no effect.[79] Similarly, phosphatidylserine and phosphatidylcholine stimulated and phosphatidylinositol 4-phosphate inhibited the low K_m cAMP phosphodiesterase activity of adipocyte microsomal (endoplasmic reticulum) and plasma membrane preparations.[58,80] The high K_m enzyme was unaffected by these same phospholipids. These initial studies suggested that phosphatidylserine and phosphatidylinositol 4-phosphate may have a counterregulatory role that could be of physiological importance in regulating enzyme activity. Farese et al.[81] have shown that the insulin-induced increase in phosphatidylserine in adipose tissue precedes PDH stimulation, a requirement for a second messenger. It seems unlikely, though, that phosphatidylserine itself is the mediator for insulin action, since the mediators discussed in this chapter are chloroform-methanol insoluble and water soluble. However, phospholipids may be involved in an alternative regulatory role for certain insulin-sensitive enzymes, such as PDH and low K_m cAMP phosphodiesterase.

Despite extensive effort over the past several years, purification and chemical identification of the mediator have yet to be achieved, because of major factors: (1) sufficient quantities of material have not been available until recently, (2) the mediator has not proved amenable to many of the separation techniques that have proved useful for other small-molecular-weight biological substances, and (3) it is possible that the mediator belongs to an as-yet undescribed class of materials of unusual chemical properties.

Recent developments in our laboratory have encouraged us to expect that purification and chemical identification may be accomplished soon. These developments include the following unpublished observations. First, we have obtained substantial quantities of mediator from liver or skeletal muscle of control and insulin-treated rats. The amount or activity of material extracted from the insulin-treated animals has been substantially greater than from controls. After acid extraction, boiling, and charcoal absorption, the active material passes through an Amicon YM2 membrane filter but is retained in a YC05 membrane filter, indicating a molecular weight between 500 and 2000. The mediator from insulin-treated rats has insulin-mimetic effects greater than the mediator from control rats by three assays: PDH, adenylate cyclase, and low K_m cAMP phosphodiesterase. The mediator can be further sized on Sephadex G-15, where a sharp peak of activity is found at an M_r of 1000. The active material is not destroyed by boiling in 6 N HCl for 24 hr but is destroyed after 72 hr, making it unlikely that the mediator is a simple peptide. The activity is destroyed after 1 hr at room temperature in 2.5 M NaOH, but is stable up to 18 hr at pH 9.0. The mediator is not retained on quaternary amine or reverse-phase columns but can be separated with several buffer systems on silica using either TLC or HPLC. This information has enabled us to reach a high state of purity with final purification within reach. The active material is assayed by all three enzymes at each stage. Enough active material is present that can be analyzed by gas chromatography–mass spectroscopy. This cautious approach, making no assumption as to the composition of the material, should prove fruitful in the near future.

VI. CONCLUSION

The data presented in this review support the contention that the low-molecular-weight material released from the plasma membrane, after insulin interacts with its receptor, is the mediator of many of insulin actions. The mediator can be produced in subcellular systems as well as intact cells. The mediator affects a variety of insulin-sensitive enzymes both in subcellular systems and in intact cells. It appears that the mediator is part of a new family of messenger compounds that may mediate the action of other hormones, as well as insulin. The chemical nature of the mediator is as yet unknown, but is likely to be novel.

The next few years should be exciting as the insulin mediator is purified and identified, the mode of generation of the mediator is determined, and the mechanism by which the mediator acts is unraveled. These projects will not be easily accomplished for several reasons. First, until recently, the quantity of mediator available has been rate limiting. Second, the presence of several mediators requires careful separations and determination of the actions of each class. Third, the chemical composition seems unusual and may include peptide, sugar, and phospholipid and/or prostaglandin. The importance of the problem, the availability of appropriate tools, and the increasing number of investigators working on the problem, however, assure success.

These studies on the insulin mediator have potentially important clinical significance in understanding the post-receptor defects of insulin resistance in obesity and type II diabetes (noninsulin-dependent diabetes mellitus). These conditions may result from abnormal release or metabolism of the mediator. The data presented in this review suggest that these defects do exist. Further studies in human subjects and experimental animals are needed to support or refute this hypothesis.

ACKNOWLEDGMENTS. We wish to thank our colleagues J.O. Macaulay, C. Penn, and J. Smith, without whom this work would not be accomplished. We thank D. Carson for typing the manuscript. The work was supported by grant AM 28144 from the National Institutes of Health and by a grant from the Reynolds Foundation. Dr. Kiechle is a Hartford fellow. Dr. S.L. Macaulay and Dr. J.C. Parker are Juvenile Diabetes Foundation postdoctoral fellows. Dr. K.L. Kelly is supported by NIH training grant AM07409.

VII. REFERENCES

1. Czech, M.P., 1977, Am. J. Med. **46**:359–384.
2. Czech, M.P., 1981, Am. J. Med. **70**:142–150.
3. Denton, R.M., 1981, Diabetologica **21**:347–362.
4. Goldfine, I.D., 1981, Biochemical Actions of Hormones, Vol. VIII, pp. 273–305, Academic Press, New York.

5. Kahn, C.R., 1979, *Proc. Soc. Exp. Biol. Med.* **162**:13–21.
6. Levine, R., 1983, *Vitam. Horm.* **39**:145–173.
7. Smith, C.J., and Rosen, O.M., 1983, in: *Diabetes Mellitus: Theory and Practice* (M. Ellenberg, and H. Rifkin, eds.), pp. 89–96, Medical Examination Publ., Garden City, New York.
8. Walaas, O., and Horn, R.S., 1981, *Trends Pharmacol. Sci.* **2**:196–198.
9. Jarett, L., and Kiechle, F.L., 1981, in: *Current Views on Insulin Receptors* (D. Adreani, R. De Pirro, R. Lauro, J. Olefsky, and J. Roth, eds.), pp. 245–253, Academic Press, London.
10. Jarett, L., Kiechle, F.L., and Parker, J.C., 1982, *Fed. Proc.* **41**:2736–2741.
11. Jarett, L., Kiechle, F.L., Parker, J.C., and Macaulay, S.L., 1983, *Am. J. Med.* **74**:31–37.
12. Jarett, L., Kiechle, F.L., Parker, J.C., and Macaulay, S.L., 1983, in: *The Adipocyte and Obesity* (A. Angel, C. Hollenberg, and D. Roncari, eds.), pp. 75–84, Raven Press, New York.
13. Jarett, L., Kiechle, F.L., Popp, D.A., and Kotagal, N., 1981, *Protein Phosphorylation*, Vol. 8, pp. 715–726, Cold Spring Harbor Laboratory, New York.
14. Larner, J., 1982, *J. Cyclic Nucleotide Res.* **8**:289–296.
15. Larner, J., 1983, *Am. J. Med.* **74**:38–51.
16. Larner, J., Cheng, K., Schwartz, C., Kikuchi, K., Tamura, S., Creacy, S., Dubler, R., Galasko, G., Pullin, C., and Katz, M., 1982, *Fed. Proc.* **41**:2724–2729.
17. Larner, J., Cheng, K., Schwartz, C., Kikuchi, K., Tamura, S., Creacy, S., Dubler, R., Galasko, G., Pullin, C., and Katz, M., 1982, *Recent Prog. Horm. Res.* **38**:511–556.
18. Seals, J.R., and Czech, M.P., 1982, *Fed. Proc.* **41**:2730–2735.
19. Jarett, L., 1974, *Methods Enzymol.* **XXXI**:60–71.
20. Jarett, L., and Smith, R.M., 1974, *J. Biol. Chem.* **249**:5195–5199.
21. Krebs, E.G., and Beavo, J.A., 1979, *Annu. Rev. Biochem.* **48**:923–959.
22. Seals, J.R., McDonald, J.M., and Jarett, L., 1978, *Biochem. Biophys. Res. Commun.* **83**:1365–1372.
23. Seals, J.R., McDonald, J.M., and Jarett, L., 1979, *J. Biol. Chem.* **254**:6991–6996.
24. Seals, J.R., McDonald, J.M., and Jarett, L., 1979, *J. Biol. Chem.* **254**:6997–7005.
25. Denton, R.M., and Hughes, W.A., 1978, *Int. J. Biochem.* **9**:545–552.
26. Seals, J.R., and Jarett, L., 1980, *Proc. Natl. Acad. Sci. USA* **77**:77–81.
27. Larner, J., Galasko, G., Cheng, K., DePaoli-Roach, A.A., Huang, L., Daggy, P., and Kellogg, J., 1979, *Science* **206**:1408–1410.
28. Jarett, L., and Seals, J.R., 1979, *Science* **206**:1407–1408.
29. Seals, J.R., and Czech, M.P., 1980, *J. Biol. Chem.* **255**:6529–6531.
30. Kiechle, F.L., Jarett, L., Popp, D.A., and Kotagal, N., 1981, *J. Biol. Chem.* **256**:2945–2951.
31. Seals, J.R., and Czech, M.P., 1981, *J. Biol. Chem.* **256**:2894–2899.
32. Saltiel, A., Jacobs, S., Siegel, M., and Cuatrecasas, P., 1981, *Biochem. Biophys. Res. Commun.* **102**:1041–1047.
33. Sakamoto, Y., Kuzuya, T., and Sato, J., 1982, *Biomed. Res.* **3**:599–605.
34. Begum, N., Tepperman, H.M., and Tepperman, J., 1982, *Endocrinology* **110**:1914–1921.
35. Begum, N., Tepperman, H.M., and Tepperman, J., 1982, *Endocrinology* **111**:1491–1497.
36. Begum, N., Tepperman, H.M., and Tepperman, J., 1983, *Endocrinology* **112**:50–59.
37. Amatruda, J.M., and Chang, C.L., 1983, *Biochem. Biophys. Res. Commun.* **112**:35–41.
38. Kiechle, F.L., Jarett, L., Popp, D., and Kotagal, N., 1980, *Diabetes* **29**:852–855.
39. Parker, J.C., Kiechle, F.L., and Jarett, L., 1982, *Arch. Biochem. Biophys.* **215**:339–344.
40. Jarett, L., Kiechle, F.L., Popp, D.A., Kotagal, N., and Gavin, J.R. III, 1980, *Biochem. Biophys. Res. Commun.* **96**:735–741.
41. Schuster, S.M., Olson, M.S., and Routh, C.A., 1975, *Arch. Biochem. Biophys.* **171**:745–752.
42. Reed, L.J., 1974, *Acc. Chem. Res.* **7**:40–46.
43. Popp, D., Kiechle, F.L., Kotagal, N., and Jarett, L., 1980, *J. Biol. Chem.* **255**:7540–7543.
44. Newman, J., Armstrong, J.McD., and Bornstein, J., 1983, *Second International Symposium on Insulin Receptors: Receptor Interaction and Insulin Action*, Rome, abst. 32.
45. Parker, J.C., and Jarett, L., 1985, *Diabetes* **34**:92–97.
46. Jarett, L., Wong, E., and Smith, J.A., and Macaulay, S.L., 1985, *Endocrinology* (in press).
47. Pershadsingh, H.A., and McDonald, J.M., 1981, *Biochem. Int.* **2**:243–248.
48. Pershadsingh, H.A., Landt, M., and McDonald, J.M., 1970, *J. Biol. Chem.* **255**:8983–8986.

49. Pershadsingh, H.A., and McDonald, J.M., 1979, *Nature* **281:**495–497.
50. Chan, K.M., and McDonald, J.M., 1982, *J. Biol. Chem.* **257:**7443–7448.
51. McDonald, J.M., Pershadsingh, H.A., Kiechle, F.L., and Jarett, L., 1981, *Biochem. Biophys. Res. Commun.* **100:**857–864.
52. Witters, L.A., 1981, *Biochem. Biophys. Res. Commun.* **100:**872–878.
53. Brownsey, R.W., and Denton, R.M., 1982, *Biochem. J.* **202:**77–86.
54. Saltiel, A.R., Doble, A., Jacobs, S., and Cuatrecasas, P., 1983, *Biochem. Biophys. Res. Commun.* **110:**789–795.
55. Loten, E.G., Assimacopoulos-Jeannet, F.D., Exton, J.H., and Park, C.R., 1978, *J. Biol. Chem.* **253:**746–757.
56. Loten, E.G., and Sneyd, J.G.T., 1970, *Biochem. J.* **120:**187–195.
57. Makino, H., and Kono, T., 1980, *J. Biol. Chem.* **255:**7850–7854.
58. Macaulay, S.L., Kiechle, F.L., and Jarett, L., 1983, *Biochim. Biophys. Acta* **760:**293–299.
59. Kiechle, F.L., and Jarett, L., 1981, *FEBS Lett.* **133:**279–281.
60. Marchmont, R.J., and Houslay, M.D., 1980, *Nature* **286:**904–906.
61. Torrez, H.N., Flawia, M.M., Hernaez, L., and Cuatrecasas, P., 1978, *J. Membr. Biol.* **43:**1–18.
62. Illiano, G., and Cuatrecasas, P., 1972, *Science* **174:**906–908.
63. Saltiel, A.R., Siegel, M.I., Jacobs, S., and Cuatrecasas, P., 1982, *Proc. Natl. Acad. Sci. USA* **79:**3513–3517.
64. Zhang, S.-R., Shi, Q.-H., and Ho, R.J., 1983, *J. Biol. Chem.* **258:**6471–6476.
65. Caro, J.F., Folli, F., Cecchin, F., and Sinha, M.K., 1983, *Biochim. Biophys. Res. Commun.* **115:**375–382.
66. Turakulov, Y.K., Gainutdinov, M.K., Lavinaa, J.I., and Akhmatov, M.S., 1977, *Rep. Acad. Sci. USSR* **234:**1471–1474.
67. Turakulov, Y.K., Gainutdinov, M.K., Lavina, I.I., and Akhmatov, M.S., 1979, *Bull. Expt. Biol. Med.* **88:**1008–1010.
68. Chen, K., Galasko, G., Huang, L., Kellogg, J., and Larner, J., 1980, *Diabetes* **29:**659–661.
69. Muchmore, D.B., Raess, B.U., Bergstrom, R.W., and de Haën, C., 1982, *Diabetes* **31:**976–984.
70. Gainutdinov, M.K., Turakulov, Y.K., Akhmatov, M.S., and Lavina, J.I., 1978, *Khim. Prirodnykh Soedinenii* **1:**141.
71. Seals, J.R., 1983, *Diabetes* **32:**57A.
72. Farese, R.V., Larson, R.E., and Sabir, M.A., 1982, *J. Biol. Chem.* **257:**4042–4045.
73. Walaas, O., Sletten, K., Horn, R.J., Lystad, E., Adler, A., and Alertsen, A.R., 1981, *FEBS Lett.* **128:**137–141.
74. Wasner, H.K., 1980, *Aktuel. Endokinol. Stoffwechsel* **1:**207–208.
75. Wasner, H.K., 1981, *FEBS Lett.* **133:**260–264.
76. Begum, N., Tepperman, H.M., and Tepperman, J., 1983, *Diabetes* **32:**32A.
77. Bereziat, G., Wolf, C., Colard, O., and Polonovski, J., 1978, *Adv. Exptl. Biol. Med.* **101:**191–199.
78. Dietze, G.J., 1982, *Mol. Cell Endocrinol.* **25:**127–149.
79. Kiechle, F.L., and Jarett, L., 1983, *Mol. Cell. Biochem.* **56:**99–105.
80. Macaulay, S.L., Kiechle, F.L., and Jarett, L., 1983, *Arch. Biochem. Biophys.* **225:**130–136.
81. Farese, R.V., Sabir, M.A., Larson, R.E., and Trudeau, W.L. III, 1983, *Biochim. Biophys. Acta* **750:**200–202.

12

Intracellular Mediators of Insulin Action

Jonathan R. Seals

I. HISTORY AND RATIONALE OF CURRENT APPROACHES TO RESEARCH ON INSULIN MEDIATOR

Controversy about the intracellular mechanism of insulin action has focused for a decade on identifying an intracellular mediator of the actions of the hormone. During this time, extensive efforts have been made to specify a cellular component that is altered in response to insulin and, in turn, directly affects enzyme activities and other cellular functions. A wide range of possibilities has been considered for this role, including cAMP, Ca^{2+}, H_2O_2, membrane potential, and cellular pH.[1,2] Interest in each possibility has been generated by data correlating changes in these parameters with changes in one or more insulin-sensitive enzyme or process. But in each case, opposing evidence has subsequently dissociated changes in the levels of the regulatory components from other insulin-sensitive functions, making it impossible to consider any of these a primary, central element of the insulin effector system. The failure of all candidates thus far proposed to satisfy all the criteria of a hormone's second messenger has led to the development of an experimental approach designed to isolate from cells components that possess the characteristics of an insulin mediator. This approach embodies two criteria of an insulin mediator, the ability to directly regulate insulin-sensitive enzymes *in vitro*, and a change in activity in response to insulin treatment of the cell source of the mediator. In practice, this has involved exposing tissues, cells or isolated plasma membranes to insulin, preparing an extract or soluble fraction from the cellular material, then testing the sample for the ability to regulate an insulin-sensitive enzyme in an isolated setting. By comparing these results with those of samples prepared from control sources, it is possible to demonstrate the existence of insulin-dependent, enzyme-regulatory activity and to use the same criteria

Jonathan R. Seals • Department of Biochemistry, University of Massachusetts Medical School, Worcester, Massachusetts.

to follow the active component through a purification protocol. Since first carried out by Larner et al.,[3] the experimental phenomenology has been repeated using a wide range of cellular sources and many different target enzymes.[4,5] Attention must now be focused on the significance of these results with regard to whether they reflect the existence of an actual chemical mediator of insulin action and whether they will ultimately lead to its identification. The critical factor that has yet to be achieved is the purification to homogeneity of a putative mediator in any of the reported experimental systems. Pending this accomplishment, it will not be possible to confirm the identity of the active component and to apply more rigorous criteria *in vivo* to evaluate its possible role as a mediator. In the interim, characterization of the phenomenon itself and attempts at characterizing the active component in unfractionated or partially purified preparations have been carried out to provide preliminary and circumstantial evidence concerning the significance of the mediator. This chapter will review results that are relevant to the possible identity of components that account for the results obtained using the mediator approach described here, and their possible role in the insulin effector system. In addition, recent progress in the purification of these components will be presented, and prospects for future progress in this area will be discussed. The results indicate that the mediator phenomenology carried out *in vitro* is most likely a reflection of some aspect of the insulin effector system, and remains one of the most promising avenues toward understanding the mechanism of insulin action. However, further conclusions will require structural identification of the putative mediator(s) and analysis of their specific functions *in vivo*.

II. EVIDENCE FOR THE EXISTENCE OF UNIDENTIFIED INTRACELLULAR INSULIN MEDIATORS

The experimental phenomenology that has been interpreted as indicating the existence of an unidentified insulin mediator has now been reproduced using a variety of cellular sources and target enzymes. All results have thus far been consistent with, but are not proof of, the model that the active factor being studied in these experiments is a mediating component in the insulin effector system. The broad range of sources and target enzymes that have been successfully used provides one of the strongest pieces of circumstantial evidence in favor of this interpretation. Insulin-dependent regulatory activity has been extracted from muscle,[3,6–8] liver[9], adipocytes[10] (also J.R. Seals and M.P. Czech, unpublished observations), cultured hepatoma cells[13] (also C.B. McDonald, C.D. Chang, N. Gibran, and J.R. Seals, unpublished observations), and adipocyte[12,13] or liver[14,15] plasma membranes. All these represent insulin-sensitive sources in terms of metabolic or growth regulation. In contrast, a cell line that does not respond to insulin, the IM-9 lymphocyte, does not exhibit the mediator phenomenon *in vitro*.[16] Enzymes and processes shown to be regulated by extracts from these sources include cAMP-dependent protein kinase,[6–8] pyruvate dehydrogenase,[17–20] glycogen synthase phosphatase,[21] low K_m phosphodiesterase,[10] adenylate cyclase,[18] plasma membrane Ca^{2+}/Mg^{2+} ATPase,[22] acetyl CoA carboxylase,[23] mitochondrial Ca^{2+} uptake,[9] nuclear RNA synthesis,[24] and nuclear DNA synthesis (C.B. McDonald, C.D. Chang, N. Gibran, and J.R. Seals, unpublished observations). All these functions have been recognized as part of the complex of alterations that occurs in response to insulin. The ability of experimental mediators to act on such a wide variety of processes suggests that the potential exists in this phenomenon to identify a component that fills the central role in the insulin effector system, linking the diverse effects of the hormone. Other results have revealed a positive correlation between the behavior of the experimental systems and the characteristics of an insulin mediator as predicted from the known parameters of the cellular effects of insulin. For example, the time course of mediator production is sufficiently rapid, occurring in seconds to minutes in cells or membranes,[12,14,15] to allow it to precede the alteration of enzyme activities throughout the cell. Thus the requirement that a mediator exhibit temporal precedence over the effects it mediates appears feasible in light of the available data. In addition, the insulin concentration dependence of mediator production occurs over the same range as does the regulation of enzymes in cells,[15,18] again providing a necessary correlation between the putative mediator and its effects. Finally, the production of mediator activity by purified plasma membranes[12,13,15,19] suggests that the mediator can provide a direct link between the initial receptor-linked and terminal enzyme-expressed levels of the effector system and is not a secondary pathway unrelated to the transfer of a regulatory signal.

Taken together, the bulk of the available evidence supports the concept that a central intracellular mediator of insulin action exists and can be studied by the methodology described here. However, the picture is far from complete or conclusive. Two major reservations about the available data should be stressed. First, the experiments have been carried out, by definition and design, in systems consisting of limited sets of cellular components *in vitro*. The behavior of any component studied in this way cannot necessarily be accurately extrapolated to the more complex intact cell. However, limited preliminary data suggest that mediators may produce insulin-like effects in intact cells and tissues. Incubation of mediator, produced from liver plasma membranes *in vitro*, with intact hepatocytes produces insulin-like effects on lipid metabolism.[25] Perfusion of liver slices with partially purified mediator from hepatocytes has been reported to inhibit gluconeogenesis and glycogenolysis.[26] Finally, injection of this partially purified mediator into alloxan-diabetic rats rapidly lowers blood levels of the animals.[26] These data are suggestive, but crude. A rigorous analysis of the role of a putative mediator in the cell will ultimately require that: (1) the mediator exist within the cell at concentrations that can be shown to regulate enzyme activity; (2) insulin increase the concentration of the mediator over a range that produces corresponding effects on enzyme activity; (3) alteration of mediator concentration over this range by nonhormonal means mimic insulin action; and (4) specifically blocking the change in mediator concentration in the presence of hormone block the expression of the effects of insulin as well.

Satisfying these criteria will confirm a component as a necessary and sufficient element of the insulin effector system according to the definition of a mediator. However, these points cannot be considered until the second major reservation about the available data has been addressed. The most obvious deficiency in the study of the mediator is the failure of any putative mediator in any system to be purified to homogeneity and identified. This achievement is critical for the interpretation of the available data, as well as for future progress in this field. It is not currently possible, for example, to conclude whether the broad range of systems being used in mediator studies all reflect the same active component, or many different ones. Much of the positive interpretation of the wide range of mediator data assumes that the same, or at least a closely related family of components are involved, since this would be compatible with the simplest model of insulin action, that of a single, central mediator. The alternative possibility—that many different components are involved—would require a much more complex model in which specific mediators are linked to individual effects. The limited analysis that has been attempted in various systems indicates general agreement on the basic characteristics of the mediator(s), consistent with the simpler model. However, because of the uncertainty of these results, as discussed in Section IV.A, definitive conclusions are precluded at this time.

Progress toward the goal of mediator purification and rigorous analysis of the mediator hypothesis has been hindered by several factors, as well as by the inherent characteristics of the system itself. In many cases, the experimental effects initially reported were relatively small.[17] However, modification of experimental conditions has made it possible to consistently demonstrate more dramatic effects. Furthermore, some experimental systems, particularly those involving purified plasma membranes, have proven difficult to reproduce consistently. In many cases, this can be attributed to the difficulty of translating complex structures, such as the plasma membrane, into an experimental form removed from the native state with adequate retention of relevant functions. The use of cells or tissues, although not as simple or direct as isolated membranes, has proven more reproducible. In addition, the use of cells or tissues may produce a greater yield of material from an equal amount of starting material than after preparation of plasma membranes (J.R. Seals and M.P. Czech, unpublished observations), which involves loss of some percentage of the membrane during fractionation. This factor is significant in attempts to purify and identify mediators. Low yields have required the use of large amounts of starting material in order to obtain sufficient mass for analysis, as described in Section IV.B. Obtaining the maximum possible yield appears to be necessary if this critical goal is to be reached.

In light of these factors, we have used isolated adipocytes as a source of mediator for the purpose of purification and identification. The adipocyte is an insulin-sensitive cell that can be prepared in relatively large homogeneous quantities. Adipocytes from 100 rats are obtained by collagenase digestion in a typical preparation. The washed cells are incubated with or without 5 nM insulin for 5 min, then extracted following removal of free insulin. The extraction is performed by homogenizing the cells in 1% trifluoroacetic acid (TFA) containing 0.5 mM each of dithiothreitol and EDTA.[27] This system was chosen empirically for producing the optimal yield of mediator activity and because TFA, a volatile acid, is readily removed by lyophilization during further processing of the samples. Bioactivity was determined as the stimulation of pyruvate dehydrogenase activity in isolated adipocyte mitochondria, prepared as described elsewhere.[13,27] This enzyme is one that has been used most extensively for the detection of mediator, including studies of mediator from adipocytes,[12,13,19] muscle,[21] liver,[14,25] and hepatoma cells.[11] The enzyme is converted from an inactive, phosphorylated

Table 1. Insulin-Dependent Regulator of
Pyruvate Dehydrogenase Extracted from
Adipocytes

Dilution of extract	PDH activity (nmol/min)	
	Control extract	Insulin extract
1/1	0.626	0.651
1/10	0.444	0.623
1/50	0.295	0.570
1/100	0.218	0.450
1/500	0.134	0.226
1/1000	0.117	0.173
1/10,000	0.125	0.136

form to an active dephosphorylated form by insulin *in vivo,* and by mediator activity *in vitro.* Initial mediator effects reported on this enzyme were relatively small. However, technical modification, such as improving factionation procedures and converting the bulk of the enzyme into the inactive form by preincubation with ATP have dramatically increased the achievable response.[27] In an attempt to detect possible mediator activity, comparable extracts from control and insulin-treated cells were added to mitochondria, and pyruvate dehydrogenase activity was assayed according to standard methods.[17,27] Normally, several dilutions are tested to determine the relative amounts of stimulatory material in the two samples. Typical results are shown in Table 1. At full strength, both samples often maximally stimulate the enzyme, but differences in concentration are revealed at lower dilutions. These results illustrate the basic mediator phenomenon. The context within which these results will need to be considered, as well as the use of this system to lead to the purification and characterization of the mediator are discussed in the following sections.

III. INSULIN MEDIATORS IN THE CONTEXT OF THE INSULIN EFFECTOR SYSTEM

A. Possibility of Multiple Insulin Mediators

Most analysis of the mediator hypothesis and data relevant to it have dealt with the simple model of a single, central component that mediates a wide range of effects. However, there is evidence that can be inferred from some characteristics of the mediator phenomenon that suggests the possible existence of more than one mediating component. Furthermore, there is no substantial evidence that rules out the existence of complex pathways within the insulin effector system involving multiple mediators. Two types of multiplicity can be considered: (1) regulatory components that have partial but agonistic roles in the total response to a single stimulus, either on the same or different enzymes, and (2) regulatory components that have antagonistic effects on the same processes. Experimental evidence exists for the latter possibility. The concentration-dependence of the action of many mediator preparations has been shown to be biphasic, with reduced or opposite effects at high concentrations.[7,13,18] This has been interpreted as indicating the action of a second, antagonistic regulator. In several cases, it has been possible to resolve the antagonistic regulators into separate monophasic preparations by partial purification.[7,18] In both cases, the concentrations of both antagonists appeared to be increased by insulin.[4,18] The possible significance of these opposing factors can only be speculative at this time. One possibility is that the antagonistic regulator acts as a limiting factor to prevent overstimulation in the presence of high concentrations of hormone. The mechanism by which the regulators produce a cumulative effect remains to be established.

The concept that the agonistic actions of a hormone may require multiple mediators for full expression also remains a possibility. As noted previously, there is no data to confirm that mediator activity observed in different enzyme systems can be attributed to a single component. It is feasible that different mediators are generated as part of a coordinated network with separate pathways leading to different enzymes. An interesting possibility raised by this model is the opportunity to integrate components such as cAMP and Ca^{2+}, which have generally been discounted as single mediators, into the overall scheme of insulin action. Several relationships can be envisioned among regulators in such a network: (1) separate pathways leading to different effector enzymes; (2) alternative pathways leading to the same enzyme; (3) antagonistic pathways leading to different effects on the same enzyme; and (4) sequential steps in the same pathway leading to one or more effects. The potential relationships between multiple regulators, specifically cAMP and Ca^{2+}, in coordinated pathways has been considered in detail and termed *synarchy*.[28] The possibility that insulin may act by producing coordinated alterations in a network of regulators should be considered as an alternative model for incorporating future data concerning insulin mediators.

B. Role of Mediators in Protein Phosphorylation Changes

A growing body of evidence supports the contention that protein phosphorylation changes serve as a unifying terminal covalent modification in the insulin effector system.[29,30] Insulin has been shown to affect the phosphorylation state of a wide range of cellular proteins,[31–33] and most, if not all, enzymes regulated by insulin are also regulated by changes in covalent phosphorylation.[34–37] Although this concept cannot be considered completely proven, it parallels an increasing recognition of protein phosphorylation as a ubiquitous regulatory phenomenon in cell function.[29] A link between mediator activity and changes protein phosphorylation would serve to strengthen the case that both are a part of the insulin effector system. Some evidence suggests that mediators may alter protein phosphorylation in producing their effects. Regulation of pyruvate dehydrogenase by mediator activity *in vitro* involves a decrease in protein phosphorylation by a process that is blocked by NaF, a protein phosphatase inhibitor.[20] In addition, inhibition of cAMP-dependent protein kinase[6–8] and stimulation of glycogen synthase phosphatase[21] would serve as a means by which a mediator might affect the phosphorylation and activity of a wide range of enzymes.

The relationship between mediator activity and protein phosphorylation changes may help clarify some puzzling aspects of insulin action on protein phosphorylation. For example, insulin appears to decrease the phosphorylation of some proteins while increasing the phosphorylation of other proteins or of other sites within the same protein. This apparently paradoxical result could be explained by the existence of multiple mediators, acting on different protein kinases and/or phosphatases with specificity for a limited population of phosphorylation sites. Resolution of this issue will require the identification of protein kinases and phosphatases involved in specific insulin effects, and determination of the ability of mediators to regulate them. This is not currently possible because no protein kinase or phosphatase involved in mediating an insulin effect has been conclusively identified.

The recent demonstration that the insulin receptor itself possesses protein kinase activity[38-40] and undergoes an autophosphorylation reaction in response to insulin binding[41-43] has raised new possibilities concerning the mechanism of insulin action. Although there is not yet evidence that activation of the receptor-associated kinase is a requisite step in the transduction of the hormonal stimulus into enzyme effects, the suggestion has been made that an enzyme cascade linked to the receptor kinase could produce enzyme phosphorylation changes without the necessity of a mediator. The components of such a system would require an unusual set of properties that have not been described in any known enzyme. The system would require that the receptor kinase, a tyrosine kinase,[38-43] ultimately regulate kinases or phosphatases that regulate enzyme targets, all of which act on phosphoserine residues. The linking of the receptor kinase to cellular substrates will therefore be a critical subject for research in evaluating this hypothesis. If receptor kinase activation can be established as a causal step in the insulin effector system, possible relationships with mediator mechanisms will require investigation. An enzyme cascade could be perceived as an alternative to a mediator pathway, or they may coexist as different elements in a coordinated system. Since the kinase is associated with the receptor, it is likely to be one of the earliest steps in the effector system, and may lead to mediator production, possibly by activating an enzyme that catalyzes this process. Another possibility is that the mediator is a substrate for the kinase, the action of which converts the mediator from an inactive to an active form. All such models would be highly speculative at this time, but it will be important to consider such possibilities in the analysis of data arising from the investigation of receptor kinase and mediator phenomena.

C. Proteolytic Involvement in Mediator Generation

An important corollary of the identification of an insulin mediator will be the identification of an insulin-dependent process that produces the mediator. This component will be required to complete the picture of mediator involvement in insulin action. The possibility that the mediator is a novel chemical component suggests that the process that produces it may also not have been recognized previously as a component of the effector system. A limited body of data has suggested that a proteolytic cleavage may be a required step in the generation of mediator by isolated membranes[13] and in cells,[44] and for the expression of insulin action on enzyme activity in cells.[44,45] Limited proteolysis has been shown to be an important regulatory mechanism in a variety of physiological processes,[46,47] but its role in hormone action has generally been limited to degradation of the ligand.[48] However, studies of mediator in adipocyte plasma membranes showed that its production was blocked if the membranes were pretreated with antiproteases, particularly some arginine esters that inhibit trypsinlike activities.[13] Furthermore, the mediator could be produced by treatment of membranes with low concentrations of trypsin in the absence of insulin.[13] One conception of the role of proteolysis is that it releases the mediator from a bound or precursor state. Alternatively, it might activate a second enzyme that catalyzes mediator production. It was recently reported that a proteolytic cleavage plays another role in insulin action. Insulin treatment of fibroblasts results in the conversion from an inactive to an active form of a protease-activated protein kinase that phosphorylates ribosome protein S6, which is phosphorylated in response to insulin in these cells,[49] implying the activation of a protease by insulin leading to this effect. These observations raise the possibility that proteolysis is a mechanistic link between two branches of the effector system, one leading to mediator production, the other to protein kinase activation. The proof of this model will require the identification of a protease activated by insulin, its substrates and products, and their effects. The coordination of these results with investigation of the mediator itself will help evaluate the significance of both phenomena.

IV. PURIFICATION AND CHARACTERIZATION OF PUTATIVE INSULIN MEDIATORS

A. Characterization in Heterogeneous Preparations

Some characteristics of mediators have been reported as part of the methodology of their preparation and observation. Mediator activity from a wide variety of sources is associated with a soluble, negatively charged (at pH 7.4) component that is more stable in acid than at neutral pH.[8,12,21,50,51] However, attempts at more detailed analysis have raised more conflicts than agreements. No insulin mediator has been successfully purified, allowing analysis in a homogeneous preparation, so attempts have been made to characterize the active components in crude preparations. These results must be interpreted with caution because of the possibility that other components may interfere with the analysis, leading to anomalous results, and because of the possibility that there may be multiple active components in the crude mixture.

One approach to characterization of the mediator has been to determine the sensitivity of the active component to specific agents, such as degradative enzymes. Several reports have indicated that mediator activity is sensitive to protease digestion, although this has been an area of controversy. Mediator activity from adipocyte plasma membranes was reported to be destroyed by trypsin or chymotrypsin treatment.[13] But other reports claim this effect to be inconsistent, although sensitivity to other proteases, including subtilisin, was found.[52] One report has been made of sensitivity of mediator from liver plasma membranes to degradation by β-galactosidase and neuraminidase,[14] but this report has not been independently confirmed. These reports suggest that the mediator may contain peptide or carbohydrate structure necessary for activity. However, the lack of agreement among the reports makes it difficult to draw a definitive conclusion about the identity of the mediator from these studies.

Another approach to identifying the mediator has been to analyze the contents of crude or partially purified preparations. Not surprisingly, many components have been identified in such studies, but in no case has a conclusive case been made to associate any of these components with mediator activity. Most preparations have been reported to contain peptide,[44,51] but sugars,[50] phospholipids,[53] and a mixture of prostaglandin E_1, inositol, and phosphate[54] have also been reported. In some cases, these analyses have been carried out on preparations asserted to be pure[50,54] but in neither case have supportive data accompanied the report. Thus, no analysis of a demonstrably pure preparation has been carried out, and no independent confirmation that a proposed structure possesses mediator properties has been presented. Thus, the question of the chemical identity of the mediator remains open pending these achievements.

The characteristics of the mediator have also been studied using chromatographic fractionation techniques. This approach serves the dual purpose of associating mediator activity with a particular behavior reflecting its characteristics and of partially purifying the mediator. The most extensively used fractionation technique has been size-exclusion chromatography both by conventional gel filtration and by high-pressure liquid chromatography. These results have generally indicated a molecular size of 1000–1500 daltons for mediator from a variety of sources,[11,12,21,44] with estimates as low as 500[24] or as high as 3000.[51] Within the range of accuracy of these estimates, it is not possible to consider them as indicating real differences.

Taken together, these data provide little basis for comparing the identities of the mediators being studied in various systems. The general similarity in size, charge, and stability, and the uncertainty of the data indicating differences in specific composition are consistent with the interpretation that the mediators reported in different systems are closely related. However, the resolution of this important question will require the purification and identification of specific mediators.

B. Status of Purification Methodology

Attempts at purifying mediator from various sources have met with limited success. These efforts face several difficult problems. First, the component to be purified is of unknown composition, making it difficult to choose conditions and approaches in advance. Second, the active component is likely to be present in small quantities, requiring extensive purification from starting material, which must be obtained in large amounts to yield sufficient purified mediator for analysis. Finally, detection of the mediator throughout the protocol is dependent on its bioactivity, requiring considerations to preserve this activity, and demanding time-consuming assay procedures at each step. In response to these considerations, a protocol has been devised to provide the most rapid and efficient purification by using high-pressure liquid chromatographic techniques.[27] This system has been applied to the purification of mediator from intact adipocytes using pyruvate dehydrogenase as the target enzyme as described in Section II. Two chromatographic approaches have been applied in gradient and isocratic modes to achieve purification, anion exchange and reversed phase partition separations. The mediator activity is resolved into two peaks, I and II, by the first anion-exchange gradient step. Each of these peaks has been purified separately. Their characteristics are compared in Table 2. Peak I activity has been purified to correspond to a single component. This component has been identified as a peptide, whose sequence revealed that it was a 50-amino acid, N-terminal fragment of bovine serum albumin, which was present during adipocyte preparation. Its identity has been confirmed as the active component in peak I by extracting adipocytes prepared without albumin, and finding that peak I is not produced. This peak is therefore an experimental artifact, but its production may provide some insights into real physiological mechanisms. The activity of this peptide may arise from a homology with a component of cellular origin. Further, the increased yield of Peak I peptide after insulin treatment may reflect an insulin-dependent process, such as an increase in cellular uptake of extracellular components, or an increased proteolytic activity, a possibility discussed in Section III.C. These possibilities will require investigation. Peak II activity has also been carried through several purification steps, but the state of purity and composition of the most purified fraction remain unknown. The activity is associated with extremely small amounts of material, and a large amount of starting material may be required to obtain enough Peak II component for analysis. Peak II is observed in cells prepared without albumin or any other exogenous components, suggesting that it is of cellular origin. When identified, it will be necessary to evaluate its physiological significance by determining its concentration in cells, response to insulin, and effects on various enzymes. These will serve as conclusive tests of the mediator hypothesis.

Table 2. Purification of Insulin-Dependent Regulators from Adipocyte Extracts

	Results observed	
Characteristic	Peak I	Peak II
Anion-exchange HPLC	Recovered at 30% eluting buffer	Recovered at 50% eluting buffer
Reverse-phase (C18) HPLC (0.1% TFA/CH$_3$CN)	Recovered at 35%	Not retained
Reverse-phase (C18) HPLC (0.1 M (NH$_4$)$_2$C$_2$H$_3$O$_2$, pH 6.0/CH$_3$OH)	—	Recovered at 20%
Reverse-phase (CN) HPLC (0.1% TFA/CH$_3$CN)	—	Not retained
Composition of purified fraction	50-amino acid peptide (albumin fragment)	π

V. PROSPECTS FOR PROGRESS IN INSULIN MEDIATOR RESEARCH

The approach to insulin mediator research described in this chapter remains one of the most promising avenues toward understanding the mechanism of insulin action. The results described in the preceding section emphasize that this approach is not immune to artifact and cannot guarantee success. However, the bulk of available data suggests that the mediator phenomenon will ultimately lead to the revelation of pathways in the insulin effector system. The immediate challenge is to purify and identify mediators from different experimental systems. This will allow more rigorous criteria concerning the role of the putative mediators *in vivo* to be applied. In addition, the relationship between the mediator phenomenon and other phenomena associated with the insulin effector system, such as receptor autophosphorylation, receptor internalization, insulin degradation and uptake, and covalent modification of enzymes by phosphorylation and dephosphorylation should be areas for fruitful investigation. The insulin effector system will undoubtedly turn out to be a complex mechanism, in which the mediator, or mediators, may play a central role. The elucidation of these mechanisms should be possible by resolving the issues raised in this chapter.

ACKNOWLEDGMENTS. The assistance of Ms. Tamera Reardon in obtaining the data described herein is gratefully recognized. Part of this work was carried out during my tenure as a postdoctoral fellow of the Juvenile Diabetes Foundation in Dr. Michael Czech's laboratory. The remainder has been supported by grants from the National Science Foundation and the American Diabetes Foundation.

VI. REFERENCES

1. Czech, M.P., 1977, *Annu. Rev. Biochem.* **46:**359–384.
2. Czech, M.P., 1981, in: *Handbook of Diabetes Mellitus* (M. Brownlee, ed.), pp. 117–149, Garland Press, New York.
3. Larner, J., Takeda, Y., Brewer, H.B., Huang, L.C., Hazen, R., Brooker, G., Murad, F., and Roach, P., 1976, in: *Metabolic Interconversions of Enzymes, 1975,* (S. Shaltiel, ed.), pp. 71–85, Springer-Verlag, Berlin.
4. Larner, J., 1982, *J. Cyclic. Nucleotide Res.* **8:**289–296.
5. Czech, M.P., Yu, K.-T., and Seals, J.R., 1984, *Diabetes Care* (in press).
6. Walkenbach, R.J., Hazen, R., and Larner, J., 1977, *Mol. Cell. Biochem.* **19:**31–41.
7. Larner, J., Lawrence, J.C., Roach, P.J., DePaoli-Roach, A.A., Walkenbach, R.J., Guinovart, J., and Hazen R.J., 1979, *Cold Spring Harbor Symp. Quant. Biol.* **6:**95–112.
8. Cheng, K., Galasko, G., Huang, L., Kellogg, J., and Larner, J., 1980, *Diabetes* **29:**659–661.
9. Turakulov, Y.K., Gainutdinov, M.K., Lavina, I.I., and Akhmatov, M.S., 1977, *Rep. Acad. Sci. USSR* **234:**1471–1474.
10. Kiechle, F.L., and Jarett, L., 1981, *FEBS Lett.* **133:**279–282.
11. Parker, J.C., Kiechle, F.L., and Jarett, L., 1982, *Arch. Biochem. Biophys.* **215:**339–344.
12. Kiechle, F.L., Jarett, L., Kotagal, N., and Popp, D.A., 1981, *J. Biol. Chem.* **256:**2945–2951.
13. Seals, J.R., and Czech, M.P., 1981, *J. Biol. Chem.* **255:**6529–6531.
14. Begum, N., Tepperman, H.M., and Tepperman, J., 1983, *Endocrinology* **112:**50–59.
15. Saltiel, A., Jacobs, S., Siegel, M., and Cuatrecasas, P., 1981, *Biochem. Biophys. Res. Commun.* **102:**1041–1047.

16. Jarett, L., Kiechle, F.L., Popp, D.A., Kotagal, N., and Gavin, J.R., 1980, *Biochem. Biophys. Res. Commun.* **96:**735–741.
17. Seals, J.R., and Jarett, L., 1980, *Proc. Natl. Acad. Sci. USA* **77:**77–81.
18. Saltiel, A.R., Siegel, M.I., Jacobs, S., and Cuatrecasas, P., 1982, *Proc. Natl. Acad. Sci. USA* **79:**3513–3517.
19. Begum, N., Tepperman, H.M., and Tepperman, J., 1982, *Endocrinology* **110:**1914–1921.
20. Popp, D.A., Kiechle, F.L., Kotagal, N., and Jarett, L., 1980, *J. Biol. Chem.* **255:**7540–7543.
21. Larner, J. Galasko, G., Cheng, K., DePaoli-Roach, A.A., Huang, L., Daggy, P., and Kellogg, J., 1979, *Science* **206:**1408–1410.
22. McDonald, J.M., Pershadsingh, H.A., Kiechle, F.L., and Jarett, L., 1981, *Biochem. Biophys. Res. Commun.* **100:**857–864.
23. Saltiel, A.R., Doble, A., Jacobs, S., and Cuatrecasas, P., 1983, *Biochem. Biophys. Res. Commun.* **110:**789–795.
24. Horvat, A., 1980, *Nature* **286:**906–908.
25. Caro, J.F., Folli, F., Cecchin, F., and Sinha, M.K., 1983, *Biochem. Biophys. Res. Commun.* **115:**375–382.
26. Gainutdinov, M.K., Abdullaev, N.K., Gizatullina, A.A., Akhmatov, M.S., and Turakulov, Y.K., 1978, *Dokl. Akad. Nauk. SSSR* **234:**1589–1592.
27. Seals, J.R., 1984, in: *Methods in Diabetes Research* (J. Larner and S. Pohl, eds.), Wiley, New York.
28. Rasmussen, H., 1981, *Calcium and cAMP as Synarchic Messengers,* Wiley, New York.
29. Cohen, P., 1982, *Nature* **296:**613–619.
30. Czech, M.P., Masague, J., Seals, J.R., and Yu, K.-T.,1984, in: *Biochemical Actions of Hormones,* Vol. 11 (G. Litwack, ed.), pp. 93–125, Academic Press, New York.
31. Avruch, J., Leone, G.R., and Martin, D.B., 1976, *J. Biol. Chem.* **251:**1511–1515.
32. Belsham, G.J., Denton, R.M., and Tanner, M.J.A., 1980, *Biochem. J.* **192:**457–467.
33. Benjamin, W.B., and Singer, I., *Biochemistry* **14:**3301–3309.
34. Seorain, V.S., Khatra, B.S., and Soderling, T.R., 1982, *Fed. Proc.* **41:**2618–2622.
35. Sobrino, F., and Hers, H.G., 1980, *Eur. J. Biochem.* **109:**239–246.
36. Ingebritsen, T.S., Geelen, M.J.H., Parker, R.A., Evenson, K.J., and Gibson, D.M., 1979, *J. Biol. Chem.* **254:**9986–9989.
37. Hughes, W.A., Brownsey, R.W., and Denton, R.M., 1980, *Biochem. J.* **192:**469–481.
38. Machicao, F., Urumow, T., and Wieland, O.H., 1982, *FEBS Lett.* **149:**96–100.
39. Shia, M.A., and Pilch, P.F., 1983, *Biochemistry* **22:**717–721.
40. Van Obberghen, E., Rossi, B., Kowalski, A., Gazzano, H., and Ponzio, G., 1983, *Proc. Natl. Acad. Sci. USA* **80:**945–949.
41. Kasuga, M., Karlsson, F.A., and Kahn, C.R., 1981, *Science* **215:**185–187.
42. Haring, H.-U., Kasuga, M., and Kahn, C.R., 1982, *Biochem. Biophys. Res. Commun.* **108:**1538–1545.
43. Kasuga, M., Zick, Y., Blithe, D.L., Crettaz, M., and Kahn, C.R., 1982, *Nature* **298:**667–669.
44. Larner, J., Cheng, K., Schwartz, C., Kikuchi, K., Tamura, S., Creacy, S., Dubler, R., Galasko, G., Pullin, C., and Katz, 1982, *Fed. Proc.* **41:**2724–2729.
45. Seals, J.R., and Czech, M.P., 1982, *Fed. Proc.* **41:**2730–2735.
46. Agarwal, M.K. (ed.), 1979, *Proteases and Hormones,* Elsevier/North-Holland, Amsterdam.
47. Reich, E., Rifkin, D.B., and Shaw, E. (eds.), 1975, *Proteases and Biological Control,* Cold Spring Harbor Laboratory, Cold Spring Harbor, New York.
48. Hammons, G.T., and Jarett, L., 1980, *Diabetes* **29:**475–486.
49. Perisic, O., and Traugh, J.A., 1983, *J. Biol. Chem.* **258:**9589–9592.
50. Gainutdinov, M.K., Turakulov, Y.K., Akhmatov, M.S., and Lavina, I.I., 1978, *Khim. Prirod. Soedin.* **1:** 141.
51. Seals, J.R., and Czech, M.P., 1981, *J. Biol. Chem.* **256:**2894–2899.
52. Larner, J., Cheng, L.K., Schwartz, C., Kikuchi, K., Tamura, S., Creacy, S., Dubler, R., Galasko, G., Pullin, C., and Katz, M., 1982, *Recent Prog. Horm. Res.* **38:**511–556.
53. Kiechle, F.L., and Jarett, L., 1983, *Mol. Cell. Biochem.* **56:**99–105.
54. Wasner, H.K., 1983, *Second International Symposium on Insulin Receptors,* August 31–September 3, 1983, Abstracts of reports, p. 33.

III

Mechanisms of Target Enzyme Modulation by Insulin Action

13

The Molecular Mechanism by Which Insulin Activates Glycogen Synthase in Mammalian Skeletal Muscle

Philip Cohen, Peter J. Parker, and James R. Woodgett

I. INTRODUCTION

One of the important metabolic effects of insulin is its ability to stimulate the synthesis of glycogen in skeletal muscle within minutes; much research has been directed toward understanding the molecular basis of this effect. Almost 25 years ago, Joseph Larner and co-workers showed that glycogen synthase, the rate-limiting enzyme in this pathway, could exist in at least two forms. One possessed little activity in the absence of glucose 6-phosphate (G6P), while the other was almost fully active in the absence of G6P.[1] Incubation of rat hemidiaphragms with insulin decreased the proportion of glycogen synthase in the G6P-dependent form which paralleled the increased rate of glycogen production.[1-3] Furthermore, the conversion of glycogen synthase to a G6P-independent form was still observed when glucose was omitted from the perfusion medium,[2,3] indicating that it was not a consequence of the increased rate of transport of glucose into muscle, which is also stimulated by insulin. Conversely, incubation of the hemidiaphragms with epinephrine increased the proportion of the G6P-dependent form and decreased the rate of glycogen synthesis.[2,3]

During the early 1960s, the conversion of glycogen synthase from a G6P-independent to a G6P-dependent form was shown to require the presence of Mg-ATP and a further protein,[4] suggesting that a phosphorylation reaction might be involved; this idea was subsequently substantiated.[5,6] The effects of insulin and epinephrine could

Philip Cohen • Department of Biochemistry, University of Dundee, Dundee, Scotland. *Peter J. Parker* • Imperial Cancer Research Fund Laboratories, London, England. *James R. Woodgett* • Salk Institute, La Jolla, California.

therefore be equated with partial dephosphorylation and phosphorylation of the enzyme, respectively, a supposition that has been confirmed by recent work.[7–13] These observations implied that insulin either causes the inhibition of a glycogen synthase kinase or the activation of a glycogen synthase phosphatase, or that it alters the conformation of glycogen synthase itself, making it a poorer substrate for a glycogen synthase kinase or a more effective substrate for a glycogen synthase phosphatase.

Following the discovery by Earl Sutherland that the stimulation of glycogenolysis by epinephrine is mediated by cyclic adenosine monophosphate (cAMP), it was observed that this second messenger activated the enzyme that converted glycogen synthase from a G6P-independent to a G6P-dependent form.[14,15] The basis for this effect became clear after the discovery of cAMP-dependent protein kinase,[16] when this enzyme was shown to catalyze the phosphorylation of glycogen synthase.[5,17] These findings indicated that epinephrine exerted its effect on glycogen synthase by activating cAMP-dependent protein kinase, raising the possibility that insulin might inhibit this protein kinase. However, measurements of cAMP in muscle failed to detect any decrease in this second messenger under conditions in which insulin stimulated glycogen synthesis.[2] This observation led Larner to suggest that insulin might inhibit cAMP-dependent protein kinase by another mechanism. It was postulated that insulin stimulated the formation of a second messenger distinct from cAMP, which bound to cAMP-dependent protein kinase and decreased its affinity for cAMP.[18] Such a molecule was subsequently reported to exist, and its levels were reported to be elevated by insulin (reviewed in ref. 19).

Our laboratory first became interested in this problem in 1973, when we noticed an unexplained discrepancy in the literature. Soderling et al.[5] had reported that the phosphorylation of glycogen synthase by cAMP-dependent protein kinase approached a plateau when approximately one phosphate had been incorporated per enzyme subunit (85,000 daltons). In contrast, Smith et al.[6] had reported that purified preparations of the G6P-dependent form of glycogen synthase contained six to seven phosphates per 85,000-dalton subunit. These studies led us to identify a glycogen synthase kinase, distinct from cAMP-dependent protein kinase, that was present as a trace contaminant in purified preparations of glycogen synthase.[20] This observation generated renewed interest in the mechanism by which insulin stimulated glycogen synthase. If more than one glycogen synthase kinase existed, then insulin might not exert its effect by inhibiting cAMP-dependent protein kinase, but rather by regulating a different protein kinase. This chapter reviews recent developments in our understanding of the structure and regulation of glycogen synthase, with particular emphasis on the mechanism by which insulin activates the enzyme.

II. THE GLYCOGEN SYNTHASE KINASES

Over the past 10 years, many glycogen synthase kinase preparations have been described that are distinct from cAMP-dependent protein kinase. Until recently, however, the relationship between these enzymes was unclear. Fortunately, the situation has been clarified during the past 2 years, and the glycogen synthase kinases listed in

Table 1. Glycogen Synthase Kinases That Have Been Identified

Class of protein kinase	Glycogen synthase kinase	Regulator	Sites phosphorylated
Cyclic nucleotide dependent	cAMP-dependent protein kinase	cAMP	$1a \gg 2 > 1b$
	cGMP-dependent protein kinase	cGMP	$1a > 2 > 1b$
Calcium ion dependent	Phosphorylase kinase	Ca^{2+}	2
	Calmodulin-dependent multiprotein kinase	Ca^{2+}	$2 > 1b$
Nonspecified	Glycogen synthase kinase-3	?	3a,3b, 3c
	Glycogen synthase kinase-4	?	2
	Glycogen synthase kinase-5	?	5
	Casein kinase-1	?	Not yet identified, but at least 6

Table 1 probably account for all the preparations that have been described (see ref. 21 and the following discussion). These enzymes can be conveniently classified into three types, as suggested by Krebs and Beavo,[22] i.e., cyclic nucleotide-dependent protein kinases, Ca^{2+}-dependent protein kinases, and protein kinases for which the physiological regulator(s) has not yet been specified.

The major sites on glycogen synthase phosphorylated by each of these enzymes have been identified, with the exception of those phosphorylated by casein kinase-1. The known sites are all contained within two cyanogen bromide peptides, termed CB-1 and CB-2, located at the N-terminal and C-terminal ends of the molecule, respectively. The amino acid sequences surrounding each phosphoserine residue are given in Fig. 1, and the organization of the sites in the polypeptide chain is shown schematically in Fig. 2.

A. Cyclic Nucleotide-Dependent Glycogen Synthase Kinases

The phosphorylation of glycogen synthase by cAMP-dependent protein kinase reaches different plateau values, depending on the concentration of protein kinase in the incubations.[7,25] Up to an incorporation of 2.5 phosphates per subunit, nearly all the radioactivity is located in sites 1a, 1b, and 2. Sites 1a and 1b are separated by only 13 residues in the primary structure, while site 2 is seven residues from the N-terminus of the protein (Fig. 1). The initial rate of phosphorylation of site 1a is 7- to 10-fold faster than at site 2 and 15- to 20-fold faster than at site 1b.

Cyclic guanosine monophosphate (GMP)-dependent protein kinase can phosphorylate glycogen synthase,[27] although the rate of phosphorylation is about 100-fold slower than that of cAMP-dependent protein kinase.[26,27] The same three sites are phosphorylated and in the same order—site 1a > site 2 > site 1b—although the difference in rate of phosphorylation between site 1a and the other two sites is not as pronounced.[26] In view of the slow rate of phosphorylation of glycogen synthase by cGMP-dependent protein kinase, as well as its very low concentration in skeletal

Figure 1. Amino acid sequences of cyanogen bromide peptides CB-1 and CB-2 of glycogen synthase. Peptide CB-1-containing site 2 is the N-terminal cyanogen bromide fragment of the protein, while CB-2-containing sites 3a, 3b, 3c, 5, 1a, and 1b is near the C-terminus (see Fig. 2). Note the large number of proline residues in the vicinity of sites 3a, 3b, and 3c, the large number of acidic residues that are C-terminal to site 5, and the pairs of adjacent basic amino acids that are N-terminal to sites 1a and 1b. These structural features are likely to be important in determining the substrate specificities of glycogen synthase kinase-3, glycogen synthase kinase-5, and cAMP-dependent protein kinase, respectively. (From Rylatt *et al.*[23] and Picton *et al.*[24])

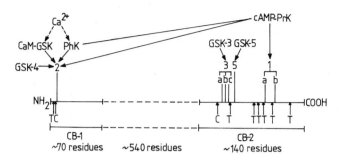

Figure 2. Organization of the phosphorylation sites in rabbit skeletal muscle glycogen synthase. Site 2 is seven residues from the N-terminus of the protein, while sites 3a, 3b, 3c, 5, 1a, and 1b are 30, 34, 38, 46, 87, and 100 residues from the N-terminus of peptide CB-2 (see Fig. 1). The positions at which trypsin (T) and chymotrypsin (C) cleave the native enzyme are indicated by arrows. Abbreviations: cAMP-PrK, cyclic adenosine monophosphate-dependent protein kinase; PhK, phosphorylase kinase; CaM, calmodulin; GSK, glycogen synthase kinase. The calmodulin-dependent glycogen synthase kinase (CaM-GSK) phosphorylates site 1b as well as site 2 (see text). (From Picton *et al.*[24])

muscle,[28] this protein kinase is not thought to be a significant glycogen synthase kinase *in vivo* and will not be considered further in this discussion.

B. Ca^{2+}-Dependent Glycogen Synthase Kinases

Roach and Larner[29,30] were the first to demonstrate that phosphorylase kinase is an effective glycogen synthase kinase; this observation has been confirmed by others.[31–33] The phosphorylation occurs at site 2.[31,32] The report of an additional phosphorylation site after prolonged incubation with high concentrations of phosphorylase kinase[34] can be explained by trace contamination with glycogen synthase kinase-5.[21] The rates of phosphorylation of glycogen phosphorylase and glycogen synthase by phosphorylase kinase are comparable *in vitro*.[31] However, the concentration of glycogen phosphorylase *in vivo* (80 μM) is much higher than that of glycogen synthase (2.8 μM).[7] If glycogen phosphorylase and glycogen synthase are mixed together at physiological concentrations, phosphorylase kinase phosphorylates glycogen phosphorylase preferentially.[35] The much higher concentration of glycogen phosphorylase *in vivo* may therefore suppress the rate of phosphorylation of glycogen synthase by phosphorylase kinase *in vivo*.

A Ca^{2+}-calmodulin-dependent protein kinase distinct from phosphorylase kinase was first identified in mammalian liver[36] and subsequently in skeletal muscle.[35] The skeletal muscle enzyme has been purified ≃5000-fold to homogeneity. It is a dodecamer with a molecular mass of 700,000 daltons formed from 12 identical 58,000-dalton subunits arranged as two hexagonal rings stacked one above the other.[37] The enzyme phosphorylates glycogen synthase at sites 2 and 1b, the initial rate of phosphorylation of site 2 being 5- to 10-fold faster than that of site 1b. The Ca^{2+}-calmodulin-dependent protein kinase has a broad substrate specificity *in vitro*[37,38] and it may mediate a number of the actions of Ca^{2+}-calmodulin *in vivo*. Accordingly, it has been termed the calmodulin-dependent multiprotein kinase.[38] It should be noted, however, that it is unable to phosphorylate glycogen phosphorylase. Furthermore, physiological concentrations of glycogen phosphorylase do not affect the rate of phosphorylation of glycogen synthase by this enzyme. The molar concentration of the calmodulin-dependent multiprotein kinase in rabbit skeletal muscle is 5- to 10-fold lower than that of phosphorylase kinase.

C. Nonspecified Glycogen Synthase Kinases

Four glycogen synthase kinases, the activities of which are unaffected by cyclic nucleotides and calcium ions, have been identified. The properties of these enzymes and their substrate specificities are summarized in Tables 2 and 3.

Glycogen synthase kinase-3 (GSK-3) phosphorylates glycogen synthase on three serine residues, termed sites 3a, 3b, and 3c, located within nine residues in the peptide chain (Figs. 1 and 2). The order of phosphorylation of these sites is not yet known. GSK-3 has been purified $\simeq$50,000-fold to homogeneity from skeletal muscle.[39-41] It is a monomeric protein that yields a molecular mass of 51,000 daltons by SDS polyacrylamide gel[39-41] and 47,000 daltons by sedimentation equilibrium centrifugation.[41] However, its apparent molecular mass on gel filtration is 75,000 daltons[21,41] (Table 2). GSK-3 can phosphorylate itself and up to four phosphates per mole enzyme can be incorporated.[39-41] The type II regulatory subunit of cAMP-dependent protein kinase is a substrate for GSK-3, and two residues (Ser-44 and Ser-47) are phosphorylated.[42] GSK-3 phosphorylates casein at a low rate (Table 3). The K-casein variant is phosphorylated preferentially, and the C-terminal cyanogen bromide fragment of this protein is labeled.[43] However, many other proteins that are phosphorylated by cAMP-dependent protein kinase are not touched by GSK-3 (Table 3). GSK-3 is identical to the enzyme termed factor F_A by Merlevede and co-workers.[39,44] This refers to the ability of GSK-3 to activate the Mg-ATP-dependent form of protein phosphatase-1, a reaction that involves the phosphorylation of a protein termed inhibitor-2 (see Section VI.A).

Glycogen synthase kinase-4 (GSK-4) phosphorylates glycogen synthase at site 2.[21] This enzyme has only been partially purified, and its subunit structure is therefore unknown. It appears to be highly specific for glycogen synthase, and no other protein tested is phosphorylated at a significant rate (Table 3). Its substrate specificity demonstrates that it is not a proteolytic fragment of phosphorylase kinase or the calmodulin-dependent multiprotein kinase that has lost its ability to be regulated by Ca^{2+}-calmodulin.[37] GSK-4 is identical to the enzyme termed $PC_{0.4}$ by Roach and co-workers.[45]

Glycogen synthase kinase-5 (GSK-5) phosphorylates glycogen synthase at site 5[21] and is identical to an enzyme termed $PC_{0.7}$[45], casein kinase-II, casein kinase G, or casein kinase TS (see ref. 21). The enzyme which has been purified and characterized from a number of mammalian tissues, has an $\alpha_2\beta_2$ structure, in which the apparent molecular masses of the α and β subunits are 43,000 and 26,000 daltons, respectively.[46-48] The

Table 2. Properties of Nonspecified Glycogen Synthase Kinases from Rabbit Skeletal Muscle[a,b]

Property	GSK-3	GSK-4	GSK-5	CK-1
Elution from DEAE-Sepharose	Buffer A[c]	Buffer A + 0.2 M NaCl	Buffer A + 0.5 M NaCl	ND
Elution from phosphocellulose	Buffer A + 0.2 M NaCl	Buffer A + 0.4 M NaCl	Buffer A + 0.7 M NaCl	Buffer A + 0.5 M NaCl
Effects of cyclic nucleotides and Ca^{2+}-calmodulin	None	None	None	None
Effect of inhibitor protein[d]	None	None	None	None
Inhibition by heparin	No	No	Yes ($K_i = 0.04$ µg/ml)	Yes ($K_i = 70$ µg/ml)
K_m for ATP (µM)	20	10	10	20
K_m for GTP (µM)	400	2000	50	ND
V (GTP/ATP)	0.65	0.42	0.85	ND
K_m for glycogen synthase (mg/ml)	0.3	0.3	1.0	ND
Apparent molecular mass on gel filtration	75,000	115,000	170,000	35,000
Activation of protein phosphatase-1[e]	Yes	No	No	No

[a] Data are taken from Cohen et al.,[21] Woodgett et al.,[37] and J.R. Woodgett (unpublished work).
[b] Abbreviations: CK-1, casein kinase-1; GSK-3, -4, -5, glycogen synthase kinases-3, -4, and -5; ND, not determined.
[c] 50 mM Tris-HCl pH 7.5 – 1.0 mM EDTA – 0.1% (v/v) 2-mercaptoethanol.
[d] The specific protein inhibitor of cyclic AMP dependent protein kinase.
[e] See Section VI.

Table 3. Substrate Specificities of Nonspecified Glycogen Synthase Kinases from Rabbit Skeletal Muscle

Protein substrate	Concentration		Relative activity (% max)			
	(mg/ml)	(μM)	GSK-3	GSK-4	GSK-5	CK-1
Glycogen synthase	0.6	7	100	100	10	12
Casein	2.0	100	3.5	2.8	70	100
Phosvitin	2.0	25	0.3–17[f]	0.8	100	100
R_{II}-subunit[d]	0.3	6	10–15	1.0	60	ND
Phosphorylase	1.5	15	<0.1	<0.1	<0.1	<0.4
Phosphorylase kinase[e]	2.0	6	<1	<1	<1	9–14
Inhibitor-1	0.2	10	<0.1	2.2	0.2	ND
L-Pyruvate kinase	0.6	10	<0.1	2.0	0.2	ND
Acetyl CoA carboxylase	0.7	3	<0.1	1.0	15	5
ATP-citrate lyase	0.6	5	<0.1	<0.1	<0.1	<1
Histone H1	0.2	10	<1	<0.1	0.8	<1
Histone H2B	0.3	20	<0.1	<0.1	<0.1	<1

[a] Data are taken from Cohen et al.,[21] Woodgett et al.,[37] and J. R. Woodgett (unpublished work).
[b] Abbreviations as in Table 2.
[c] The assays for CK-1 were carried out as described for GSK-3, GSK-4, and GSK-5,[21] except that the concentration of NaCl in the assays was 100 mM rather than 50 mM.
[d] Type II regulatory subunit of cAMP-dependent protein kinase.
[e] Although the β subunit of phosphorylase kinase has been reported to be phosphorylated 10- to 15-fold faster than the α subunit (Singh et al.[75]), our group found the phosphorylation of the β subunit to be only 3-fold faster than the α subunit (J.R. Woodgett, unpublished experiments).
[f] The rate varies 50-fold depending on the batch of phosvitin used. Only a less highly phosphorylated form of phosvitin appears to serve as a substrate for GSK-3.[21]

43,000-dalton component appears to be the catalytic subunit,[46,49] while incubation with Mg-ATP results in phosphorylation of the 26,000-dalton subunit.[46–48] This enzyme has a number of striking properties, including up to 40-fold activation by polyamines at low Mg^{2+} concentrations, potent inhibition by heparin, and ability to use GTP as a substrate almost as effectively as ATP[21,46] (Table 2). The enzyme has a broad substrate specificity in vitro[21,46,50] (Table 3).

Huang and co-workers have described a glycogen synthase kinase, which they originally termed glycogen synthase kinase-1.[51,52] It has recently become clear that this enzyme is distinct from the other nonspecified protein kinases (J.R. Woodgett and P. Cohen, unpublished work) and that it is identical to casein kinase-1.[46] It is also the enzyme termed $PC_{0.6}$[45] (P.J. Roach, personal communication). Casein kinase-1 is a monomeric enzyme with a molecular mass of ≃37,000 daltons.[46,52] If glycogen synthase is incubated for prolonged periods with high concentrations of casein kinase-1, up to six phosphates can be incorporated per enzyme subunit. The sites of phosphorylation have not yet been identified, although they all appear to be located in CB-1 and CB-2 (J.R. Woodgett and P. Cohen, unpublished observations).

III. EFFECT OF PHOSPHORYLATION ON THE ACTIVITY OF GLYCOGEN SYNTHASE

Although phosphorylation of glycogen synthase decreases the activity, the changes in kinetic parameters are complex and different phosphorylation sites have different effects. The K_m for UDPG increases progressively as the extent of phosphorylation increases, but the effects of phosphorylation can be reversed by the allosteric activator G6P, which decreases the K_m for UDPG. However, phosphorylation also increases the K_a for G6P and decreases the K_i for inhibitors, such as inorganic phosphate. The effect of inhibitors is antagonized by G6P.[53,54] Because of this complexity, activity measurements are therefore usually performed at a single UDPG concentration, ≈ 5 mM, in the presence and absence of a saturating concentration of G6P. The *activity ratio*, $\mp$G6P, can then be used to monitor the effect of phosphorylation on activity. These measurements have demonstrated that phosphorylation of sites 3a + 3b + 3c decreases the activity ratio and increases the K_a for G6P to a greater extent than phosphorylation at site 2.[21,55] Similarly, phosphorylation of site 2 decreases the activity ratio to a greater extent than at site 1a.[26,55] However, the effects are additive, so that even larger decreases in the activity ratio and increases in the K_a for G6P are observed when all five sites are combined.[55] Phosphorylation of sites 1b and 5 appear to have little or no effect on the activity ratio, $\mp$G6P.[21,26] The role of site 5 phosphorylation is considered further in Section V.

Phosphorylation of glycogen synthase with casein kinase-1 to 4 moles per subunit decreases the activity ratio to a level that appears to be lower than that obtained after phosphorylation of sites 3a + 3b + 3c).[51,52] However, as discussed earlier, the sites phosphorylated by casein kinase-1 have not yet been identified.

IV. ASSESSMENT OF THE RELATIVE ACTIVITIES OF GLYCOGEN SYNTHASE KINASES IN SKELETAL MUSCLE EXTRACTS

In studies reported by Cohen et al.,[21] skeletal muscle extracts were acidified to pH 6.1, centrifuged to pellet the protein–glycogen complex, and fractionated from 33–55% ammonium sulfate. This material was chromatographed on phosphocellulose, and nonspecified glycogen synthase kinases, which are retained by the column (Table 2), were eluted by stepwise increases in NaCl concentration. Assays were carried out in the presence of EGTA to inhibit Ca^{2+}-dependent protein kinases and the inhibitor protein (Table 2) to inactivate cAMP-dependent protein kinase. These experiments indicated that GSK-3 accounted for 84–92%, GSK-4 for 5–7%, and GSK-5 for 0.2–10% of the glycogen synthase kinase activity eluted from the column, depending on the particular assay conditions employed.[21] Casein kinase-1 was not detected in these studies, as it represents no more than 1–2% of the glycogen synthase kinase activity eluted from the column.

Precipitation at pH 6.1 and fractionation with ammonium sulfate were included to remove phosphorylase kinase prior to chromatography on phosphocellulose. However, recent work has demonstrated that this procedure not only eliminates phosphorylase kinase, but 90% of the GSK-4 as well. If muscle extracts are applied directly to phosphocellulose, comparable amounts of GSK-3 and GSK-4 activity are recovered from the column (J.R. Woodgett, unpublished work). It therefore appears that the potential activities of GSK-3 and GSK-4 in muscle extracts are similar, while that of casein kinase-1 is much lower. However, despite the low activity of casein kinase-1 toward glycogen synthase, prolonged incubation with high concentrations of this protein kinase results in the phosphorylation of many sites, decreasing the activity ratio to a low value (see Section III).

The activity of GSK-5 is also much lower than that of GSK-3 and GSK-4. Nevertheless, site 5 is almost fully phosphorylated *in vivo* under all conditions so far examined (see Section VII). This may reflect the very slow dephosphorylation of site 5 by the protein phosphatases present in skeletal muscle (Section VI).

V. ROLE OF SITE 5 PHOSPHORYLATION

Although site 5 does not affect the activity ratio of glycogen synthase, it has recently been demonstrated that phosphorylation of this residue is a prerequisite for phosphorylation of sites 3a + 3b + 3c by GSK-3.[56] Purified preparations of the G6P-independent form of glycogen synthase usually contain ≈0.5 moles phosphate per mole subunit in site 5,[12] which can be removed by incubation with potato acid phosphatase.[56] This treatment abolishes the phosphorylation of glycogen synthase by GSK-3 without affecting phosphorylation by either cAMP-dependent protein kinase, phosphorylase kinase, the calmodulin-dependent multiprotein kinase, or GSK-4. Rephosphorylation of site 5 restores the ability of GSK-3 to phosphorylate the enzyme.[56] The presence of 0.5 moles phosphate per mole subunit at site 5 may explain why phosphorylation by GSK-3 reaches a plateau near 1.5 moles phosphate per subunit *in vitro*,[39,55] rather than 3 moles per subunit. Roach and co-workers recently reported phosphorylation by GSK-3 in excess of 2 moles per subunit if the phosphate present at site 5 was first increased by preincubation with GSK-5 and Mg-ATP.[57] These results suggest that the role of GSK-5 is a novel one in forming the recognition site for another protein kinase (GSK-3). This may explain why site 5 is fully phosphorylated *in vivo* (see Section VII).

Similar observations have recently been made for the type II regulatory subunit of cAMP-dependent protein kinase.[42] This protein as normally isolated contains 1.5–1.8 moles phosphate per subunit, of which 1.3 moles is located in the thermolytic peptide containing Ser-74 and Ser-76, the sites phosphorylated by GSK-5. Removal of these phosphate groups by incubation with potato acid phosphatase prevents GSK-3 from phosphorylating Ser-44 and Ser-47. The amino acid sequence following Ser-76, i.e., Glu-Asp-Glu-Glu-Asp, is very similar to the sequence following site 5 of glycogen synthase, which is Glu-Asp-Glu-Glu-Glu[42] (Fig. 1).

VI. THE GLYCOGEN SYNTHASE PHOSPHATASES

Four protein phosphatases—1, 2A, 2B, and 2C—are capable of dephosphorylating proteins involved in the regulation of glycogen metabolism.[58] Three of these—1, 2A, and 2C—have broad substrate specificities *in vitro* and can dephosphorylate glycogen synthase.[59] At near physiological pH and Mg^{2+} concentrations, protein phosphatase-1 (PrP-1) accounts for 65–75% of the glycogen synthase phosphatase activity toward sites 1a, 2 and 3a + 3b + 3c. Protein phosphatase-2A (PrP-2A) accounts for the remainder of the activity toward these sites, since the contribution of protein phosphatase-2C, at only 1–2%, is negligible.[60] PrP-1 and PrP-2A dephosphorylate sites 1a, 2, and 3a + 3b + 3c at comparable rates *in vitro*.[61] However, if glycogen synthase is phosphorylated at sites 1a, 1b, and 2, using cAMP-dependent protein kinase, PrP-1 dephosphorylates site 2 5- to 10-fold faster than site 1a and more than 100-fold more rapidly than site 1b. The dephosphorylation of site 1a occurs more rapidly once site 2 is dephosphorylated, and dephosphorylation of site 1b only takes place at a significant rate after sites 2 and 1a are both dephosphorylated.[26]

Approximately 50% of the PrP-1 activity in skeletal muscle is bound to the protein–glycogen complex, the functional particles on which glycogen synthase and other enzymes of glycogen metabolism are located.[60] In contrast, neither PrP-2A nor PrP-2C is associated with this fraction.[60]

These observations suggest that PrP-1 may be the most important glycogen synthase phosphatase in skeletal muscle, although PrP-2A accounts for a significant percentage of the activity. Site 1b and especially site 5 are dephosphorylated very slowly by protein phosphatases 1, 2A, and 2C; no other phosphatases capable of dephosphorylating these sites have so far been detected. The large number of acidic residues immediately C-terminal to site 5 may act as a negative specificity determinant for protein phosphatases.[61] The very weak phosphatase activity toward this site may explain its high phosphorylation state *in vivo* (see Section VII).

A. Structure and Regulation of Protein Phosphatase-1 and Protein Phosphatase-2A

Protein phosphatase-1 can be isolated as the active 37,000-dalton catalytic subunit[62] or in an inactive form termed the Mg-ATP-dependent protein phosphatase.[63] The latter consists of a 1 : 1 complex between the catalytic subunit and a protein termed inhibitor-2.[62,64–66] Activation of this complex requires Mg-ATP and a further protein, termed factor F_A[63] that has been identified as GSK-3[39] (see also Section II.C). Activation is triggered by the phosphorylation of inhibitor-2 on a specific threonine residue.[64–66] This represents the only known example of a protein phosphatase that is activated by a protein kinase.

The identity of factor F_A and GSK-3 is of considerable interest, but rather surprising. The problem is that GSK-3 catalyzes two apparently opposing reactions, i.e., the phosphorylation of glycogen synthase and the activation of PrP-1, one of whose functions is to dephosphorylate glycogen synthase. Further information about the physiological role of the Mg-ATP dependent protein phosphatase is required to understand this anomaly. In this context, it would be relevant to measure the phosphorylation state of inhibitor-2 under different metabolic conditions (Section IX).

Protein phosphatase-1 is also inhibited by nanomolar concentrations of a further protein termed inhibitor-1, which only functions as an inhibitor if it is first phosphorylated by cAMP-dependent protein kinase.[67] The complete primary structure of inhibitor-1 has been determined,[68] and the site of phosphorylation is Thr-35. The rate of phosphorylation of inhibitor-1 *in vitro* is similar to that of glycogen synthase,[7] and its concentration in skeletal muscle (1.8 μM)[69] is higher than that of PrP-1 ($\approx$0.5 μM).[62]

The state of phosphorylation of inhibitor-1 *in vivo* is under hormonal control. Administration of epinephrine *in vivo* increases the level of phosphorylation three- to fourfold.[70] Using a perfused rat hemicorpus system, the phosphorylation state of inhibitor-1 was found to be exquisitely sensitive to β-adrenergic agonists, half-maximal effects being observed at 1.0 nM isoproterenol. A concentration of 0.5 nM isoproterenol produced a 40% increase in cAMP and a two-fold rise in inhibitor-1 phosphorylation, i.e., 15% to 30% phosphorylation. Both effects were prevented by nanomolar concentrations of insulin added together with isoproterenol. However, insulin had no effect on cAMP levels or inhibitor-1 phosphorylation either at high levels of isoproterenol (10 nM), or in the presence of the β-adrenergic antagonist L-propranolol, where the state of phosphorylation was 10%.[71]

These experiments suggest that the inhibition of protein phosphatase-1 by inhibitor-1 is likely to be important in the hormonal control of glycogen metabolism; the effects of epinephrine on the phosphorylation of glycogen synthase *in vivo* are explained most simply by this mechanism (Section VII.C). (The ability of insulin to suppress the effects of low concentrations of β-adrenergic agonists is considered further in Sections VII and IX.)

Protein phosphatase-2A can be resolved into three forms—PrP-2A$_0$, PrP-2A$_1$, and PrP-2A$_2$—by chromatography on DEAE-cellulose. These species, which have apparent molecular masses of 210,000, 210,000, and 150,000 daltons, respectively, each contain the same 36,000-dalton catalytic subunit.[59] PrP-2A$_1$ contains two subunits, 60,000 and 55,000 daltons, in addition to the catalytic subunit, while PrP-2A$_2$ contains only the 60,000-dalton component complexed with the catalytic subunit.[58,59,72,73] Rechromatography of PrP-2A$_1$ on DEAE-cellulose results in partial dissociation to PrP-2A$_2$.[59] This observation suggests that PrP-2A$_2$ may be generated from PrP-2A$_1$ during purification and that PrP-2A$_1$ is the form that exists *in vivo*.

PrP-2A$_0$, -2A$_1$ and -2A$_2$ are dissociated to the 36,000-dalton catalytic subunit by treatment with 80% ethanol at room temperature or by freezing and thawing in the presence of 0.2 M 2-mercaptoethanol. This treatment results in severalfold activation of PrP-2A$_1$ and a smaller increase in PrP-2A$_2$,[59] suggesting that the 55,000-dalton subunit may hold the catalytic subunit in a less active conformation. PrP-2A$_0$ accounts for 10–20% of the potential PrP-2A activity in skeletal muscle and liver.[59] This species is completely inactive, and its activity is only expressed following dissociation to the catalytic subunit. PrP-2A$_0$ has not yet been obtained in pure form, and its subunit structure is unknown. Mechanisms for regulating the activity of PrP-2A *in vivo* have not yet been identified. The catalytic subunits of PrP-1 and PrP-2A are distinct gene products,[62] and PrP-2A activity is unaffected by inhibitor-1 and inhibitor-2.[61]

VII. PHOSPHORYLATION STATE OF GLYCOGEN SYNTHASE IN VIVO

A. Methodology

In order to determine the phosphorylation state of a protein in vivo, it is essential that the tissue be extracted and that the purification be carried out under conditions that prevent further phosphorylation or dephosphorylation from taking place. This objective is achieved by including EDTA and sodium fluoride in the buffers. The EDTA chelates Mg^{2+} and so inhibits protein kinases, while sodium fluoride inhibits protein phosphatases. In the case of glycogen synthase, the effectiveness of this procedure has been validated by measurements of the activity ratio, $\mp$G6P, and of K_a for G6P in the muscle extracts and at each stage of purification.[7,12,13] The phosphate bound to the pure enzyme can then be determined by standard procedures.

However, for enzymes that are regulated by multisite phosphorylation the phosphorylation of each site must be measured in order to obtain useful information. This complex analysis was greatly facilitated by the finding that incubation of glycogen synthase with trypsin cleaves the molecule between sites 3c and 5, while chymotrypsin cleaves just N-terminal to site 3a (Fig. 2). This approach facilitated assessment of the phosphate content of sites 2, 3a + 3b + 3c, and 5 + 1a + 1b.[12,13] The development of further methodology for separating the tryptic peptides containing each phosphorylation site[12,13] was required to estimate the phosphate content of sites 5, 1a, and 1b individually; these procedures permitted evaluation of the phosphorylation of sites 2 and 3a + 3b + 3c by a second independent method.

B. Effect of Insulin[13]

In these experiments, animals were deprived of food for 24 hr in order to reduce the level of insulin in the circulation. Fifteen minutes prior to sacrifice, rabbits were given an intravenous injection of L-propranolol (2 mg/kg body weight) or propranolol plus insulin (16.5 μg/kg). In the latter case, a further injection of insulin (16 μg/kg) was given 7.5 min prior to sacrifice. The purpose of injecting propranolol was twofold: (1) to minimize animal-to-animal differences that might be caused by the presence of variable amounts of epinephrine in the circulation, and (2) to ensure that any effect of insulin would not be related to its ability to suppress the rise in cAMP produced by low concentrations of epinephrine (Section VI). (This important point is considered further in Sections VII–IX.)

In muscle extracts prepared from animals that had been injected with propranolol alone, the activity ratio, $\mp$G6P, of glycogen synthase was 0.18 ± 0.02, and the K_a for G6P was 1.2 ± 0.1 mM (±SEM for 11 preparations). Injection of propranolol plus insulin increased the activity ratio to 0.35 ± 0.02 and decreased the K_a for G6P to 0.6 ± 0.05 mM (±SEM for 12 preparations) after 15 min. Similar results were generally, but not quite so reproducibly, observed when the amount of insulin injected was decreased four-fold and the animals sacrificed after 5 min.

Glycogen synthase isolated from animals injected with propranolol alone contained 2.75 ± 0.09 moles phosphate per mole subunit (±SEM for 14 preparations). This value decreased to 2.33 ± 0.09 moles phosphate per mole subunit (±SEM for 14 preparations) after injection of propranolol plus insulin. This confirms that insulin decreases the phosphorylation state of glycogen synthase, but only by 15%.

The tryptic phosphopeptides containing the known phosphorylation sites (Fig. 2) were isolated from animals injected with propranolol or propranolol plus insulin. The results, summarized in Fig. 3, demonstrated that covalently bound phosphate was associated with every site. The sum of the phosphate in each peptide (3.08 moles per mole subunit in animals injected with propranolol and 2.61 moles per mole subunit in animals injected with propranolol plus insulin[13]) was in good agreement with the total amount of phosphate in glycogen synthase. Insulin decreased the phosphate content of the tryptic peptide-containing sites (3a + 3b + 3c) by ≈0.4 mole per mole. Changes in the phosphate contents of other sites were not statistically significant, although the possibility that small changes in the phosphorylation of these other sites take place is not excluded. The increase in activity ratio, ∓G6P, from 0.18 to 0.35 and the decrease in K_a for G6P from 1.2 to 0.6 mM would be expected from a loss of 0.4–0.5 moles phosphate from sites 3a + 3b + 3c *in vitro*.[13,55]

Lawrence and co-workers[74] recently incubated rat hemidiaphragms with [^{32}P]-P$_i$ both in the presence and the absence of insulin. Following extraction of the tissue, glycogen synthase was immunoprecipitated, digested with cyanogen bromide, and sub-

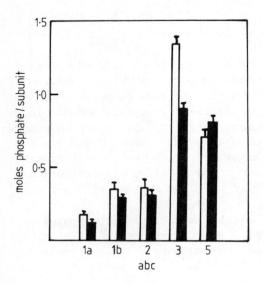

Figure 3. Effect of insulin on the phosphorylation of the seven phosphoserine residues of rabbit skeletal muscle glycogen synthase. Open bars show the phosphate contents of the various sites in normally fed animals injected with propranolol; the filled bars represent the phosphate contents after injection of propranolol plus insulin. (Data are replotted from Parker *et al.*[13])

jected to SDS polyacrylamide gel electrophoresis. These experiments demonstrated that peptides CB-1 (containing site 2) and CB-2 (containing sites 3a, 3b, 3c, 5, 1a, and 1b) are the only cyanogen bromide peptides phosphorylated *in vivo*. In the absence of insulin, ^{32}P-radioactivity in CB-2 was 6.9-fold greater than in CB-1, in agreement with our own studies. Incubation with insulin (25 mU/ml) for 30 min reduced the ^{32}P radioactivity associated with glycogen synthase by 44%—a larger decrease than that observed after injection of insulin *in vivo*.[13] In the presence of insulin, ^{32}P radioactivity was 6.4-fold greater in CB-2 than in CB-1. From these results, it can be calculated that 89% of the decrease in ^{32}P radioactivity occurred in CB-2. This is also in agreement with our studies, although the results of Lawrence and co-workers indicate that the phosphate content of CB-1 (containing site 2) is significantly decreased by insulin under their conditions.

C. Effect of Epinephrine[7,12]

An understanding of the molecular mechanism by which insulin stimulates glycogen synthase activity first requires an explanation of the effect of epinephrine on the enzyme. Intravenous injection of a maximally effective dose of epinephrine (70 μg/kg body weight) 30 sec prior to sacrifice decreased the activity ratio, $\mp$G6P, from 0.21 ± 0.01 to 0.04 ± 0.01 and increased the K_a for G6P from 1.38 ± 0.11 mM to 5.95 ± 0.74 mM ($\pm$SEM for nine preparations). The phosphate content was increased from 2.94 ± 0.10 to 5.08 ± 0.14 mole/mole subunit ($\pm$SEM for 18 preparations). Similar results have been reported by other laboratories.[9,11]

The effect of epinephrine on the distribution of phosphate is summarized in Fig. 4. The sum of the phosphate in each tryptic peptide (3.13 mole/mole subunit in the presence of propranolol and 5.25 mole/mole subunit after injection of epinephrine[12]) was again in good agreement with the total phosphate content of glycogen synthase. Epinephrine increased the phosphorylation of sites 1a and 1b by 0.25–0.3 mole/mole subunit, site 2 by 0.6 mole/mole subunit, and sites 3a + 3b + 3c by 1.2 mole/mole subunit (Fig. 4). The inactivation produced by epinephrine is therefore caused by increased phosphorylation at sites 3a + 3b + 3c and 2.

The considerable increase in phosphorylation at sites 3a + 3b + 3c was surprising. The effects of epinephrine are thought to be mediated by cAMP, yet cAMP-dependent protein kinase does not phosphorylate sites 3, and the activity of GSK-3 is unaffected by cAMP. Furthermore, attempts to demonstrate the phosphorylation and activation of GSK-3 by cAMP-dependent protein kinase *in vitro* or the activation of GSK-3 following administration of epinephrine *in vivo* have been unsuccessful (B.A. Hemmings and J.R. Woodgett, unpublished work). The possibility that increased phosphorylation at sites 1a, 1b, and 2 alters the conformation of glycogen synthase in such a way that sites 3a + 3b + 3c are phosphorylated more rapidly by GSK-3 has also been excluded.[56]

The simplest explanation for the increased phosphorylation of sites 3a + 3b + 3c in response to epinephrine is that PrP-1 is inhibited through the phosphorylation of inhibitor-1 (Section VI). However, the high levels of phosphorylase *a* formed in response to epinephrine may also decrease the rate of dephosphorylation of sites

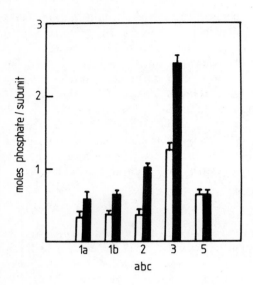

Figure 4. Effect of epinephrine on the phosphorylation of the seven phosphoserine residues of rabbit skeletal muscle glycogen synthase. Open bars show the phosphate contents of the various sites in animals starved for 24 hr that were injected with propranolol; filled bars represent the phosphate contents after injection of epinephrine. (Data are replotted from Parker et al.[12])

3a + 3b + 3c by simple competition for PrP-1 and PrP-2A, since these two enzymes account for all the phosphorylase phosphatase activity exhibited in skeletal muscle.[58–61] In view of the high concentration of phosphorylase (80 μM) relative to glycogen synthase (2.8 μM) *in vivo*, this additional mechanism may be important.

The increased phosphorylation of sites 1a and 1b is presumably catalyzed by cAMP-dependent protein kinase. Increased phosphorylation of site 2 could take place, however, through (1) phosphorylation by cAMP-dependent protein kinase; (2) phosphorylation by phosphorylase kinase, which is activated by cAMP-dependent protein kinase; or (3) inhibition of PrP-1 through the two mechanisms discussed above. These three possibilities might explain why site 2 becomes maximally phosphorylated in response to epinephrine (Fig. 4).

The high level of phosphorylation in the absence of epinephrine (3 mole/mole subunit) is presumably a reflection of the activity of GSK-3, GSK-4, and GSK-5 (and perhaps casein kinase-1) under these conditions. This high basal level of phosphorylation makes it extremely difficult, if not impossible, to study the effects of low concentrations of epinephrine on glycogen synthase. In the perfused rat hemicorpus, 0.5 nM isoproterenol increased the phosphate content of inhibitor-1 from 0.15 to 0.3 mole/mole, doubling its activity (Section VI). A similar effect on glycogen synthase would only increase the phosphate content by 5%, which would not register as a significant change in the activity or phosphate content of glycogen synthase, in view of the technical limitations of these experiments. This presumably explains why the activity ratio, ∓G6P, and K_a for G6P are not significantly affected if injection with propranolol prior to sacrifice is omitted[12] (Section VIII).

The ability of insulin to suppress the rise in cAMP produced by epinephrine is only observed in skeletal muscle at low concentrations of β agonists.[71] It therefore follows that it is extremely difficult to detect antagonism between insulin and epinephrine by measuring the activity ratio $\mp$G6P, or phosphate content of glycogen synthase.

VIII. INSULIN AND THE REGULATION OF cAMP-DEPENDENT PROTEIN KINASE

The original suggestion that insulin might activate glycogen synthase by forming a chemical mediator that binds to cAMP-dependent protein kinase and decrease its affinity for cAMP[18] was made before other glycogen synthase kinases had been identified. At the time, it appeared to represent the simplest explanation for the failure to detect significant changes in cAMP levels in response to insulin. However, it now seems appropriate to review this postulate in the light of recent findings.

It is clear that this mechanism cannot account for the activation by insulin observed in the presence of propranolol. An inhibition of cAMP-dependent protein kinase should result in decreased phosphorylation of sites 1a, 1b, and 2, but these sites are essentially unaffected under these conditions (Fig. 3). It could be argued, however, that the injection of propranolol selects against detecting such an effect of insulin, since this β-antagonist decreases cAMP to basal levels and so inhibits cAMP-dependent protein kinase in the absence of insulin. Therefore, if propranolol were omitted, the activation of glycogen synthase might occur through two mechanisms, one of which involved inhibition of cAMP-dependent protein kinase. This possibility seems most unlikely according to the following rationale: Although propranolol was injected to simplify interpretation of the results, its omission does not significantly affect the activity ratio, $\mp$G6P, or K_a for G6P of glycogen synthase. When saline was substituted for propranolol, the activity ratio was 0.22 ± 0.03, and the K_a for G6P was 1.4 ± 0.1 mM ($\pm$ SEM for three preparations). After injection of insulin, in the absence of propranolol the activity ratio, G6P, was 0.34 ± 0.02 and the K_a for G6P was 0.8 ± 0.1 mM ($\pm$ SEM for three preparations).

Recent work appears to argue against the idea that activation of glycogen synthase by insulin involves an inhibition of cAMP-dependent protein kinase.

IX. MOLECULAR MECHANISM BY WHICH INSULIN DEPHOSPHORYLATES GLYCOGEN SYNTHASE

The results reviewed in Section VII.B demonstrate that the acute effect of insulin to activate glycogen synthase within minutes is explained by partial dephosphorylation of the tryptic peptide-containing sites 3a + 3b + 3c. This indicates that the action of insulin must involve inhibition of GSK-3 and/or activation of PrP-1 and PrP-2A (Fig. 5). PrP-1 and PrP-2A are not specific for the dephosphorylation of sites 3a + 3b + 3c *in vitro,* and site 2 is dephosphorylated at a comparable rate.[23,61] It is possible, however, that these phosphatases are more specific for sites 3 *in vivo* or that dephosphorylation of site 2 by insulin is counterbalanced by activation of one of the glycogen synthase kinases that act on this site. It should also be remembered that

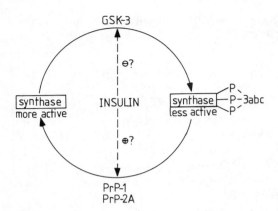

Figure 5. Insulin regulates the phosphorylation state of sites 3a + 3b + 3c of glycogen synthase by inhibiting glycogen synthase kinase-3 (GSK-3) and/or activating protein phosphatase-1 (PrP-1) and/or protein phosphatase 2A (PrP-2A).

the sites phosphorylated by casein kinase-1 have not yet been identified (see Section II.C). Therefore, the possibility that this enzyme might also phosphorylate the tryptic peptide-containing sites (3a + 3b + 3c) cannot yet be discounted.

In order to simplify interpretation of the results animals were injected with propranolol prior to sacrifice. Under these conditions, the levels of phosphorylation of glycogen phosphorylase and inhibitor-1 were extremely low (<5%) even in the absence of insulin, and administration of insulin did not affect the phosphorylation state of these proteins. This excludes the possibility that insulin acts by decreasing the phosphorylation of these proteins, leading to activation of PrP-1 and PrP-2A (Section VII.C).

It does not seem possible that insulin could act by decreasing the intracellular concentration of cAMP. In the presence of propranolol, the phosphorylation state of inhibitor-1 is very low, indicating that cAMP-dependent protein kinase is essentially inactive even before the addition of insulin. Furthermore, insulin does not significantly affect the sites on glycogen synthase phosphorylated by cAMP-dependent protein kinase (sites 1a, 1b, and 2). The activities of GSK-3, casein kinase-1, and PrP-2A are unaffected by cAMP, and the only mechanism by which PrP-1 is known to be regulated by this second messenger (phosphorylation of inhibitor-1) is inoperative under these conditions. Finally, insulin does not influence cAMP levels in skeletal muscle in the presence of propranolol.[71]

It is also inconceivable that insulin could act by altering the intracellular concentration of calcium ions. First, the effects of insulin are observed in resting muscle, in which the cytoplasmic concentration of calcium ions ($\leqslant 10^{-7}$ M) should be insufficient to activate Ca^{2+}-dependent enzymes. Second, the Ca^{2+}-dependent glycogen synthase kinases do not phosphorylate sites (3a + 3b + 3c) (Fig. 2). Third, the only known Ca^{2+}-dependent protein phosphatase (2B) does not dephosphorylate glycogen synthase.[58,61] The Ca^{2+}-dependent glycogen synthase kinases presumably play a role in suppressing the activity of glycogen synthase during muscle contraction. However, the *in vivo* experiments needed to test this hypothesis have yet to be performed.

At present, it is only possible to speculate about the mechanism by which insulin might regulate GSK-3, casein kinase-1, PrP-1, and PrP-2A. However, two possibilities can be envisaged. First, interaction of insulin with its receptor might generate a second messenger that would regulate one or more of these enzymes. Alternatively, a second messenger for insulin might not exist, and the tyrosine-specific protein kinase associated with the insulin receptor (see Chapters 1, 4, and 5, this volume) might mediate the stimulation of glycogen synthase by this hormone. The binding of insulin to the α-subunit of the receptor, located on the outer surface of the plasma membrane, activates the tyrosine protein kinase that appears to be associated with the β subunit. The tyrosine kinase might then phosphorylate one of the serine-specific protein kinases or phosphatases that act on sites 3 *in vivo*.

As discussed in Section VI, protein phosphatase-1 can be isolated either as a free catalytic subunit or in an inactive form consisting of a 1 : 1 complex between the catalytic subunit and inhibitor-2. It is particularly intriguing that this species can only be activated by GSK-3, a reaction that involves the phosphorylation of inhibitor-2 on a threonine residue. It was therefore of interest to investigate whether the insulin receptor kinase was capable of phosphorylating homogeneous preparations of one or more of these proteins. These experiments, carried out in collaboration with Dr. Raphael Nemenoff and Dr. Joseph Avruch, have demonstrated that the insulin receptor is capable of phosphorylating inhibitor-2 on a tyrosine residue, even when this protein is associated with PrP-1. The 37,000-dalton catalytic subunit of PrP-1[62] and the 51,000-dalton catalytic unit of GSK-3 are not phosphorylated by the insulin receptor. In contrast to phosphorylation by GSK-3, however, phosphorylation of inhibitor-2 by the insulin receptor does not lead to activation of PrP-1. In order to ascertain whether the phosphorylation of inhibitor-2 has any physiological significance, it will be necessary to study the effects of hormones on the phosphorylation of inhibitor-2 and determine whether this protein contains phosphotyrosine *in vivo*.

ACKNOWLEDGMENTS. This work was supported by a Programme Grant from the Medical Research Council and the British Diabetic Association. P.J.P. acknowledges a postdoctoral fellowship and J.R.W. a postgraduate studentship from the Medical Research Council.

X. REFERENCES

1. Villar-Palasi, C., and Larner, J., 1960, *Biochim. Biophys. Acta* **39:**171–173.
2. Craig, J.W., and Larner, J., 1964, *Nature* **202:**971–973.
3. Danforth, W.H., 1965, *J. Biol. Chem.* **240:**588–593.
4. Friedman, D.L., and Larner, J., 1963, *Biochemistry* **2:**669–675.
5. Soderling, T.R., Hickenbottom, J.P., Reiman, E.M., Hunkeler, F.L., Walsh, D.A., and Krebs, E.G., 1970, *J. Biol. Chem.* **245:**6317–6328.
6. Smith, C.H., Brown, N.E., and Larner, J., 1971, *Biochim. Biophys. Acta* **242:**81–88.
7. Cohen, P., 1978, *Curr. Top. Cell. Regul.* **14:**117–196.
8. Roach, P.J., Rosell-Perez, M., and Larner, J., 1977, *FEBS Lett.* **80:**95–98.
9. Sheorain, V.S., Khatra, B.S., and Soderling, T.R., 1981, *FEBS Lett.* **127:**94–96.
10. Chiasson, J.L., Aylward, S.G., Shikama, H., and Exton, J.H., 1981, *FEBS Lett.* **127:**97–100.

11. Uhing, R.J., Shikama, H., and Exton, J.H., 1981, *FEBS Lett.* **80:**95–98.
12. Parker, P.J., Embi, N. Caudwell, F.B., and Cohen, P., 1982, *Eur. J. Biochem.* **124:**47–55.
13. Parker, P.J., Caudwell, F.B., and Cohen, P.,1983, *Eur. J. Biochem.* **130:**227–234.
14. Rosell-Perez, M., and Larner, J., 1964, *Biochemistry* **3:**81–88.
15. Huijing, F., and Larner, J., 1966, *Biochem. Biophys. Res. Commun.* **23:**259–263.
16. Walsh, D.A., Perkins, J.P., and Krebs, E.G., 1968, *J. Biol. Chem.* **243:**3763–3765.
17. Schlender, K.K., Wei, S.H., and Villar-Palasi, C., 1969, *Biochim. Biophys. Acta* **191:**272–278.
18. Larner, J., 1973, *Diabetes* **21:**428–438.
19. Larner, J., 1982, *J. Cyclic Nucleotide Res.* **8:**289–296.
20. Nimmo, H.G., and Cohen, P., 1974, *FEBS Lett.* **47:**162–167.
21. Cohen, P., Yellowlees, D., Aitken, A., Donella-Deana, A., Hemmings, B.A., and Parker, P.J., 1982, *Eur. J. Biochem.* **124:**21–35.
22. Krebs, E.G., and Beavo, J.A., *Annu. Rev. Biochem.* **48:**923–959.
23. Rylatt, D.B., Aitken, A., Bilham, T., Condon, G.D., Embi, N., and Cohen, P., 1980, *Eur. J. Biochem.* **107:**529–537.
24. Picton, C., Aitken, A., Bilham, T., and Cohen, P., 1982, *Eur. J. Biochem.* **124:**37–45.
25. Soderling, T.R., 1975, *J. Biol. Chem.* **250:**5407–5412.
26. Embi, N., Parker, P.J., and Cohen, P., 1981, *Eur. J. Biochem.* **115:**405–413.
27. Lincoln, T.M., and Corbin, J.D., 1977, *Proc. Natl. Acad. Sci. USA* **74:**3239–3243.
28. Takai, Y., Kishimoto, A., and Nishizuka, Y., 1982, in: *Calcium and Cell Function*, Vol. II (W.Y. Cheung, ed.), pp. 385–412, Academic Press, New York.
29. Roach, P.J., DePaoli-Roach, A.A., and Larner, J., 1978, *J. Cyclic Nucleotide Res.* **4:**245–257.
30. DePaoli-Roach, A.A., Roach, P.J., and Larner, J., 1979, *J. Biol. Chem.* **254:**4212–4219.
31. Embi, N., Rylatt, D.B., and Cohen, P., 1979, *Eur. J. Biochem.* **100:**339–347.
32. Soderling, T.R., Srivastava, A.K., Bass, Y.A., and Khatra, B.S., 1979, *Proc. Natl. Acad. Sci. USA* **76:**2536–2540.
33. Walsh, K.Y., Millikin, D.M., Schlender, K.K., and Reimann, E.M., 1979, *J. Biol. Chem.* **254:**6611–6616.
34. Soderling, T.R., Shearain, V.S., and Ericsson, L.H., 1979, *FEBS Lett.* **106:**181–184.
35. Woodgett, J.R., Tonks, N.K., and Cohen, P., 1982, *FEBS Lett.* **148:**5–11.
36. Payne, M.E., and Soderling, T.R.,1980, *J. Biol. Chem.* **255:**8054–8056.
37. Woodgett, J.R., Davison, M.T., and Cohen, P., 1983, *Eur. J. Biochem.* **136:**481–487.
38. McGuinness, T.L., Lai, Y., Greengard, P., Woodgett, J.R., and Cohen, P., 1983, *FEBS Lett.* **163:**329–337.
39. Hemmings, B.A., Yellowlees, D., Kernohan, J.C., and Cohen, P., 1981, *Eur. J. Biochem.* **115:**443–451.
40. Hemmings, B.A., and Cohen, P., 1983, *Methods Enzymol.* **99:**337–345.
41. Woodgett, J.R., and Cohen, P., 1984, *Biochim. Biophys. Acta* **788:**339–347.
42. Hemmings, B.A., Aitken, A., Cohen, P., Rymond, M., and Hofmann, F., 1982, *Eur. J. Biochem.* **127:**473–481.
43. Donella-Deana, A., Pinna, L.A., Hemmings, B.A., and Cohen, P., 1983, *Biochim. Biophys. Acta* **745:**149–153.
44. Vandenheede, J.R., Yang, S.D., Goris, J., and Merlevede, W., 1980, *J. Biol. Chem.* **255:**11768–11774.
45. DePaoli-Roach, A.A., Roach, P.J., and Larner, J., 1979, *J. Biol. Chem.* **254:**12062–12068.
46. Hathaway, G., and Traugh, J.A., 1982, *Curr. Top. Cell. Regul.* **21:**101–127.
47. DePaoli-Roach, A.A., Ahmad, Z., and Roach, P.J., 1981, *J. Biol. Chem.* **256:**8955–8962.
48. Huang, K.P., Itarte, E., Singh, T.J., and Akatsuka, A., 1982, *J. Biol. Chem.* **257:**3236–3242.
49. Feige, J., Cochet, C., Pirollet, F., and Chambaz, E.M., 1983, *Biochemistry* **22:**1452–1459.
50. DePaoli-Roach, A.A., Roach, P.J., Pham, K., Kramer, G., and Hardesty, B., 1981, *J. Biol. Chem.* **256:**8871–8874.
51. Itarte, E., Robinson, J.C., and Huang, K.P., 1977, *J. Biol. Chem.* **252:**1231–1234.
52. Itarte, E., and Huang, K.P., 1979, *J. Biol. Chem.* **254:**4052–4057.
53. Roach, P.J., Takeda, Y., and Larner, J., 1976, *J. Biol. Chem.* **251:**1913–1919.
54. Roach, P.J., 1982, *Curr. Top. Cell. Regul.* **20:**45–105.
55. Embi, N., Rylatt, D.B., and Cohen, P., 1980, *Eur. J. Biochem.* **107:**519–527.
56. Picton, C., Woodgett, J.R., Hemmings, B.A., and Cohen, P., 1982, *FEBS Lett.* **150:**191–196.

57. DePaoli-Roach, A.A., Ahmad, Z., Camici, M., Lawrence, J.C., and Roach, P.J., 1983, *J. Biol. Chem.* **258**:10702–10709.
58. Ingebritsen, T.S., and Cohen, P., 1983, *Science* **221**:331–338.
59. Ingebritsen, T.S., Foulkes, J.G., and Cohen, P., 1983, *Eur. J. Biochem.* **132**:263–274.
60. Ingebritsen, T.S., Stewart, A.A., and Cohen, P., 1983, *Eur. J. Biochem.* **132**:297–307.
61. Ingebritsen, T.S., and Cohen, P., 1983, *Eur. J. Biochem.* **132**:255–261.
62. Lim-Tung, H.Y., Resink, T., Hemmings, B.A., Shenolikar, S., and Cohen, P., 1984, *Eur. J. Biochem.* **138**:635–641.
63. Goris, J., Dopere, F., Vandenheede, J.R., and Merlevede, W., 1980, *FEBS Lett.* **117**:117–121.
64. Hemmings, B.A., Resink, T., and Cohen, P., 1982, *FEBS Lett.* **150**:319–324.
65. Resink, J., Hemmings, B.A., Lim-Tung, H.Y., and Cohen, P., 1983, *Eur. J. Biochem.* **133**:455–461.
66. Ballou, L.M., Brautigan, D.L., and Fischer, E.H., 1983, *Biochemistry* **22**:3393–3399.
67. Huang, F.L., and Glinsmann, W.H., 1976, *Eur. J. Biochem.* **70**:419–426.
68. Aitken, A., Bilham, T., and Cohen, P., 1982, *Eur. J. Biochem.* **126**:235–246.
69. Foulkes, J.G., and Cohen, P., 1980, *Eur. J. Biochem.* **105**:195–203.
70. Foulkes, J.G., and Cohen, P., 1979, *Eur. J. Biochem.* **97**:251–256.
71. Foulkes, J.G., Cohen, P., Strada, S.J., Everson, W.V., and Jefferson, L.S., 1982, *J. Biol. Chem.* **257**:12493–12496.
72. Tamura, S., and Tsuiki, S., 1980, *Eur. J. Biochem.* **114**:217–224.
73. Tamura, S., Kikuchi, H., Kikuchi, K., Hiraga, A., and Tsuiki, S., 1980, *Eur. J. Biochem.* **104**:347–355.
74. Lawrence, J.C., Hiken, J.F., DePaoli-Roach, A.A., and Roach, P.J., 1983, *J. Biol. Chem.* **258**:10710–10719.
75. Singh, T.J., Akatsuka, A., and Huang, K.P., 1982, *J. Biol. Chem.* **257**:13379–13384.

14

Skeletal Muscle Glycogen Synthase: Hormonal Regulation

Thomas R. Soderling and Virender S. Sheorain

I. INTRODUCTION

Studies by Larner and associates[1,2] during the early 1960s established the hormonal responsiveness of glycogen synthase in skeletal muscle and other tissues. Epinephrine treatment of rat diaphragm results in inactivation of glycogen synthase whereas insulin promotes activation of the enzyme. Since similar kinetic effects on glycogen synthase can be obtained in muscle extracts by conditions that favor protein phosphorylation–dephosphorylation, this was postulated to be the molecular mechanism underlying the hormonal regulation.

Extensive biochemical studies have characterized purified rabbit skeletal muscle glycogen synthase, synthase kinases, and synthase phosphatase. Seven sites of phosphorylation catalyzed by six to seven protein kinases have been identified[3,4] (Fig. 1) (for reviews, see refs. 5 and 6; see also Chapter 13, this volume). Phosphorylation of synthase is known to inactivate the enzyme by increasing both the K_m value for substrate (UDP-glucose) and the K_a value for the allosteric activator glucose 6-phosphate (G6P). Standard enzyme assays are conducted at 4.8 mM UDP-glucose which, in the absence of G6P, is close to V_{max} for dephosphosynthase but not for phosphosynthase. Maximal activity of phosphosynthase is determined in the presence of 4.8 mM G6P, which activates synthase by decreasing the K_m value for UDP-glucose. Thus, the synthase activity ratio (activity minus G6P/activity plus G6P) has an inverse relationship to the K_m value for UDP-glucose and varies from about 0.8–0.9 for dephosphosynthase to 0.02–0.05 for phosphosynthase. The synthase activity ratio is quite sensitive to changes in phosphate content of 0–2 moles phosphate per subunit (90,000 daltons) of synthase,

Thomas R. Soderling and Virender S. Sheorain ● Howard Hughes Medical Institute and Department of Physiology, Vanderbilt Medical Center, Nashville, Tennessee.

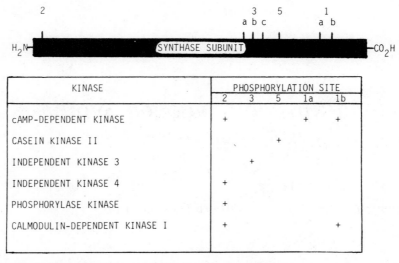

Figure 1. Specificities of glycogen synthase kinases. The approximate localization of the seven phosphorylation sites in the synthase subunit and the specificities of kinases for these sites are indicated.[3,4] (See note added in proof.)

but becomes less accurate at higher phosphorylation states. The K_a value for G6P is 5–10 μM for dephosphosynthase, increasing to several millimolar for phosphosynthase. Phosphorylation of sites 1a, 2, and 3 has significant effects on the K_m value for UDP-glucose (and activity ratio). Phosphorylation of all sites, except site 5, influences the K_a value for G6P. Phosphorylation of site 5 has no known effect on synthase kinetic parameters,[7] but reportedly influences the ability of site 3 to be phosphorylated.[8]

During the past five years, several groups have established that hormonal regulation of rabbit muscle glycogen synthase does indeed occur through phosphorylation–dephosphorylation.[9–11] This chapter summarizes results from our laboratory on the mechanisms of epinephrine, diabetes, and insulin on glycogen synthase of rabbit skeletal muscle.

II. GENERAL APPROACH AND CONSIDERATIONS

Rabbit skeletal muscle was the tissue of choice for these studies on hormonal regulation, since almost all biochemical studies have used purified synthase, kinases, and phosphatase from this tissue. The approach involved subjecting the rabbits to various hormonal treatments, purifying the enzyme from the skeletal muscle, and characterizing the purified synthase in terms of kinetic properties, phosphate content, and occupancy of the phosphorylation site. It was crucial to the purpose of this investigation that the phosphate content of the synthase not be altered by handling of the rabbits, method of anesthesia, excision and homogenization of the muscle, or purification of the enzyme. Extensive investigations indicated that such was the case.[9]

In all studies except those involving epinephrine and short-term insulin therapy, the rabbits were anesthetized by intravenous injection of propranolol over a 5-min period. This treatment was found to stabilize the phosphorylase activity ratio (about 0.1), which with pentobarbital (Nembutal) varied up to 0.5 due to the excitability of the rabbits and the release of endogenous catecholamines. In these preliminary studies, it was evident that the synthase activity ratio was not as variable as the phosphorylase activity ratio. Additional experiments established that the kinetic properties of synthase were the same whether the muscle was frozen *in situ* or quickly excised from the rabbit prior to homogenization. Of great importance was the fact that these kinetic parameters did not change during purification of the synthase.

III. EXPERIMENTAL GROUPS

Rabbits were divided into the following experimental groups:

1. *Control:* fed *ad libidum*
2. *Epinephrine-treated, nondiabetic:* given IV epinephrine (66 μg/kg) 30 sec before sacrifice
3. *Insulin-treated (short term), nondiabetic:* given IV glucose (0.5 g/kg) 10 min prior to IV insulin (160 mU/kg), sacrificed 10 min later
4. *Insulin-treated (long term) nondiabetic:* given 1 unit NPH insulin per day for 4 days
5. *Diabetic:* fasted overnight, given IV alloxan (160 mg/kg), maintained 5–7 days before sacrifice
6. *Insulin-treated diabetics:* 5–7-day diabetics maintained additional 4 days with 8–10 units NPH insulin twice daily

The various experimental groups responded as expected in terms of serum glucose and muscle glycogen levels. Diabetic (group 5) and epinephrine-treated (group 2) rabbits were found to have glycogen levels of 4.5–5.0 mg/g compared with a control value of 9.0. Serum glucose of untreated diabetic rabbits averaged 412 mg/dl, and insulin administration reduced this level to 131 mg/dl. For other experimental details, see Sheorain *et al.*[9]

IV. KINETIC PARAMETERS AND PHOSPHATE CONTENT

Glycogen synthase was purified from the skeletal muscles of rabbits under conditions in which the kinetic values were constant. We interpret this to indicate that the phosphate content was not altered and reflected the *in vivo* value. The apparent kinetic constants for the groups are listed in Table 1. The K_m values for UDP-glucose and the K_a values for G6P were elevated in synthase from epinephrine-treated and

Table 1. Effects of Hormones on the Kinetic Parameters of Glycogen Synthase[a]

Groups	Activity ratio ($-$G6P/$+$G6P)	K_m (UDP-glucose)[b] (mM)	K_a (glucose 6-phosphate) (mM)
Control (6)	0.24 ± 0.03	2.43 ± 0.35	0.17 ± 0.04
Insulin, long-term (5)	0.39 ± 0.05[c]	1.10 ± 0.15[c]	0.09 ± 0.02[c]
Insulin, short-term (5)	0.40 ± 0.05[c]	1.18 ± 0.29[c]	0.09 ± 0.03[c]
Diabetic (5)	0.11 ± 0.02[c]	3.44 ± 0.52[c]	0.44 ± 0.07[c]
Diabetic + insulin (5)	0.29 ± 0.01[d]	2.56 ± 0.36[d]	0.14 ± 0.04[d]
Epinephrine (5)	0.07 ± 0.02[c]	4.24 ± 0.48[c]	0.77 ± 0.12[c]

[a] Glycogen synthase, purified from the indicated treatment groups, was assayed to determine the specific kinetic parameters. Number of samples in each group is given in parentheses.
[b] Determined at 0.25 mM glucose 6-phosphate (G6P).
[c] Significantly different from control value.
[d] Significantly different from diabetic value.

diabetic animals. Diabetic subjects given insulin therapy were found to have significantly lowered values for both parameters. Insulin therapy of nondiabetics also lowered both the UDP-glucose K_m and G6P K_a values. These changes in synthase kinetic values suggest that epinephrine and diabetes produce an increased phosphorylation state of the enzyme, whereas insulin treatment of diabetics and nondiabetics results in net dephosphorylation.

Analysis of the total phosphate content of the purified synthase confirmed this hypothesis (Table 2). Synthase from both diabetic (group 5) and epinephrine-treated rabbits had about 4 moles phosphate per mole synthase subunit (90,000 daltons) compared with a value of 2.4 for synthase from controls. Insulin administration to the diabetics reduced the synthase phosphate content to the same level found in the control group.

Table 2. Effects of Diabetes, Insulin, and Epinephrine on the Phosphorylation State of Glycogen Synthase[a]

Groups	Phosphate content (mole/mole subunit)		
	Total	Trypsin-sensitive	Trypsin-insensitive
Control (6)	2.35 ± 0.29	0.95 ± 0.10	1.40 ± 0.12
Insulin, long-term (5)	2.01 ± 0.13	1.20 ± 0.17	0.81 ± 0.09[b]
Insulin, short-term (5)	2.36 ± 0.11	1.08 ± 0.06	1.28 ± 0.06
Diabetic (5)	3.91 ± 0.25[b]	1.08 ± 0.12	2.84 ± 0.13[b]
Diabetic + insulin (5)	2.25 ± 0.12[c]	1.07 ± 0.07	1.18 ± 0.09[c]
Epinephrine (5)	3.85 ± 0.26[b]	2.09 ± 0.14[b]	1.75 ± 0.09

[a] Glycogen synthase, purified from the indicated treatment groups, was analyzed for total phosphate content and localization of phosphate in "trypsin-sensitive" and "trypsin-insensitive" regions. See text for details.
[b] Significantly different from control value.
[c] Significantly different from diabetic value.

These results are in general agreement with other published studies. Using rabbits, Cohen and associates found epinephrine[10] to elevate the phosphate content of glycogen synthase to about 5 moles/mole subunit from a control value of about 2.5–2.7 and insulin[11] to decrease the phosphate level to 2.3. Glycogen synthase from rat muscle perfused with saline showed 3.1–3.4 moles phosphate per 85,000 g protein; perfusion with epinephrine increased this value, and insulin decreased it to 4.9 and 2.7, respectively.[12,13]

V. OCCUPANCY OF THE PHOSPHORYLATION SITE

Biochemical studies on purified synthase and synthase kinases have provided us with the specificities of the kinases for the phosphorylation sites in synthase.[4] Similar studies are currently being undertaken with synthase phosphatases. If diabetes and insulin therapy alter the phosphate content of only one or two sites, this would allow us to formulate a hypothesis as to which kinase or phosphatase activity may be altered. On the basis of this expectation, we have analyzed occupancies of phosphorylation sites of the purified glycogen synthase using three different techniques.

A. Trypsin-Sensitivity

Several studies have shown that limited trypsinization (10 μg/ml trypsin for 5–15 min) of native synthase reduces the subunit M_r from 90,000 to about 73,000 by generating several small peptides from the C-terminus and a tetrapeptide from the N-terminus.[14,15] These small trichloroacetic acid (TCA)-soluble peptides contain phosphorylation sites 1a, 1b, and 5. Sites 2, 3a, 3b, and 3c remain in the TCA-precipitable 73,000-dalton peptide. Therefore, by limited proteolysis and TCA precipitation, one can resolve sites 1a, 1b, and 5 from sites 2, 3a, 3b, and 3c. We refer to the former sites as trypsin sensitive and the latter sites as trypsin insensitive.[16]

Using this approach, we determined that 76% of the additional phosphorylation of synthase in response to epinephrine was localized in trypsin-sensitive sites (Table 2). This finding is consistent with the interpretation that epinephrine is acting in skeletal muscle primarily through activation of the cyclic adenosine monophosphate (cAMP-dependent protein kinase, since this kinase *in vitro* phosphorylates sites 1a, 1b, which are trypsin sensitive, and site 2, which is trypsin insensitive. Assays of tissue extracts showed that only the epinephrine-treated group exhibited activated cAMP-dependent protein kinase (Table 3).

In contrast to epinephrine treatment, the increased phosphate in the diabetic group was almost exclusively (92%) in trypsin-insensitive sites 2 or 3 or both (Table 2). Insulin therapy of the diabetic animals (group 6) specifically reduced the phosphate content of these sites, as did insulin therapy of nondiabetics (groups 3 and 4). These data indicate that insulin deficiency and administration alter the phosphate contents of sites 2 or 3 or both in glycogen synthase of skeletal muscle.

Table 3. Activity Ratio of cAMP-Dependent
Protein Kinase

Group	Activity ratio ($-$cAMP/ $+$cAMP)[a]
Control (7)	0.08 ± 0.01
Epinephrine (6)	0.33 ± 0.01[+]
Diabetic (8)	0.08 ± 0.02
Diabetic + insulin (7)	0.10 ± 0.02
Insulin, long-term (8)	0.10 ± 0.02

[a] Muscle samples were frozen *in situ* and powdered, and the homogenates were assayed for protein kinase activity in the absence and presence of cAMP using the synthetic peptide Kemptide (L-R-R-A-S-L-G) as substrate. Only the epinephrine-treated group was significantly different (+) from control group. Numbers in parentheses indicate the number of samples analyzed.

B. In Vitro Phosphorylation

If a particular site has been phosphorylated *in vivo*, one would expect decreased ^{32}P incorporation into that site during subsequent *in vitro* phosphorylation of the purified synthase using purified kinases. A potential caveat of this approach is that phosphorylation of one site might affect subsequent phosphorylation of another site. In fact, it has been reported that phosphorylation of site 3 is enhanced by the presence of phosphate in site 5.[8] Since the phosphate content of site 5 appears to be quite high and does not change significantly with the experimental treatments examined (see below), this particular site–site interaction may not be a problem. The synthase preparations used for *in vitro* phosphorylation were also purified over phosphocellulose to remove trace contamination by kinases and phosphatases.[17]

The results presented in Table 4 show a threefold decrease in ^{32}P incorporation catalyzed by synthase kinase 3 into site 3 of synthase purified from diabetic rabbits, as compared with synthase from control animals. Insulin therapy of the diabetics brought site-3 phosphorylation back to the control level. With synthase from the epinephrine-treated group, there was a small but significant decrease in ^{32}P incorporation at site 3 as well. High-performance liquid chromatography (HPLC) peptide mapping (see Section V.C) of tryptic digests of the [^{32}P]synthase confirmed that greater than 90% of the ^{32}P was in site 3. *In vitro* phosphorylation of site 2 was also markedly depressed in synthase from diabetics and was slightly elevated in synthase from insulin-treated rabbits. These results suggest an increased phosphate content *in vivo* in sites 2 and 3 from diabetic rabbits and reversal of this effect by insulin.

When the cAMP-dependent protein kinase was the catalyst *in vitro*, synthase from the epinephrine-treated group was the poorest substrate. Phosphorylation of sites 1a, 1b, and 2 was all depressed. Again, these results strongly argue that the cAMP-dependent protein kinase is the major mediator of epinephrine action in this enzyme system in skeletal muscle.

Table 4. In Vitro ^{32}P Incorporation into Purified Glycogen Synthase[a]

Source of purified glycogen synthase	^{32}P incorporation (mole/mole)			
	Site 3	Site 1a	Site 1b	Site 2
Control (5)	0.97 ± 0.04	0.20 ± 0.02	0.23 ± 0.03	0.24 ± 0.02
Insulin, long-term (5)	1.20 ± 0.11	0.20 ± 0.01	0.38 ± 0.04[b]	0.33 ± 0.02[b]
Diabetic (6)	0.34 ± 0.07[b]	0.20 ± 0.01	0.25 ± 0.02	0.06 ± 0.01[b]
Diabetic + insulin (5)	1.03 ± 0.07[c]	0.19 ± 0.01	0.26 ± 0.01	0.15 ± 0.01[c]
Epinephrine (5)	0.74 ± 0.08[b]	0.04 ± 0.02[b]	0.14 ± 0.01[b]	0.04 ± 0.01[b]

[a] Purified glycogen synthase from the indicated group, was further phosphorylated in vitro for 60 min using [γ-^{32}P]-ATP and synthase kinase 3 (for site 3) or cAMP-dependent protein kinase (for sites 1a, 1b, and 2). The ^{32}P-synthase was subjected to extensive trypsinization (1 mg/ml for 5 hr), and the ^{32}P-labeled peptides were resolved by reverse-phase high-performance liquid chromatography to determine distribution of ^{32}P in each site.
[b] Significantly different from control value.
[c] Significantly different from diabetic value.

It is important to point out that the phosphorylation stoichiometries gained from these in vitro phosphorylations are qualitative rather than quantitative in nature. That is to say, one cannot deduce that site 1a contained 0.8 mole phosphate in vivo, since only 0.2 mole ^{32}P was incorporated in vitro into that site using synthase from control animals. This is clearly seen when one uses dephosphosynthase as substrate. Although this synthase contains little or no phosphate in sites 1a, 1b, or 2, only about 0.4–0.5 mole ^{32}P is incorporated into each of these sites under the conditions used in this study. Partial-site phosphorylation is a common observation with in vitro phosphorylation of glycogen synthase as well as some other proteins.[18]

C. HPLC Peptide Mapping

If an extensive tryptic digest (1 mg/ml trypsin for 5 hr) of synthase is chromatographed on reverse-phase high-performance liquid chromatography (RP-HPLC) under suitable gradient conditions, the various phosphopeptides can be resolved (Fig. 2A).[4] The separated phosphopeptides obtained from synthase phosphorylated in vivo are lyophilized and assayed for inorganic phosphate to quantitate the phosphorylation stoichiometry for each site. Figure 2B shows the profile for a sample of synthase purified from epinephrine-treated rabbits. Although sites 3a, 3b, and 3c are in a single tryptic peptide, referred to collectively as site 3, one usually obtains up to three peaks corresponding to site 3 on HPLC (see fig. 5 of ref. 4). Since the phosphate content of a peptide can markedly affect its retention time on RP-HPLC,[19] these multiple peaks probably represent mono-, di-, and triphosphorylated site 3 peptide.

Five preparations of glycogen synthase per group from control, epinephrine-treated, and diabetic rabbits were analyzed by this method. Figure 3 shows the results of HPLC peptide mapping analysis. Epinephrine treatment produced 2-fold elevations of phosphate in sites 2 and 1b and a 1.5-fold increase in site 3. These results are in agreement with the other analyses described above except that we also expected an

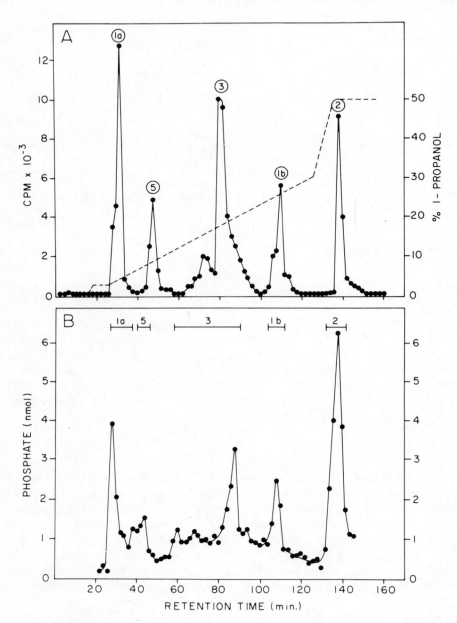

Figure 2. Separation of tryptic phosphopeptides derived from glycogen synthase by HPLC. (A) A mixture of ^{32}P-labeled peptides obtain by tryptic digestion (1 mg/ml for 5 hr) of synthase phosphorylated *in vitro* was chromatographed on reverse-phase (C_{18}) HPLC. The column was developed using a 1-propanol gradient (0% for 10 min, 0–2.5% for 1 min, 2.5% for 9 min, 2.5–15% for 100 min and 15–50% for 10 min) in 0.1% trifluoroacetic acid. (B) 2.4 mg synthase, purified from epinephrine-treated rabbits, was digested with trypsin and chromatographed as in (A). Fractions were analyzed for inorganic phosphate.

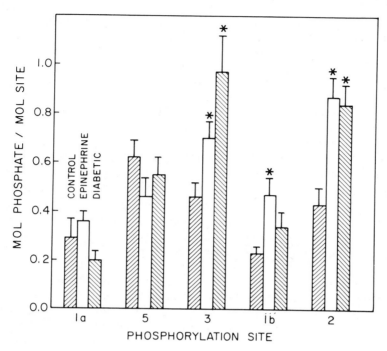

Figure 3. Effects of epinephrine and insulin on glycogen synthase phosphorylation site occupancies determined by HPLC. Glycogen synthase purified from control, diabetic, and epinephrine-treated rabbits was digested with trypsin (1 mg/ml) for 5 hr. The tryptic peptides were resolved by HPLC and assayed for the inorganic phosphate content of each phosphorylation site. The asterisk (*) denotes statistically significant difference from the control value.

increase in site 1a. The reason for the failure to detect an increase in site 1a is not certain, but it is probably a technical problem. Site 1a elutes at very low concentrations of 1-propanol (less than 5%), and some site 1a is probably lost in the flowthrough fractions when large amounts (~10 nmole) of peptide mixtures are loaded to the RP-HPLC column.

Synthase from diabetic rabbits had 2-fold elevations of phosphate in sites 3 and 2. These results are consistent with the findings from the *in vitro* phosphorylation analyses and the limited trypsinization experiments.

VI. CONCLUSIONS

The effects of diabetes, insulin, and epinephrine treatment in intact rabbits on skeletal muscle glycogen synthase have been investigated. Epinephrine treatment and diabetes produced inactivation of the enzyme due to increases in both the K_m value for UDP-glucose and the K_a value for the allosteric activator glucose 6-phosphate.

Insulin administration decreased both of these kinetic parameters thereby activating glycogen synthase. Similar effects of diabetes and insulin on the kinetic values of rat muscle synthase have been reported.[20]

The effects of these hormonal treatments on the phosphate content of glycogen synthase and distribution of phosphate among the various phosphorylation sites were determined by three methods, summarized in Table 5.

Epinephrine produced significant increases in the phosphate contents of sites 1a, 1b, 2, and 3. The effects on sites 1a, 1b, and 2 can be explained by the known activation of the cAMP-dependent protein kinase in skeletal muscle. The mechanism for phosphorylation of site 3 is less clear (see note added in proof). There is no reported effect of the cAMP-dependent protein kinase on synthase kinase 3. It is known that epinephrine, acting through the cAMP-dependent protein kinase, can increase the phosphorylation state of phosphatase inhibitor I.[21,22] This may result in increased inhibition of protein phosphatase I, which selectively dephosphorylates sites 3 and 2,[11] thereby elevating the phosphate content of these two sites. However, the ability of phosphatase inhibitor I to inhibit native forms of protein phosphatase I was recently questioned.[23]

All three methods showed that insulin deficiency in the alloxan-treated diabetic rabbits produced an increased phosphate content in sites 2 and 3. Insulin treatment of the diabetics reduced the phosphate content of these two sites. These results could be most readily explained if insulin were to activate protein phosphatase I to dephosphorylate sites 2 and 3 selectively. Loss of synthase phosphatase activity in the diabetic and restoration by insulin have been reported for liver[24] and heart muscle.[25] This possibility for skeletal muscle is currently under investigation.

It is possible that insulin could affect the activity of some undiscovered synthase kinase that phosphorylates sites 2 and 3. For example, protease-activated kinase II is reported to mediate the increased phosphorylation of ribosomal protein S6 in response to insulin.[26] This and/or other kinases may phosphorylate glycogen synthase as well.

Table 5. Comparative Results from Methods for Analysis of Phosphorylation Site Occupancies

	Method		
Group	Limited trypsinization	*In vitro* phosphorylation	HPLC peptide mapping
Insulin, long-term	Decrease in sites 2 and/or 3	Decrease in sites 2 and 1b	ND[a]
Diabetic	Increase in sites 2 and/or 3	Increase in sites 2 and/or 3	Increase in sites 2 and/or 3
Diabetic + insulin	Decrease in sites 2 and/or 3	Decrease in sites 2 and/or 3	ND[a]
Epinephrine	Increase in sites 1a, 1b, and/or 5	Increase in sites 1a, 1b, 2, and 3	Increase in sites 1b, 2, and 3

[a] ND, not determined.

Cohen and associates, on the other hand, report that insulin administration results in selective dephosphorylation of only site 3 and suggest that the hormone may in some way attenuate synthase kinase 3 activity.[11] Since these workers were measuring an acute 15-min effect of insulin whereas we were treating the rabbits for several days, the two sets of results are not necessarily at conflict.

Note added in proof: Recent experiments have shown that physiological concentrations (0.5 to 1.0 μM) of cAMP-dependent protein kinase can catalyze phosphorylation of site 3 in addition to sites 1a, 1b, and 2. Thus, activation of the cAMP-dependent kinase in response to epinephrine can account for phosphorylation of sites 1a, 1b, 2, and 3 as observed (V. S. Sheorain, J. D. Corbin, and T. R. Soderling, *J. Biol. Chem.*, in press.)

VII. REFERENCES

1. Craig, J.W., and Larner, J., 1964, *Nature* **202**:971–973.
2. Craig, J.W., Rall, T.W., and Larner, J., 1969, *Biochim. Biophys. Acta* **177**:213–219.
3. Picton, C., Aitken, A., Bilham, T., and Cohen, P., 1982, *Eur. J. Biochem.* **124**:37–45.
4. Juhl, H., Sheorain, V.S., Schworer, C.M., Jett, M.F., and Soderling, T.R., 1983, *Arch. Biochem. Biophys.* **222**:518–526.
5. Soderling, T.R., and Khatra, B.S., in: *Calcium and Cell Function,* Vol. 3 (W.Y. Cheung, ed.), pp. 189–221, Academic Press, New York.
6. Roach, P.J., 1981, *Curr. Top. Cell. Regul.* **20**:45–104.
7. DePaoli-Roach, A.A., Ahmad, Z., and Roach, P.J., 1981, *J. Biol. Chem.* **256**:8955–8962.
8. Picton, C., Woodgett, J., Hemmings, B., and Cohen, P., 1982, *FEBS Lett.* **150**:191–196.
9. Sheorain, V.S., Khatra, B.S., and Soderling, T.R., 1982, *J. Biol. Chem.* **257**:3462–3470.
10. Parker, P.J., Embi, N., Caudwell, F.B., and Cohen, P., 1982, *Eur. J. Biochem.* **124**:47–55.
11. Parker, P.J., Caudwell, F.B., and Cohen, P., 1983, *Eur. J. Biochem.* **130**:227–234.
12. Chiasson, J.L., Aylward, J.H., Shikama, H., and Exton, J.H., 1981, *FEBS Lett.* **127**:97–100.
13. Uhing, R.J., Shikama, H., and Exton, J.H., 1981, *FEBS Lett.* **134**:185–188.
14. Takeda, Y., and Larner, J., 1975, *J. Biol. Chem.* **250**:8951–8956.
15. Soderling, T.R., 1976, *J. Biol. Chem.* **251**:4359–4364.
16. Soderling, T.R., Jett, M.F., Hutson, N.J., and Khatra, B.S., 1977, *J. Biol. Chem.* **252**:7517–7524.
17. Soderling, T.R., Srivastava, A.K., Bass, M.A., and Khatra, B.S., 1979, *Proc. Natl. Acad. Sci. USA* **76**:2536–2540.
18. Soderling, T.R., 1979, *Mol. Cell. Endocrinol.* **16**:157–179.
19. Soderling, T.R., and Walsh, K., 1982, *J. Chromatogr.* **253**:243–251.
20. Komuniecki, P.R., Kochan, R.G., Schlender, K.K., and Reimann, E.M., 1982, *Mol. Cell. Biochem.* **48**:129–134.
21. Foulkes, J.G., and Cohen, P., 1979, *Eur. J. Biochem.* **97**:251–256.
22. Khatra, B.S., Chiasson, J.L., Shikama, H., Exton, J.H., and Soderling, T.R., 1980, *FEBS Lett.* **114**:253–256.
23. Khatra, B.S., and Soderling, T.R., 1983, *Arch. Biochem. Biophys.* **227**:39–51.
24. Gold, A.H., 1970, *J. Biol. Chem.* **245**:903–906.
25. Miller, T.B., 1978, *J. Biol. Chem.* **253**:5389–5394.
26. Perisic, O., and Traugh, J.A., 1983, *J. Biol. Chem.* **256**:9589–9592.

15

Insulin Regulation of Glycogen Synthase in Rat Heart and Liver

Thomas B. Miller, Jr.

I. INTRODUCTION

This chapter touches on selected aspects of the regulation of glycogen synthase by insulin in rat heart and liver. An update on this area would seem timely due to reports during the past 10 years of reproducible insulin effects on glycogen synthase in perfused rat hearts and isolated rat liver preparations. While considerable work has been carried out in rat heart and liver preparations *in vitro*, the mechanistic knowledge of glycogen synthase regulation remains far behind that reported for rabbit skeletal muscle. The purpose of this chapter is to attempt to summarize recent knowledge related to the insulin regulation of rat heart and liver glycogen synthase and to present recent data pertinent to insulin action from our laboratory.

II. ACUTE ACTION OF INSULIN ON GLYCOGEN SYNTHASE IN RAT HEART

As early as 1961, it was demonstrated that addition of insulin to perfused rat hearts produced an increase in cardiac glycogen.[1] An effect of insulin to promote the activation of glycogen synthase (conversion from a Glc-6-P-dependent form to a Glc-6-P-independent form) has been well documented in the rat heart, *in vivo*,[2-6] as well as in the perfused rat heart.[7-11] As is true for other species and tissues, glycogen synthase activity in rat heart appears to be regulated by phosphorylation–dephosphorylation reactions of specific sites on the synthase molecule.[12] There-

Thomas B. Miller, Jr. ● The Department of Biochemistry, University of Massachusetts Medical School, Worcester, Massachusetts.

fore, it is credible to postulate that the regulation of rat heart glycogen synthase by insulin is mediated through a change in the phosphorylation state of the synthase subunit, involving kinases and/or phosphatases acting on synthase.

A. Insulin and cAMP

Current literature makes it doubtful that the acute action of insulin in the perfused rat heart is mediated through a change in cyclic adenosine monophosphate (cAMP) or the cAMP-dependent protein kinase. At least five reports on perfused rat hearts have shown no measurable effect of insulin on cAMP, cAMP-dependent protein kinase activity, phosphorylase kinase activity, or phosphorylase activity.[7–10,13] It has also been suggested that the acute activation of synthase by insulin is independent of calcium metabolism.[9] That another protein kinase may be involved remains to be determined. There are at least two reports in the literature showing the existence of cAMP-independent kinase(s) in rat heart,[12,14] and a recent report on data obtained in rabbit skeletal muscle suggests that a cAMP-independent synthase kinase activity may be decreased by insulin, leading to net dephosphorylation of synthase.[15] Therefore, an effect of insulin to reduce cardiac synthase kinase will have to remain speculative until further work is carried out.

B. Insulin Effects on Synthase Phosphatase

An ideal candidate for insulin-mediated synthase activation in rat heart is a synthase phosphatase. Data obtained from the hearts of diabetic animals make the prospect of synthase phosphatase regulation by insulin even more attractive. In a whole-animal study, it was reported that hearts from fed diabetic rats contained decreased synthase phosphatase activity as compared with normal rats and that insulin treatment for 1 hr could restore the phosphatase activity to normal levels.[16] Another study showed that insulin injection into normal rats resulted in activation of rat heart glycogen synthase by 5 min, whereas insulin was without effect on synthase in the hearts of diabetic rats over a 15-min time course.[6] Concomitant with a lack of effect of insulin on synthase in the diabetic hearts was an approximate 70% decrease in total synthase phosphatase activity associated with the diabetic state. The same study reported synthase phosphatase activity to be fully restored in diabetic hearts 30 min after injection of insulin in vivo.

In a study involving isolated perfused rat hearts,[8] perfusion of hearts from normal rats with insulin resulted in glycogen synthase activation to maximal levels between 5 and 15 min, whereas perfusion of hearts from alloxan diabetic rats with insulin resulted in no synthase activation. It was also demonstrated that the impairment of synthase activation in diabetic hearts was coincident with a 60–70% decrease in total synthase phosphatase activity. When insulin was injected into the diabetic animal 1–6 hr before perfusion, synthase phosphatase activity was restored, as was acute insulin activation of synthase. Since restoration of synthase phosphatase activity and the acute effect of insulin to activate synthase could be blocked by simultaneous treatment of diabetic rats with cycloheximide and insulin, it was suggested that insulin plays a dual role in the regulation of glycogen synthase. One effect of insulin appears to be a chronic effect on synthesis of synthase phosphatase. The other effect is an acute (minutes) response of synthase activation (synthase D to I conversion) by insulin. If these assumptions are correct, it would imply that the acute effect of insulin to activate glycogen synthase in perfused hearts is mediated through an acute effect to activate synthase phosphatase. However, the perfused heart study[8] did not demonstrate an acute effect of insulin on synthase phosphatase. A similar study,[6] performed *in vivo*, reported synthase phosphatase activity in diabetic rat heart to be normalized 30 min after insulin injection, while insulin appeared to be without effect on synthase phosphatase in hearts from normal fed rats during the same time period. Therefore, a reproducible insulin effect to activate synthase phosphatase acutely concomitant with or prior to synthase activation remains to be demonstrated.

C. Rat Heart Phosphoprotein Phosphatases

Recent reviews[17,18] on phosphoprotein phosphatases in many tissues and species emphasize the existence of multiple forms of phosphatases. At least two studies[19,20] support their existence in rat heart. Studies carried out in our laboratory by R.K. McPherson on rat heart support the existence of multiple phosphatase activities. Figure 1 shows the elution profile of synthase-directed and phosphorylase-directed phosphatase

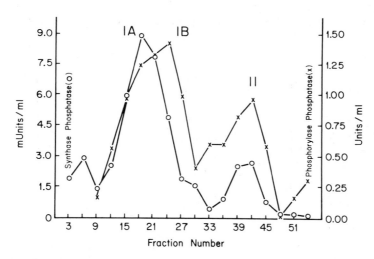

Figure 1. Rat heart phosphoprotein phosphatase activities. Rat hearts (50 g) were homogenized in four volumes of 50 mM Tris, 50 mM β-mercaptoethanol, 1 mM EDTA, 50 mM NaCl, 250 mM sucrose at pH 7.4. All procedures except assays were carried out at 2°C. Homogenate was centrifuged at $10,000 \times g$ for 10 min, and that supernatant was centrifuged at $110,000 \times g$ for 1 hr. The supernatant was brought to 60% saturation with saturated $(NH_4)_2SO_4$ and centrifuged at $10,000 \times g$ for 10 min. After overnight dialysis against the above buffer containing 6 mM $MgCl_2$ or 6 mM $MnCl_2$, the dialysate was applied to a 2.6×13 cm DE-53 column at a flow rate of 0.6 ml/min. After the optical density returned to baseline, the protein was eluted with a linear 50–250-mM NaCl gradient between fractions 1–54. Synthase phosphatase and phosphorylase phosphatase activites were assayed using purified rabbit skeletal muscle glycogen synthase D and phosphorylase a as substrates. Synthase phosphatase was assayed with 6 mM $MgCl_2$, and phosphorylase phosphatase was assayed with 6 mM $MnCl_2$.

activities from a rat heart extract using a 50–250 mM NaCl gradient over DEAE. There appear to be two major phosphatase activity peaks. Peak I (fractions 12–27) contains activities directed toward both synthase and phosphorylase, although the peak for synthase-directed activity resides in fraction 18 (peak IA), while the peak for phosphorylase-directed activity resides in fraction 24 (peak IB). Peak II (fractions 36–45) contains phosphatase activities directed toward synthase and phosphorylase, although in this case the peak activities appear to reside in the same fraction. Figure 1 is in agreement with the existence of multiple phosphatase activities in rat heart. Figures 2 and 3 represent elution profiles of similar extracts from normal and diabetic heart using a 50–250-mM NaCl gradient over DEAE with synthase phosphatase activity assayed in the presence of either manganese (Fig. 2) or magnesium (Fig. 3). In Fig. 2, the synthase phosphatase activity peaks for normal and diabetic reside in the same fractions. Activity in fraction 10 is decreased approximately 60%, and in fraction 20 it is decreased approximately 50%, in the diabetic. Synthase phosphatase activity assayed in the presence of magnesium (Fig. 3) shows a virtual absence of activity in the diabetic, peaking in fraction 14 in the normal subject. Therefore, it appears that there are multiple forms of phosphoprotein phosphatase activity in rat heart, all of which act to a greater or lesser degree on synthase D. Which, if any, of these phosphatase activities is regulated acutely by insulin remains to be determined. Diabetes in rats decreases two of the rat heart phosphatase activities by 50–60% (Fig. 2) and one almost totally (Fig. 3). Since acute insulin activation of synthase in the perfused rat heart is not decreased 50–60% by diabetes, but rather, is virtually abolished, it

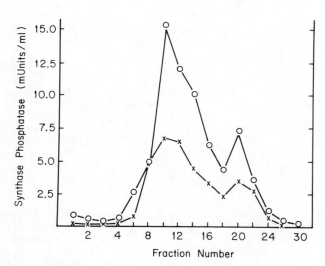

Figure 2. Manganese stimulatable synthase phosphatase from normal and diabetic rat heart. Procedures for preparations of rat heart extracts to be applied to DE-53 were the same as described in Fig. 1, except that 6 mM MnCl$_2$ replaced MgCl$_2$. Dialysate was applied to a 1.5 × 11-cm DE-53 column and washed to baseline with the same buffer described for Fig. 1. Protein was eluted from the column with a linear 50–250-mM NaCl gradient between fractions 1–32. Synthase phosphatase activity was assayed with 6 mM MNCl$_2$ using purified rabbit skeletal muscle synthase D as substrate. (O) Normal; (X) diabetic.

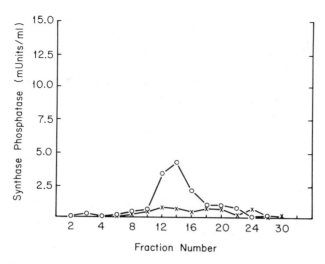

Figure 3. Magnesium-stimulatable synthase phosphatase from normal and diabetic rat heart. Procedures were the same as described in Fig. 2, except that 6 mM MgCl₂ replaced MnCl₂. (O) Normal; (X) diabetic.

will be interesting to carry out further studies on the magnesium-stimulatable phosphatase activity in fraction 14 (Fig. 3) to determine whether it may be acutely regulated by insulin. Whether protein phosphatase inhibitor-1 and -2 are involved in regulation of one or more rat heart phosphatases also remains to be determined. A recent study on the dephosphorylation of rat heart synthase by *Escherichia coli* alkaline phosphatase[21] supports the existence of low-molecular-weight inhibitors of phosphatase activity in rat heart extracts.

Whether there is a general phosphatase activity in rat heart that physiologically regulates both glycogen synthase and phosphorylase remains to be determined. There is circumstantial evidence that suggests two separate phosphatase activities, one acting on synthase and the other on phosphorylase. One set of studies[8] in perfused hearts of normal and 3–4-day diabetics has demonstrated a diabetes related 60–70% decrease in synthase phosphatase activity. In another report[22] from the same laboratory, under the same conditions, 3–4-day diabetes resulted in no detectable decrease in phosphorylase phosphatase activity. Another study using perfused rat hearts from spontaneously diabetic rats along with control nondiabetic littermates[11] found synthase phosphatase activity to be significantly reduced, while phosphorylase phosphatase activity was unchanged in hearts from the spontaneously diabetic rats only 48 hr after cessation of insulin therapy. However, these studies are circumstantial in nature; more data at the molecular level will have to be obtained to determine whether rat heart synthase phosphatase and phosphorylase phosphatase are separate enzymes.

D. Insulin and Synthase Phosphorylation–Dephosphorylation

Glycogen synthase activity has been referred to simply as the *D* (Glc-6-P-dependent) or *I* (Glc-6-P-independent) forms. Such designations represent an extreme oversimplification used only for convenience. As is true for skeletal muscle,[15] rat heart glycogen synthase may also contain seven or more phosphorylatable sites. The degree of phosphorylated species, specific regulation by phosphorylation–dephosphorylation, and insulin-mediated changes in specific phosphorylation sites in rat heart synthase have not yet been determined. One study in rat heart suggests that four to five phosphates/subunit can be incorporated into synthase,[23] while another investigation[12] has demonstrated the existence of at least two CNBr-generated phosphopeptides of glycogen synthase that appeared to be preferentially phosphorylated by either cAMP-dependent or cAMP-independent kinases. Another report[9] from the same laboratory on insulin activation of glycogen synthase in perfused rat heart has demonstrated that insulin lowers the $A_{0.5}$ of synthase for Glc-6-P. Insulin was also shown to decrease the $S_{0.5}$ of synthase for UDP-Glc. These reports are consistent with multiple phosphorylation sites on rat heart synthase responsible for multiple changes in kinetic parameters, as has been shown for rabbit skeletal muscle synthase.[15] Further work will be required to fill in the details.

In summary, rat heart glycogen synthase activity is regulated by insulin. There is an acute effect of insulin to promote synthase D to I conversion and a chronic effect to maintain the responsiveness of the D to I conversion system to insulin. The effect of insulin is probably independent of cAMP-mediated events. Current evidence suggests that insulin action is mediated through changes in synthase phosphatase activity. While the acute effects of insulin are thought to be due to alteration of the phosphorylation state of synthase, data on such an effect have not been reported. Conclusive evidence of the involvement of a cAMP-independent synthase kinase, synthase phosphatase, and/or synthase phosphatase inhibitors in the acute effects of insulin remains to be demonstrated.

III. REGULATION OF RAT LIVER GLYCOGEN SYNTHASE BY INSULIN

Although numerous studies through the years have reported no insulin effect on synthase in rat liver preparations, only the reports of positive effects are dealt with here. As early as 1959, it was demonstrated that within a few hours after administration of insulin to diabetic rats, glycogen rapidly accumulated in the liver without any measurable increase in liver Glc-6-P.[24] A subsequent study from the same laboratory[25] using alloxan-treated diabetic rats demonstrated an increase in liver glycogen and a sevenfold increase in the Glc-6-P-independent form of liver glycogen synthase without any change in liver Glc-6-P concentrations, all within 2 hr of insulin administration. The effect of insulin to activate rat liver glycogen synthase *in vivo* was confirmed in several other laboratories during the late 1960s and early 1970s.[26–30]

A study in isolated livers from rats[31] demonstrated that perfusion with insulin plus acetylcholine resulted in greater glycogen accumulation than did perfusion with acetylcholine alone. In 1973, two separate studies using red blood cell (RBC)-supplemented Krebs-Henseleit bicarbonate buffer systems demonstrated insulin activation of glycogen synthase in isolated perfused rat livers.[30,32] One study[30] demonstrated that insulin added simultaneously with glucagon promoted glycogen synthase activation, a decrease in glucagon-induced glycogen breakdown, and increased [^{14}C]glucose incorporation into glycogen. The same study reported that insulin alone, in the absence of exogenous glucagon, produced a slight but significant increase in glycogen synthase activity and [^{14}C]glucose incorporation into glycogen. In another study,[32] isolated rat livers were perfused with glucagon or epinephrine for 30–45 min, which inactivated glycogen synthase, after which they were perfused with insulin, which reactivated glycogen synthase. Together, these reports documented an acute effect of insulin (maximum effects within 20 min) to activate glycogen synthase in the isolated perfused rat liver. Further documentation of a direct effect of insulin alone on glycogenesis was obtained in rat liver slices.[33] When insulin was added, both [^{14}C]glucose incorporation into glycogen and total liver glycogen levels were increased by 25% over a 2-hr incubation period. Since 1972, numerous publications have reported an insulin effect on glycogen synthesis and/or glycogen synthase in isolated rat hepatocytes,[34–45] on perfused rat liver,[30–32,46,47] and on cultured rat hepatocytes.[48–52]

A. Potential Mechanisms of Insulin Action on Synthase

Glycogen synthase in rat liver has the potential of being regulated by allosteric effectors, protein kinases and phosphoprotein phosphatases. There is ample evidence that rat liver synthase, as is true for rabbit muscle synthase, is regulated through phosphorylation and dephosphorylation.[53–59] A logical conclusion is that insulin effects the phosphorylation state of rat liver synthase.

B. Insulin Antagonism of Rat Liver cAMP

One of the best known and most reproducible effects of insulin in rat liver is the ability to lower hormonally induced increases in cAMP. It was demonstrated in perfused rat liver that insulin could antagonize the effects of glucagon[30,32] and epinephrine[32] on glycogen synthase inactivation. In this situation,[30] insulin produced a 40% decrease in glucagon-elevated cAMP and a 30% decrease in cAMP-dependent synthase kinase activity. Synthase phosphatase activity was not measured in either study,[30,32] and it was suggested that insulin activated synthase by lowering synthase kinase activity subsequent to its effect to decrease liver cAMP. Similar studies on insulin counteraction of the effects of glucagon on glycogen synthase and phosphorylase have been reported for isolated hepatocytes.[44,60,61] In a study carried out in rat liver glycogen pellets isolated from rats injected with glucagon,[62] glucagon injection was reported to produce an acute decrease in synthase phosphatase activity. In perfused rat liver,[63] glucagon was reported to produce a time- and concentration-dependent inactivation of glycogen synthase phosphatase that was partially reversed by insulin. The effect of insulin to reverse glucagon-inactivated synthase phosphatase was statistically significant by 10 min after exposure to insulin. Therefore, besides the effect of insulin to counteract the actions of cAMP-mediated hormones on the synthase inactivation by cAMP-dependent synthase kinase, it appears that rat liver synthase phosphatase may also be regulated through insulin–glucagon antagonism. Whether the effect is mediated through a phosphorylatable phosphatase inhibitor in rat liver remains to be determined.

C. Insulin Effects on Basal cAMP-Dependent Protein Kinase

There is some evidence in the literature that insulin produces a decrease in rat liver cAMP-dependent protein kinase activity from the basal state. For example, in the livers of fed rats perfused without or with insulin for 0–30 min, insulin activated glycogen synthase and was found to inactivate synthase kinase between 6–15 min.[30] In isolated rat hepatocytes, it was reported that insulin inactivated cAMP-dependent protein kinase from the basal state.[64] Another study in isolated hepatocytes[44] reported that insulin decreased basal cAMP levels in the absence of added glucagon. It has also been reported that insulin stimulated the synthesis of an adenylate cyclase- and cAMP-dependent protein kinase inhibitor in isolated rat hepatocytes.[65] Therefore, it remains a possibility that insulin may be able to affect glycogen synthase in rat liver by reducing cAMP-dependent protein kinase activity below basal levels.

D. Effects of Insulin Alone in Rat Hepatocyte Preparations

With the advent of a viable isolated hepatocyte preparation,[66] studies on insulin effects in liver have proliferated. Many of these studies have reported an effect of insulin alone on glycogen synthesis and/or glycogen synthase.[34–45] In studies from our laboratory using hepatocytes isolated from normal fed rats, an insulin effect on [^{14}C]glucose incorporation into glycogen and glycogen synthase activation has been observed. Figure 4 shows incorporation of [^{14}C]glucose into glycogen in isolated

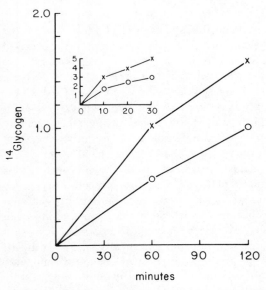

Figure 4. Effects of insulin on ^{14}C-glycogenesis in isolated hepatocytes. Isolated hepatocytes were prepared by perfusion of livers from normal fed rats with collagenase using conventional procedures. Dispersed hepatocytes were incubated at 37°C in a shaker bath in Krebs-Henseleit bicarbonate buffer containing 15 mM glucose, 10 mM glutamine, 1 mM palmitate, and 2.7% albumin (1 ml packed hepatocytes plus 9 ml buffer). After 10-min incubation, 1 μCi of [14C]glucose was added to the constantly oxygenated flasks (95% O_2–5% CO_2). At the same time, either insulin (final concentration of 10^{-8}) or insulin diluent was added to the flasks. Cells were incubated for 120 min or for 30 min (insert). One-ml aliquots were withdrawn for determination of [^{14}C]glycogen at 0, 60, and 120 min, or at 0, 10, 20, and 30 min, respectively. The ordinate represents [^{14}C]glycogen as counts per minute × 10^{-3}, while the abscissa represents minutes of incubation both without and with insulin. At 60 and 120 min, the difference in the means has a p value of less than 0.01, using paired statistics. Each mean represent six determinations. (O) Control; (X) insulin.

hepatocytes from normal rats for 0–120 min in the absence and presence of 10^{-8} M insulin. An insulin effect to increase ^{14}C-glycogenesis is observed at 60 and 120 min. The insert in Fig. 4 shows the mean of two experiments on ^{14}C-glycogenesis in isolated hepatocytes incubated in the absence and presence of 10^{-8} M insulin for 0–30 min. Although statistical analysis was not performed, the insert suggests that the insulin effect on ^{14}C-glycogenesis is acute. Figure 5 shows the results of synthase assays on hepatocytes incubated for 10 min in the presence and absence of insulin using exactly the same conditions described for Fig. 4, except without [^{14}C]glucose. Glycogen synthase activity assayed under standard conditions (6.7 mM UDP-Glc $\pm$ 10mM Glc-6-P) was 25% in the controls and 27% in the hepatocytes incubated for 10 min with insulin (data not shown). Also shown in Fig. 5 is the relative velocity (observed velocity/V_{max} at infinite Glc-6-P) of synthase in these hepatocytes incubated both without and with insulin for 10 min assayed using 6.7 mM UDP-Glc with varying concentrations of Glc-6-P. Under these assay conditions, no effect of insulin is observed. Figure 6 shows the synthase activity data when synthase is assayed using a more physiological concentration of UDP-Glc (0.2 mM) with increasing concentrations of Glc-6-P—an assay previously used for rat skeletal muscle.[67] Whereas no effect of insulin was observed in Fig. 5 when synthase was assayed under standard conditions (6.7 mM UDP-Glc), synthase activity expressed as the relative velocity using 0.2 mM

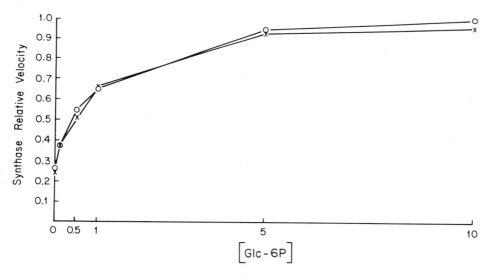

Figure 5. Hepatocyte synthase relative velocity assayed using 6.7 mM UDP-Glc as substrate. Isolated hepatocyte were prepared and incubated using the same buffer components described in Fig. 4. The relative velocity of synthase (activity determined at a given Glc-6-P concentration/maximal velocity of activity at infinite Glc-6-P concentration) is plotted versus final Glc-6-P concentration in the assay. Assays were carried out for 10 min at pH 7.8 using 6.7 mM UDPGlc. The V_{max} at infinite Glc-6-P was calculated using the Michaelis-Menten equation with weighted linear regression analysis. Each mean represents data from six different experiments. (O) Control; (X) insulin.

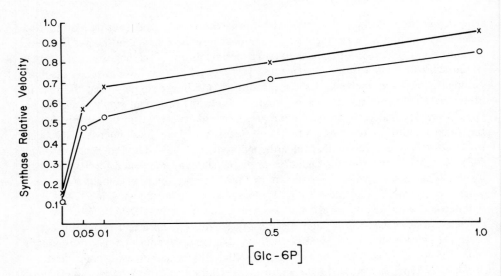

Figure 6. Hepatocyte synthase relative velocity using 0.2 mM UDP-Glc as substrate. Assays and analysis of data were carried out on the same hepatocytes described in Fig. 5. Synthase activity was assayed at pH 6.9 for 5 min using 0.2 mM UDP-Glc as substrate. Each mean represent data from six different experiments. (○) Control; (X) insulin. For control versus insulin at [Glc-6-P] of 0, 0.05, 0.1, 0.5, and 1.0 mM, $p < 0.05$, 0.025, 0.001, 0.05, and 0.05, respectively.

UDP-Glc and increasing Glc-6-P concentrations shows that insulin increases the relative velocity of synthase at every Glc-6-P concentration tested.

When these data are calculated using a combination of the Hill equation and Michaelis-Menten weighted linear regression analysis (Table 1), the calculated maximal velocity at infinite Glc-6-P concentrations (V_{inf} Glc-6-P) is unchanged by insulin, using either 6.7 or 0.2 mM UDP-Glc. While the concentration of Glc-6-P necessary

Table 1. Effect of Insulin on Maximum Velocity at Infinite Glc-6-P and the $A_{0.5}$ for Glc-6-P[a]

UDP-Glc (mM)	pH	V_{inf} Glc-6-P[b] (nmoles/g per min)		$A_{0.5}$ for Glc-6-P[c] (mM Glc-6-P)	
		Control	Insulin	Control	Insulin
6.7	7.8	780 ± 145	695 ± 104	0.598 ± 0.052	0.660 ± 0.100
0.2	6.9	58.7 ± 13.4	53.6 ± 7.1	0.074 ± 0.015	0.028[d] ± 0.007

[a] Data were obtained from the hepatocytes described in Figs. 5 and 6.
[b] V_{inf} G6P (V_{max} at infinite Glc-6-P) was calculated using weighted linear regression analysis of the Michaelis-Menten equation. Each mean represents six data points.
[c] The concentration of Glc-6-P required for half-maximal activation of synthase ($A_{0.5}$) was calculated using the Hill equation[67]. Each mean represents six data points.
[d] $p < 0.0125$ versus control without insulin.

for half-maximal activation of synthase ($A_{0.5}$) is unaffected by insulin with 6.7 mM UDP-Glc, the $A_{0.5}$ for Glc-6-P is reduced from 0.074 mM in the 10-min control to 0.028 mM in the hepatocytes incubated for 10 min with insulin. Data on a similar insulin effect in isolated hepatocytes have also been reported[43] by another group. Although the absolute values obtained in these two studies differ somewhat, both studies demonstrate a similar insulin-induced increase in synthase relative velocity and a decrease in the $A_{0.5}$ of synthase for Glc-6-P. These data, sh. wing an effect of insulin in isolated hepatocytes to lower the $A_{0.5}$ of synthase for Glc 6-P, would support the hypothesis that insulin effects a net dephosphorylation of a site on synthase, regulating affinity for the binding and allosteric stimulation by Glc-6-P. Studies on phosphorylation–dephosphorylation of rat liver synthase[54,58,59] are consistent with an insulin effect mediated through dephosphorylation of a site affecting the $A_{0.5}$ for Glc-6-P. Only further work will determine the true mechanism of the effect.

E. Insulin Regulation of Basal Synthase Phosphatase

Since it appears that multiple phosphorylation sites are present on synthase[68] and rat liver glycogen synthase phosphatase has not been extensively studied, there are as yet no mechanistic studies on insulin regulation of rat liver synthase phosphatase. However, studies in normal versus diabetic rat liver are consistent with the possibility that synthase phosphatase is regulated by insulin. Numerous reports have conclusively demonstrated a defect in glycogen synthase activation by glucose in liver preparations from diabetic rats.[26,28,69–75] Restoration of synthase activation by glucose in perfused diabetic rat liver can be achieved by insulin injection *in vivo*.[69–72] It appears that glycogen synthase phosphatase, which is either missing or altered in livers of diabetic rats, may be chronically regulated by insulin.[28,71–73,75–79] Two reports on restoration of synthase phosphatase in diabetic rat liver by insulin demonstrate that the insulin effect can be blocked by cycloheximide,[72,73] suggesting a role for protein synthesis.

A recent set of experiments in our laboratory suggests that insulin regulates glycogen synthase and synthase phosphatase in primary cultures of normal rat hepatocytes. Hepatocytes prepared from normal rats were plated on collagen-coated dishes, and primary cultures were maintained for 3 days in serum-free medium containing MEM-Earl's salts, 0.3% crystalline albumin, 5x physiological concentration of fed rat amino acids, 50 μg/ml hydrocortisone hemisuccinate and 50 ng/ml triiodothyronine. Cells were fed with fresh medium each day. One-half the culture dishes contained 10^{-7} M insulin. [^{14}C]glucose (1 μCi) was added to culture dishes only 18 hr before [^{14}C]glycogen determination. Table 2 shows the data obtained on hepatocytes in primary culture for 18, 42, and 66 hr incubated both without and with insulin. During the first 18 hr, ^{14}C-glycogenesis was not different between hepatocytes maintained either without or with insulin. Glycogen synthase phosphatase activity was also unchanged by the presence of insulin, although insulin did appear to effect glycogen synthase activation. After 42 hr, insulin facilitated a fourfold increase in the 18-hr [^{14}C]glycogen deposition, a 42% increase in synthase *I*, and a twofold increase in synthase phosphatase activity. After 66 hr, insulin promoted a fivefold increase in 18-hr [^{14}C]glycogen deposition, a 42% increase in synthase *I* activity, and a 2.3-fold

Table 2. Effects of Insulin on Primary Hepatocyte Cultures[a]

Culture time (hr)	[^{14}C]glycogen cpm/μg DNA		Glycogen synthase % active		Synthase phosphatase pmoles/μg DNA	
	− Ins	+ Ins	− Ins	+ Ins	− Ins	+ Ins
18	436 ± 82	454 ± 80	34 ± 4	48[c] ± 3	72.5 ± 31.4	84.4 ± 29.0
42	492 ± 128	2054[d] ± 469	38 ± 3	54[c] ± 5	80.5 ± 15.5	170.6[e] ± 43.2
66	727 ± 229	3605[f] ± 706	46 ± 5	66[d] ± 4	134.6 ± 28.1	310.7[c] ± 61.1

[a] Isolated hepatocytes were cultured for 66 hr, as described in the text. Glycogen synthase activity was assayed using the standard ± Glc-6-P filter paper assay and is expressed as Glc-6-P-independent/Glc-6-P-dependent × 100. Synthase phosphatase activity was determined using purified rabbit skeletal muscle synthase D as substrate and is expressed as picomoles of [^{14}C]glucose incorporation into glycogen from [^{14}C]-UDP-Glc/μg DNA. Each mean represents at least six different data points.
[b] $p < 0.01$ versus no insulin.
[c] $p < 0.0005$ versus no insulin.
[d] $p < 0.05$ versus no insulin.
[e] $p < 0.0025$ versus no insulin.

increase in synthase phosphatase activity. Although the data are not shown, cells incubated with insulin had 50–80% greater total synthase activity. Taken together, these data show that chronic exposure to insulin of primary cultures of rat hepatocytes results in ^{14}C-glycogenesis, synthase activation, and increased synthase phosphatase activity. Whether the insulin-mediated increase in synthase phosphatase is a result of a net increase in the phosphatase protein or of an activity change in the existing enzyme remains to be determined. In either case, the data appear to demonstrate that synthase phosphatase activity can be regulated by insulin. Data supporting the existence of synthase phosphatase inhibitors have been reported,[77,80] although the involvement of inhibitors in liver glycogen synthase regulation remains unknown.

IV. CONCLUSIONS AND SPECULATIONS

The mechanisms whereby insulin regulates rat heart and liver synthase are yet to be determined. In the overall scheme, extrapolation of data obtained in rabbit skeletal muscle[15] to data cited here suggests that similar, if not the same, mechanisms may be involved. For rat heart and liver, however, it remains to be determined (1) how many sites per synthase monomer are phosphorylated, (2) which phosphoprotein phosphatase is specific for synthase and whether it is involved in insulin action on synthase, (3) whether phosphoprotein phosphatase inhibitors-1 and -2 are involved in insulin regulation of synthase, and (4) whether insulin affects the activity of a cAMP-independent synthase kinase. Many references have been cited in this chapter that support inferences to many or all of these possibilities, although conclusive findings on any of these points for rat heart and liver are lacking.

That an effect of insulin on rat heart and liver synthase exists *in vitro* has been established and well documented during the past 10–15 years. In rat heart, insulin activates glycogen synthase and initiates glycogen synthesis most likely in the absence of changes in cAMP. In rat liver, insulin can activate glycogen synthase by counteracting the effects of glucagon and epinephrine on cAMP-mediated inactivation of synthase. Insulin has also been shown to activate glycogen synthase in various *in vitro* rat liver preparations independent of change in cAMP-mediated systems, the mechanism of which possibly involves a specific synthase phosphatase. That insulin can activate rat liver synthase in the absence of glucagon suggests that mechanistically such an effect might be important under certain metabolic circumstances.

This chapter is intended to be an update of insulin effects on synthase regulation in rat liver and heart rather than a general comprehensive review. For the most part, references have been made only to those studies that specifically involve insulin activation of synthase in rat liver and heart. A comprehensive review of the mechanism of insulin action on synthase regulation in a variety of species and tissues is provided in Chapter 16. Finally, the mechanism(s) of insulin activation of rat heart and liver synthase and the effects of diabetes on the affected systems are yet to be determined.

ACKNOWLEDGMENTS. Studies carried out on rat heart phosphoprotein phosphatases were supported by research grant HL-20476 and work on rat hepatocyte preparations by research grant AM-18269, from the National Institutes of Health.

V. REFERENCES

1. Williamson, J.R., and Krebs, H.A., 1961, *Biochem. J.* **80**:540–551.
2. Williams, B.J., and Mayer, S.E., 1966, *Mol. Pharmacol.* **2**:454–464.
3. Huijing, F., Nuttall, F.Q., Villar-Palasi, C., and Larner, J., 1969, *Biochim. Biophys. Acta* **177**:204–212.
4. Sacristan, A., and Rosell-Perez, M., 1971, *Rev. Esp. Fisiol.* **27**:331–342.
5. Nuttall, F.Q., Gannon, M.C., and Bergstrom, W.J., 1975, *Am. J. Physiol.* **228**:1815–1820.
6. Nuttall, F.Q., Gannon, M.C., Corbett, V.A., and Wheeler, M.P., 1976, *J. Biol. Chem.* **251**:6724–6729.
7. Adolfsson, S., Isaksson, O., and Hjalmarson, A., 1972, *Biochim. Biophys. Acta* **279**:146–156.
8. Miller, T.B., Jr., 1978, *J. Biol. Chem.* **253**:5389–5394.
9. Ramachandran, C., Angelos, K.L., and Walsh, D.A., 1982, *J. Biol. Chem.* **257**:1448–1457.
10. Ramachandran, C., Angelos, K.L., Sivaramakrishnan, S., and Walsh, D.A., 1983, *Fed. Proc.* **42**:9–13.
11. Miller, T.B., Jr., 1983, *Am. J. Physiol.* **245**:E379–E383.
12. McCullough, T.E., and Walsh, D.A., 1979, *J. Biol. Chem.* **254**:7336–7344.
13. Keely, S.L., Corbin, J.D., and Park, C.R., 1975, *J. Biol. Chem.* **250**:4832–4840.
14. Schlender, K.K., and Reimann, E.M., 1977, *J. Biol. Chem.* **252**:2384–2389.
15. Parker, P.J., Caudwell, F.B., and Cohen, P., 1983, *Eur. J. Biochem.* **130**:227–234.
16. Das, I., 1973, *Can. J. Biochem.* **51**:637–641.
17. Lee, E.Y.C., Silberman, S.R., Ganapathi, M.K., Petrovic, S., and Paris, H., 1980, *Adv. Cyclic Nucleotide Res.* **13**:95–131.
18. Li, H.-C., 1982, *Curr. Top. Cell. Regul.* **20**:129–174.
19. Kinohara, N., Usui, H., Imazu, M., Imaoka, T., and Takeda, M., 1982, *J. Biochem.*, **91**:177–190.
20. Yang, S-D., Vandenheede, J.R., Goris, J., and Merlevede, W., 1980, *FEBS Lett.* **111**:201–204.
21. Lau, K-H.W., Chen, I-I.G., and Thomas, J.A., 1982, *Mol. Cell. Biochem.* **44**:149–159.
22. Miller, T.B., Jr., Praderio, M., Wolleben, C., and Bullman, Jr., 1981, *J. Biol. Chem.* **256**:1748–1753.
23. Larner, J., Villar-Palasi, C., and Brown, N.E., 1969, *Biochim. Biophys. Acta* **178**:470–479.
24. Steiner, D.F., and Williams, R.H., 1959, *J. Biol. Chem.* **234**:1342–1349.

25. Steiner, D.F., Rauda, V., and Williams, R.H., 1961, *J. Biol. Chem.* **236**:299–304.
26. Kreutner, W., and Goldberg, N.D., 1967, *Proc. Natl. Acad. Sci. USA* **58**:1515–1519.
27. Friedmann, B., Goodman, E.H., Jr., and Weinhouse, S., 1967, *Endocrinology* **81**:486-496.
28. Gold, A.H., 1970, *J. Biol. Chem.* **245**:903–905.
29. Blatt, L.M., and Kim, K-H., 1971, *J. Biol. Chem.* **246**:7256–7264.
30. Miller, T.B., Jr., and Larner, J., 1973, *J. Biol. Chem.* **248**:3483–3488.
31. Mondon, C.E., and Burton, S.D., 1971, *Am. J. Physiol.* **220**:724–734.
32. Hostmark, A.T., 1973, *Acta Physiol. Scand.* **88**:248–255.
33. Davidson, M.B., and Berliner, J.A., 1974, *Am. J. Physiol.* **227**:79–87.
34. Johnson, M.E.M., Das, N.M., Butcher, F.R., and Fain, J.N., 1972, *J. Biol. Chem.* **247**:3229–3235.
35. Wagle, S.R., Ingebretsen, W.R., and Sampson, L., 1973, *Biochem. Biophys. Res. Commun.* **53**:937–943.
36. Seglen, P.O., 1973, *FEBS Lett.* **30**:25–28.
37. Akpan, J.O., Gardner, R. and Wagle, S.R., 1974, *Biochem. Biophys. Res. Commun.* **61**:222–229.
38. Witters, L.A., Alberico, L., and Avruch, J., 1976, *Biochem. Biophys. Res. Commun.* **69**:997–1003.
39. Witters, L.A., and Avruch, J., 1978, *Biochemistry* **17**:406–410.
40. Baldwin, D., Jr., Terris, S., and Steiner, D.F., 1980, *J. Biol. Chem.* **255**:4028–4034.
41. Beynen, A.C., and Geelen, M.J.H., 1981, *Indian J. Exp. Biol.* **19**:46–48.
42. Beynen, A.C., and Geelen, M.J.H., 1981, *Horm. Metab. Res.* **13**:376–378.
43. Ciudad, C.J., Bosch, F., and Guinovart, J., 1981, *FEBS Lett.* **129**:123–126.
44. Nyfeler, F., Fasel, P., and Walter, P., 1981, *Biochim. Biophys. Acta* **675**:17–23.
45. Venkatesan, N., and Davidson, M.B., 1983, *Life Sci.* **32**:467–474.
46. Storer, G.B., Trimble, R.P., and Topping, D.L., 1980, *Biochem. J.* **192**:219–222.
47. Storer, G.B., Topping, D.L., and Trimble, R.P., 1981, *FEBS Lett.* **136**:135–137.
48. Plas, C., Menuelle, P., Moncany, M.L.J., and Fulchignoni-Lataud, M.C., 1979, *Diabetes* **28**:705–712.
49. Schudt, C., 1980, *Biochim. Biophys. Acta* **629**:499–509.
50. Menuelle, P., and Plas, C., 1981, *Diabetologia* **20**:647–653.
51. Plas, C., and Desbriquois, B., 1982, *Biochem. J.* **202**:333–341.
52. Reed, G.B., 1983, *Exp. Mol. Pathol.* **38**:183–192.
53. Jett, M.F., and Soderling, T.R., 1979, *J. Biol. Chem.* **254**:6739–6745.
54. Guinovart, J.J., Salavert, A., Massaque, J., Ciudad, C.J., Salsas, E., and Itarte, E., 1979, *FEBS Lett.* **106**:284–288.
55. Dopere, F., Vanstapel, F., and Stalmans, W., 1980, *Eur. J. Biochem.* **104**:137–146.
56. Hiraga, A., Kikuchi, K., Tamura, S., and Tsuiki, S., 1981, *Eur. J. Biochem.* **119**:503–510.
57. Bosch, F., Ciudad, C.J., and Guinovart, J., 1983, *FEBS Lett.* **151**:76–78.
58. Akatsuka, A., Singh, T.J., and Huang, K-P., 1983, *Arch. Biochem. Biophys.* **220**:426–434.
59. Huang, K-P., Akatsuka, A., Singh, T.J., and Blake, K.R., 1983, *J. Biol. Chem.* **258**:7094–7101.
60. Van de Werbe, G., Hue, L., and Hers, H.G., 1977, *Biochem. J.* **162**:135–142.
61. Massague, J., and Guinovart, J.J., 1977, *FEBS Lett.* **82**:317–320.
62. Gilboe, D.P., and Nuttall, F.Q., 1978, *J. Biol. Chem.* **253**: 4078–4081.
63. Miller, T.B., Jr., Garnache, A., and Vicalvi, J.J., Jr., 1981, *J. Biol. Chem.* **256**:2851–2855.
64. Mor, M.A., Vila, J., Ciudad, C.J., and Guinovart, J.J., 1981, *FEBS Lett.* **136**:131–134.
65. Wasner, H.K., 1981, *FEBS Lett.* **133**:260–264.
66. Berry, M.N., and Friend, D.S., 1969, *J. Cell. Biol.* **43**:506–520.
67. Kochan, R.G., Lamb, D.R., Reimann, E.R., and Schlender, K.K., 1981, *Am. J. Physiol.* **240**:E197–E202.
68. Soderling, T.R., and Khatra, B.S., 1982, *Calcium and Cell Function* (W.Y. Cheung, ed.), Vol. III, pp. 189–221, Academic Press, New York.
69. Miller, T.B., Jr., Hazen, R., and Larner, J., 1973, *Biochem. Biophys. Res. Commun.* **53**:466–474.
70. Whitton, P.D., and Hems, D.A., 1975, *Biochem. J.* **150**:153–165.
71. Miller, T.B., Jr., 1978, *Am. J. Physiol.* **234**:E13–E19.
72. Miller, T.B., Jr., 1979, *Biochim. Biophys. Acta* **583**:36–46.
73. Haverstick, D.M., Dickemper, D., and Gold, A.H., 1979, *Biochem. Biophys. Res. Commun.* **87**:177–183.
74. Miller, T.B., Jr. Vicalvi, J.J., Jr., and Garnache, A.K., 1981, *Am. J. Physiol.* **240**:E539–E543.
75. Bollen, M., Hue, L., and Stalmans, W., 1983, *Biochem. J.* **210**:783–787.

76. Golden, S., Wals, P.A., Okajima, F., and Katz, J., 1979, *Biochem. J.* **182:**727–734.
77. Khandelwal, R.L., Zimman, S.M., and Zebrowski, E.J., 1977, *Biochem. J.* **168:**541–548.
78. Tan, A.W.H., and Nuttall, F.Q., 1976, *Biochim. Biophys. Acta* **445:**118–130.
79. Appel, M.C., Like, A.A., Rossini, A.A., Carp, D.B., and Miller, T.B., Jr., 1981, *Am. J. Physiol.* **240:**E83–E87.
80. Shenolikar, S., Strada, S.J., and Steiner, A.L., 1983, *Fed. Proc.* **42:**1801.

16

Protein Phosphorylations As a Mode of Insulin Action

Joseph Avruch, Raphael A. Nemenoff, Mark Pierce, Yan C. Kwok, and Perry J. Blackshear

I. INTRODUCTION

Within minutes after the interaction of insulin with its receptor, marked changes occur in the metabolic pattern of the cell. In classic target tissues, insulin inhibits catabolic pathways (e.g., glycogenolysis and lipolysis), concurrently activating anabolic pathways (such as glycogen synthesis and *de novo* fatty acid synthesis). These effects are due to alterations in the activity of a variety of intracellular enzymes and transcellular transport processes. Over a longer period of time, insulin also participates in regulating the expression of specific genes and, in certain cell types, insulin modulates cell division and growth.

Given that regulation of cell function by hormones is a multistep process that may involve multiple parallel and perhaps branching pathways, it seems likely that the earliest, most rapidly expressed physiological actions of the hormone will involve the fewest steps and will be most directly coupled to the signals generated by the hormone–receptor interaction. On the basis of this reasoning, considerable effort has been applied to understanding the mechanisms by which insulin achieves its rapid effects on metabolism. The approach most commonly taken parallels the classic studies of Sutherland and co-workers,[1] which elucidated the mechanism of glucagon–catecholamine activation of hepatic glycogenolysis and led to the discovery of cyclic adenosine monophosphate (cAMP). The hormone-regulated step within an insulin-

Joseph Avruch, Raphael A. Nemenoff, Mark Pierce, Yan C. Kwok, and Perry J. Blackshear ● Howard Hughes Medical Institute Laboratory, Harvard Medical School, the Diabetes Unit and Medical Services, Massachusetts General Hospital; and the Department of Medicine, Harvard Medical School, Boston, Massachusetts.

sensitive pathway is identified, and it is determined whether the altered flux through this reaction is caused by a stable alteration in the enzyme activity. Thus, if the altered enzyme activity induced by insulin is assayable after cell disruption and extensive dilution or partial purification, it is inferred that the altered activity is most likely due to a modification of the enzyme molecule itself, rather than to an indirect effect of the hormone on enzyme activity mediated by altered substrates, products, or allosteric ligands. This approach was first employed in the study of insulin action by Larner and Villar-Palasi[2] and has led to the identification of a number of enzymes that undergo such stable insulin-induced modifications of activity; glycogen synthase[3-7] and pyruvate dehydrogenase (in adipose tissue[8,9]) are perhaps the two best studied examples. In both instances, extensive evidence indicates that the insulin-induced activation is due to dephosphorylation of the enzyme, which, in the case of glycogen synthase, occurs in a relatively site-specific fashion. In addition, there are a number of enzymes whose activity is stably modified by insulin, for which evidence supporting dephosphorylation is available but incomplete and/or indirect.[10-15] Thus, a large body of evidence indicates that many of the rapid effects of insulin on metabolism are mediated by alterations in the phosphorylation of key rate-limiting enzymes in a variety of pathways. Understanding the mechanism of insulin action may then be restated in terms of defining the mechanisms by which insulin regulates the phosphorylation of these enzymes.

II. INSULIN REGULATION OF PROTEIN PHOSPHORYLATION IN INTACT CELLS

In an attempt to identify the enzymes mediating insulin-directed alterations in regulatory protein phosphorylation, we initiated an approach to the study of insulin action that differed from that described above. We hypothesized, by analogy to the actions of glucagon and β-adrenergic agents,[16] that altered protein phosphorylation was likely to serve as an intermediate step in many actions of insulin in addition to the already well described effects of the hormone on glycogen synthase and pyruvate dehydrogenase. Furthermore, since certain actions of insulin were clearly mediated independently of alterations in cellular cAMP levels (e.g., activation of glucose transport[18] and PDH activity[19]), we surmised that the primary effector pathway for insulin regulation of protein phosphorylation was independent of the cAMP cascade. This hypothesis predicted that insulin altered the phosphorylation of a set of cellular proteins through a cAMP-independent mechanism. In order to examine this prediction experimentally, we attempted to define the overall patterns by which insulin altered regulatory protein phosphorylation in intact cells. Cells were equilibrated with extracellular $^{32}P_i$ to achieve steady-state labeling of nucleotide pools and cellular phosphoproteins; insulin and other hormones could then be added under conditions in which altered ^{32}P incorporation into protein would reflect changes in the activity of protein kinases and/or protein phosphatases. By this approach, we hoped to identify (1) a pattern of altered protein phosphorylation unique to insulin and not due to changes in cAMP levels or in the activity of the cAMP-dependent protein kinase, and (2) specific proteins that were substrates for the insulin-regulated kinases or phosphatases operative in intact cells; these proteins could be purified, identified, and used in *in vitro* assays to detect the insulin-regulated kinases and/or phosphatases and in turn permit the purification and characterization of these species.

We have used this approach to examine the effects of insulin on cellular protein phosphorylation in isolated rat adipocytes[20–25] and hepatocytes[26,27] as well as in 3T3-L1 fatty fibroblasts[28] and H_4 hepatoma cell cultures (P.J. Blackshear, unpublished observations). Comparable studies have been carried out in a number of laboratories, employing these and other cell types.[29–40] The patterns of response to insulin have been qualitatively very consistent; four distinguishable patterns have been observed.

The most prominent response has been the rapid stimulation of ^{32}P incorporation into serine and threonine residues on a subset of major and minor ^{32}P peptides. The peptides most consistently detected are described later. Protein dephosphorylation has been observed but is much less evident; for example, dephosphorylation was not observed in our initial studies in analysis of one-dimensional gels of ^{32}P-labeled adipocytes[23] and hepatocytes,[26] although Hughes *et al.* were able to detect dephosphorylation of the α subunit of pyruvate dehydrogenase in their studies in adipose tissue.[9,34] The difficulty in visualizing protein dephosphorylation in response to insulin is surprising, as this clearly would be the expected result. Thus, while both increases and decreases in protein phosphorylation are observed in response to insulin added as the sole hormone, the greater number and prominence of the peptides undergoing increased phosphorylation are striking, as analyzed by this technique.

Despite the relative scarcity of ^{32}P-labeled peptides dephosphorylated by insulin as the sole hormone, insulin antagonism of cAMP-stimulated protein phosphorylation was readily visualized. While this observation was entirely expected on the basis of earlier studies and did not suggest new regulatory mechanisms (i.e., insulin regulation of cAMP phosphodiesterase and/or adenylcyclase), it did serve to strongly support the likelihood that the technique was capable of detecting physiologically significant patterns of regulatory protein phosphorylation in response to insulin.

Recently Kasuga *et al.*[41] made the important observation that insulin stimulated the phosphorylation of its own receptor in a number of intact cell systems. Although the large majority of insulin-stimulated ^{32}P incorporated into the receptor is on serine and threonine residues, a small fraction is recovered as [^{32}P] tyrosine after acid hydrolysis.[42] This observation has several important implications. First, it served to highlight an analogy between an early step (i.e., tyrosine phosphorylation) in insulin action and in the action of certain growth factors, i.e., epidermal growth factor (EGF) and platelet-derived growth factor (PDGF) and retroviruses. Second, it served to identify an entirely new pattern of insulin-stimulated protein phosphorylation; since the specificity of serine/threonine kinases and tyrosine kinases show essentially no overlap, it is clear that two classes of protein kinase must be involved in the insulin stimulated receptor phosphorylation. Given the demonstration by Cohen and co-workers[43] that the EGF receptor itself was a tyrosine protein kinase, the idea arose immediately that the insulin receptor itself might be the tyrosine kinase responsible for receptor tyrosine phosphorylation. Several lines of evidence indicate strongly that this in fact is the case.

To date, insulin-stimulated phosphorylation of both serine/threonine and tyrosine residues is of unknown physiological significance. Moreover, the relationship of these processes to one another is unknown. We anticipate that a regulatory role for both serine/threonine and for tyrosine phosphorylation will emerge and that both phenomena are involved in propagating the signals for the actions of insulin. Consequently, an understanding of the mechanisms underlying these responses will be important for understanding not only certain aspects of insulin action, but the role of similar modifications in the actions of growth factors and retroviruses as well.

The insulin-induced dephosphorylations are clearly regulatory modifications of identifiable physiological significance. The unanswered questions here relate to the steps by which insulin-receptor interaction achieves these dephosphorylations. Before examining the insulin-stimulated protein phosphorylations, we will briefly review data from our laboratory and others on the mechanisms of insulin induced dephosphorylation.

III. INSULIN-INDUCED DEPHOSPHORYLATION

A. Insulin Antagonism of cAMP-Stimulated Protein Phosphorylation

Insulin inhibits certain protein phosphorylations induced by β-catecholamines in adipocytes[23,44] and glucagon in liver cells.[27] The peptides exhibiting this response include hormone-sensitive lipase (84,000 M_r) and an unidentified 69,000-M_r peptide in adipocytes, and pyruvate kinase (59,000 M_r) and glycogen phosphorylase (94,000 M_r) in hepatocytes. The phosphorylation of these proteins in response to cAMP and insulin shows the following features. In the absence of either hormone, a fairly low level of ^{32}P incorporation is observed. In the case of hormone-sensitive lipase, this corresponds to absence of detectable ^{32}P in the tryptic peptide, which bears the phosphorylation site of the cAMP-dependent protein kinase (P. Belfrage, personal communication). A similar situation has been reported for muscle phosphorylase b kinase; i.e., in the absence of epinephrine, the preferred phosphorylation site of the cAMP-dependent protein kinase is unfilled.[45] These observations suggest that in the basal state (i.e., absence of catabolic hormones) the activity of the cAMP-dependent protein kinase is low (or lacking entirely). The calculations of Beavo and co-workers[46] on the degree of activation of the cAMP-dependent kinase in resting skeletal muscle support this idea. Direct measurements of the activation ratio cAMP-dependent protein kinase indicate that 15–20% of the total enzyme is active in the absence of catabolic hormones;[47] the activity of the relevant phosphatases in the basal state is unknown. Insulin, added as the sole hormone, has little or no effect on the basal phosphorylation of the ^{32}P proteins indicated above. When added in the presence of a submaximally stimulating concentration of epinephrine and/or glucagon, insulin partially antagonizes the cAMP-stimulated phosphorylation of these enzymes; at maximal concentrations of the catabolic hormones, insulin is without effect.[27] This antagonistic effect of insulin is most economically explained by the ability of insulin to inhibit cAMP generation in response to catabolic hormones.[17,47] Since insulin does not alter basal levels of cAMP, the hormone does not modify the basal phosphorylation of these proteins, which is either very low, or mediated largely by cAMP-independent kinases. At submaximally stimulated levels of intracellular cAMP, levels that yield only submaximal activation of the cAMP-dependent protein kinase, a lowering of cAMP by insulin will be reflected in a deactivation of the kinase and a decrease in ^{32}P incorporation. At high levels of these catabolic hormones, intracellular cAMP levels are attained that greatly exceed those required for full activation of the kinase. Although insulin can still lower intracellular cAMP levels under these circumstances, even these inhibited levels are likely to remain higher than those required for full activation of the kinase.

The ability of insulin to lower agonist-stimulated cAMP levels, while undoubtedly a very important physiological mechanism, does not eliminate the existence of other sites of insulin action in antagonizing cAMP-stimulated phosphorylation. Thus, Larner and associates[48] have reported insulin antagonism of cAMP action under circumstances in which cAMP levels are unaffected by insulin (i.e., in skeletal muscle); these workers have proposed that insulin leads to the generation of a factor that directly inhibits the cAMP-dependent protein kinase.[49] This provides a fully adequate explanation for insulin antagonism of cAMP directed phosphorylation. Finally, an equally plausible mechanism is insulin activation of a phosphatase that removes ^{32}P from the sites phosphorylated by the cAMP-dependent protein kinase. Evidence in support of this pathway is discussed below.

B. Insulin Regulation of Protein Phosphatase Inhibitor-1

The ability of insulin to dephosphorylate cellular peptides when added as the sole hormone requires a somewhat different set of mechanisms than those considered above. Since the activity of the cAMP-dependent kinase is low in the basal state, and its major sites of phosphorylation largely unfilled, the ability of insulin to promote dephosphorylation under these circumstances must involve sites phosphorylated primarily by cAMP-independent protein kinases. Certainly the sites on pyruvate dehydrogenase dephosphorylated in response to insulin are phosphorylated by a cAMP-independent protein kinase (i.e., PDH kinase). The data of Cohen and colleagues[6] on the insulin-induced dephosphorylation of glycogen synthase in rabbit skeletal muscle indicate strongly that a substantial portion of the dephosphorylation represents sites phosphorylated by the cAMP-independent protein kinase known as glycogen synthase kinase-3. Thus, the potential loci for insulin action include inhibition of one or more cAMP-independent protein kinases or activation of one or more protein phosphatases. We have undertaken studies of insulin-regulated protein phosphatase activity.

Characterization of protein phosphatases has proceeded more slowly than for protein kinases, largely because of difficulties in purifying unproteolyzed forms of phosphatases and lack of agreement as to the properties and substrate specificites of phosphatases purified in different laboratories. Several laboratories now concur that a major component of the phosphatase activity directed against enzymes whose phosphorylation state is regulated by insulin can be accounted for by a protein phosphatase of broad substrate specificity, which Cohen and co-workers have designated as protein phosphatase-1 and which is probably identical to the so-called Mg ATP-dependent phosphatase.[50] It is likely that an understanding of the mechanism of regulation of phosphatase-1 would lead to an understanding of the mechanism of dephosphorylation of many of the proteins dephosphorylated by insulin.

Protein phosphatase-1 is subject to regulation *in vitro* by a class of protein modulators, which are acid-soluble, heat-stable proteins of 15,000–40,000 M_r.[51] Two of these proteins—phosphatase inhibitor-1 and inhibitor-2—have been characterized, and their inhibitory activity has been shown to be regulated by phosphorylation–dephosphorylation reactions.[52,53] An attractive model is the possibility that insulin-regulated protein phosphatase activity may be mediated by these heat-stable modulators.

Inhibitor-1 was first indentified in rabbit skeletal muscle as a 26,000-M_r protein.[51] Inhibitor-1 is a substrate of the cAMP-dependent protein kinase *in vitro*; upon phosphorylation by the cAMP-dependent protein kinase, it becomes a potent inhibitor of phosphatase-1. Removal of the phosphate from the inhibitor protein abolishes its inhibitory activity. Thus, inhibitor-1 provides a mechanism whereby an increase in cAMP level results in an inhibition of protein phosphatase activity. Elevations of cAMP would activate the cAMP-dependent protein kinase, thereby increasing the phosphorylation of, for example, phosphorylase *b* kinase, while simultaneously inhibiting dephosphorylation of the same enzyme via inhibitor-1. Such a mechanism would amplify the signal generated by elevated levels of cellular cAMP. Several groups have shown that in perfused skeletal muscle, administration of catecholamines results in an increase in inhibitor-1 activity, presumably by stimulating its phosphorylation via the cAMP-dependent protein kinase.[54,55] In the basal state, there is very little inhibitor-1 activity in the perfused muscle preparation. Insulin, acting as a sole hormone, was not observed to cause any detectable change in the measured inhibitor-1 activity although insulin did activate glycogen synthase under these conditions.[54] Insulin, however, was able to antagonize activation of inhibitor-1 by catecholamines, apparently by lowering cAMP levels.[55]

Our laboratory has studied inhibitor-1 in rat adipose tissue, another insulin-sensitive system.[56] This tissue permits convenient study of the effects of insulin on inhibitor-1 phosphorylation and phosphatase inhibitory activity in an isolated cell preparation. We have purified inhibitor-1 from rat adipose tissue; the protein has a molecular weight of 32,000, slightly larger than the skeletal muscle analog. In contrast to perfused muscle, inhibitor-1 isolated from adipose tissue not exposed to catabolic hormones shows appreciable inhibitory activity, approximately 50% of maximal. As expected, β-adrenergic agents cause an increase in the phosphorylation state and inhibitory activity, presumably by activating the cAMP-dependent protein kinase. In contrast to perfused skeletal muscle, in which insulin as the sole hormone has no effect on the activity of inhibitor-1, in rat adipose tissue insulin alone causes a substantial decrease in both phosphorylation state and phosphatase inhibitory activity (measured as inhibition of phosphorylase *a* phosphatase). Thus, insulin may regulate protein phosphatase activity in adipose tissue by dephosphorylating and inactivating inhibitor-1, resulting in an increased phosphatase-1 activity. As to the mechanism whereby insulin dephosphorylates inhibitor-1, the hormone could inhibit a protein kinase, or activate a protein (i.e., inhibitor-1) phosphatase. Aside from the cAMP-dependent protein kinase, little is known of the protein kinases and phosphatases that act on inhibitor-1. The high basal phosphorylation of inhibitor-1 in adipose tissue strongly suggests that cAMP-independent kinases may be acting on this protein, analogous to those characterized for glycogen synthase. Thus, insulin might inhibit such a cAMP-dependent protein kinase, resulting in dephosphorylation of inhibitor-1 and activation of phosphatase-1. Alternatively, insulin may activate an as yet unidentified protein phosphatase which dephosphorylates inhibitor-1, thereby activating phosphatase-1; such a mechanism would then be a cascade of phosphatases analogous to cascades of kinase responsible for the activation of glycogen phosphorylase.

In conclusion, the ability of insulin to induce protein dephosphorylation is an important pathway of insulin action, since insulin regulates the activity of a number of enzymes by promoting their dephosphorylation. Some of the sites undergoing dephosphorylation are phosphorylated by the cAMP-dependent protein kinase, whereas others are not substrates for this kinase; any scheme for insulin-induced dephosphorylation must explain this observation. Clearly, insulin can inhibit the cAMP-dependent protein kinase by lowering the intracellular concentration of the activating ligand, cAMP, probably through the activation of a low-K_m cAMP phosphodiesterase.[57–60] Some data indicate that insulin may inhibit the cAMP-dependent protein kinase through other mechanisms. The effect of insulin on the important class of cAMP-independent kinases has not been investigated. Insulin could induce protein dephosphorylation by activation of one or more protein phosphatases. The ability of insulin to diminish phosphatase inhibitor-1 activity in adipose tissue by dephosphorylating it may underlie certain of the other dephosphorylations reduced by insulin. Nevertheless, the mechanism of insulin induced dephosphorylation of inhibitor-1 is itself unknown. The ability of insulin to regulate the activity of inhibitor-2 is largely unexplored and is clearly a topic worthy of further study. Finally, it will be necessary to show how phosphorylation and dephosphorylation of different proteins by insulin flow from the same initial event—insulin binding to its receptor.

IV. INSULIN-STIMULATED (SERINE/THREONINE) PROTEIN PHOSPHORYLATION

A. General Features

The observation that insulin promotes intracellular protein phosphorylation was entirely unexpected; there had been no example of a regulatory response to insulin associated with increased protein phosphorylation. Therefore, a series of studies was carried out to characterize the properties of insulin-stimulated phosphorylation and the pathway mediating this effect, in an effort to determine whether it had features compatible with a signaling mechanism elicited in response to the hormone. The general features of this response are as follows. Insulin-stimulated protein phosphorylation is observed very rapidly after exposure of rat adipose and liver cells to insulin; in our studies in hepatocytes, it is detectable within 1–3 min after hormone exposure and reaches plateau values within 5–15 minutes.[26] In studies by Belfage and co-workers in adipocytes, half-maximal stimulation was observed in less than 1 min.[44] The response is freely reversible if insulin is removed by washing or by adding insulin antibody; the stimulated phosphorylations returned toward basal levels within several minutes.[61] The response is furthermore a rather ubiquitous expression of insulin action, observed not only in the classic target tissues, but in a variety of cultured cells as well. The response appears to be initiated by interaction of insulin with its high-affinity receptor; thus, the half-maximal stimulation of these phosphorylations is observed at 10–50 μU/ml[25,26] (70–350 pM) insulin. Moreover, incubation of adipocytes with antisera containing antibodies directed against the insulin receptor, which have the ability to mimic several of insulin's well-defined metabolic actions, elicits an array of stimulated intracellular protein phosphorylations that is indistinguishable from those observed in response to insulin itself.[36] Finally, several lines of evidence indicate that insulin-stimulated phosphorylation is elicited through a pathway that does not involve cAMP or the cAMP-dependent protein kinase. Insulin stimulates protein phosphorylation as the sole hormone, under conditions in which no change in the levels of cAMP is detected.[23,26] Insulin stimulation of phosphorylation is evident and quantitatively unaltered in the presence of maximally stimulated levels of cAMP (in response either to glucagon or to the addition of dibutyryl-cAMP).[26] Agents other than insulin that are able to antagonize catecholamine-stimulated cAMP accumulation (e.g., propranolol, nicotinic acid, and concanavalin A) fail to stimulate protein phosphorylation, although they do inhibit cAMP-stimulated phosphorylation much as insulin.

In summary, the properties of insulin-stimulated phosphorylation are entirely compatible with those expected of a signaling pathway. The ultimate verification of this hypothesis requires the identification of several representative proteins that undergo this modification, as well as the direct demonstration that the altered phosphorylation serves a regulatory role. This verification has not yet been produced, thus the physiological significance of the pathway of insulin-stimulated protein phosphorylation remains unknown. A more detailed examination of the best characterized insulin-stimulated phosphoprotein is presented below.

B. ATP Citrate Lyase

ATP citrate lyase is a cytosolic enzyme required for the production of cytosolic acetyl CoA in non ruminants. In nutritional states associated with high rates of *de novo* fatty acid synthesis in the rat, ATP citrate lyase constitutes approximately 3–5% of total cytosolic protein in both liver and adipose tissue.[27] The native enzyme is a tetramer of identical 123,000-M_r subunits. In addition to a phosphohistidine enzyme intermediate,[62] lyase is phosphorylated on both serine and threonine[63] residues. A portion of these residues turn over very slowly, in that estimates of the basal phosphate content measured as alkali-labile phosphate (i.e., serine plus threonine) of the enzyme purified from rat hepatocytes is much higher (i.e., 0.24 moles P/per mole 123,000-M_r subunit) than estimated by ^{32}P labeling of intact hepatocytes (i.e., 0.08 mole P/subunit).[63] The phosphorylation of lyase in intact cells is stimulated by insulin and glucagon (i.e., cAMP); the effects of these hormones have been studied in most detail (the phosphorylation of the enzyme in intact liver cells is also stimulated by vasopressin and EGF). The enzyme can be purified from basal and hormone stimulated hepatocytes with 80–90% recovery of initial ^{32}P content (estimated by immunoprecipitating the enzyme from homogenates). Analysis of acid hydrolysates indicates that greater than 90% of the ^{32}P is in phosphoserine residues before and after hormone stimulation. In response to a maximal dose of insulin, the ^{32}P content increases two- to threefold[63]; a similar increase in ^{32}P content is elicited in response to glucagon[64] and in the presence of both hormones simultaneously an essentially additive increment is observed. (This additivity has not been detected in the studies of Swergold et al. in 3T3-L1 cells.[65]) Thus, insulin and glucagon each stimulate the incorporation of an additional 0.15 moles of phosphate per subunit and in the presence of both hormones near 0.5 moles of phosphate per subunit can be observed. Identical estimates of the hormone-stimulated component of phosphorylation are obtained by measurement of the alkali-labile P content of the purified enzyme. ATP citrate lyase can be phosphorylated in vitro by the cAMP-dependent protein kinase with incorporation of ~0.5 moles of phosphate per subunit; if lyase is slightly denatured, an additional 0.5 moles of phosphate per subunit may be incorporated in vitro into the same site, to a total stoichiometry of 1 mole per subunit.[64,66] This phosphorylation proceeds in vitro at approximately 2% of the rate observed for phosphorylase b kinase (as substrate). Phosphorylation by the cAMP-dependent protein kinase occurs on exquisitely tryptic-sensitive region of lyase, and the tryptic peptide containing the cAMP kinase phosphorylation site has been purified and sequenced (Thr-Ala-Ser-p-Phe-Ser-Glu-Ser-Arg).[66]

Similarly, lyase has been purified to homogeneity from insulin-[67] and glucagon[66]-stimulated ^{32}P-labeled hepatocytes and the phosphorylation site has been characterized in these preparations. A single major tryptic peptide has been isolated from the enzyme after both insulin and glucagon stimulation; this peptide contains greater than 80% of the hormone-induced increment in ^{32}P content and is indistinguishable in chromatographic properties from the single major tryptic ^{32}P-labeled peptide obtained from lyase purified from unstimulated hepatocytes. Most importantly, the lyase tryptic ^{32}P-labeled peptide purified from insulin- or glucagon-treated cells exhibits the identical chromatographic behavior, composition, peptide sequence, and phosphorylated amino acid residue as the tryptic peptide (shown above) that bears the phosphorylation site of the cAMP-dependent protein kinase. Thus, insulin and glucagon, acting on intact hepatocytes, both direct the phosphorylation of lyase on the same serine residue as is phosphorylated by the cAMP-dependent protein kinase in vitro. It is important to note that phosphorylation of lyase by the cAMP-dependent protein kinase to a stoichiometry between 0.5–1.0 moles P per lyase subunit has no detectable effects on enzyme activity.[68,69] Similarly, glucagon and insulin stimulation of lyase phosphorylation in intact cells does not alter the activity of the enzyme measured in homogenates[70] (also observed by M.C. Alexander, M. Adelman, R.A. Nemenoff, and J. Avruch, unpublished results), although de novo fatty acid synthesis in intact cells is stimulated by insulin and inhibited by glucagon.[71]

On the basis of the low stoichiometry of hormone-stimulated lyase phosphorylation in intact cells, the lack of an effect of phosphorylation on enzyme activity, and most importantly the observation that insulin and glucagon phosphorylate the same site (although they regulate the pathway in which lyase is situated in a mutually antagonistic fashion), we surmise that phosphorylation of lyase is not a regulatory modification and that it is incidental to the program of hormone-directed phosphorylation initiated by both glucagon and insulin. Thus, lyase phosphorylation occurs primarily because of the abundance of the protein and because of the presence within its structure of a exposed segment whose sequence or conformation approximates the recognition requirements of several protein kinases. Other explanations for the role of lyase phosphorylation remain feasible and cannot be discounted with present evidence. Thus, hormone-directed phosphorylation may modify the rate of intracellular proteolysis of lyase or provide a recognition signal for association of lyase with membranes or other macromolecules. No evidence is available, however, in support of such formulations.

The notion that a post-translational modification may occur in a "mistaken" or incidental fashion is not new. Microheterogeneity of protein glycosylation is well known; examples (e.g., ribonuclease) exist wherein a protein that is normally not glycosylated can be detected in a glycosylated form. If one accepts the notion that lyase represents an incidental phosphorylation, it provides an important caveat for the study of hormone regulated phosphorylation by the ^{32}P labeling/SDS gel technique; i.e., that substoichiometric phosphorylation (i.e., <0.2 mole P/subunit even with maximal hormone stimulation) of abundant proteins may not be physiologically relevant modifications. The relatively slow rate of lyase phosphorylation by the cAMP-dependent protein kinase *in vitro* was likely a clue to the lack of significance of this modification, despite the good stoichiometry (0.5–1.0 mole P/subunit) that could ultimately be achieved in the test tube.[64]

Finally, it should be emphasized that even if the hormone-stimulated phosphorylation of lyase is itself an irrelevant event, it remains a valid indicator of the hormone activation of protein kinase activity occurring in intact cells. For example, the irrelevance of glucagon-stimulated lyase phosphorylation in no way impeaches the significance of glucagon/cAMP-stimulated protein phosphorylation as a regulatory pathway; the same proviso is likely to apply for insulin.

C. Acetyl CoA Carboxylase

Acetyl CoA carboxylase is considered in detail in Chapters 17 and 18; however, several points may be briefly noted here. It is clear that carboxylase activity in intact adipose and liver cells is regulated by β-catecholamines[72] and glucagon (which inhibit the enzyme) and by insulin[71,73] (which activates it). It is also clear that each hormone individually stimulates a modest increment in the ^{32}P content of the enzyme, of the order of 10–20%.[73,75,77] Since the [^{32}P]phosphate content in the basal state appears to be quite high, (3–4 moles per 240,000-dalton peptide subunit[76]) even this modest incremental increase in response to hormones is likely to represent 0.3–1.2 moles of phosphate added per subunit of protein. Thus, the hormone-stimulated phosphorylation of carboxylase is quantitatively much greater than that observed with ATP citrate lyase. Data from Denton and co-workers[73] and Witters et al.[77] clearly indicate that the major sites of phosphorylation in response to insulin and epinephrine reside on different tryptic peptides. With regard to the effects of phosphorylation on the function of the enzyme, it is clear that the ability of the epinephrine to inhibit the enzyme in intact cells can be mimicked by phosphorylation of the enzyme *in vitro* by the cAMP-dependent protein kinase[78–80]; the enzyme purified from epinephrine-treated cells retains both the increment in phosphate and the inhibition of activity detected in the original homogenate.[77] The major area of uncertainty is the relationship of insulin stimulated phosphorylation to insulin activation. Although carboxylase can be purified from insulin-treated liver cells with the increment in ^{32}P intact, the apparent stimulation of activity observed in the homogenate is lost progressively during purification.[76,80] No protein kinase has yet been identified whose phosphorylation of carboxylase is associated with a stimulation of carboxylase activity.[84] In sum, the functional role of insulin-stimulated carboxylase phosphorylation is at present unknown.

D. Ribosomal Protein S6

S6 is the major phosphoprotein of the 40S subunit of the eukaryotic ribosome.[85] The phosphorylation of this protein is strongly stimulated by insulin in 3T3-L1 cells, as first shown by Smith *et al.*,[39] as well as in liver cells[86] (Fig. 1), adipose cells, H4 hepatoma cells[87] and other cultured cells.[41] The maximally-stimulated phosphorylation of S6 is quantitatively substantial; S6 separated by two-dimensional electrophoresis after maximal phosphorylation shows up to 5 discrete spots of more anodic mobility, indicating the presence of 5 phosphorylation sites.[85–91] Two sites of cAMP-stimulated phosphorylation on S6 have been sequenced.[92] While an early report indicated that the sites of insulin and cAMP-stimulated S6 phosphorylation in Hela cells resided on different tryptic peptides,[93] more recent studies by Thomas and coworkers,[94] in 3T3 cells reveal a more complex situation. Using partial tryptic digestion, they can detect 10 or 11 tryptic phosphopeptides which probably represent 5–6 phosphorylation sites. Upon addition of serum an ordered, sequential phosphorylation of these sites on S6 occurs. The more recent studies of Martin Perez *et al.*[94,95] indicate that these sites are common to cAMP as well as insulin and the growth factors; i.e., cAMP can promote only the "earliest" steps in the sequential phosphorylation, whereas insulin and the growth factors can promote more distal steps as well. However, the combination of agents is required to elicit the full sequence of phosphorylation observed in response to serum. It is not clear whether S6 phosphorylation stimulated by serum or growth factors requires the sequential operation of a series of different protein kinases, each with specificity for one or two sites, or whether a single protein kinase mediates the incremental phosphorylation; whether the earlier phosphorylations facilitate the subsequent phosphorylations is also unknown. Moreover, it is not entirely certain whether the augmented S6 phosphorylation is due to an activation of kinase(s), a redistribution of kinase(s) to the ribosome, or an enhanced availability of the substrate S6 itself, which may undergo an ordered series of changes determined by the sequential assembly of the 80S ribosome, i.e., the process of initiation of protein synthesis.

While suggestions have been put forth that S6 is involved specifically in mRNA binding or other partial reactions in the initiation of protein synthesis, direct evidence is lacking. Both the function of S6 within the 40S and the functional consequences of hormone-directed phosphorylation are unknown.[85,96]

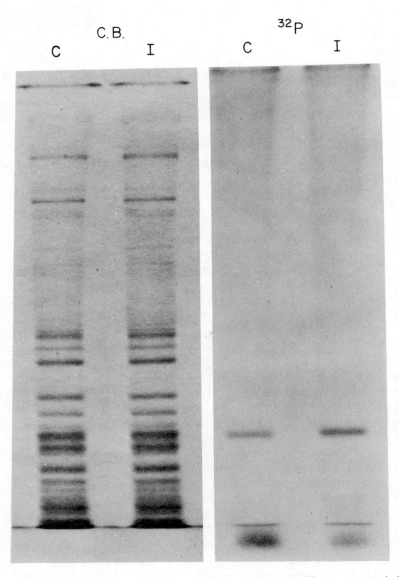

Figure 1. Insulin-stimulated phosphorylation of ribosomal S6 in rat hepatocytes. Hepatocytes were isolated from 48-hr starved rats and labeled *in vitro* by incubation with $^{32}P_i$. After 2 hr, a portion of the cells were exposed to insulin (7 nm) for 10 min and homogenized; the supernatant of a $1.95 \times 10^5 \times g$ min) centrifugation was recentrifuged at $4.68 \times 10^6 \times g$ min). The pellet was resuspended in 1% deoxycholate–0.5% Nonidet and centrifuged over a cushion of 0.5 M sucrose at $1.5 \times 10^7 \times g$ min. The pellets from control and insulin cells were resuspended, matched for protein (optical density 260/280 = 1.82), and subjected to SDS gel electrophoresis. (Left) Protein pattern. (Right) Autoradiograph. [^{32}P]-S6 (confirmed by two-dimensional electrophoresis) is the major ribisomal ^{32}P-labeled protein.

E. M_r 22,000 Protein

A 22,000 dalton insulin-stimulated phosphopeptide has been detected in rat adipocytes,[25,37,97] 3T3-L1 cells,[28] H_4 hepatoma cells[87] (Fig. 2) and freshly isolated rat liver cells[87] (Fig. 3). This low abundance cytoplasmic peptide exhibits several interesting properties: it is stable to boiling and soluble in 1% trichloroacetic acid, it migrates as a doublet on SDS gel electrophoresis but as a single band on electrophoresis in a cationic detergent, and on isoelectric focusing, 2–3 ^{32}P-labeled spots can be detected between 4.5–5.0.

If 3T3-L1 cells are labeled with ^{35}S methionine, approximately half of this peptide can be detected at M_r 21,000 near pH 4.7; exposure of the cells to maximal insulin

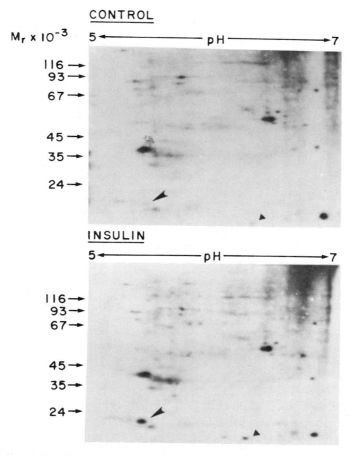

Figure 2. Insulin-stimulated protein phosphorylation in cultured rat hepatoma cells. Rat hepatoma cells (H_4IIE) were incubated with [32]P$_i$ in control conditions or given insulin (7 nm × 10 min); autoradiographs of two-dimensional gels of soluble proteins are shown. Large arrows point to a protein of 22,000 M_r; small arrows point to a protein of 19,000 M_r, the phosphorylations of which are stimulated by insulin.

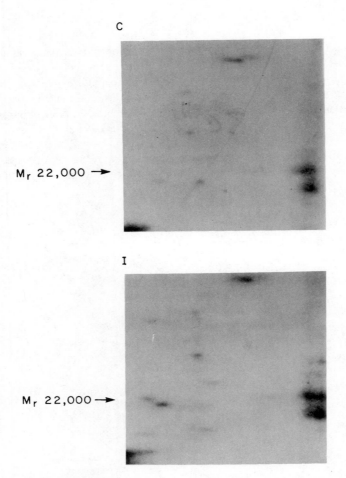

Figure 3. Insulin-stimulated phosphorylation of a heat-stable 22,000-M_r protein-from rat hematocytes. Freshly isolated rat hepatocytes were exposed to $^{32}P_i$ and control conditions (C) or insulin (I), (7 nM × 10 min). Heat-stable extracts from these cells were prepared as described in Ref. 25 and subjected to two-dimensional electrophoresis and autoradiography. The anode is to the left of the figure. The arrow points to a heat-stable, 22,000-M_r protein the phosphorylation of which is stimulated by insulin in these cells. This protein is similar or identical to the 22,000-M_r protein of H$_4$II E cells shown in Fig. 2 and to the 22,000-M_r protein of rat adipose tissue[25] and 3T3-L1 adipocytes.[28]

results in the shift of almost all of this ^{35}S labeled peptide to a more acidic spot at M_r 22,000, indicating a near stoichiometric extent of charge modification (i.e., phosphorylation) in response to insulin. The structure of the peptide containing the site of insulin-stimulated phosphorylation is unknown, but the data of Blackshear *et al.*[28] indicate that threonine is the residue preferentially phosphorylated in response to insulin (as well as EGF). In the 3T3-L1 cell, the M_r 22,000 ^{32}P peptide is also phosphorylated in response to EGF, PDGF, and serum, but unaffected or slightly inhibited by beta adrenergic catecholamines and dibutryl cAMP. An acidic M_r 22,000 ^{32}P peptide, whose phosphorylation is inhibited by cAMP, can also be detected in H_4 rat hepatoma cells, hepatocytes and cardiac myocytes[87]; it is not clear whether this peptide is identical to the insulin-stimulated 22,000 peptide of adipose cells and 3T3-L1 cells. To date there is no evidence on the identity of this heat resistant, acid soluble protein; it is not protein phosphatase inhibitor 1 or 2, the Walsh inhibitor of the cAMP-dependent protein kinase, calmodulin or myosin light chain, which all share these peculiar solubility characteristics.

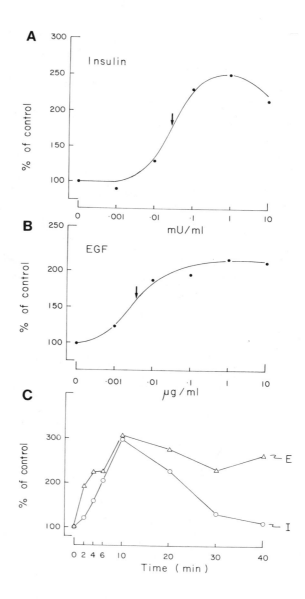

Figure 4. Insulin- and EGF-stimulated phosphorylation of the 46,000-M_r phosphoprotein in isolated rat hepatocytes. (A, B) Freshly isolated rat hepatocytes were exposed to [32]P$_i$ and insulin or epidermal growth factor (EGF) (15 min) at the concentrations indicated. The hormone-stimulated phosphorylation of the 46,000-M_r protein shown in Fig. 5 was determined by scanning the autoradiograph; the results are expressed as % of control phosphorylation. Each data point represents the average of two separate incubation samples. (C) Time course of phosphorylation of the 46,000-M_r protein is shown in response to insulin (1 mU/ml) or EGF (1 μg/ml). Each data point represents the average of determinations from two separate incubation samples.

F. 46,000-M_r Peptide

The 46,000-M_r protein has been detected thus far only in freshly isolated rat liver cells[26,94–97]; it is not identifiable as an insulin-stimulated phosphopeptide in adipose tissue or cultured cells. Its phosphorylation in the intact hepatocyte is stimulated by insulin, glucagon, dibutyryl cAMP, vasopressin, EGF (Fig. 4), and the α-adrenergic agents. The protein behaves as a monomer on gel filtration and ultracentrifugation (J. Avruch, unpublished observations). On two-dimensional gel analysis, the protein can be resolved into a series of five to six ^{32}P-labeled spots with an approximate isoelectric point of ~4.5.[96] Insulin leads to an increase in ^{32}P in all spots, with a modest shift in the distribution of ^{32}P from basic to acidic species; this response suggests that the underlying charge heterogeneity may be due to some modification other than phosphorylation or that the protein may already be highly phosphorylated in the basal state (Fig. 5). On acid hydrolysis of the protein partially purified from insulin-treated hepatocytes, [^{32}P]serine is the only phosphoaminoacid recovered (J. Avruch, unpublished observations), tryptic digestion of the 46,000-M_r protein from basal ^{32}P-labeled he-

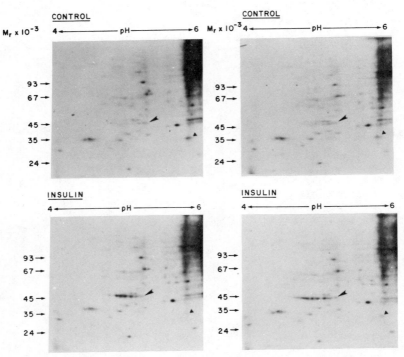

Figure 5. Insulin-stimulated protein phosphorylation in isolated rat hepatocytes. Freshly isolated hepatocytes prepared from 48-hr starved rats were exposed to ^{32}Pi and control conditions or insulin (7 nm × 10 min). Shown are autoradiographs of replicate two-dimensional gels of cytosolic ^{32}P-labeled proteins. Large arrows point to a 46,000-M_r protein whose phosphorylation is stimulated by insulin, appearing as a series of ^{32}P-labeled spots. Small arrow points to a protein of 34,000 M_r the phosphorylation of which is inhibited by insulin.

patocytes yields a relatively limited number of ^{32}P peptides, and a qualitative identical pattern of the tryptic peptide is obtained after insulin or glucagon stimulation (Fig. 6); thus, both hormones lead to the stimulation of the phosphorylation on the same peptide segments. A similar conclusion has been reached by LeCam.[96] We have succeeded in partially purifying the protein from isolated hepatocytes 300–600-fold such that it is the sole ^{32}P-labeled peptide and represents 5–10% of the Coomassie blue-stained protein in the isolate; this implies that an approximate 10,000-fold purification would

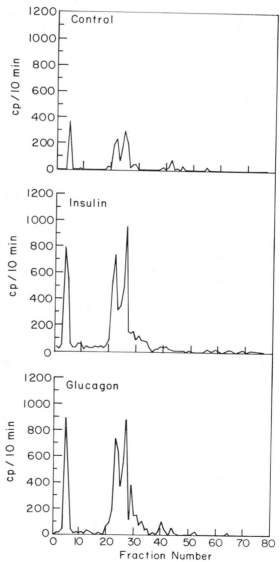

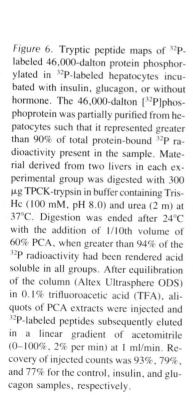

Figure 6. Tryptic peptide maps of ^{32}P-labeled 46,000-dalton protein phosphorylated in ^{32}P-labeled hepatocytes incubated with insulin, glucagon, or without hormone. The 46,000-dalton [^{32}P]phosphoprotein was partially purified from hepatocytes such that it represented greater than 90% of total protein-bound ^{32}P radioactivity present in the sample. Material derived from two livers in each experimental group was digested with 300 μg TPCK-trypsin in buffer containing Tris-Hc (100 mM, pH 8.0) and urea (2 m) at 37°C. Digestion was ended after 24°C with the addition of 1/10th volume of 60% PCA, when greater than 94% of the ^{32}P radioactivity had been rendered acid soluble in all groups. After equilibration of the column (Altex Ultrasphere ODS) in 0.1% trifluoroacetic acid (TFA), aliquots of PCA extracts were injected and ^{32}P-labeled peptides subsequently eluted in a linear gradient of acetomitrile (0–100%, 2% per min) at 1 ml/min. Recovery of injected counts was 93%, 79%, and 77% for the control, insulin, and glucagon samples, respectively.

be required for purification to homogeneity, indicating that this is a relatively scarce component of liver cytosol. The ^{32}P content of this protein may be very high, i.e., 10 moles phosphate/subunit or greater. The function of this protein is unknown.

In addition to the peptides described above, the insulin receptor may also be considered among the insulin-stimulated (serine/threonine) phosphoproteins.[41] We have no information concerning the stoichiometry of serine (or tyrosine) phosphoryl-ation of the receptor in the intact cell or the functional significance of these modifi-cations. Denton and co-workers[37] and Benjamin and Clayton[31] have each reported the detection of additional ^{32}P-labeled peptides whose phosphorylation is stimulated by insulin in adipocytes. We have detected a further five to ten minor insulin-stimulated ^{32}P-labeled peptides by analysis of two-dimensional gels, all phosphorylated on serine or threonine, or both (P.J. Blackshear, J. Avruch, unpublished observations). The characterization of these minor species has not yet been carried further.

G. Methodological Considerations

On the basis of the data summarized above, it is clear that the ^{32}P-labeling/SDS-PAGE approach has a variety of advantages and limitations. The chief advantage is the ability to exemplify patterns of altered protein phosphorylation in response to hormones or other stimuli. This permits detection of entirely unanticipated responses, e.g., insulin-stimulated phosphorylation. This method can also provide a very char-acteristic array of altered ^{32}P-labeled peptides for each hormone or class of hormones, i.e., subsets of cellular peptides that exhibit increased or decreased ^{32}P content; the greater the number of the ^{32}P-labeled peptides detected, the more specific a fingerprint of the hormone effect is obtained. Thus, in contrast to examinations of a single insulin-regulated function such as glucose transport in fat cells, or inhibition cAMP accu-mulation, the ^{32}P-labeling approach permits simultaneous examination of a large num-ber of hormonally regulated cellular ^{32}P-labeled peptides. A further advantage is that this technique may detect cellular ^{32}P-labeled peptides, providing a convenient model in the study of hormone-regulated kinases/phosphatases; an example is the 40S ribo-somal protein S6, whose phosphorylation in intact cells is regulated by insulin, cAMP, and a variety of growth factors. Although the functional significance of S6 phospho-rylation is unknown, S6 phosphorylation in cell homogenates provides a potentially useful assay for insulin- and growth factor-regulated protein kinases.

Several major drawbacks to this approach limit its utility. The most serious is that the ^{32}P-labeled peptides are detected after detergent gel electrophoresis, a procedure that denatures and defunctionalizes the protein. Thus, it is necessary to provide in-dependent evidence on the functional identity of the ^{32}P-labeled peptides detected and the functional consequences of altered phosphorylation. This approach requires the independent purification of an activity, using conventional or immunological tech-niques, a goal that has been accomplished for only a minority of the hormone-regulated ^{32}P-labeled peptides. A second limitation is the bias of the method toward detecting abundant ^{32}P-labeled peptides to the disadvantage of scarce but functionally important proteins. It is important to recall that sites of hormone regulation are often rate-limiting enzymatic steps, which in turn are generally those enzymes with the lowest V_{max} in a given pathway; they therefore tend to be rather low abundance proteins. To some degree, this bias can be overcome if one examines tissues specialized for the function of specific interest; here, even the rate-limiting enzymes may be relatively abundant. However, even in adipose tissue, for example, which is dedicated to the synthesis, storage, and hydrolysis of triglyceride, all under rapid hormonal control, an enzyme such as hormone-sensitive lipase, which has a very high intrinsic turnover number, is a very low-abundance species and is easily overlooked in one-dimensional SDS gels of ^{32}P-labeled adipocytes, without prior knowledge of its electrophoretic mobility.

A final shortcoming of the ^{32}P-labeling technique is it inability to aid in the identification of the site of hormone action on protein phosphorylation. Thus, a hormone may act to alter phosphorylation by changing kinase activity, phosphatase activity, or both. A potentially powerful approach toward distinguishing these sites of action is by changing the ^{32}P specific activity of the [^{32}P]-ATP precursor pool concomitant with the introduction of the hormone stimulus. Unfortunately, the rate of equilibration of extracellular ^{32}P into intracellular [^{32}P]-ATP is quite slow (on the order of hours) and is a hormone-sensitive process, whereas hormone-induced changes in phosphorylation are elicited in minutes. Thus, in the absence of some novel technique for rapid re-equilibration of intracellular ATP, the ^{32}P-labeling technique in intact cells is unable to provide information as to the site of hormone action. Finally, an increase in ^{32}P incorporation induced by a hormone might be due to initial dephosphorylation of previously unlabeled sites, followed by refilling of these sites by ^{32}P. Thus, independent evidence for an increase in net phosphate content must be obtained; most convenient is the demonstration that increased ^{32}P incorporation is accompanied by an anodic shift of the protein on isoelectric focusing (run under denaturing conditions). Direct measurement of the alkali labile or total phosphate content of the purified protein is of course most definitive.

H. Comparisons among Insulin-, cAMP-, and Growth Factor-Regulated Protein Phosphorylation

In the absence of direct evidence as to the significance of insulin-stimulated phosphorylation, it is informative to compare this phenomenon with that of cAMP-stimulated phosphorylation, a pathway of known regulatory significance. Perhaps most relevant are the observations of Steinberg and Coffino,[100,101] who attempted to classify the cAMP-mediated protein phosphorylations in S49 lymphoma cells and other cell lines, as detected by two-dimensional gel analysis of extracts from [^{35}S]methionine- and ^{32}P-labeled cells. If one considers only the post-translational phosphorylations induced by dibutyryl cAMP, they observed two general types of substrates for the endogenous cAMP-dependent protein kinase: most of the substrates were peptides of very low abundance (less than 0.005% of cellular protein) that underwent a nearly complete conversion to a phosphorylated form. In addition, however, these workers observed two relatively abundant (0.5–1.0% of cell protein) cellular phosphoproteins that underwent cAMP-stimulated phosphorylation but that did not undergo greater than 10% conversion to a phosphorylated form. Since these proteins were not identified, it is not possible to conclude as to the functional significance of these two patterns of cAMP-directed phosphorylation. Nevertheless we would suggest that the quantitatively minor, highly phosphorylated peptide reflect the response expected for the modification of important regulatory enzyme proteins. Conversely, it is unlikely that this is true of the slightly phosphorylated, highly abundant peptides; these species are probably analogous to ATP citrate lyase.[67] Thus, as assessed by one- and two-dimensional analysis of ^{32}P-labeled cells, insulin and cAMP elicit strikingly similar kinds of changes in protein phosphorylation. The predominant response to both agents is an increase in ^{32}P content of a subset of cellular peptides that includes highly abundant slightly phosphorylated peptides, and low-abundance peptides phosphorylated to a stoichiometric extent. cAMP has recently been shown to induce dephosphorylation of peptides in several cells.[97,100,101] These shared features support the notion that insulin-stimulated phosphorylation, like cAMP-stimulated phosphorylation, is a regulatory pathway. Nevertheless, many other features clearly indicate that these two pathways are initiated by different mechanisms and mediate different functions. These include the noncongruity of many substrates, the presence of distinct phosphorylation sites on shared substrates (e.g., acetyl CoA carboxylase), and the antagonism of certain cAMP-stimulated phosphorylations by insulin. By contrast, the patterns of protein phosphorylation elicited by the growth factors EGF and PDGF exhibit none of the features that distinguish cAMP-stimulated phosphorylation from insulin-stimulation phosphorylation. The ability of EGF and PDGF to stimulate serine/threonine phosphorylation in target cells has perhaps been underemphasized in comparison with growth factor-stimulated tyrosine phosphorylation. Nevertheless, it is clear that the growth factors can elicit a major increment in serine/threonine protein phosphorylation in target cells. As with cAMP and insulin, growth factor-stimulated protein phosphorylation is rapid in onset and is observed in all cells bearing appropriate receptors. The striking feature of growth factor-stimulated protein phosphorylation in our view is that the array of ^{32}P undergoing phosphorylation in response to growth factors and serum overlaps to a very great extent with that observed in response to insulin. Thus, in freshly isolated rat liver cells, EGF stimulates the phosphorylation of ATP citrate lyase, acetyl CoA carboxylase, M_r 46,000, M_r 22,000; i.e., most of the major insulin-stimulated phosphopeptides. In 3T3-L1 cells, a similar congruity of stimulated substrates is observed.[28] It is not certain as to whether these common substrates are phosphorylated on the same or different sites in response to insulin and the growth factors (although preliminary data indicate that a single site is more likely for lyase, S6, 22,000 and 46,000-dalton peptides P.J. Blackshear, M.W. Pierce, J. Avruch, unpublished observations), nor is the extent of the overlap in substrates known. It is clear, however, that no antagonism exists between the phosphorylations stimulated by insulin and these agents.

Since the functional significance of growth factor-stimulated protein phosphorylation is unknown, its identification adds little direct information to our understanding of insulin-stimulated phosphorylation. Nevertheless, if the insulin- and growth factor-stimulated phosphorylations reflect regulatory responses, the extensive overlap in the phosphorylation responses to insulin and the growth factors suggests that the reactions being regulated and the sense of the regulation might be similar or identical for these agents. This overlap also suggests that the mechanisms involved in carrying out insulin-

and growth factor-stimulated protein phosphorylation, i.e., the kinases and phosphatases, may in some instances be shared or at least have closely related substrate specificity. A much more detailed examination of the biological actions and protein phosphorylation patterns elicited by these agents in several cell types will be required to judge the validity of this idea.

I. Mechanism of Insulin-Stimulated (Serine/Threonine) Protein Phosphorylation

On the basis of the premise that insulin-stimulated phosphorylation reflects a regulatory signal, several laboratories have attempted to uncover the mechanism by which the insulin–receptor interaction initiates this response. In general, three mechanisms are possible: (1) stimulation of one or more protein kinases, (2) inhibition of protein phosphatases, or (3) a ligand-induced modification of the substrate that alters its susceptibility to phosphorylation or dephosphorylation. The previous identification of cAMP-, calcium/calmodulin-, and growth factor-stimulated protein kinases has led most investigators to consider an insulin-mediated kinase activation as the most likely mechanism. The recent demonstration of insulin-stimulated tyrosine phosphorylation on its receptor suggested that insulin must activate one or more tyrosine protein kinases in addition to serine/threonine protein kinase(s). The identification of the insulin receptor itself as an insulin-activated tyrosine kinase has been rapidly forthcoming from several laboratories, including our own (see Section 5.1). The demonstration of an insulin-activated serine/threonine kinase has been much more problematic and will be briefly reviewed first.

One goal of the ^{32}P-labeled approach was to identify suitable substrates for the detection of the insulin-regulated kinases and phosphatases. Thus, despite the continuing lack of understanding as to the functional role of insulin-stimulated phosphoproteins, it is quite feasible to use these peptides in an assay for the detection of their respective kinases. The most convincing evidence supporting the identification of such a kinase is to demonstrate that treatment of cells or tissue with hormone leads to the generation of an activated kinase, which can be assayed after cell disruption. In general, these assays are carried out by adding Mg γ [^{32}P]-ATP immediately after homogenization and monitoring the transfer of ^{32}P to the substrate chosen. The acceptor protein may be the insulin-stimulated phosphoprotein endogenous to the tissue homogenate; it is preferable, however, to add an exogenous (at least partially) purified protein substrate to permit a more reliable estimate of kinase activity *per se*.

Employing endogenous S6 as a substrate, Smith *et al.*[39] reported that after insulin treatment of intact 3T3-L1 cells, the subsequent homogenate exhibited an increase in the phosphorylation of the endogenous S6. This observation has been confirmed in our laboratory (Fig. 7). Rosen *et al.*[102] also reported that a cell-free particulate fraction prepared from 3T3-L1 cells, which contained insulin binding activity, ribosomes, and S6 kinase, showed a 1.5–3-fold stimulation of phosphorylation of endogenous S6 on addition of insulin *in vitro*. This finding implies that all the components required for insulin simulation of S6 protein kinase are present in this particulate fraction; S6 kinase activation proceeds, for example, on supplementation with Mg ATP and insulin.

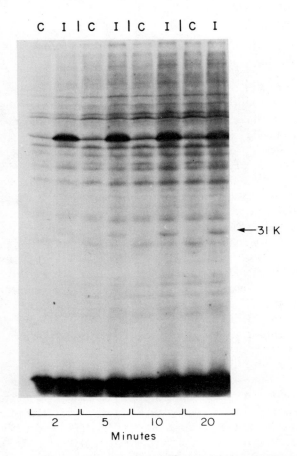

Figure 7. Homogenates from insulin treated 3T3-L1 cells show enhanced phosphorylation of endogenous ribosomal S6. 3T3-L1 fatty fibroblasts were induced and maintained as in (28). On the day of the experiment, the medium was replaced by Krebs Ringer bicarbonate with 25 mm Hepes (pH 7.4) 5 mM glucose, 1% crystalline bovine serum albumin for 2 hr. Insulin (0.18 μM) was added to one set; after 10 min, the Krebs buffer was aspirated and the cells scraped into sucrose (0.25 M), Tris (10 mM pH 7.0), $MgCl_2$ (1.5 mM), benzamidine (25 mM), at 0°C and disrupted by polytron homogenization. The suspension was centrifuged (3 min $\times$ 3000 $\times$ g) to remove unbroken cells and nuclei. The supernatants from the control and insulin-treated cells were incubated at 0°C with the addition of Mg (10 mM) and γ ^{32}P ATP (10 μM, ~1000 CPM/pmole). Aliquots of the mixture were taken for detergent gel electrophoresis at the times indicated. The stimulated phosphoprotein at M_r 31,000 is ribosomal S_6; the prominent phosphorylation at M_r 60,000 is rapidly discharged by the subsequent addition of CoASH (0.3 mM) and nonradioactive ATP (1 mM) and is likely a phosphoenzyme intermediate rather than a protein kinase mediated phosphorylation.

Most recently, Cobb and Rosen[103] reported the partial purification of an insulin-stimulated S6 kinase, starting with salt extracts of ribosomal fractions from insulin-treated cells; these workers conclude that the enzyme shares several properties with casein kinase I. By contrast, Perisic and Traugh[106] have reported the partial purification from 3T3-L1 cells of an insulin-stimulated S6 kinase, employing DEAE cellulose chromatography, and have tentatively identified this kinase as a trypsin-activated kinase.

Using adipocytes, Denton and Brownsey[105] observed that after treatment with insulin, the high-speed cell supernatant, which contains endogenous ATP-citrate lyase and acetyl CoA carboxylase, shows an enhanced rate of autophosphorylation of these endogenous peptides, as well as of added partially purified ATP-citrate lyase and acetyl CoA carboxylase. In these studies, the total cellular kinase directed at these substrates is apparently not increased by insulin, merely the fraction of the kinase activity found in the cytosol. This response is reminiscent (albeit in reverse) of the phorbol ester-induced translocation of protein kinase C from the cytosol to the surface membrane. Thus, direct data supporting the existence of insulin-stimulated (serine/threonine) protein kinase are available, but incomplete and conflicting.

It is not clear whether insulin gives a net increase in kinase activity, redistribution of kinase or both. It is unclear whether a single kinase of wide specificity mediates the insulin-stimulated serine/threonine phosphorylations, or whether several kinases are involved. Those laboratories that have detected insulin-stimulated S6 kinase activity disagree as to the properties and identity of the enzyme. We have been unable to detect an insulin increment in total cellular kinase active against purified added ATP-citrate lyase, acetyl CoA carboxylase, 40S ribosomal subunits, casein, histone, or phosvitin, even in extracts that exhibit an insulin-stimulated autophosphorylation of endogenous S6.

A second approach to the identification of insulin-regulated kinases is to purify the protein kinases that phosphorylate an insulin-stimulated phosphopeptide *in vitro* on the same site as is phosphorylated in response to insulin in the intact cells. It is anticipated that the kinase mediating the effect of insulin will be within the set of cAMP-independent kinases that show the appropriate site specificity. This hypothesis may be difficult to verify if the regulatory properties of these kinases are altered during purification, e.g., by proteolysis or loss of regulatory subunits. Nevertheless, we have attempted to purify the hepatic ATP-citrate lyase kinases that can phosphorylate the major site phosphorylated by insulin in intact cells. By this approach, Palmer subjected liver cytosol to phosphocellulose chromatography and identified a zone of Ca^{2+} and cAMP-independent lyase kinase activity (J.C. Palmer and J. Avruch, unpublished observations). This complex peak was further subfractionated on DEAE-Sephadex with resolution of five distinguishable peaks of lyase-kinase activity. The two lyase kinase peaks that eluted earliest from DEAE-Sephadex showed little phosphorylation of the insulin-directed site, but rather showed substantial phosphorylation of threonine and serine residues on other tryptic peptides. A kinase(s) with similar specificity has been partially purified by Benjamin and co-workers.[106,107] Given the relatively poor phosphorylation of the insulin-directed site by these kinases, we do not consider them likely candidates for mediating the insulin-stimulated phosphorylation of ATP-citrate lyase. The other three kinases separated on DEAE-Sephadex all show selective phosphorylation of the insulin-stimulated site on lyase. One of these has been identified as the catalytic subunit of the cAMP-dependent protein kinase on the basis of its inhibition by the Walsh inhibitor. The remaining two peaks of lyase-kinase phosphorylate ATP-citrate lyase at the insulin-directed site in a Ca^{2+}- and cAMP-independent fashion and are also able to phosphorylate acetyl CoA carboxylase (to the extent of 1–2 moles of phosphate incorporated/subunit) without an alteration in carboxylase activity (J.L. Palmer and J. Avruch, unpublished observations). These two kinases are distinct from casein kinase I and II. However, their relationship to other cAMP-independent protein kinases is unknown and awaits further purification.

V. INSULIN-STIMULATED TYROSINE PROTEIN PHOSPHORYLATION

A. Identification of the Insulin Receptor As a Tyrosine Protein Kinase

In contrast to the uncertainty as to the existence of insulin-stimulated serine kinases, the situation with respect to an insulin-stimulated tyrosine kinase is quite different. Within a brief interval after the initial report of insulin-stimulated tyrosine phosphorylation in intact cells,[41] several laboratories succeeded in documenting the existence of insulin-stimulated tyrosine kinase in cell-free extracts.[108–111] Initial studies demonstrated that detergent extracts of membranes rich in insulin receptors catalyzed the transfer of ^{32}P from [λ-^{32}P]-ATP onto the 95,000-dalton subunit of the insulin receptor endogenous to these extracts. The insulin-induced increment in ^{32}P incorporation appeared to be largely if not entirely onto tyrosine residues. It should be emphasized that three conditions are important to the facile demonstration of this reaction in crude membrane preparations: (1) inclusion of a nonionic detergent, (2) the use of Mn rather than Mg, since Mn ATP appears to have a greater affinity for the kinase, and (3) adequate concentrations of ATP (i.e., 0.1–0.2 mM), since the kinase exhibits a rather high apparent K_m for Me^{2+} ATP in comparison, for example, with the EGF receptor tyrosine kinase.[111,112] Rigorous proof of an insulin-stimulated tyrosine protein kinase followed from the demonstration of insulin-stimulated tyrosine phosphorylation of exogenous phosphate acceptor proteins.[110] We have found histone subfraction 2b to serve as a particularly convenient substrate; under the conditions we have employed, H_2b is phosphorylated on tyrosine only—a reaction stimulated three- to fivefold by insulin.[112]

Several lines of evidence support the conclusion that the insulin-binding site and the insulin-stimulated tyrosine protein kinase are both functions of the insulin receptor itself. Three groups have demonstrated the presence of an ATP-binding site on the 95,000-dalton subunit of the insulin receptor by photoaffinity labeling.[108–116] We have compared the recovery of the insulin-stimulated tyrosine kinase activity with the recovery of the insulin-binding function during purification of these two moieties. We observed that the kinase and the binding copurify in constant ratio through sequential adsorption and elution from wheat germ agglutinin Sepharose and DEAE cellulose chromatography, with approximately 20-fold purification. The partially purified kinase is bound to insulin Sepharose in parallel with the insulin-binding function; both functions remain bound after extensive washing and are quite resistant to elution under nondenaturing conditions. However, the tyrosine kinase is easily assayed when bound to insulin Sepharose. In this immobilized state, the kinase, after extensive washing, exhibits at least 80% of the activity observed for the enzyme applied to the insulin Sepharose beads. Elution of these well-washed beads with SDS yields two major silver-stained peptides, of 130,000 and 95,000 M_r. The 95,000-M_r peptide contains [^{32}P]tyrosine if the beads are incubated with Me^{2+} and [γ-^{32}P]-ATP prior to SDS elution.[112] The rate of autophosphorylation of the immobilized receptor is quite comparable to that observed in detergent solution, indicating that this is, in fact, an intramolecular reaction. These observations indicate that the insulin-binding function of the receptor remains tightly coupled to the tyrosine kinase, although both are in an immobilized state. Furthermore, these two functions copurify in an essentially constant stoichiometric ratio to the state at which only two peptides (130,000 and 95,000 M_r) are recovered. Thus, it appears that the kinase is not a trace contaminant of the receptor, but a function of the same molecule responsible for the insulin-binding. Moreover, our estimate of the turnover number for the insulin receptor tyrosine kinase is approximately 10–30 mole P transferred to histone 2b per min per mole of insulin bound. This measure of intrinsic activity is comparable to that of the cGMP-dependent protein kinase, and perhaps 10–20% that of the catalytic subunit of the cAMP-dependent protein kinase.[118] Thus, the tyrosine kinase activity of the insulin receptor is not a trivial activity.

B. Activation of the Receptor Kinase by Autophosphorylation

The co-identity of the binding and kinase function on the same receptor protein is entirely compatible with the kinetic features observed for the kinase: binding of insulin activates the kinase with a 3 to 5 fold increase in V_{max} for the protein substrate H2b, and little change in K_m. The extent of activation is directly proportional to the extent of insulin binding. The major unorthodox feature of the kinase is related to its requirements for metal/ATP. As the kinase was purified free of competing ATPases, protein phosphatases, etc., it became evident that either Mn^{2+} or Mg^{2+} could support the reaction (as opposed to an apparent absolute requirement for Mn^{2+} in crude extracts). However with Mg^{2+} as the sole divalent cation, the enzyme displayed an unusually low apparent affinity for ATP, and nonlinear double reciprocal plots for ATP. Even at high ATP concentrations (>0.5 mM) a significant curvilinearity in rate was observed, with Mg^{2+} as the sole divalent cation, before maximal rates of H2b phosphorylation were achieved.[112] This curvilinearity in the progress curves suggested that the kinase might be undergoing an autoactivation during the reaction, as originally proposed by Rosen et al.[112a] We examined the question of activation using the immobilized insulin receptor (Y. Kwok, R.A. Nemenoff, M.W. Pierce, and J. Avruch, unpublished data). In this state the receptor can be conveniently preincubated under any conditions and extensively washed prior to assay of histone tyrosine kinase, without loss of receptor kinase activity. We observed that preincubation of the immobilized receptor with Me^{2+}/ATP abolished the curvelinearity in the phosphorylation of H2b and markedly increased the apparent rate of H2b phosphorylation when measured at low concentrations of ATP (<0.1 mM). The preactivated kinase exhibited orthodox saturation kinetics for Mg^{2+} ATP with a K_m of 0.11 mM and a K_m for Mn^{2+} ATP of 0.008 mM. Activation occurred to an equal extent with Mn^{2+} or Mg^{2+} in the preincubation, although 10-fold lower concentrations of Mn^{2+} ATP (~10 μM for half maximal activation) are required, consistent with the higher affinity of the kinase for Mn^{2+} ATP. This feature, together with the inability of AMP PNP or AMP PCP to yield activation suggested that autophosphorylation mediated the activation. In fact the extent of activation and autophosphorylation of the 95,000 subunits are closely congruent both temporally as well as over a wide range of ATP concentrations in the preactivation incubation. This phosphorylation is exclusively onto tyrosine residues on the M_r 95,000 subunit. Moreover, in confirmation of the results of Rosen et al.,[112a] treatment of the preactivated receptor with bacterial alkaline phosphatase leads to a partial dephosphorylation and deactivation of the histone kinase; both are reversed by further incubation with metal ATP. Thus, insulin binding triggers an intramolecular phosphorylation of the 95,000 subunit. This phosphorylation dramatically enhances the affinity of the receptor kinase for Me^{2+}/ATP, at least one consequence of which is a stimulation of H2b kinase activity when measured at subsaturating concentrations of ATP.

Given present information we envision the operation of the receptor kinase as follows: the intracellular nucleotide substrate is undoubtedly Mg ATP, with ATP levels of 2–3 mM. The data Kasuga et al.[41] indicate that in the intact cell, the receptor is largely dephosphorylated in the absence of insulin; tyrosine protein kinase activity is commensurately low. The extent of basal receptor tyrosine phosphorylation in the cell is probably maintained at a low level by the action of tyrosine phosphatases. Binding of insulin facilitates autophosphorylation through an intramolecular reaction in proportion to receptor occupancy. Consequent to autophosphorylation, in the presence of Mg ATP the tyrosine kinase function is converted from an essentially inactive to active form.

Although insulin binding greatly facilitates the activating autophosphorylation, binding of hormone (at least to solubilized receptors) is probably not absolutely required; Mn ATP, presumably because of its (at least 10-fold) higher affinity for the receptor than Mg ATP, can support significant receptor autophosphorylation and consequent protein kinase activity, even in the absence of insulin. The higher affinity of the insulin receptor kinase for Mn ATP is advantageously applied for in vitro assay of receptor kinase; inclusion of Mn is essential when employing ATP concentrations ≤0.1 mM. These considerations appear to explain the apparrent requirement for Mn^{2+} ATP observed for insulin receptor autophosphorylation in crude extracts.

The physiologic implications of this autoactivation via autophosphorylation are worthy of consideration. The superimposition of another level of regulation of receptor kinase activity beyond that of binding of the activating ligand (i.e., insulin) serves to emphasize in a general way the likelihood that this kinase function is of substantial importance in physiologic regulation. It is conceivable that cellular tyrosine kinase other than the insulin receptor kinase itself may be capable of phosphorylating the insulin receptor so as to activate it. Although all of the tyrosine kinases are subject to autophosphorylation, there is no evidence to suggest that this modification underlies a functional activation for either the retroviral or the growth factor receptor kinase. Thus, this form of regulation may be unique to the insulin receptor. The ability of autophosphorylation to lower the K_m for ATP together with the presence of the product inhibition of the kinase by ADP, couples the insulin receptor tyrosine kinase to the cellular adenylate charge in a steeply positive feed-forward manner, and conversely is likely to lead to an abrupt fall in the kinase activity if intracellular ATP levels fall, even in the presence of insulin. In fact it is well known that a variety of insulin's actions are quite sensitive to modest depletion of intracellular ATP[112b,c,d] to levels (0.1 to 0.2 mM total cellular ATP) which are near or above the K_m for most ATP requiring actions. For example, cAMP mediated processes are activated quite readily in such cells.[112d] However, at present, there is of course, no evidence that a fall in tyrosine kinase underlies the ability of modest ATP depletion to inhibit insulin action.

Finally, it should be noted that the autophosphorylation–activation phenomenon is the only functional change yet detected as a consequence of tyrosine phosphorylation *in vitro* or *in vivo*. If this event occurs *in vivo* consequent to insulin binding to its receptor, it is then the only "physiologically relevant," i.e., regulatory, tyrosine phosphorylation yet identified.

C. Role of Insulin Receptor Tyrosine Kinase in Insulin Action

Insulin-stimulated tyrosine phosphorylation of the 95,000-M_r receptor subunit is demonstrable in crude membrane fractions from virtually all tissues of the rat containing specific insulin-binding sites, including brain, heart, skeletal muscle (Fig. 8), spleen, liver, and adipose tissue, as well as in membrane fractions from amphibian liver. Thus tyrosine kinase activity appears to be a general property of insulin receptors. The demonstration that insulin can promote tyrosine phosphorylation of the 95,000-M_r subunit of its own receptor in a variety of intact cells[41,117] is good evidence that the insulin tyrosine kinase function is operative in intact cells (although site analysis is necessary to confirm this conclusion). The critical problem is the elucidation of the role of this kinase in the initiation of insulin actions; a variety of models can be formulated (Table 1). Clearly, the most attractive possibility envisions the tyrosine kinase as phosphorylating other membrane or intracellular proteins, which in turn transmit and amplify the signal that serves to reorient cellular metabolism and/or growth. The validation of this idea requires the detection and identification of the relevant insulin stimulated phosphotyrosine-containing proteins. While such proteins have not yet been detected, analogous studies have been underway for sometime in Rous sarcoma virus transformed cells, and in EGF and PDGF treated cells, and it is instructive to briefly review the progress in these systems.

Hunter and colleagues[118–120] have shown that [32P]tyrosine constitutes <0.1% of the total [32P]phosphoamino acids in acid hydrolysates of 32P-labeled chick embryo fibroblasts. Transformation with Rous sarcoma virus (RSV) leads to a 10-fold increase

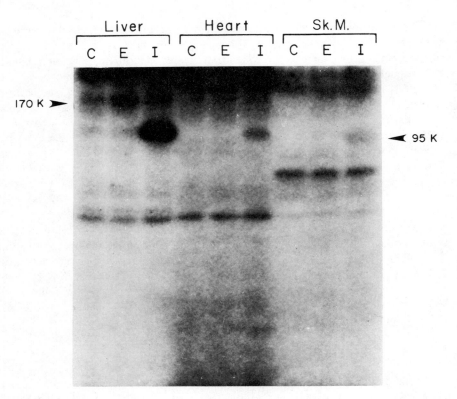

Figure 8. Insulin and EGF-promoted receptor autophosphorylation in detergent extracts of microsomes from rat liver, heart, and skeletal muscle. Detergent extracts of rat liver, heart, and skeletal muscle microsomes were prepared and phosphorylated in the presence of control conditions (C), EGF (E), or insulin (I), as described by Avruch *et al.*[111] Autoradiograph of an alkali-treated gel of the reaction products indicates the 170,000-M_r EGF receptor and the 95,000-M_r β subunit of the insulin receptor.

Table 1. Possible Functions of the Insulin Receptor Tyrosine Kinase

1. Signaling, i.e., activation of cellular response
   ? Growth effects
   ? Metabolic effects
   Substrate(s): other proteins in surface membrane ± intracellular proteins
2. Modification of receptor interaction with unknown effector unit
   Substrate: insulin receptor
3. Modification of receptor life cycle, e.g., internalization, proteolysis
   Substrate: insulin receptor
4. Other, e.g., nonprotein substrate

in the content of [^{32}P]tyrosine, which nevertheless remains a trace constituent in cellular proteins.

Because of the low abundance of [^{32}P]tyrosine, the background from [^{32}P]serine must be diminished if [^{32}P]tyrosine peptides are to be detected. Hunter and colleagues have used base treatment of polyacrylamide gels to achieve this end. This procedure (2 hr at 55°C in 1 M KOH) is adequate to discharge approximately 90% [^{32}P]serine; however individual phosphoserine residues in proteins vary in their susceptibility to base hydrolysis, and the large majority of ^{32}P-labeled peptides remaining after this treatment contain [^{32}P]serine and [^{32}P]threonine exclusively, as determined after acid hydrolysis of the excised bands (P.J. Blackshear, J. Avruch, unpublished observations). By this approach, Hunter and colleagues have detected four to five cellular peptides which show an increase in [^{32}P]tyrosine after cellular transformation by RSV. These workers have identified the function of several of these peptides, by immunoprecipitation and/or purification, as vinculin[121] and the glycolytic enzymes enolase, phosphoglycerate mutase, and lactate dehydrogenase.[122] The surprising feature of the tyrosine phosphorylation of these proteins is the low stoichiometry of the modification in RSV-transformed cells. The [^{32}P]tyrosine content in none of these proteins exceeded 0.05 mole P/mole protein; this low value virtually excludes a significant (i.e., >10%) effect on catalytic function. Since these enzymes catalyze near-equilibrium reactions, they are unlikely candidates for regulators of glycolytic flux in the first instance. This situation is strikingly reminiscent of that observed for the insulin-stimulated (serine) phosphorylation of ATP citrate lyase,[67] i.e., the presence of stoichiometrically trivial amounts of phosphate on a relatively abundant protein most likely occurring as an incidental consequence of the presence of a large amount of active (RSV-associated) tyrosine kinase in the transformed cells. This type of systematic survey of ^{32}P-labeled proteins in RSV- and growth factor-treated cells has thus far failed to identify a [^{32}P]tyrosine-labeled protein of clear-cut regulatory significance; the peptides enumerated thus far are either trivially phosphorylated or remain unidentified as to function.

A complementary approach has attempted to enrich for [^{32}P]tyrosine-containing proteins by immunological means. These include polyclonal and monoclonal antibodies capable of binding to [^{32}P]tyrosine residues in proteins,[123,124] as well as antibodies raised against synthetic peptides identical to the region surrounding the tyrosine phosphorylation site in the *src* gene product.[125] Such reagents are able to partially purify [^{32}P]tyrosine-containing peptides, as identified on SDS gels; among these are viral-transforming antigens and the growth factor receptors themselves, as would be anticipated from the ability of these species to undergo autophosphorylation on tryosine residues *in vitro*. The identity of the other [^{32}P]tyrosine-containing peptides and the stoichiometry of their (tyrosine) phosphorylation are unknown. It is of interest that virtually all these [^{32}P]tyrosine peptides detected also contain substantial amounts of [^{32}P]serine/threonine, as is true of the insulin receptor; the significance of this concurrence, if any, is unknown.

We have initiated studies aimed at identifying insulin-stimulated ^{32}P-containing peptides by methods analogous to those described above. Nevertheless, given the relatively limited insights afforded thus far by careful and extensive earlier studies, we have become persuaded that the identification of physiologically relevant substrates of the insulin receptor kinase, if in fact they exist beyond the receptor itself, may best be achieved by an approach that examines the relationship of tyrosine phosphorylation to a specific functional consequence of insulin action (i.e., intelligent guessing). In this regard, insulin appears to offer substantial advantages over the retrovirus' and growth factors; the best studied biological actions of these agents are related to cell growth, a very complex and delayed effect, the earliest unequivocal manifestations of which (i.e., traverse of specific cell cycles) are many steps (and hours) removed from the tyrosine kinase reaction. By contrast, the rapid metabolic effects of insulin can, in some cases, be localized to specific proteins, which can be evaluated as potential substrates for the insulin receptor tyrosine kinase *in vivo* and *in vitro*. To identify a potentially relevant substrate by this approach, it is necessary to demonstrate that the protein undergoes a rapid stoichiometrically significant phosphorylation *in vitro* by the purified insulin receptor kinase, coupled with the demonstration that this phosphorylation alters the function of the protein in a manner consistent with the biological action of the hormone. Finally, it will be necessary to demonstrate that such a modification occurs in intact cells.

In considering the likely candidates for such a modification, we have generated the working hypothesis that tyrosine phosphorylation will regulate serine/threonine phosphorylation; among the physiologically relevant substrates for the insulin receptor tyrosine kinase will be one or more serine/threonine protein kinases, and perhaps serine protein phosphatases and/or the modulatory proteins for these enzymes. In view of the great scarcity of phosphotyrosine containing cellular proteins, it seems likely that the major amplification of the hormonal signal occurs at a more distal step; we propose that this amplification occurs primarily through the modification of the activity of serine kinases. This hypothesis is readily extended to the actions of the growth factors, and the retroviruses whose transforming antigen is a tyrosine kinase. While there is no direct evidence for such a scheme, several recent observations are entirely compatible with this hypothesis. Decker[126] observed that serum stimulated the serine phosphorylation of ribosomal S6 in chick embryo fibroblasts. If the cells were transformed with RSV, S6 phosphorylation was present at a high level and was no longer stimulated in the presence of serum. If transformation was accomplished with a temperature-sensitive variant of RSV, wherein tyrosine kinase was expressed at the permissive temperature only, then the increment in S6 serine phosphorylation paralleled the temperature sensitivity. This suggests that an active *src* tyrosine kinase is required for increased (serine) S6 phosphorylation. Even stronger support for this conclusion is provided by recent experiments of Erickson and Maller,[127] who showed that microinjection of the *src* gene product (tyrosine kinase, produced from cloned *src*) into *Xenopus* oocytes led to an increased phosphorylation of S6 serine residues. These experiments, although compatible with a tyrosine kinase-induced activation of a S6 serine kinase, do not eliminate the possibility of more indirect mechanisms, e.g., *src*-induced activation of protein synthesis leading to increased S6 phosphorylation as a passive consequence. The direct demonstration that a serine kinase is activated through tyrosine phosphorylation still remains to be done. This might involve tyrosine phosphorylation of the serine kinase catalytic moiety itself, of a modulator protein, or of an (inhibitory) protein that serves as a cellular binding site. We have also entertained the possibility that insulin-induced dephosphorylation might be a direct consequence of tyrosine phosphorylation. The insulin receptor tyrosine kinase might act directly on one or more of the proteins involved in the expression of (serine) protein phosphatase activity. In collaboration with Philip Cohen and colleagues, we have observed that the insulin receptor kinase is capable of phosphorylating highly purified protein phosphatase inhibitor-2, both alone or when complexed to the catalytic subunit of protein phosphatase-1, forming the so-called Mg ATP-dependent phosphatase. The significance of this preliminary observation is as yet unknown.

VI. SUMMARY

In conclusion, the insulin–receptor interaction leads to the rapid phosphorylation and dephosphorylation of serine/threonine residues in target cells. The dephosphorylation reactions are clearly related to a variety of rapid metabolic actions exerted by insulin on substrate metabolism. Many of the anticatabolic effects are adequately explained by the ability of insulin to inhibit the cAMP-dependent protein kinase by lowering cAMP levels. The locus of this action is most likely through an activation of cAMP phosphodiesterase, an effect for which the mechanism is as yet unknown. Certain anabolic actions of the hormone require dephosphorylation through pathways that do not involve cAMP, i.e., inhibiton of cAMP-independent protein kinases and/or activation of protein phosphatases; little is known of the insulin regulation of these elements. The function of insulin-stimulated serine/threonine phosphorylation remains unknown. This response behaves as expected for a rapid signaling pathway, and a very similar pathway is activated by EGF and PDGF; nevertheless, the specific cellular functions modified remain unknown. The identification of the insulin receptor as an insulin-activated tyrosine protein kinase, analogous to the receptor for EGF and PDGF, suggests that tyrosine phosphorylation is an early step in cellular activation for all three agents. However, the scarcity of $[^{32}P]$tyrosine-containing proteins has impeded efforts to identify potentially relevant substrates for this modification. We propose that the signal generated by the insulin receptor tyrosine kinase is amplified and transmitted by either specific serine protein kinases/protein phosphatases or modulatory proteins of these enzymes or both.

ACKNOWLEDGMENTS. This work was supported in part by grant AM1776 from the National Institutes of Health.

VII. REFERENCES

1. Robison, G.A., Butcher, R., and Sutherland, E.W., in 1971, *cAMP*, pp. 1–23, Academic Press, New York.
2. Larner, J., and Villar-Palasi, C., 1971, *Curr. Top. Cell. Regul.* **3**:195–236.
3. Roach, P.J., Rosell, M., Rosell Perez, M., and Larner, J., 1977, *FEBS Lett.* **80**:95–98.
4. Sheorain, V.S., Khatra, B.S., and Soderling, T.R., 1982, *J. Biol. Chem.* **257**:3462–3470.

5. Uhing, R.J., Shikama, H., and Exton, J., 1981, *FEBS Lett.* **134:**185–188.
6. Parker, P.J., Caudwell, F.B., and Cohen, P., 1983, *Eur. J. Biochem.* **130:**227–234.
7. Lawrence, J.C., Hiken, J.F., DePaoli-Roach, A.A., and Roach, P.J., 1983, *J. Biol. Chem.* **258:**10710–10719.
8. Mukherjee, C., and Jungas, R.L., 1975, *Biochem. J.* **148:**229–235.
9. Denton, R.M., Hughes, W.A., Bridges, B.J., Brownsey, R.W., McCormack, S.G., and Stansbie, D., 1978, *Horm. Cell Regul.* **2:**191–208.
10. Engstrom, L., 1980, *Mol. Aspects Cell Regul.* **1:**11–30.
11. Claus, T.H., and Pilkis, S.J., 1981, *Biochem. Actions Horm.* **8:**209–271.
12. Fredrikson, G., Stralfors, P., Nilsson, N.O., and Belfrage, P., 1981, *J. Biol. Chem.* **256:**6311–6320.
13. Ingebritsen, T.S., Geelen, M.J.H., Evenson, K.J., and Gibson, D.M., 1979, *J. Biol. Chem.* **254:**9986–9989.
14. Vande Werve, G., Hue, L., and Hers, H.G., 1977, *Biochem. J.* **162:**135–142.
15. Marchmont, R.J., and Houslay, M.D., 1980, *Nature,* **286:**904–906.
16. Krebs, E.G., 1972, *Curr. Top. Cell. Regul.* **5:**99–131.
17. Kono, T., and Barham, T.W., 1983, *J. Biol. Chem.* **248:**7417–7426.
18. Sneyd, J.G.T., Corbin, J.D., and Park, C.R., 1968, in *Pharmacology of Hormonal Polypeptides and Proteins* (H. Back, L. Martini, and R. Paoletti, eds.), p. 376, Plenum Press, New York.
19. Sica V., and Cuatrecasas, 1973, *Biochemistry,* **12:**2782–2291.
20. Avruch, J., Leone, G., and Martin, D.B., 1974, *J. Clin. Invest.* **53:**3a.
21. Avruch, J., Leone, G., and Martin, D.B., 1974, *Diabetes* **23:**348.
22. Avruch, J., Leone, G., and Martin, D.B., 1976, *J. Biol. Chem.* **251:**1505–1510.
23. Avruch, J., Leone, G., and Martin, D.B., 1976, *J. Biol. Chem.* **251:**1511–1515.
24. Crapo, L., Graham, W., and Avruch, J., 1980, *Biochem. Biophys. Res. Commun.* **94:**1331–1336.
25. Blackshear, P.J., Nemenoff, R.A., and Avruch, J., 1982, *Biochem. J.* **204:**817–824.
26. Avruch, J., Witters, L.A., Alexander, M.C., and Bush, M.A., 1978, *J. Biol. Chem.* **253:**4754–5761.
27. Alexander, M.C., Kowaloff, E.M., Witters, L.A., Dennihy, D.T., and Avruch, J., 1979, *J. Biol. Chem.* **254:**8052–8056.
28. Blackshear, P.J., Nemenoff, R.A., and Avruch, J., 1983, *Biochem. J.* **214:**11–19.
29. Benjamin, W.B., and Singer, I., 1974, *Biochim. Biophys. Acta* **351:**28–42.
30. Benjamin, W.B., and Singer, I., 1975, *Biochemistry* **14:**3301–3309.
31. Benjamin, W.B., and Clayton, N.L., 1978, *J. Biol. Chem.* **253:**1700–1707.
32. Ramakrishna, B., and Benjamin, W.B., 1979, *J. Biol. Chem.* **254:**9232–9236.
33. Forn, J., and Greengard, P., 1976, *Arch. Biochem. Biophys.* **176:**721–733.
34. Hughes, W.A., Brownsey, R.W., and Denton, R.M., 1977, in *Phosphorylated Proteins and Related Enzymes* Pinna, LA (ed.), pp. 17–32, Information Retrieval, London.
35. Hughes, W.A., Brownsey, R.W., and Denton, R.M., 1980, *Biochem. J.* **192:**469–481.
36. Belsham, G.J., Brownsey, R.W., Hughes, W.A., and Denton, R.M., 1980, *Diabetologia* **18:**307–312.
37. Belsham, G.J., Denton, R.M., and Tanner, M.J.A., 1980, *Biochem. J.* **192:**457–467.
38. Smith, C.J., Wejksnora, P.J., Warner, J.R., Rubin, C.S., and Rosen, O.M., 1979, *Proc. Natl. Acad. Sci. USA* **76:**2725–2729.
39. Smith, C.J., Rubin, C.S., and Rosen, O.M., 1980, *Proc. Natl. Acad. Sci. USA* **77:**2641–2645.
40. Haselbacher, G.K., Humbel, R.E., and Thomas, G., 1979, *FEBS. Lett.* **100:**185–187.
41. Kasuga, M., Karlsson, F.A., and Kahn, R.C., 1982, *Science* **215:**185–187.
42. Kasuga, M., Zick, Y., Blithe, D.L., Karlsson, F.A., Haring, H.U., and Kahn, C.R., 1982, *J. Biol. Chem.* **257:**9891–9894.
43. Cohen, S., Carpenter, G., and King, L., 1980, *J. Biol. Chem.* **255:**4834–4842.
44. Nilsson, N.D., Stralfors, P., Fredrikson, G., and Belfrage, P., 1980, *FEBS Lett.* **111:**125–130.
45. Nimmo, H.G., and Cohen, P., 1977, *Adv. Cyclic Nucleotide Res.* **8:**145–266.
46. Beavo, J.A., Bechtel, P.J., and Krebs, E.G., 1974, *Proc. Natl. Acad. Sci. USA* **71:**3580–3583.
47. Soderling, T.R., Corbin, J.D., and Park, C.R., 1973, *J. Biol. Chem.* **248:**1822–1829.
48. Walkenbach, R.J., Hazen, R., and Larner, J., 1978, *Mol. Cell Biochem.* **19:**31–41.
49. Larner, J., Galasko, G., Cheng, K., DePaoli-Roach, A.A., Huang, L., Duggy, P., and Kellog, I., 1979, *Science,* **206:**1408–1410.

50. Ingebritsen, T.S., and Cohen, P., 1983, *Science* **221:**331–338.
51. Huang, F.L., and Glinsman, W.H., 1976, *Eur. J. Biochem.* **70:**419–426.
52. Foulkes, J.G., and Cohen, P., 1979, *Eur. J. Biochem.* **97:**251–256.
53. Hemmings, B.A., Resink, T.J., and Cohen, P., 1982, *FEBS. Lett.* **150:**319–324.
54. Khatra, B.S., Chiasson, J.L., Shikima, H.J., Exton, J.A., and Soderling, T.R., 1980, *FEBS. Lett.* **114:**253–256.
55. Foulkes, J.G., Cohen, P., Strada, S.J., Everson, W.V., and Jefferson, L.S., 1982, *J. Biol. Chem.* **257:**12493–12496.
56. Nemenoff, R.A., Blackshear, P.B., and Avruch, J., 1983, *J. Biol. Chem.* **258:**9437–9443.
57. Manganiello, V., and Vaughn, M., 1973, *J. Biol. Chem.* **248:**7164–7170.
58. Makino, H., and Kono, T., 1980, *J. Biol. Chem.* **255:**7850–7854.
59. Loten, E.G., Assimacopoulos-Jeannet, F.D., Exton, J.A., and Park, C.R., 1978, *J. Biol. Chem.* **253:**746–757.
60. Heyworth, C.M., Wallace, A.V., and Houslay, M.D., 1983, *Biochem. J.* **214:**99–110.
61. Belsham, G.J., Brownsey, R.W., and Denton, R.M., 1982, *Biochem. J.* **204:**345–352.
62. Cottam, G.L., and Srere, P.A., 1969, *Biochem. Biophys. Res. Commun.* **35:**895–900.
63. Alexander, M.C., Palmer, J.C., Pointer, R.H., Koumjian L., and Avruch, J., 1982, *J. Biol. Chem.* **257:**2049–2055.
64. Alexander, M.C., Palmer, J.C., Pointer, R.H., Koumjian, L., and Avruch, J., 1981, *Biochim. Biophys. Acta* **674:**37–47.
65. Swergold, C.D., Rosen, O.M., and Rubin, C.S., 1982, *J. Biol. Chem.* **257:**4207–4215.
66. Pierce, M.W., Palmer, J.L., Keutmann, H.T., Hall, T.A., and Avruch, J., 1981, *J. Biol. Chem.* **256:**8867–8870.
67. Pierce, M.W., Palmer, J.L., Keutmann, H.T., Hall, T.A., and Avruch, J., 1982, *J. Biol. Chem.* **257:**10681–10686.
68. Avruch, J., Alexander, M.C., Palmer, J.L., Pointer, R.H., Nemenoff, R.A., and Pierce, M.W., 1981, *Cold Spring Harbor Conf. Cell Prolif.* **8:**759–770.
69. Ranganathan, N.S., Linn, T.C., and Srere, P.A., 1982, *J. Biol. Chem.* **257:**698–702.
70. Janski, A.M., Srere, P.A., Cornell, N.W., and Veech, R.L., 1979, *J. Biol. Chem.* **254:**9365–9368.
71. Witters, L.A., Moriarty, P., and Martin, D.B., 1979, *J. Biol. Chem.* **254:**6644–6649.
72. Brownsey, R.W., Hughes, W.A., and Denton, R.W., 1979, *Biochem. J.* **184:**23–32.
73. Brownsey, R.W., and Denton, R.M., 1982, *Biochem. J.* **202:**77–86.
74. Witters, L.A., 1981, *Biochem. Biophys. Res. Commun.* **100:**872–878.
75. Witters, L.A., Kowaloff, E.M., and Avruch, J., 1979, *J. Biol. Chem.* **254:**245–248.
76. Witters, L.A., and Vogt, B., 1981, *J. Lipid Res.* **22:**364–369.
77. Witters, L.A., Tipper, J.P., and Bacon, G.W., 1983, *J. Biol. Chem.* **258:**5643–5648.
78. Brownsey, R.W., and Hardie, D.G., 1980, *FEBS Lett.* **120:**67–70.
79. Hardie, D.G., and Grey, P.S., 1980, *Eur. J. Biochem.* **110:**167–177.
80. Tipper, J., and Witters, L.A., 1982, *Biochim. Biophys. Acta* **715:**162–169.
81. Tipper, J., Bacon, G.W., and Witters, L.A., 1983, *Arch. Biochem. Biophys.* **227:**386–396.
82. Wool, I.G., 1979, *Annu. Rev. Biochem.* **48:**719–754.
83. Wettenhall, R., Cohen, P., Caudwell, B., and Holland, R., 1982, *FEBS Lett.* **148:**207–213.
84. Gressner, A., and Wool, I.G., 1974, *J. Biol. Chem.* **249:**6917–6925.
85. Lastick, S.M., Neilson, O.J., and McConkey, E.H., 1977, *Mol. Gen. Genet.* **152:**223–230.
86. Thomas, G., Siegmann, M., Kubler, A.M., Gordon, J., and Jimenez de Asua, L., 1980, *Cell* **19:**1015–1023.
87. Thomas, G., Martin-Perez, J., Siegmann, M., and Otto, A., 1982, *Cell* **30:**235–242.
88. Wettenhall, R.E.H., and Cohen, P., 1982, *FEBS Lett.* **140:**263–269.
89. Lastick, S.M., and McConkey, E.H., 1981, *J. Biol. Chem.* **256:**583–585.
90. Martin-Perez, J., and Thomas, G., 1983, *Proc. Natl. Acad. Sci. USA* **80:**926–930.
91. Martin-Perez, J., Siegmann, M., and Thomas, G., 1984, *Cell* **36:**287–294.
92. Leader, D.P., 1980, in *Molecular Aspects of Cell Regulation,* (P. Cohen, ed.), Vol. 1, pp. 203–234, Elsevier–North-Holland, Amsterdam.
93. Belsham, G.J., and Denton, R.M., 1980, *Biochem. Soc. Trans.* **8:**832–833.

94. Garrison J.C., 1978, *J. Biol. Chem.* **253**:7091–7100.
95. Garrison, J.C., Borland, M.K., Florio, V.A., and Twible, D.N., 1979, *J. Biol. Chem.* **254**:7147–7156.
96. LeCam, A., 1982, *J. Biol. Chem.* **257**:8376–8385.
97. Vargas, A.M., Halestrap, A.P., and Denton, R.M., 1982, *Biochem. J.* **208**:221–229.
98. Steinberg, R.A., and Coffino, P., 1979, *Cell* **18**:719–726.
99. Steinberg, R.A., 1981, *Cold Spring Harbor Conf. Cell Prolif.* **8**:179–193.
100. Garrison, J.C., Borland, M.K., Moyland, R.D., and Ballard, B.J., 1981, *Cold Spring Harbor Conf. Cell Prolif.* **8**:529–545.
101. Blackshear, P.J., Nemenoff, R.A., Bonventre, J.V., Cheung, J.Y., and Avruch, J., 1984, *Amer. J. Physiol. (Cell Physiol.)* **246**:C439–C449.
102. Rosen, O.M., Rubin, C.S., Cobb, M.H., and Smith, C.J., 1981, *J. Biol. Chem.* **256**:3630–3633.
103. Cobb, M.H., and Rosen, O.M., 1983, *J. Biol. Chem.* **258**:12472–12481.
104. Perisic, O., and Traugh, J.A., 1983, *J. Biol. Chem.* **258**:9589–9592.
105. Brownsey, R.W., Edgell, N.J., Hopkirk, T.J., Denton, R.M., 1984, *Biochem. J.* **218**:733–743.
106. Ramakrishna, S., and Benjamin, W.B., 1981, *FEBS Lett.* **124**:140–144.
107. Ramakrishna, S., Pucci, D.L., and Benjamin, W.B., 1981, *J. Biol. Chem.* **256**:10213–10216.
108. Van Obberghen, E., and Kowalski, 1982, *FEBS Lett.* **143**:179–182.
109. Kasuga, M., Zick, Y., Blithe, D.L., Crettaz, M., and Kahn, C.R., 1982, *Nature* **298**:667–669.
110. Petruzelli, L.M., Ganguly, S., Smith, C.J., Cobb, M.H., Rubin, C.S., and Rosen, O.M., 1982, *Proc. Natl. Acad. Sci. USA* **79**:6790–6796.
111. Avruch, J., Nemenoff, R.A., Blackshear, P.J., Pierce, M.W., and Osathanondh, R., 1982, *J. Biol. Chem.* **257**:15162–15166.
112. Nemenoff R.A., Kwok, Y.C., Shulman, G.I., Blackshear, P.J., Osathanondh, R., and Avruch, J., 1984, *J. Biol. Chem.* **259**:5058–5065.
112a. Rosen, O.M., Herrera, R., Olowe, Y., Petruzelli, L.M., and Cobb, M.H., 1983, *Proc. Natl. Acad. Sci. U.S.A.* **80**:3237–3240.
112b. Kono, T., Robinson, F.W., Sarver, J.A., Vega, F.V., and Pointer, R.H., 1977, *J. Biol. Chem.* **252**:2226–2233.
112c. Siegel, J., and Olefsky, J.M., 1980, *Biochemistry* **19**:2183–3190.
112d. Häring, H.U., Bierman, E., and Kemmler, W., 1981, *Amer. J. Physiol.* **240**:E556–E565.
113. Roth, D.A., and Carrell, D.J., 1983, *Science* **219**:299–301.
114. Van Obberghen, E., Rossi, B., Kowalski, A., Gazzaro, H., and Ponziog, G., 1983, *Proc. Natl. Acad. Sci. USA* **809**:945–949.
115. Shia, M., and Pilch, C., 1983, *Biochemistry* **22**:717–721.
116. Sugden, P.H., Holladay, L.A., Reimann, E.M., and Corbin, J.B., 1976, *Biochem. J.* **159**:409–422.
117. Haring, H.U., Kasuga, M., and Kahn, C.R., 1983, *Biochem. Biophys. Res. Commun.* **108**:1538–1545.
118. Hunter, T., and Sefton, B.M., 1982, in *Molecular Aspects of Cellular Regulation* (Cohen, P., and Van Heyningen, S., eds.), Vol. 2, pp. 337–370, Elsevier–North-Holland, Amsterdam.
119. Hunter, T., and Cooper, J.A., 1981, *Cell* **24**:741–752.
120. Cooper, J.A., Bowen-Pope, D.G., Raines, E., Ross, R., and Hunter, T., 1982, *Cell* **31**:263–273.
121. Sefton, B.M., Hunter, T., Boll, E.H., and Singer, S.J., 1981, *Cell* **24**:165–174.
122. Cooper, J.A., Reiss, N.A., Schwartz, R.J., and Hunter, T., 1983, *Nature* **302**:218–223.
123. Frackelton, A.R., Ross, A.H., and Eisen, H.N., 1983, *Mol. Cell Biol.* **3**:1341–1352.
124. Ross, A.H., Baltimore, D., and Eisen, H.N., 1981, *Nature* **294**:654–656.
125. Wong, T.W., and Goldberg, A.R., 1981, *Proc. Natl. Acad. Sci. USA* **78**:7412–7416.
126. Decker, S., 1981, *Proc. Natl. Acad. Sci. USA* **78**:4112–4115.
127. Spivak, J.G., Erikson, R.L., and Maller, J.L., 1984, *Molecular and Cell Biol.*, in press.

17

Role of Phosphorylation in the Regulation of Acetyl CoA Carboxylase Activity

Roger W. Brownsey and Richard M. Denton

I. INTRODUCTION

Insulin increases the rate of fatty acid synthesis in white fat cells,[1-3] brown fat cells,[4] liver cells,[5-8] and the lactating mammary gland[9,10] within a few minutes. Proportionally, the most impressive effects of insulin appear to occur in white and brown fat cells. In contrast, at least in white adipose tissue and liver, hormones that increase the concentration of cyclic adenosine monophosphate (cAMP) result in inhibition of fatty acid synthesis, esepcially in the presence of insulin.[3,11] To date, this has not been clearly demonstrated for brown fat and mammary tissue—probably because of the lack of adequate *in vitro* preparations of these tissues.

This chapter reviews the evidence that changes in activity of acetyl CoA carboxylase[2] represent an important element in the opposing short-term effects of insulin and hormones that increase cAMP, followed by a detailed discussion of our understanding of the role of phosphorylation of acetyl CoA carboxylase in these hormonal effects. Our own studies have concentrated mainly on rat epididymal fat cells, a bias that is inevitably reflected in this presentation. Our loyalty to this tissue is based on two important advantages: the ease with which large and reproducible effects of insulin and other hormones can be obtained *in vitro*, and the low levels of protease activity in cell extracts. Acetyl CoA carboxylase is extremely susceptible to limited proteolysis. We begin our account with an outline of the current understanding of the structure of acetyl

Roger W. Brownsey and Richard M. Denton • Department of Biochemistry, University of Bristol Medical School, Bristol, England. *Present address of R.W.B.:* Department of Biochemistry, University of British Columbia, Vancouver, Canada.

CoA carboxylase from mammalian and avian sources and its regulation by allosteric regulators. These aspects have been described in greater detail in a number of review articles.[3,11–13]

II. ACETYL CoA CARBOXYLASE: STRUCTURE AND REGULATION BY ALLOSTERIC EFFECTORS

The active form of acetyl CoA carboxylase purified from avian or mammalian sources is a long filamentous polymer ($M_r \sim 10^7$) composed of protomeric units, which are dimers of apparently identical subunits. Each subunit contains 1 mole/mole of biotin and contains active sites both for the ATP-dependent carboxylation of biotin and for the transfer of the carboxyl group to acetyl CoA-producing malonyl CoA. The molecular weight of the subunit as determined by sodium dodecylsulfate polyacrylamide gel electrophoresis (SDS-PAGE) has been reported to be in the range of 215,000–260,000.[11–16] Some of the variability may be explained in part by the use of different molecular-weight standards, but differences in size may also occur, since acetyl CoA carboxylase is also extremely sensitive to limited proteolysis especially during its isolation from liver.[15,16] Very importantly, proteolysis may result in considerable alterations in kinetic properties. For example, proteolytic clipping, which decreases the apparent molecular weight on SDS-PAGE by 10,000–30,000 gives a form of the enzyme that has a much lower K_m for acetyl CoA and a fivefold increase in specific activity from ~ 1.5–2.0 unit/mg protein to ~ 10–15 units/mg protein. Proteolysis may also remove potential phosphorylation sites. To minimize proteolysis during preparation, it is necessary to add a suitable cocktail of protease inhibitors and to employ a rapid procedure. Useful steps include ammonium sulfate and/or polyethylene glycol fractionation,[17] high-speed differential centrifugation in the presence of citrate,[18] and the use of biotin-treated avidin-Sepharose.[16,19,20] In our hands, the enzyme purified from rat epididymal adipose tissue has a molecular weight close to 230,000 on SDS-PAGE acrylamide gel electrophoresis using as standards myosin (200,000 M_r), fat cell fatty acid synthetase (230,000 M_r), and perhaps most usefully the two heavy chains of human red cell spectrin (250,000 and 225,000 M_r). This enzyme has a specific activity of ~ 1.5 units/mg and a relatively high K_m for acetyl CoA. Its migration on SDS-PAGE is also identical to that of [^{32}P]acetyl CoA carboxylase in a fresh acid extract of fat cells previously incubated with medium containing [^{32}P]phosphate.

Acetyl CoA carboxylase from many animal sources, including rat adipose tissue, has been shown to be activated by citrate and inhibited by fatty acyl CoA esters. These effects were first recognized during the early 1960s[21,22]; subsequently it was shown that the effects of citrate were associated with conversion of the enzyme into the polymerized form, whereas those of fatty acyl CoA were associated with depolymerization of the enzyme to inactive dimers.[3,12,23] Although superficially these appeared to be examples of feed-forward activation (by citrate) and of end-product inhibition (by fatty acyl CoA esters), it has been very difficult to establish the physiological role if any, in the regulation of acetyl CoA carboxylase by changes in concentrations of these two potential effectors. On the whole, few examples of parallel changes in rates of fatty acid synthesis and whole tissue or cytoplasmic concentrations of citrate have been found.[24–27] In particular, there is no evidence that insulin increases the cytoplasmic concentration of citrate in either liver[28] or adipose tissue.[24,29,30] Moreover, the whole-tissue concentration of citrate is markedly increased on incubation of epididymal fat pads with fluoroacetate or pyruvate, but this does not alter the rate of fatty acid synthesis.[30]

The potential role for fatty acyl CoA esters has obtained some support from a number of reports of inverse changes in their whole-tissue concentration and the rate of fatty acid synthesis.[24,25,29,31,32] These include the finding that insulin treatment can cause a marked decrease in the level of fatty acyl CoA esters in fat cells.[24,29] The interpretation of such whole-tissue measurements is particularly hazardous because of the uncertainties in the distribution, not only between different intracellular compartments (as with citrate), but between aqueous and hydrophobic phases and specific binding sites within each compartment as well. Nevertheless, the possibility that changes in fatty acyl CoA esters play a role, perhaps supplementing the effects of phosphorylation, should not be forgotten (see Section 4).

A number of other potential regulators of acetyl CoA carboxylase activity have been reported. These include free coenzyme A, which may markedly diminish the K_m for acetyl CoA,[13,33] and a number of adenine and guanine nucleotides,[34] including adenosine monophosphate (AMP).[35] The physiological role of these effectors has yet to be established.

III. SHORT-TERM EFFECTS OF INSULIN AND OTHER HORMONES ON ACTIVITY OF ACETYL CoA CARBOXYLASE

The first direct evidence of insulin-regulated activity of acetyl CoA carboxylase was the finding by Halestrap and Denton in 1973 that acetyl CoA carboxylase activity was increased in fresh extracts of rat epididymal adipose tissue previously exposed to insulin.[36] The increase was found to persist despite extensive dilution or incubation of tissue extracts in the presence of serum albumin, but was no longer evident after incubation of extracts with sufficient citrate to cause maximum activation.[29,36] Subsequently, similar observations have been made by others with fat cells[37,39] and liver cells.[6,7] Activation can also be demonstrated in vivo following manipulation of circulating plasma insulin concentrations with, for example, injections of anti-insulin serum or glucose, in both white and brown adipose tissue[4] and perhaps less convincingly in liver.[40] Some examples are given in Fig. 1.

In contrast to these effects of insulin, treatment of epididymal adipose tissue or liver cells with hormones that give rise to increases in cAMP and inhibition of fatty acid synthesis results in decreased activity of acetyl CoA carboxylase (Fig. 1). This effect has been demonstrated with epididymal adipose tissue pieces, with isolated epididymal fat cells with β-adrenergic agonists[29,37,41] and glucagon,[39] and with isolated liver cells with glucagon.[6,7,42]

The changes in kinetic properties of acetyl CoA carboxylase brought about by exposure of adipose tissue to insulin are not strictly the inverse of those seen with epinephrine.[18,41] Whereas the effects of epinephrine are still apparent after incubation of tissue extracts with citrate, the effects of insulin are abolished. Conversely, the

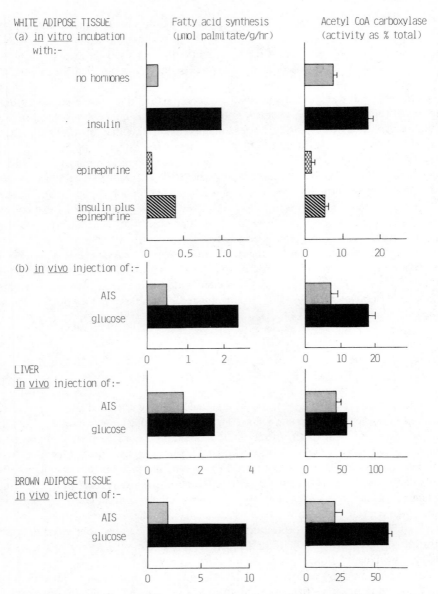

Figure 1. Effects of insulin and epinephrine on rates of fatty acid synthesis and activities of acetyl CoA carboxylase. Rates of fatty acid synthesis *in vitro* were calculated from the incorporation of ¹⁴C from [U-¹⁴C]glucose; rates *in vivo* were calculated from incorporation of ³H from ³H₂O. Initial activity of acetyl CoA carboxylase was measured in extracts of tissue previously frozen in liquid nitrogen and expressed as percentage of the activity observed following incubation with citrate. In the *in vivo* experiments, circulating plasma insulin concentrations were manipulated by injections of anti-insulin anti-serum (AIS) or glucose using fed animals. (Data from refs. 4, 29, 40, and 41, where further details can be found.)

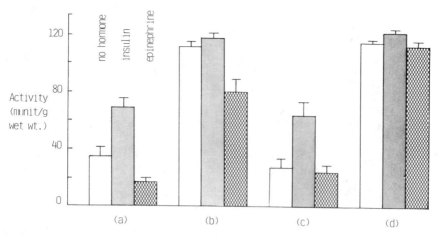

Figure 2. Persistence of changes in acetyl CoA carboxylase activity in extracts of rat epididymal adipose tissue previously exposed to no hormones (unshaded), insulin (shaded) or epinephrine (cross-hatched). Acetyl CoA carboxylase activity was determined after incubation of tissue extracts for 20–30 min at 30°C with (a) no additions, (b) citrate (20 mM), (c) MgCl$_2$ (5 mM) plus Ca^{2+} (~ 50 μM) or (d) as for (c) followed by citrate (20 mM) for a further 20 min. (Data taken from Brownsey and Denton.[18])

effects of epinephrine disappear if tissue extracts are incubated with Mg^{2+} and Ca^{2+}, while those of insulin remain (see Fig. 2). As will be seen, under these latter conditions, partial dephosphorylation of acetyl CoA carboxylase occurs. These observations suggest that epinephrine and insulin may act through different modes of regulation.

Changes in activity in fresh extracts are likely to be the result of changes in phosphorylation or some other covalent modification, but they may also be the result of changes in the residual tight binding of some allosteric ligand, e.g., fatty acyl CoA esters. However, changes in the activity of the adipose tissue enzyme with insulin have been found to persist at least, in part, during 100-fold purification by ammonium sulfate fractionation and differential high-speed centrifugation.[18] This finding strongly suggests that covalent modification is involved in the action of insulin on the enzyme. Recently, Witters *et al.*[38] reported that they could find no persistence of the insulin effect through ammonium sulfate fractionation. However, for reasons that are not apparent, these workers recovered only a small fraction (apparently less than 20%) of acetyl CoA carboxylase activity after precipitation with ammonium sulfate, whereas we and others routinely obtain recoveries in the range of 60–100%.[18]

IV. PHOSPHORYLATION OF ACETYL CoA CARBOXYLASE: EFFECTS OF INSULIN AND OTHER HORMONES

The finding of covalently bound phosphate in acetyl CoA carboxylase purified from rat liver provided the first clue that reversible phosphorylation may play a role in the regulation of this enzyme.[14] In 1973, Carlson and Kim[43] reported the inactivation and incorporation of ^{32}P from [γ-^{32}P]-ATP into a partially purified preparation of rat liver acetyl CoA carboxylase. The kinase involved was apparently distinct from cAMP-dependent protein kinase.[13,44,45]

The phosphorylation of acetyl CoA carboxylase in intact cells was first demonstrated in 1977 by Brownsey et al.[46] In these studies, fat cells from epididymal adipose tissue were incubated with medium containing ^{32}P$_i$ and the acetyl CoA carboxylase was rapidly separated from other cell proteins by either specific immunoprecipitation or affinity-chromatography on Sepharose-avidin. Both techniques resulted in the isolation of ^{32}P-labeled protein, which comigrated with the purified enzyme (230,000 M_r) on SDS-PAGE (see Fig. 3). These studies showed that all the ^{32}P incorporated into protein of 230,000 M_r in a whole-cell extract could be assigned to acetyl CoA carboxylase, which was in fact one of the major phosphoproteins in fat cells. There had been reports that fatty acid synthase, which has a subunit of almost identical molecular weight, may also be subject to regulation by phosphorylation,[47] but we found no evidence of ^{32}P incorporation into fatty acid synthase immunoisolated from fat cells previously incubated in medium containing ^{32}P$_i$.[46]

Incorporation of ^{32}P into fat cell acetyl CoA carboxylase reaches a steady-state value within 60 min. Addition of either insulin or epinephrine at this point results in modest increases in phosphorylation.[18,41,46] Within 15 min, the level of phosphorylation as judged by ^{32}P$_i$ incorporation was found to be increased by 15 ± 6% in the presence of insulin and by 40 ± 6% with epinephrine (mean ± SEM of more than 20 independent observations). This was somewhat surprising, initially, as under these conditions insulin was causing a doubling in the initial activity of acetyl CoA carboxylase, whereas epinephrine resulted in a more than 30% decline in activity (Fig. 4). Quantitatively similar results have been obtained subsequently by others using similar techniques, both in fat cells and with liver cells exposed to either glucagon or insulin.[42,48]

Since the studies of Hardie and Cohen[11,49,50] on purified mammary gland enzyme had suggested that phosphorylation of the enzyme could occur at multiple sites, we explored the extent of multisite phosphorylation fo acetyl CoA carboxylase in fat cells.[18,51] Cells were first incubated with ^{32}P$_i$ and [^{32}P]acetyl CoA carboxylase, separated by immunoprecipitation as described above, and were then digested with trypsin; the peptides were separated in a two-dimensional system.[18] A number of ^{32}P-labeled peptides were evident from autoradiographs, and these were assigned to groups C, A, and I (Fig. 4). Incorporation into the A group was increased by epinephrine about twofold, but was unaffected by insulin. The most striking finding was a fivefold increase in the incorporation into the I peptide in insulin-treated tissue. Epinephrine also increased incorporation into this site, but to a lesser extent. Neither hormone apparently affected phosphorylation of the C group of peptides. These results indicated that exposure of fat cells to insulin and epinephrine led essentially to the phosphorylation of different sites on acetyl CoA carboxylase and that these different patterns of phosphorylation might be associated with the opposite changes in activity. Recently, general confirmation of our findings has come from the studies of Witters et al.,[38] in which the enzyme was isolated by use of ammonium sulfate plus Sepharose-avidin and the tryptic peptides were separated by reverse-phase high-performance liquid chromatog-

Protein
(coomassie blue stained)

^{32}p

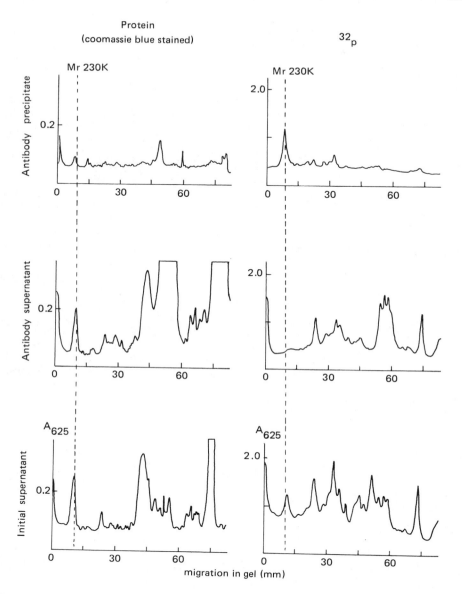

Figure 3. Immunoprecipitation of fat cell [^{32}P]acetyl CoA carboxylase. Fat cells were incubated with medium containing ^{32}P$_i$ for 60 min and a 100,000 × *g* initial supernatant was prepared. This preparation was then incubated with antiserum to acetyl CoA carboxylase for 30 min at 30°C before being centrifuged at 80,000 × *g* to give antibody precipitate and supernatant fractions. Proteins in equivalent samples were then separated by Sodium dodecylsulfate polyacrylamide gel electrophoresis; and densitometric scans of Coomassie blue-stained proteins and of autoradiographs were compared. It should be noted that the protein band of 230,000 M_r in the initial supernatant represents both acetyl CoA carboxylase and fatty acid synthase, as the component remaining in the antibody supernatant could be precipitated following incubation with antisera to fatty acid synthase. For further details, see refs. 41 and 46.

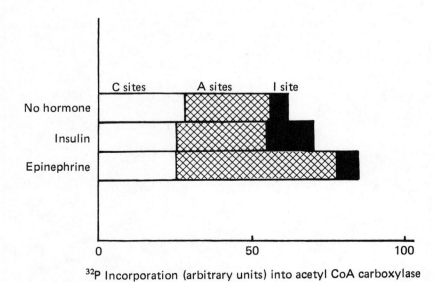

Figure 4. Effects of insulin and epinephrine on the phosphorylation of specific sites on acetyl CoA carboxylase in rat epididymal fat cells. Cells were incubated with [^{32}P]phosphate and appropriate hormones; acetyl CoA carboxylase was isolated by immunoprecipitation, and overall ^{32}P incorporation was determined on the basis of the specific activity of the medium phosphate. Peptides released by trypsin digestion were then separated by two-dimensional thin-layer chromatography and ^{32}P-labeled peptides located by autoradiography and assigned to three groups A, C, and I, according to mobility. The incorporation into these groups indicated in the figure involves the assumption that these sites represent all the phosphorylation sites on acetyl CoA carboxylase. (Data calculated from results of Brownsey and Denton.[18])

raphy (RP-HPLC). The changes in phosphorylation appear to be somewhat smaller in this later study, although this may be related to the rather low recovery of the enzyme through the isolation procedure, which apparently was less than 20%.[38]

V. BRIEF COMMENT ON INCREASES IN PHOSPHORYLATION OF OTHER SPECIFIC PROTEINS IN FAT CELLS EXPOSED TO INSULIN

Acetyl CoA carboxylase is by no means the only protein that exhibits increased phosphorylation in fat cells exposed to insulin (see Table 1 and Chapter 16). Other examples of increased phosphorylation by proteins of known identity include ATP-citrate lyase, the ribosomal protein S6, and the β subunit of the insulin receptor. In addition, at least three further proteins of unknown function exhibit increased phosphorylation with subunit values of 63,000, 61,000, and 22,000 M_r on SDS-PAGE. The smaller of these proteins shows the greatest percentage increase in phosphoryla-

Table 1. Proteins Exhibiting Increased Phosphorylation in Fat Cells Exposed to Insulin

Band No.[a]	Approx. subunit M_r ($\times 10^{-3}$)	Cell location	Identity	Extent of increase in overall phosphorylation	Comments	References
1	230	Cytoplasm	Acetyl CoA carboxylase	10%	Increase in "1-site" phosphorylation about 5-fold	18,38,46
2	130	Cytoplasm	ATP-citrate lyase	2–3-fold	No change in activity	52–55
	95	Plasma membrane	Insulin-receptor β-subunit		Increase in serine and tyrosine phosphorylation	56–58
5	63	Cytoplasm/fat cake	?	2-fold	—	59,60
	61	Plasma membrane	?	30%	—	61
7	30–35	Microsomes	S₆	2–3-fold	Relationship to changes in protein synthesis?	62–65
9	22	Cytoplasm	?	>4-fold	Acid and heat stable	60,66–68

[a] Refers to the numbers allocated in this laboratory to the main bands of ^{32}P-labeled phosphoproteins in a whole-cell extract, separated by sodium dodecylsulfate polyacrylamide electrophoresis. [69]

tion.[66] This protein is thermostable and is not precipitated by 1–2% trichloroacetic acid (TCA), properties shown by a number of similar-size proteins that have important regulatory functions, including inhibitor-1, myosin light chains, and calmodulin.[60,66–68] Increased phosphorylation of this 22,000-M_r protein is closely associated with an apparent decrease in the phosphorylation of another heat- and acid- stable protein band with an apparent molecular weight of 20,000. It seems possible that these two protein bands represent species of the same protein that differ with respect to the number of sites phosphorylated, the species migrating with an apparent molecular weight of 22,000 containing more sites phosphorylated than the species with the apparent molecular weight of 20,000.[60,68,70] The role of this protein remains an intriguing problem. We have speculated previously that increased phosphorylation of this protein may play a role in initiating the dephosphorylation of other proteins in cells exposed to insulin, including glycogen synthase, triglyceride lipase, and hydroxymethyl glutaryl CoA reductase, perhaps through activation of protein phosphatase activity.[69] We have not yet obtained convincing evidence either for or against this proposal.

The increases in phosphorylation summarized above were all deduced from changes in the steady-state level of ^{32}P incorporation from ^{32}P$_i$ in the incubation medium. It is clearly important to be sure that the elevated ^{32}P labeling of this series of phosphoproteins does represent a true increase in phosphate content—especially as it is sometimes argued[71,72] that a common element in the mechanism of insulin action is the dephosphorylation of proteins. A fairly straightforward and certainly rigorous method of proving that increased steady-state ^{32}P incorporation does represent a net increase in phosphorylation is to demonstrate reversal of the increases by subsequent removal of insulin. Application of the approach to small pieces of epididymal fat pads has shown quite clearly that the increased phosphorylation of ATP-citrate lyase, the cytoplasmic 63,000-M_r protein, and the thermostable cytoplasmic 22,000-M_r protein were all reversed on removing insulin by the addition of anti-insulin antiserum.[60]

A number of the proteins listed in Table 1, including those of 130,000, 35,000, and 22,000 M_r also exhibit increased phosphorylation following exposure of fat cells to epinephrine. In the case of ATP-citrate lyase, it appears that the same site is phosphorylated in response to both hormones.[55] However, with the ribosomal S6 protein (30,000–35,000 M_r), it appears that a different spectrum of sites may be phosphorylated in HeLa and other cells exposed to insulin than those in cells in which cAMP has been increased.[73,74] Only in the case of acetyl CoA carboxylase and possibly S6[74] is an increase in phosphorylation associated with a change in catalytic or other functional activity.

A final point worth emphasizing is that where established, the increases in phosphorylation summarized in Table 2 appear to be confined to serines (e.g., acetyl CoA carboxylase, ATP-citrate lyase, S6, and the 22,000-M_r protein) or threonines (22,000-M_r protein). Only the increased phosphorylation of the β subunit of the insulin receptor has been reported to involve an enhanced amount of tyrosine phosphorylation, but even in this case much of the increased phosphorylation observed with intact tissue may occur on serine and threoninne residues.[57]

Table 2. Effects of Insulin Treatment of Rat Epididymal Adipose Tissue on the Rate of Phosphorylation of Acetyl CoA Carboxylase and ATP-Citrate Lyase by High-Speed Supernatant Fractions[a,b]

| | | Rate of phosphorylation[c] | | | |
| | | Endogenous proteins | | Added proteins | |
Protein	M_r	%[d]	N^f	%[d]	N^f
Acetyl CoA carboxylase	230,000	127 ± 6[e]	20	184 ± 30[e]	6
ATP-citrate lyase	125,000	177 ± 12[e]	20	196 ± 26[e]	6
—	78,000	121 ± 4[e]	20	—	
—	43,000	122 ± 8[e]	20	—	

[a] Data from Brownsey et al.[78]
[b] Experiments were carried out as in Fig. 5, with the further addition of purified acetyl CoA carboxylase (200 μg/ml) or ATP-citrate lyase (100 μg/ml) when appropriate.
[c] Rates of phosphorylation in extracts from insulin-treated tissue were calculated as percentage of that found in extracts from control tissue.
[d] Mean ± SEM.
[e] $p < 0.02$.
[f] Number of separate experiments.

VI. STUDIES ON PROTEIN KINASES INVOLVED IN THE ACTION OF EPINEPHRINE AND INSULIN ON ACETYL CoA CARBOXYLASE ACTIVITY

Acetyl CoA carboxylase from rat epididymal adipose tissue,[38,75] rat liver,[20] and rabbit mammary gland[49] has been shown to be phosphorylated on addition of the catalytic subunit of cAMP-dependent protein kinase. Phosphorylation of the mammary gland and adipose tissue enzymes under these conditions appears to occur mainly at the same sites (A sites) as those exhibiting increased phosphorylation in fat cells exposed to epinephrine (Brownsey and Hardie[51]; R.W. Brownsey, unpublished observations, 1983). It is also associated with a similar degree of inactivation.[18,41,76] These effects of the kinase are prevented by addition of the specific protein inhibitor of cAMP-dependent protein kinase (Walsh inhibitor) and are reversed on subsequent dephosphorylation.[41,75,76] These results give strong support to the view that the effects of epinephrine on the activity of acetyl CoA carboxylase in fat cells are brought about by an increase in cAMP concentration and hence activation of cAMP-dependent protein kinase. It seems very likely that the diminution of acetyl CoA carboxylase activity in liver cells exposed to glucagon is brought about by the same mechanism. It should be noted that other cyclic nucleotide-independent protein kinases have been reported that phosphorylate and inactivate acetyl CoA carboxylase[13] and these kinases may also play a role in the action of hormones such as epinephrine and glucagon. For example, Lent and Kim[77] recently argued that the effect of cAMP-dependent protein kinase on liver acetyl CoA carboxylase may not be direct, but may occur via an independent kinase that appears to be activated when phosphorylated by cAMP-dependent protein kinase.

The finding that insulin increases the phosphorylation of a specific site on acetyl CoA carboxylase raises the question of the protein kinase involved. Purified fat cell plasma membranes were found to display substantial protein kinase activity that was capable of phosphorylating acetyl CoA carboxylase with concomitant activation of the enzyme[75] (Fig. 5). The kinase was found to employ Mg-ATP, to be insensitive to cyclic nucleotides and the specific (Walsh) inhibitor protein, and apparently to be fully active in the absence of Ca^{2+}.[75] The changes in catalytic activity observed were similar to those observed following exposure of fat cells to insulin—in particular, increases occurred on initial activity and were no longer evident after treatment of the enzyme with citrate. Under these conditions, the maximum level of phosphorylation that could be achieved was about 0.25–0.5 mole/mole of subunit of which about 50% occurred at I sites as defined in Fig. 4, with the remainder at C sites (R.W. Brownsey, unpublished observations, 1983). Fat cell plasma membranes also contain considerable amounts of a similar protein kinase activity capable of phosphorylating ATP-citrate lyase (R.W. Brownsey and T.J. Hopkirk, unpublished observations, 1983), but further studies are needed to establish whether this is the same enzyme that phosphorylates

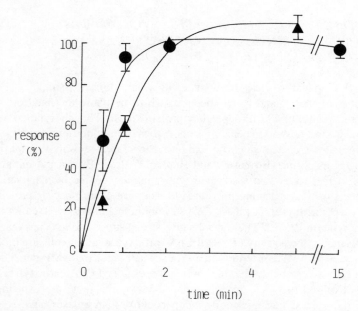

Figure 5. Parallel phosphorylation and activation of fat cell acetyl CoA carboxylase upon incubation with fat cell plasma membranes and [γ-^{32}P]-ATP. Acetyl CoA carboxylase (▲) activity was assayed in the absence of citrate and ^{32}P incorporation (●) into acetyl CoA carboxylase determined after separation of the enzyme from other labeled proteins by sodium dodecylsulfate polyacrylamide gel electrophoresis. No changes were found in acetyl CoA carboxylase activity nor did appreciable phosphorylation occur in the absence of either added ATP or plasma membranes. (Data taken from Brownsey et al.[75])

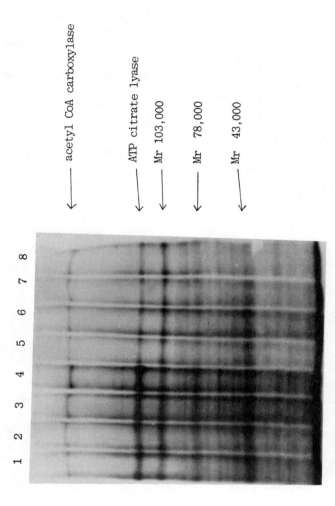

Figure 6. Increased phosphorylation of acetyl CoA carboxylase and ATP-citrate lyase in supernatant fractions of rat epididymal adipose tissue previously exposed to insulin. Tissue was extracted in a sucrose-based medium containing EGTA and protease inhibitors, and $100,000 \times g$ supernatant fractions were rapidly prepared, then incubated for 2 min at 37°C with $MgCl_2$ (5 mM) and $[\gamma^{-32}P]$-ATP (20 μM and 1200 dpm/pmole). Proteins were then separated by Sodium dodecylsulfate polyacrylamide gel electrophoresis. Autoradiograph shows results of four separate experiments. Tracks 1–4 were obtained with extracts from insulin-treated tissue, while tracks 5–8 were the matched control extracts.

and activates acetyl CoA carboxylase. Despite many attempts, we have been unable to find any consistent changes in protein kinase activity using either acetyl CoA carboxylase or ATP-citrate lyase as substrates following the direct exposure to insulin of fat cell membrane preparations.

On the other hand, exposure of intact fat cell preparations to insulin does result in a clear-cut increase in protein kinase activity in subsequently prepared high-speed supernatant fractions[78] (see Fig. 6 and 7 and Table 2). These fractions were prepared as rapidly as possible by centrifugation (Beckman Airfuge). Substrates for the kinase include both endogenous and added purified acetyl CoA carboxylase and ATP-citrate lyase. In addition, increased phosphorylation of two unknown proteins of 78,000 and 43,000 M_r is evident in the high-speed supernatant from insulin-treated adipose tissue. Increased phosphorylation of proteins of these apparent sizes has not been reported within whole cells, but this may simply be due to inadequate separation from other labeled phosphoproteins. The properties of this protein kinase appear to be similar to the cAMP-independent activity associated with fat cell plasma membranes. The activity apparently is not altered by Ca^{2+}, cyclic nucleotides, or the Walsh inhibitor. Supernatant fractions from fat cells exhibit appreciable cAMP-dependent protein kinase activity that phosphorylated both acetyl CoA carboxylase and ATP-citrate lyase on addition of cAMP. However, this activity also phosphorylates further endogenous

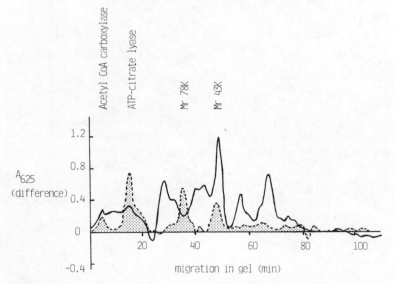

Figure 7. Difference in densitometric traces illustrating the specificity of incorporation of ^{32}P from [γ-^{32}P]-ATP into proteins in high-speed supernatant fractions following treatment of rat epididymal adipose tissue with insulin or treatment of tissue extracts with cAMP. The shaded area represents the (insulin-control) difference densitometric trace from an experiment carried out as in Fig. 6. The unshaded area is the (cAMP-protein kinase inhibitor protein) trace from the same experiment in which the high-speed supernatant fraction from control tissue was incubated with [γ-^{32}P]-ATP for 2 min at 37°C in the presence of either cAMP (50 μm) or protein kinase inhibitor (70 unit/ml). (Data from Brownsey *et al.*[78])

proteins that are not substrates for the protein kinase whose activity is increased in the fractions from insulin-treated adipose tissue (Fig. 7).

The activity of phosphoprotein phosphatase in the high-speed supernatant is very modest (only about 10% of that of the kinase), and no evidence was found for any decrease in activity in the supernatant fractions from insulin-treated tissue that might have contributed to the observed increases in incorporation of ^{32}P. Thus, there seems little doubt that insulin leads to a substantial increase in cytoplasmic protein kinase activity able to phosphorylate at least two of the proteins exhibiting enhanced phosphorylation in cells exposed to insulin. We have been unable to detect any increase in activity associated with crude particulate or purified plasma membranes prepared from adipose tissue previously incubated with insulin. A problem with these studies is the presence of ATPase activity in both supernatant and other subcellular fractions. Although the amount of ATPase activity is not altered by insulin treatment, it does limit the extent of protein phosphorylation that can be achieved. For example, it has not proved possible, to date, to achieve sufficient phosphorylation of acetyl CoA carboxylase with supernatant fractions to cause a detectable change in activity.

VII. CONCLUSIONS

Acetyl CoA carboxylase is activated in fat and other cells exposed to insulin; it seems reasonable to conclude that this activation together with the parallel activations of glucose transport and pyruvate dehydrogenase brings about the increased conversion of glucose into fatty acids. Activation of acetyl CoA carboxylase in insulin-treated fat cells is associated with increased phosphorylation of a specific site on the enzyme. The phosphorylation of this site appears to be brought about by a cyclic nucleotide-independent protein kinase, much of which appears to be bound to plasma membranes in fat cells. However, after exposure of fat cells to insulin, increased acetyl CoA carboxylase and ATP-citrate lyase protein kinase activity is evident in the supernatant fraction. Further studies are clearly necessary to characterize these protein kinase activities, particularly the interrelationship between the plasma membrane and supernatant forms.

The increase in protein kinase activity in the high-speed supernatant fraction of insulin-treated tissue is very persistent and is essentially unaffected by dialysis or gel filtration.[78] Although it is possible for a tightly bound regulator to remain attached during such procedures, it is unlikely. The more likely explanations are either (1) that the kinase has undergone some covalent modification (such as a change in phosphorylation), or (2) that translocation has occurred from other sites in the cell (such as the plasma membrane). Our findings are quite similar to those of a number of laboratories studying the increased phosphorylation of the ribosomal protein S6 in 3T3-L1 and other cells.[64,65,73,74] As with acetyl CoA carboxylase, it appears that the sites phosphorylated are different from those phosphorylated by cAMP-dependent protein kinase. Moreover, the activity of the cyclic-nucleotide-independent protein kinase apparently involved in the action of insulin is elevated in subcellular fractions from insulin-treated cells. Further studies will no doubt be concerned with exploring the relationship between this protein kinase and that capable of phosphorylating acetyl CoA carboxylase and ATP-citrate lyase found in our studies. The above observations are compatible with the working hypothesis we first put forward in 1981[69] that the binding of insulin to its plasma membrane receptors may lead to dissociation of the kinase from the inner face of the plasma membrane. It is becoming increasingly evident that a characteristic of insulin action of cells may be the rapid translocation of components between plasma membrane locations. Examples include both the recruitment of glucose transporters from intracellular sites[79,80] and the rapid internalisation of insulin-receptor complexes.[81–84]

Another attraction of our hypothesis is that it does offer a means of amplification of the insulin signal. It has been suggested that the effect of insulin on acetyl CoA carboxylase activity is brought about by a mediator, presumably a peptide, acting directly on the enzyme.[85] However, there are about 20 million acetyl CoA carboxylase subunits in a fat cell, which means that approximately 10,000 times the number of insulin receptors need to be occupied in order to obtain maximum activation. It is very difficult to visualize how the binding of insulin to a receptor could result in the release of such a large number of mediator molecules, especially if the mediator was a peptide generated by proteolysis from a larger precursor. This argument also applies to other proposed targets of putative insulin mediator, such as cAMP-dependent protein kinase and pyruvate dehydrogenase.[69]

The important studies of Kasuga and colleagues[57,86] and subsequently of other laboratories[87–93] have demonstrated that solubilized, and even purified, preparations of insulin receptors display protein-tyrosine kinase activity that is greatly enhanced on addition of insulin. The physiological substrates of this kinase are unknown except for the β subunit of the receptor itself. It is an attractive possibility that dissociation and activation of the protein-serine kinase involved in the increased phosphorylation of acetyl CoA carboxylase and perhaps other proteins is initiated by the phosphorylation of tyrosine residues by the insulin receptor kinase. The close association of the insulin receptor and of a plasma membrane protein-serine kinase is suggested by the finding that the increased phosphorylation of the β subunit of the receptor in cells exposed to insulin is predominantly on serine residues.[57]

ACKNOWLEDGMENT. We thank the Medical Research Council (U.K.), the British Diabetic Association, and the Percival Waite Salmond Bequest for support.

VIII. REFERENCES

1. Cahill, G.F., LeBoeuf, B., and Renold, A.E., 1959, *J. Biol. Chem.* **234:**2540–2543.
2. Flatt, J.B., and Ball, E.G., 1964, *J. Biol. Chem.* **239:**675–685.
3. Volpe, J.J., and Vagelos, P.R., 1976, *Physiol. Rev.* **56:**339–417.
4. McCormack, J.G., and Denton, R.M., 1977, *Biochem. J.* **166:**627–630.
5. Haft, D.E., 1968, *Diabetes* **17:**251–255.
6. Geelen, M.J.H., Beynen, A.C., Christiansen, R.Z., Lepreau-Jose, M.J., and Gibson, D.M., 1978, *FEBS Lett.* **95:**326–330.
7. Witters, L.A., Moriarty, D. and Martin, D.B., 1979, *J. Biol. Chem.* **254:**6644–6649.

8. Assimacopoulos-Jeannet, F., McCormack, J.G., Prentki, M., Jeanrenaud, B., and Denton, R.M., 1982, *Biochim. Biophys. Acta* **717**:86–90.
9. Munday, M.R., and Williamson, D.H., 1981, *Biochem. J.* **196**:831–837.
10. McNellie, E.M., and Zammit, V.A., 1982, *Biochem. J.* **204**:273–280.
11. Hardie, D.G., 1980, in *Recently Discovered Systems of Enzyme Regulation by Reversible Phosphorylation* (Cohen, P., ed.), Vol. 1, pp. 33–62, Elsevier/North Holland Biomedical Press, Amsterdam.
12. Lane, M.D., Moss, J., and Polakis, S.E., 1974, *Curr. Top. Cell Regul.* **8**:139–195.
13. Kim, K-H., 1983, *Curr. Top. Cell Regul.* **22**:143–176.
14. Inoue, H. and Lowenstein, J.M., 1972, *J. Biol. Chem.* **247**:4825–4832.
15. Tanabe, T., Wada, K., Okazaki, T., and Numa, S., 1975, *Eur. J. Biochem.* **57**:15–24.
16. Song, C.S., and Kim, K.H., 1981, *J. Biol. Chem.* **256**:7786–7788.
17. Hardie, D.G., and Cohen, P., 1978, *Eur. J. Biochem.* **92**:25–34.
18. Brownsey, R.W., and Denton, R.M., 1982, *Biochem. J.* **202**:77–86.
19. Gravel, R.A., Lam, K.F., Mahuran, D., and Kronis, A., 1980, *Arch. Biochem. Biophys.* **201**:669–673.
20. Tipper, J.P., and Witters, L.A., 1982, *Biochim. Biophys. Acta* **715**:162–169.
21. Martin, D.B. and Vagelos, P.R., 1962, *Biochem. Biophys. Res. Commun.* **7**:101–106.
22. Bortz, W.M., and Lynen, F. 1963, *Biochem. Z.* **337**:505–509.
23. Ogiwara, H., Tanabe, T., Nikawa, J., and Numa, S., 1978, *Eur. J. Biochem.* **89**:33–41.
24. Denton, R.M., and Halperin, M.L., 1968, *Biochem. J.* **110**:27–38.
25. Guynn, R.W. Velaso, D., and Veech, R.L., 1972, *J. Biol. Chem.* **247**:7325–7331.
26. Clarke, S.D., Watkins, P.A., and Lane, M.D., 1979, *J. Lipid Res.* **20**:974–985.
27. Watkins, P.A., Tarlow, D.M., and Lane, M.D., 1977, *Proc. Natl. Acad. Sci. USA* **74**:1496–1501.
28. Saggerson, E.D., and Greenbaum, A.L., 1970, *Biochem. J.* **119**:221–242.
29. Halestrap, A.P., and Denton, R.M., 1974, *Biochem. J.* **142**:365–377.
30. Brownsey, R.W., Bridges, B.J., and Denton, R.M., 1977, *Biochem. Soc. Trans.* **5**:1286–1288.
31. Nishikori, K., Iritani, N., and Numa, S., 1973, *FEBS Lett.* **32**:19–21.
32. Goodridge, A.G., 1973, *J. Biol. Chem.* **248**:4318–4326.
33. Yeh, A.L., and Kim, K-H., 1980, *Proc. Natl. Acad. Sci. USA* **77**:3351–3355.
34. Witters, L.A., Friedman, S.A., Tipper, J.P., and Bacon, G.W., 1981, *J. Biol. Chem.* **256**:8573–8578.
35. Chen, S.L., and Kim, K-H., 1982, *J. Biol. Chem.* **257**:7325–7331.
36. Halestrap, A.P., and Denton, R.M., 1973, *Biochem. J.* **132**:509–517.
37. Lee, K.H., Thrall, T., and Kim, K-H., 1973, *Biochem. Biophys. Res. Commun.* **54**:1133–1140.
38. Witters, L.A., Tipper, J.P., and Bacon, G.W., 1983, *J. Biol. Chem.* **258**:5643–5648.
39. Zammit, V.A., and Corstophine, C.G., 1982, *Biochem. J.* **208**:783–788.
40. Stansbie, D., Brownsey, R.W., Crettaz, M., and Denton, R.M., 1976, *Biochem. J.* **160**:413–416.
41. Brownsey, R.W., Hughes, W.A., and Denton, R.M., 1979, *Biochem. J.* **184**:23–32.
42. Wittters, L.A., Kowaloff, E.M., and Avruch, J., 1979, *J. Biol. Chem.* **254**:245–248.
43. Carlson, C.A., and Kim, K-H., 1973, *J. Biol. Chem.* **248**:378–380.
44. Carlson, C.A., and Kim, K-H., 1974, *Arch. Biochem. Biophys.* **164**:478–489.
45. Carlson, C.A., and Kim, K-H., 1974, *Arch. Biochem. Biophys.* **164**:490–501.
46. Brownsey, R.W., Hughes, W.A., Denton, R.M., and Mayer, R.J., 1977, *Biochem. J.* **168**:441–445.
47. Qureshi, A.A., Jenik, R.A., Kim, M., Lornitzo, F.A., and Porter, J.W., 1975, *Biochem. Biophys. Res. Commun.* **66**:344–351.
48. Witters, L.A., 1981, *Biochem. Biophys. Res. Commun.* **100**:872–878.
49. Hardie, D.G., and Cohen, P., 1978, *FEBS Lett.* **91**:1–7.
50. Hardie, D.G., and Cohen, P., 1978, *Eur. J. Biochem.* **92**:25–34.
51. Brownsey, R.W., and Hardie, D.G., 1980, *FEBS Lett.* **120**:67–70.
52. Linn, T.C., and Srere, P.A. 1979, *J. Biol. Chem.* **254**:1691–1698.
53. Ramakrishna, S., and Benjamin, W.B., 1979, *J. Biol. Chem.* **254**:9232–9236.
54. Alexander, M.C., Kowaloff, E.M., Witters, L.A., Denniby, D.T., and Avruch, J., 1979, *J. Biol. Chem.* **254**:8052–8056.
55. Pierce, M.W., Palmer, J.L., Keutmann, H.T., Hall, T.A., and Avruch, J., 1982, *J. Biol. Chem.* **257**:10681–10686.
56. Häring, H.U., Kasuga, M., and Kahn, C.R., 1982, *Biochem. Biophys. Res. Commun.* **108**:1538–1545.

57. Kasuga, M., Zick, Y., Blith, D.L., Karlson, F.A., Häring, H.V., and Kahn, C.R., 1982, *J. Biol. Chem.* **257:**9891–9894.

58. Plehwe, W.A., Williams, P.F., Caterson, I.D., Harrison, L.C., and Turtle, J.R., 1983, *Biochem. J.* **214:**361–366.

59. Benjamin, W.B., and Clayton, N., 1978, *J. Biol. Chem.* **253:**1700–1709.

60. Belsham, G.J., Brownsey, R.W., and Denton, R.M., 1982, *Biochem. J.* **204:**345–352.

61. Belsham, G.J., Denton, R.M., and Tanner, M.J.A., 1980, *Biochem. J.* **192:**457–467.

62. Hughes, W.A., Brownsey, R.W., and Denton, R.M., 1977, in *Phosphorylated Proteins and Related Enzymes* (L. Pinna, ed.), pp. 17–33, Information Retrieval, London.

63. Hughes, W.A., Brownsey, R.W. and Denton, R.M., 1980, *Biochem. J.* **192:**469–481.

64. Smith, C.J., Wejksnora, P.J., Warner, J.R., Rubin, C.S., and Rosen, O.M., 1979, *Proc. Natl. Acad. Sci. USA* **76:**2725–2729.

65. Smith, C.J., Rubin, C.S., and Rosen, O.M., 1980, *Proc. Natl. Acad. Sci. USA* **77:**2641–2645.

66. Belsham, G.J., and Denton, R.M., 1980, *Biochem. Soc. Trans.* **8:**382–383.

67. Blackshear, P.J., Nemenoff, R.A., and Avruch, J., 1982, *Biochem. J.* **204:**817–824.

68. Blackshear, P.J., Nemenoff, R.A., and Avruch, J., 1983, *Biochem. J.* **214:**11–19.

69. Denton, R.M., Brownsey, R.W., and Belsham, G.J., 1981, *Diabetologia* **21:**347–362.

70. Vargas, A.M., Halestrap, A.P., and Denton, R.M., 1982, *Biochem. J.* **208:**221–229.

71. Cohen, P., 1980, in *Recently Discovered Systems of Enzyme Regulation by Reversible Phosphorylation* (P. Cohen, ed.), Vol. 1, pp. 1–10, 255–258. Elsevier Biomedical Press, Amsterdam.

72. Ingebritsen, T.S., and Gibson, D.M., 1980, in *Recently Discovered Systems of Enzyme Regulation by Reversible Phosphorylation* (P. Cohen, ed.), Vol. 1, pp. 63–94, Elsevier Biomedical Press, Amsterdam.

73. Lastick, S.M., and McConkey, E.H., 1981, *J. Biol. Chem.* **256:**583–585.

74. Perisic, O., and Traugh, J.A., 1983, *J. Biol. Chem.* **258:**9589–9592.

75. Brownsey, R.W., Belsham, G.J., and Denton, R.M., 1981, *FEBS Lett.* **124:**145–150.

76. Hardie, D.G., and Guy, P.S., 1980, *Eur. J. Biochem.* **110:**167–177.

77. Lent, B.A., and Kim, K-H., 1983, *Arch. Biochem. Biophys.* **225:**972–978.

78. Brownsey, R.W., Edgell, N.J., Hopkirk, T.J., and Denton, R.M., 1984, *Biochem. J.* **218:**733–743.

79. Karnieli, E., Zarnowski, M.J., Hissin, P.J., Simpson, I.A., Salans, L.B., and Cushman, S.W., 1981, *J. Biol. Chem.* **256:**4772–4777.

80. Kono, T., Robinson, F.W., Blevina, T.L., and Ezaki, O., 1982, *J. Biol. Chem.* **257:**10942–10947.

81. Olefsky, J.M., and Kao, M. 1982, *J. Biol. Chem.* **257:**8667–8673.

82. Desbuquois, B., Lopez, S., and Burlet, H. 1982, *J. Biol. Chem.* **257:**10852–10860.

83. Khan, M.N., Posner, B.I., Khan, R.J., and Bergeron, J.J.M., 1982, *J. Biol. Chem.* **257:**5969–5976.

84. Caro, J.F., Muller, G., and Glennon, T.A., 1982, *J. Biol. Chem.* **257:**8459–8466.

85. Saltiel, A.R., Doble, A., Jacobs, S. and Cuatrecasas, P., 1983, *Biochem. Biophys. Res. Commun.* **110:**789–795.

86. Kasuga, M., Karlson, F.A., and Kahn, C.R., 1982, *Science* **215:**185–187.

87. Avruch, J., Nemenoff, R.A., Blackshear, P.J., Pierce, M.W., and Osathanonad, R., 1982, *J. Biol. Chem.* **257:**15162–15166.

88. Machicao, F., Urumow, T., and Wieland, O.H., 192, *FEBS Lett.* **149:**96–100.

89. Petruzzelli, L.M., Ganguly, S., Smith, C.J., Cobb, M.H., Rubin, C.S., and Rosen, O.M., 1982, *Proc. Natl. Acad. Sci. USA* **79:**6792–6796.

90. Roth, R.S., and Cassell, D.J., 1983, *Science* **219:**299–301.

91. Zick, Y., Whittacker, J., and Roth, R.A., 1983, *J. Biol. Chem.* **258:**3431–3434.

92. Shia, M.A., and Pilch, P.F., 1983, *Biochemistry* **22:**717–721.

93. Van Obberghen, E., and Kowalski, A., 1982, *FEBS Lett.* **143:**179–182.

18

Regulation of Acetyl CoA Carboxylase by Insulin and Other Hormones

Lee A. Witters

Acetyl CoA carboxylase (E.C. 6.4.1.2.) catalyzes the first committed step of fatty acid synthesis, namely, the biotin-dependent conversion of acetyl CoA to malonyl CoA by the fixation of bicarbonate. Given this pivotal position, the modulation of its activity serves as an important locus for the regulation of the rate of fatty acid synthesis. It seems clear that it is not the only potential locus, for several other more proximal steps leading to the generation of cytoplasmic acetyl CoA are also subject to acute regulation. These include the uptake of glucose and other precursor molecules, as well as the reactions catalyzed by phosphofructokinase, pyruvate kinase, and pyruvate dehydrogenase.[1] However, the study of the regulation of acetyl CoA carboxylase (ACC) has proved an excellent reporter system for detailing the mechanisms by which several hormones, including insulin, regulate enzyme activity. This chapter first reviews the current information regarding the complexities of the overall regulation of ACC activity and then focuses on the specific issue of acute regulation by hormones.

In all mammalian tissues studies to date, ACC activity is distributed throughout several cellular fractions. While its intracellular location has traditionally been viewed as being entirely cytosolic, substantial activity has been detected in crude microsomes and mitchondrial fractions,[2] depending on the nutritive state of the animal and the conditions of tissue homogenization. ACC can be isolated in a protomeric form of 500,000 M_r consisting of two identical subunits of 215,000–260,000 M_r, depending on the method of isolation and the species studied.[3-11] *In vitro*, ACC can be converted to a large-molecular-weight polymer of 5–20×10^6 daltons by incubation with citrate and other tricarboxylic acids [12,13] associated with an increase in the specific activity.

Lee A. Witters ● Diabetes Unit and Medical Services, Massachusetts General Hospital; and the Department of Medicine, Harvard Medical School, Boston, Massachusetts.

It is not clear as to which form of ACC is predominant in the cell during varying states of lipogenesis.

I. REGULATION OF ACETYL CoA CARBOXYLASE ACTIVITY

The rate of fatty acid synthesis and the activity of ACC are subject to both long-term and short-term regulation. Long-term regulaton (hours to days) is quantitatively much greater in amplitude; the transition from the fasting state to the refed state in rats, for example, is associated with a 50-fold increase in hepatic fatty acid synthesis and a 15–25-fold increase in ACC activity. The increase in fatty acid synthesis is due in part to the marked increase in the cellular content of several key lipogenic enzymes including ACC, ATP-citrate lyase, malic enzyme, and fatty acid synthetase, probably all due to an increase in their synthetic rate.[14,15] The precise roles of nutrients and several hormones in this inducton have not been fully clarified. Short-term regulation (minutes) permits fine tuning of the rate of fatty acid synthesis in response to immediate metabolic demand; this regulatory mechanism represents the focus of this chapter.

Interconversion between the protomeric and polymeric forms of ACC rapidly alters the catalytic activity of ACC. Indeed, it is not entirely clear that the protomer represents an active form of ACC. Polymerization and activation of ACC can be achieved *in vitro* by incubation of ACC protomer with citrate and other tricarboxylic acids[12,13]; protomerization and inactivation of ACC polymer occur on incubation with fatty acyl CoA derivatives.[16] The shift between protomer and polymer is also sensitive to pH, salt concentration, and buffer composition.[17] Allosteric regulation of ACC by citrate and fatty acyl CoA has been an attractive mechanism for the acute modulation of enzyme activity. Other possibly important regulatory ligands have also been identified, including CoA, polyphosphoinositides, and guanine nucleotides.[18–20]

As stated above, ACC activity is not exclusively confined to the cytosol. Furthermore, the activity associated with membranous fractions increases commensurate with increases in fatty acid synthesis. Our work has indicated that the specific activity of the microsomal ACC is sevenfold higher than that of cytosolic ACC and that the percentage of total cellular ACC activity present in microsomes rises from 7% to 43% during the transition from low rates to high rates of fatty acid synthesis.[2] The nature of the microsomal ACC has not been fully clarified; it is possible that this represents an enzyme polymer or covalently modified enzyme that specifically associates with a membranous compartment or that it represents an isozymic form of ACC. The translocation of ACC from the cytosolic to microsomal compartment may represent another mechanism for the regulation of ACC activity and of fatty acid synthesis.

Much current interest has focused on the role of covalent enzyme phosphorylation–dephosphorylation as a key event in the regulation of ACC activity. ACC is a multiply-phosphorylated enzyme containing 3–6 moles phosphate/moles subunit, when isolated from various tissues.[5–9,11] *In vitro* studies indicate that a number of protein kinases and phosphatases can stoichiometrically phosphorylate or dephosphorylate ACC.[6,7,11,21–26] Much work is needed to clarify the overall role of phosphorylation in ACC regulation, but several aspects have recently been elucidated.

Incubation of ACC, isolated from rat or rabbit mammary gland[6,7] or rat liver[8] with the catalytic subunit of the cAMP-dependent protein kinase results in the incorporation of 0.5–2 moles of phosphate/mole subunit associated with parallel inactivation of ACC. The inactivation and phosphorylation are promptly reversed upon incubation with a protein phosphate preparation.[7,8] Three different protein phosphatases, as isolated from rabbit skeletal and liver (protein phosphatases 1, 2A, and 2C) can reverse the phosphorylation and inactivation catalyzed by the kinase.[26] Enzyme inactivation is due to an upward shift in the K_a for citrate; at maximally activating citrate concentrations, the effects on activity of this phosphorylation are not apparent. This point emphasizes a key relationship between allosteric regulation and changes in enzyme phosphorylation. These are not two mutually exclusive modes of enzyme regulation; there are several examples, including ACC, where the phosphorylation or dephosphorylation of an enzyme alters the K_a/K_i for an allosteric activator/inhibitor or the K_m for a substrate. Conversely, there are examples wherein the allosteric regulator may alter the rate of phosphorylation or dephosphorylation.

Several other protein kinases are present in rat liver or rat/rabbit mammary gland that will stoichiometrically phosphorylate ACC. These include casein kinase I, casein kinase II, and several (?) cAMP-independent protein kinases, which have been only partially characterized.[6–8,11,21–25] Phosphorylation of ACC by some, but not all, of these kinases has been associated with enzyme inactivation. Phosphorylase kinase and glycogen synthase kinase-3 phosphorylate ACC only at very low rates (L.A. Witters and J.P. Tipper, unpublished observations). There does not appear to be any unique ACC protein phosphatase; phosphatase activity directed against ACC can be entirely accounted for by the multifunctional protein phosphatases 1, 2A, and 2C.[26,27]

It is important to emphasize that ACC, like several other important enzymes, is multisite-phosphorylated; thus, as has proved to be the case with glycogen synthase,[28] the relationship of this covalent modification to alterations in enzyme activity may be quite complex. The nature of the phosphate already present in ACC, as isolated, is uncertain. While 3–6 moles P_i/mole subunit of alkali-labile phosphate are present, it is not clear whether this phosphate is distributed in a small number of sites or in a large number in substoichiometric amounts. At least some of this phosphate appears to reside in a site(s) mediating inactivation of ACC, in that incubation of ACC with three different protein phosphatases will remove up to 2 moles of this phosphate associated with enzyme activation.[26] Whether this phosphate is present in sites that are phosphorylatable by the cAMP-dependent protein kinase or another inactivating protein kinase is unknown. Casein kinase I and II appear to phosphorylate different phosphoserine residues than those phosphorylated by the cAMP-dependent protein kinase.[23]

Studies of the effect of altered phosphorylation on ACC activity must be interpreted with caution. ACC is quite sensitive to proteolytic modification and can be readily activated upon incubation with small concentrations of trypsin. The proteolysis of this relatively small domain of the polypeptide chain does not result in a major cleavage of ACC, yet results in enzyme activation.[29,30] Clustering of phosphorylation sites for three of the ACC kinases (cAMP-dependent kinase, casein kinase I and II) on a 16,000-dalton M_r cyanogen bromide peptide that is readily released to an acid-soluble fraction after brief incubation with low concentrations of trypsin has been noted.[25] Putative deposphorylation or changes in ACC activity observed on incubation of ACC in crude extracts might well be entirely explained by the action of endogenous proteases, which cleave only a small peptide segment from either the NH_2- or COOH-terminus of the polypeptide chain. In addition, failure to observe phosphorylation with any kinase might be due to loss of the phosphorylaton site during enzyme isolation. Any studies of the *in vitro* or *in vivo* phosphorylation of ACC must take into account this important regulatory modification.

In addition to the studies cited above indicating a role for covalent phosphorylation based on *in vitro* studies with purified substrate and kinase/phosphatase, there is abundant evidence that phosphorylation occurs *in vivo* and that the phosphorylation sites are turning over the phosphate at reasonable rates. Incubation of rat adipose tissue or liver with $^{32}P_i$ results in the incorporation of $[^{32}P]$phosphate into $[^{32}P]$phosphoserine; traces of $[^{32}P]$phosphothreonine, but no $[^{32}P]$phosphotyrosine are also detectable (L.A. Witters and J.P. Tipper, unpublished observations). On the basis of the measured specific activity of $[^{32}P]$-ATP in liver cells, one can calculate that up to 3 moles of $[^{32}P]$phosphate are incorporated under cell-labeling conditions.[8,31] These data compare well with the measured amount of alkali-labile phosphate and indicate that the bulk of the covalently bound phosphate is turning over during relatively brief periods of tissue labeling. Like the *in vitro* labeled phosphorylation sites, those labeled in the intact cell are also quite sensitive to proteolysis.[8] As will be discussed in Section 2, the phosphate is distributed among several tryptic phosphopeptides of ACC, and the phosphate content of various sites is rapidly altered in response to hormonal stimuli.

In summary, there appear to be at least two important regulatory mechanisms that rapidly alter ACC activity: protomer–polymer interconversion and changes in covalent phosphorylation. It seems likely that these are interdependent mechanisms; changes in covalent phosphorylation (either site-specific phosphorylation or dephosphorylation) may alter the rate of interconversion between polymer and protomer, either due to a direct effect of the phosphorylation event of the structure of ACC or through an alteration in the interaction of ACC with an important allosteric ligand, such as citrate or fatty acyl CoA. The latter could occur without any concomitant change in the intracellular concentration of such a regulatory ligand. Conversely, the protomer–polymer state of ACC, as determined in part by these ligands, might alter the rates of phosphorylation–dephosphorylation by a number of kinases/phosphatases by exposing or masking of phosphorylation sites. Another potential mechanism for rapid regulation of activity might involve translocation of ACC between various cellular compartments. Conditions that favor enzyme polymerization lead to the association of ACC with microsomes, while conditions favoring protomerization lead to its dissociation to the cytosolic compartment.[2] The full nature of these events needs further exploration. Another regulatory concept also deserves comment. It was postulated several years ago that the polymeric structure of ACC might serve as a microfilamentous backbone for the assembly of the important enzymes of fatty acid synthesis, permitting coordinated flow of substrates and products from acetyl CoA to fatty acid.[32] It is of interest that, while in the rat liver cytosol the molar ratios of ATP-citrate lyase and fatty acid synthetase to ACC are 5 : 20 : 1, in the microsomes (based on certain assumptions of specific activity), these ratios approach unity (L.A. Witters and J.P. Tipper, unpublished observations). Whether such a complex exists intracellularly remains to be determined.

II. ACUTE REGULATION OF ACETYL CoA CARBOXYLASE ACTIVITY BY HORMONES

The rates of fatty acid synthesis, as measured by ^3H_2O or [^{14}C]acetate incorporation, is rapidly altered following hormone exposure in several tissues, including rat adipose tissue and rat and chicken liver.[31,33–40] Insulin in rat adipose tissue and liver leads to an increase in the rate, while glucagon (liver) epinephrine (adipose tissue) and α-adrenergic catecholamines (liver) lead to a decrease in the rate. In rat liver, rates of 3-β-hydroxysterol synthesis are unchanged during these periods of exposure to hormone.[35,39] Because cytoplasmic acetyl CoA is a common precursor for the synthesis of both fatty acid and 3-β-hydroxysterol, these data suggest that the principal locus of regulation lies distal to the generation of cytoplasmic acetyl CoA. In addition, alterations in the rate of fatty acid synthesis induced by hormones *in vivo* correlate closely with the measured intracellular levels of malonyl CoA.[35,38] Taken together, these data suggest that the principal locus of hormonal regulation lies at the conversion of acetyl CoA to malonyl CoA—the step catalyzed by ACC.

In rat liver and adipose tissue, the activity of ACC, as measured in crude extracts, changes rapidly in response to hormones.[23,31,34,37–45] Insulin leads to activation, while glucagon, epinephrine, and α-agonists lead to inactivation. Under these conditions, in rat liver and adipose tissue, the activity of the next enzyme in the synthesis cascade, fatty acid synthetase, is unchanged.[39,41] In chicken liver, however, Lane and others have reported that the activity of ACC in crude extracts after glucagon exposure is unchanged.[35,36] While the reasons for this discrepancy are not entirely apparent, we belive that attention must be drawn to the conditions used to assay ACC in such extracts. Prolonged incubation of ACC, either in crude extracts or as the isolated enzyme, with citrate result in maximal enzyme activity. These conditions likely result in nearly complete enzyme polymerization and perhaps enzyme proteolysis and dephosphorylation (especially in crude extracts); they might well mask or alter the effects of a hormone in the intact cell. For example, the effects of phosphorylation by the cAMP-dependent protein kinase *in vitro* or epinephrine *in vivo* on enzyme activity are completely reversed on prolonged incubation with citrate, i.e., the effects of this phosphorylation are on the citrate K_a rather than on the V_{max} of the enzyme.[7,11] We and others have generally employed an assay of short duration in the absence or presence of low concentrations of citrate to minimize these enzyme modifications during the course of the assay after tissue homogenization and extract preparation. Denton and colleagues have expressed the data as a ratio of the initial activity to that activity achieved after prolonged incubation with citrate.[42,44] A similar approach has been taken by Zammit and co-workers in the measurement of ACC in crude extracts of rat mammary tissue.[45]

There has been some controversy as to whether the measurement of ACC activity in crude liver extracts by $H^{14}CO_3$ fixation accurately reflects the activity of ACC: Is all the ^{14}C-labeled product of the assay indeed [^{14}C]malonyl CoA?[46,47] It is particularly important to exclude a major contribution of pyruvate carboxylase to overall bicarbonate fixation. Unfortunately, other assays for ACC that rely on measurement of NADPH oxidation (in the assay linked with that of fatty acid synthetase) are not useful in crude extracts because of high levels of nonspecific NADPH oxidation. Using the $H^{14}CO_3$ fixation assay and a specific high-performance liquid chromatography (HPLC) system for assessing the products of the carboxylation reaction, we have documented in both rat liver and adipose tissue extracts that greater than 80% of the ^{14}C-labeled products of the assay are recoverable as [^{14}C]malonyl CoA. Changes in apparent ACC activity measured by total $H^{14}CO_3^-$ fixation after exposure to insulin, glucagon, and epinephrine are accurately reflected by direct measurement of [^{14}C]malonyl CoA generation during the assay (L.A. Witters and J.P. Tipper, unpublished observations).

The effects of insulin, glucagon, and epinephrine on ACC activity in crude extracts prepared from rat liver or adipose tissue appear to be due predominantly to a shift in the citrate K_a rather than a change in V_{max}, the K_a increasing after glucagon exposure and decreasing after insulin exposure.[39] However, in rat adipose tissue, the predominant change in ACC isolated after epinephrine exposure appears to be on V_{max} rather than on the K_a,[23] while glucagon in rat liver leads to both an increase in the K_a and a reduction in the V_{max}.[31] The reasons for this discrepancy are not apparent. It is possible that phosphorylation events initiated through the α-agonist properties of epinephrine lead to V_{max} changes that are not mimicked by the β-adrenergic cascade. The apparent K_i for fatty acyl CoA is also altered by hormones, increasing after insulin and decreasing after glucagon exposure.[39]

Data generated from several laboratories would indicate that both major regulatory mechanisms for the acute modification of ACC activity may be operative in the rapid regulation of ACC by hormones. Perturbations of the protomer–polymer equilibrium have been implicated in the inactivation of ACC by glucagon in the chicken liver,[48] by epinephrine in rat adipose tissue,[43] and by glucagon in the rat liver (L.A. Witters and J.P. Tipper, unpublished observations). Protomerization of ACC, as measured by sucrose density-gradient fractionation, has been documented for both rat adipose tissue and liver enzyme after exposure to epinephrine and glucagon (L.A. Witters and J.P. Tipper, unpublished observations; Meredith and Lane[48]). Lane and colleagues have used an additional novel approach in addressing this issue.[48] On light digitonin lysis of chick liver cells, ACC is released more slowly than that of other cytosolic enzymes, such as lactate dehydrogenase. Inclusion of citrate (which favors ACC polymerization) in the lysis medium slows the rate of efflux, while inclusion of malonyl CoA (which promotes protomerization) increases the efflux rate. Glucagon in these cells leads to an increase in the efflux rate consistent with enzyme protomerization. To our knowledge, there have been no reports of the direct measurement of the protomer–polymer state of ACC after insulin exposure by either of the above techniques. Denton and colleagues reported that, in rat adipose tissue, insulin led to an increase in the amount of ACC activity in a high-speed pellet fraction and interpreted this as being ACC polymer.[42] In our hands, with either sucrose density gradients or the digitonin lysis technique, we have been unable to document a change in apparent ACC size after exposure to insulin in rat liver cell (L.A. Witters and J.P. Tipper, unpublished observations). However, given the sensitivity of the protomer–polymer equilibrium to changes on dilution, buffer composition, and salt concentration, one might interpret these experiments cautiously.

Changes in the cytoplasmic concentration of citrate or fatty acyl CoA in response to hormonal stimuli might promote the conversion between the protomeric and polymeric form of ACC. The repeated observation in several laboratories that the effects of hormones survive cell breakage and extensive dilution argue against, but do not disprove, the major effects of hormones being explained by allosteric regulation. The measurement of the concentration of these two ligands in broken tissues is complicated by the fact that citrate is distributed in both the mitochondrial and cytosolic compartments and that the bulk of fatty acyl CoA is protein bound. No major change in the concentration of whole-cell citrate in rat adipose tissue and liver after exposure to epinephrine, insulin, and glucagon has been detectable[31,35,38,42] (also observed by L.A. Witters and J.P. Tipper, unpublished results). We have also found no change in cytoplasmic citrate measured after glucagon exposure and light digitonin lysis of rat liver cells (L.A. Witters and J.P. Tipper, unpublished observations). In contrast, Lane and colleagues have demonstrated a marked fall in total cell citrate and cytoplasmic citrate in chicken liver cells after exposure to glucagon.[36] This may reflect a true species difference. Denton and colleagues have shown a small decrease in total fatty acyl CoA after insulin exposure of rat adipose tissue[42]; we have been unable to confirm this observation in rat liver cells (L.A. Witters and J.P. Tipper, unpublished observations). The effects of both epinephrine in adipose tissue and glucagon in the liver persist after isolation of the ACC to homogeneity, indicating that a major part of the inactivation is due to covalent modification (see Section 2). However, the effects of insulin do not persist through enzyme isolation, so at least part of the effects of insulin to activate ACC might be explained by allosteric regulation. It is also possible that the effects of hormones are exerted through changes in the concentration of other regulatory ligands, such as the guanine nucleotides, CoA, and polyphosphoinositides.[18–20] To recant, however, it is important to recognize the close relationship between allosteric regulation and covalent phosphorylation changes in the interpretation of these data.

There is substantial evidence that hormonally directed changes in ACC phosphorylation are a major mechanism responsible for the acute regulation of activity by hormones. This evidence has been derived from the measurement of ACC phosphorylation after ^{32}P labeling of rat adipose tissue and liver by direct immunoprecipitation of [^{32}P]ACC, analysis by polyacrylamide gel electrophoresis, and isolation of [^{32}P]ACC to homogeneity.[23,31,44,49–51] While the total ^{32}P content can be assessed by these techniques, it has also become apparent that it is necessary to measure the ^{32}P content of individual phosphorylation sites in this multiply-phosphorylated enzyme in order to best characterize these changes. This has been accomplished by enzymatic digestion followed by separation of ^{32}P-labeled-peptides by either HPLC, two-dimensional TLC/TLE, or isoelectric focusing.[23,31,52–53]

The evidence that changes in ACC phosphorylation underlie the acute changes in enzyme activity on hormone exposure is strongest for epinephrine in adipose tissue and glucagon in liver. Isolation of [^{32}P]ACC from rat adipose tissue after brief epinephrine exposure has been achieved rapidly through the use of avidin-Sepharose affinity chromatography.[23] As compared with control tissue, the epinephrine-exposed [^{32}P]ACC is inactivated (principally through a change in V_{max}) and shows a small increase in total ^{32}P content. On fractionation of [^{32}P]tryptic phosphopeptides by HPLC, there is a marked increase in ^{32}P content of a single phosphopeptide, which comigrates in three HPLC systems with the major peptide phosphorylated *in vitro* by the cAMP-dependent protein kinase. Denton and Hardie have also documented the phosphorylation of a tryptic phosphopeptide with an isoelectric point of 6–7 in response to epinephrine *in vivo* and by this kinase *in vitro*.[52] In rat adipose tissue, epinephrine increases total ^{32}P content of ACC and leads to the phosphorylation of a unique set of phosphopeptides (separated by TLC/TLE) generated by the tryptic digestion of immunoprecipitates of [^{32}P]ACC.[44,53] Nearly identical results have been obtained with rat liver cells after exposure to glucagon and ACC isolation.[31,54] The *in vivo* modification of both side-specific ACC phosphorylation and altered activity can be completely reversed on incubation with a protein phosphatase (L.A. Witters and J.P. Tipper, unpublished observations; Brownsey and Denton[53]). Thus, it appears that the effects of epinephrine in adipose tissue and glucagon in the liver to regulate ACC can be accounted for by the expected epinephrine/glucagon-induced activation of the cAMP-dependent protein kinase.

Kim and co-workers have studied the effects of α-agonists on ACC phosphorylation and activity in rat liver cells by measurement of activity in crude preparations of enzyme and phosphorylation by immunoprecipitation.[40] These data indicate that α-agonists lead to an increase in total ^{32}P content associated with inactivation. This effect is not blocked with a β-antagonist and appears to be dependent on the inclusion of Ca^{2+} in the medium. The nature of the protein kinase(s) mediating these changes as well as the site of phosphorylation are unknown.

The evidence that changes in ACC phosphorylation underlie the activation of ACC in response to insulin, as measured in crude extracts, is less clear. On the basis of the observation that several protein phosphatase preparations lead to the activation of ACC, some workers speculated that the effects of insulin were due to ACC dephosphorylation. However, measurement of total ^{32}P content of ACC after insulin exposure of rat adipose tissue and rat liver revealed an unexpected small increase in ^{32}P content.[23,53,55] Furthermore, on fractionation of tryptic [^{32}P]phosphopeptides by HPLC or two-dimensional TLC/TLE, it has been observed that this insulin-stimulated increase occurs on a peptide(s) different from that phosphorylated in response to epinephrine or glucagon *in vivo* and by the cAMP-dependent protein kinase *in vitro*.[23] Another important observation is that insulin does not lead to the dephosphorylation of the major regulatory site phosphorylated in response to epinephrine, glucagon, or the cAMP-dependent protein kinase. In fact, no dephosphorylation of any sites labeled during these periods of tissue incubation is observed after exposure to insulin.

In my view, this insulin-stimulated increase in ACC phosphorylation cannot be at present related to an insulin-induced activity change. Despite this phosphorylation change, the activity of ACC isolated to homogeneity from insulin-exposed adipose tissue does not differ from that of the control.[23] Brownsey and Denton have also observed a severe decrement in the marked insulin-stimulated increase seen in crude extracts through a partial purification.[53] It is possible that the various isolation techniques are associated with the modification of an insulin-induced change in covalent modification, such as dephosphorylation or proteolysis. Alternatively, it is also possible that the assays employed do not accurately recreate the milieu in which this insulin-stimulated change is expressed in crude extracts; for example, other critical regulatory ligands, conditions favoring enzyme polymerization, and so forth may be absent. The linkage of insulin-induced stimulation of ACC phosphorylation and alteration in ACC activity remains an important focus for future work.

Irrespective of the problems associated with the changes in ACC activity, the insulin-stimulated increase in ACC phosphorylation at a specific site may serve as an important reporter system for unraveling the components that mediate this change both in ACC and in the other reported substrates for insulin-stimulated phosphorylation (see Chapter 16). This insulin-stimulated increase could be due to an increase in protein kinase activity, a decrease in protein phosphatase activity, or a ligand-mediated conformational change in ACC that changes its reactivity for kinase/phosphatase. Efforts have been principally directed at the first explanation, i.e., stimulation of a cAMP-dependent protein kinase in response to insulin. Brownsey and Denton have reported that adipose tissue plasma membranes lead to the activation of partially purified ACC in the presence of ATP/Mg^{2+}; incorporation of ^{32}P into ACC is observed by gel electrophoresis and autoradiography.[56] These workers have reported that this kinase is insensitive to cyclic nucleotides and is active in the absence of Ca^{2+}. At least some of the ^{32}P is incorporated into a tryptic phosphopeptide that is phosphorylated in response to insulin in the intact fat cell.[57] We have been unable to confirm the ATP-dependent activation of either crude or homogeneous ACC on incubations with these membranes (L.A. Witters and J.P. Tipper, unpublished observations). Furthermore, the data, as presented, do not permit calculation of the stoichiometry of phosphate incorporation, nor has it been demonstrated that this activation–phosphorylation is reversible on incubation with a purified protein phosphatase.

Two other explanations are still tenable for the observed ATP-dependent activation. ATP/Mg^{2+} could lead to the activation of the ATP/Mg^{2+}-dependent protein phosphatase, which has been shown to be capable of both dephosphorylating and activating ACC.[58] Alternatively, the activation could be due to the action of an ATP-dependent protease, which leads ACC proteolysis and activation; substoichiometric phosphate incorporation might simply be an epiphenomenon unrelated to the ACC activation. Recreation of this activation with purified kinase and substrate coupled with the demonstration of stoichiometric phosphorylation and reversibility on incubation with a protein phosphatase will be necessary to firm up these interesting observations. Finally, it has yet to be demonstrated that insulin directs a net increase in phosphate content at a specific phosporylation site. It is possible that the observed increases in ^{32}P content might simply reflect the turnover of $^{31}P_i$ for $^{32}P_i$ in a site undergoing net dephosphorylation.

We have been fractionating a number of cyclic nucleotide-independent ACC kinases present in rat liver cytosol and microsomal fractions. Two of these kinases lead to the stoichiometric phosphorylation of the same tryptic phosphopeptide that is phosphorylated in response to insulin in the intact cell, as judged by comigration on HPLC fractionation.[23] These two kinases are identical to casein kinase I and casein kinase II, as originally isolated and characterized in rabbit reticulocytes.[59] Despite the incorporation of up to 2 moles phosphate/subunit with casein kinase I and 0.6 mole/subunit with casein kinase II, no alteration in ACC activity has been observed under the conditions of our assay.[25] We have no information as to whether either of these kinases leads to the phosphorylation of the same phosphoserine residue that is phosphorylated in response to insulin. It is clear that specific sequence information will be required, especially in view of the multiple phosphorylation sites on ACC and their clustering. We have recently developed techniques for the measurement of casein kinase I and II activities in crude extracts of adipose tissue and have further partially purified both activities from rat adipose tissue after exposure to insulin. With either analytical technique to date, we have been unable to detect any change in the activity of either of these kinases after insulin exposure (L.A. Witters and J.P. Tipper, unpublished observations). The limitations of these experiments are obvious, but at present, they cannot support any role for these kinases either in the insulin-stimulated phosphorylation or activation of ACC. More recently, we have isolated a third cAMP-independent ACC kinase from rat liver cytosol (L.A. Witters, R.A. Nemenoff, and J. Avruch, unpublished observations). This kinase phosphorylates the same tryptic phosphopeptide as casein kinase I and II but, in contrast to those kinases, it also phosphorylates the phosphopeptide on ATP-citrate lyase that is phosphorylated in response to insulin in vivo. Ongoing studies are examining the potential role of this kinase in insulin-directed phosphorylation.

If activation of a serine protein kinase is an important element of the activation of ACC by insulin in intact cells, it is interesting to speculate on the possible nature of the activation. One attractive model is that a serine kinase is autophosphorylated on a tyrosine residue by the insulin receptor tyrosine kinase leading to serine kinase activation or that autophosphorylaton of the insulin receptor on insulin binding leads to the dissociation of the serine kinase from the membranes to the cytosolic fraction. We have explored the former possibility with respect to casein kinase I or II. Neither of these kinases is phosphorylated or undergoes activity changes after incubation with purified human placental insulin receptor in the presence of ATP/Mg^{2+} (L.A. Witters, R.A. Nemenoff, and J. Avruch, unpublished observations). Preliminary studies with the third site-specific cAMP-independent kinase have revealed an ATP-dependent activation of ACC kinase activity on incubation with the partially purified receptor that has been variably stimulated by the addition of insulin.

Another model would be the producton of a low-molecular-weight regulatory ligand that itself activates the serine kinase, analogous to the production of cAMP by adenylate cyclase with the subsequent activation of the cAMP-dependent protein kinase. Cuatrecasas and co-workers have recently reported the production of a mediator by liver plasma membranes in response to insulin, which leads to the activation of crude ACC preparations.[60] This "mediator" appears to copurify with that previously described substance that leads to the activation of pyruvate dehydrogenase in intact mitochondria. It has not been reported as to whether ACC phosphorylation is stimulated in response to this extract.

Although insulin-stimulated protein phosphorylation of a number of intracellular proteins has been reported, in our view the relationship between this event and physiological insulin action remains to be established. ACC is an important substrate in these studies because it is the only one of the substrates for which phosphorylation is stimulated in response to insulin and for which insulin has been shown to effect a change in enzyme activity. It is possible that direct phosphorylation by a specific serine protein kinase could lead to ACC activation. Alternatively, this phosphorylation could initiate the dephosphorylation of a second inactivating phosphoserine residue leading to ACC activation. It is also possible that this phosphorylation could lead to the binding of ACC to membranous fractions or its association with other lipogenic enzymes. Any of these events might be accompanied by ACC polymerization at prevailing intracellular allosteric regulator concentrations. The complex nature of ACC regulation, including allosteric regulation initiated through changes in the protomer–polymer equilibrium and multiple-site covalent phosphorylation, will continue to provide fertile soil for the investigation of insulin action, as mirrored in the rapid activation of ACC and the stimulation of fatty acid synthesis.

ACKNOWLEDGMENTS . The work cited from the author's laboratory has been supported by a grant from the United States Public Health Service, AM-19720 and a Research Career and Development Award from the United States Public Health Service, AM-00520. Portions of this work were completed in collaboration with Dr. D. Grahame Hardie, while the author was a Fellow of the John Simon Guggenheim Foundation. The author would also like to acknowledge the contributions of Dr. Jennifer P. Tipper and Geoffrey W. Bacon to much of the work.

III. REFERENCES

1. Hardie, G., 1981, *Trends Biochem. Sci.* **6**:75–77.
2. Witters, L.A., Friedman, S.A., and Bacon, G.W., 1981, *Proc. Natl. Acad. Sci. USA* **78**:3639–3643.
3. Moss, J., Yamagishi, M., Kleinschmidt, A.K., and Lane, M.D., 1972, *Biochemistry* **20**:3779–3786.
4. Tanabe, T., Wada, K., Okazaki, T., and Numa, S., 1975, *Eur. J. Biochem.* **57**:15–24.
5. Ahmad, F., Ahmad, P.M., Pieretti, L., and Watters, G.T., 1978, *J. Biol. Chem.* **253**:1733–1737.
6. Hardie, D.G., and Cohen, P., 1978, *FEBS Lett.* **91**:1–7.
7. Hardie, D.G., and Guy, P.S., 1980, *Eur. J. Biochem.* **110**:167–177.
8. Witters, L.A., and Vogt, B., 1981, *J. Lipid. Res.* **22**:364–369.
9. Song, C.S., and Kim, K-H., 1981, *J. Biol. Chem.* **256**:7786–7788.
10. Beaty, N., and Lane, M.D., 1982, *J. Biol. Chem.* **257**:924–929.
11. Tipper, J.P., and Witters, L.A., 1982, *Biochem. Biophys. Acta* **715**:162–169.
12. Vagelos, P.R., Alberts, A.W., and Martin, D.B., 1963, *J. Biol. Chem.* **238**:533–540.
13. Moss, J., and Lane, M.D., 1972, *J. Biol. Chem.* **247**:4944–4951.
14. Gibson, D.M., Lyons, R.T., Scott, D.F., and Muto, Y., 1972, *Adv. Enzyme Regul.* **10**:187–204.
15. Brunengraber, H., Boutry, M., and Lowenstein, J.M., 1973, *J. Biol. Chem.* **248**:2656–2669.
16. Ogiwara, H., Tanabe, T., Nikawa, J., and Numa, S., 1978, *Eur. J. Biochem.* **89**:33–41.

17. Gregolin, C., Ryder, E., Warner, R.C., Kleinschmidt, A.K., Chang, H-C., and Lane, M.D., 1968, *J. Biol. Chem.* **243**:4236–4245.
18. Yeh, L-A., Song, C-S., and Kim, K-H., 1981, *J. Biol. Chem.* **256**:2289–2296.
19. Blytt, H.J., and Kim, K-H., 1982, *Arch. Biochem. Biophys.* **213**:523–529.
20. Witters, L.A., Friedman, S.A., Tipper, J.P., and Bacon, G.W., 1981, *J. Biol. Chem.* **256**:8573–8578.
21. Shiao, M-S., Drong, R.F., and Porter, J.W., 1981, *Biochem. Biophys. Res. Commun.* **98**:80–87.
22. Lent, B., and Kim, K-H., 1982, *J. Biol. Chem.* **257**:1897–1901.
23. Witters, L.A., Tipper, J.P., and Bacon, G.W., 1983, *J. Biol. Chem.* **258**:5643–5648.
24. Allred, J.B., Harris, G.J., and Goodson, J., 1983, *J. Lipid Res.* **24**:449–455.
25. Tipper, J.P., and Witters, L.A., 1983, *Arch. Biochem. Biophys.* **227**:386–396.
26. Ingebritsen, T.S., Blair, J.B., Guy, P., Witters, L.A., and Hardie, D.G., 1983, *Eur. J. Biochem.* **132**:275–281.
27. Ingebritsen, T.S., and Cohen, P., 1983, *Eur. J. Biochem.* **132**:255–261.
28. Cohen, P., 1982, *Nature* **296**:613–620.
29. Iritani, N., Nakanishi, S., and Numa, S., 1969, *Life Sci.* **8**:1157–1165.
30. Guy, P.S., and Hardie, D.G., 1981, *FEBS Lett.* **132**:67–70.
31. Holland, R., Witters, L.A., and Hardie, D.G., 1984, *Eur. J. Biochem.* **140**:325–333.
32. Gregolin, C., Ryder, E., Kleinschmidt, A.K., Warner, R.C., and Lane, M.D., 1966, *Proc. Natl. Acad. Sci. USA* **56**:148–155.
33. Denton, R.M., and Halperin, M.L., 1968, *Biochem J.* **110**:27–38.
34. Klain, G.T., and Weiser, P.C., 1973, *Biochem. Biophys. Res. Commun.* **55**:76–83.
35. Cook, G.A., Nielsen, R.C., Hawkins, R.A., Mehlman, M.A., Lakshmanan, M.R., and Veech, R.L., 1977, *J. Biol. Chem.* **252**:4421–4424.
36. Watkins, P.A., Tarlow, D.M., and Lane, M.D., 1977, *Proc. Natl. Acad. Sci. USA* **74**:1497–1501.
37. Geelen, M.J.H., Beynen, A.C., Christiansen, R.Z., Lepreau-Jose, M.J., and Gibson, D.M., 1978, *FEBS Lett.* **95**:326–330.
38. Beynen, A.C., Vaartjes, W.J., and Geelen, M.J.H., 1979, *Diabetes* **28**:828–835.
39. Witters, L.A., Moriarity, D., and Martin, D.B., 1979, *J. Biol. Chem.* **254**:6644–6649.
40. Ly, S., and Kim, K-H., 1981, *J. Biol. Chem.* **256**:11585–11590.
41. Halestrap, A.P., and Denton, R.M., 1973, *Biochem J.* **132**:509–517.
42. Halestrap, A.P., and Denton, R.M., 1974, *Biochem. J.* **142**:365–377.
43. Lee, K-H., and Kim, K-H., 1978, *J. Biol. Chem.* **253**:8157–8161.
44. Brownsey, R.W., Hughes, W.A., and Denton, R.M., 1979, *Biochem J.* **208**:783–788.
45. Zammit, V.A., and Chorstorphini, C.G., 1982, *Biochem. J.* **208**:783–788.
46. Davies, D.R., Van Schaftingen, E., and Hers, H-G., 1982, *Biochem J.* **202**:559–560.
47. Allred, J.B., and Goodson, J., 1982, *Biochem J.* **208**:247–248.
48. Meredith, M.J., and Lane, M.D., 1978, *J. Biol. Chem.* **253**:3381–3383.
49. Witters, L.A., Kowaloff, E.M., and Avruch, J.A., 1979, *J. Biol. Chem.* **254**:245–248.
50. Lee, K-H., and Kim, K-H., 1979, *J. Biol. Chem.* **254**:1450–1453.
51. Brownsey, R.W., Hughes, W.A., Denton, R.M., and Mayer, R.J., 1977, *Biochem J.* **168**:441–445.
52. Brownsey, R.W., and Hardie, D.G., 1980, *FEBS Lett.* **120**:67–70.
53. Brownsey, R.M., and Denton, R.M., 1982, *Biochem. J.* **207**:77–86.
54. Holland, R., Witters, L.A., and Hardie, D.G., 1983, in *Symposium on Isolation, Characterization and Use of Hepatocytes* (R.A. Harris and N.W. Cornell, eds.), pp. 585–590, Elsevier Biomedical Press, New York.
55. Witters, L.A., 1981, *Biochem. Biophys. Res. Commun.* **100**:872–878.
56. Brownsey, R.W., Belsham, G.J., and Denton, R.M., 1981, *FEBS Lett.* **124**:145–150.
57. Denton, R.M., and Brownsey, R.W., 1983, *Phil. Trans. R. Soc. Lond. B.* **302**:33–45.
58. Stewart, A.A., Hemmings, B.A., Cohen, P., Goris, J., and Merlevede, W., 1981, *Eur. J. Biochem.* **115**:197–205.
59. Hathaway, G.M., and Traugh, J.A., 1982, *Curr. Top. Cell. Regul.* **21**:101–127.
60. Saltiel, A., Doble, A., Jacobs, S., and Cuatrecasas, P., 1983, *Biochem. Biophys. Res. Commun.* **110**:787–795.

Control of Gene Expression by Insulin

Direct Regulation of Nuclear Functions by Insulin: Relationship to mRNA Metabolism

I.D. Goldfine, F. Purrello, R. Vigneri, and G.A. Clawson

I. INTRODUCTION

Insulin is a major anabolic hormone for most mammalian species. The hormonal potency of insulin is largely the result of its ability to regulate target cells at a variety of cellular sites. The effects of insulin on membrane transport, enzyme activity, and protein synthesis have been studied extensively. Most likely, many of these effects result from the direct interaction of insulin with its plasma membrane receptor. Insulin also regulates nuclear functions such as DNA and RNA synthesis, but just how insulin influences these processes is unknown. The presence of specific binding sites for insulin on nuclei and nuclear envelopes has been documented and characterized. These binding sites have biochemical characteristics that are different from insulin binding sites on the plasma membrane. Moreover, direct *in vitro* effects of insulin on messenger RNA (mRNA) metabolism have now been reported. These effects include (1) stimulation of mRNA efflux from intact nuclei; (2) stimulation of nuclear envelope nucleoside triphosphatase (NTPase), the enzyme that regulates mRNA efflux; and (3) inhibition of ^{32}P incorporation into nuclear envelopes. Thus, significant insight is now being gained concerning the actions of insulin on nuclear functions.

I.D. Goldfine ● Cell Biology Research Laboratory, Harold Brunn Institute, Mount Zion Hospital and Medical Center, San Francisco, California; and the Departments of Medicine and Physiology, University of California, San Francisco, California. *F. Purrello and R. Vigneri* ● Cell Biology Research Laboratory, Harold Brunn Institute, Mount Zion Hospital and Medical Center, San Francisco, California; and the Department of Physiology, University of California, San Francisco, California. *G.A. Clawson* ● Department of Pathology, University of California, San Francisco, California.

II. ACTIONS OF INSULIN ON DNA AND RNA SYNTHESIS

Insulin *in vitro* stimulates the growth of cells in tissue culture.[1] In most studies, however, higher than physiological concentrations of insulin are necessary for this effect,[2,3] and it is likely that in many instances insulin is interacting with receptors for the various insulinlike growth factors, such as somatomedin, MSA, or nonsuppressible insulinlike activity (NSILA-s).[2–4] In a few instances, however, physiological concentrations of insulin increase cell division. Insulin has been shown to be necessary for the regeneration of liver in partially hepatectomized rats and other animals.[5,6] Also, in cultured cells from regenerating rat liver, there is evidence indicating that insulin stimulates DNA synthesis.[7] Finally, physiological concentrations of insulin stimulate DNA synthesis in H35 hepatoma cells.[8]

Insulin regulates RNA levels in many tissues. It is possible that insulin has multiple effects on RNA metabolism, including the regulation of transcriptional and post-transcriptional events. Effects of insulin on transcription have been reported in liver, pancreas, adipose tissue, and mammary gland. Insulin, *in vivo,* increases both RNA polymerase activity and template activity of liver.[9,10] Inhibition of RNA synthesis by actinomycin D has been reported to block the insulin-stimulated synthesis of several enzymes in diabetic rats, including fatty acid synthetase, glycogen synthetase, hexokinase, phosphofructokinase, and pyruvate kinase.[9,11–13] In liver of diabetic rats, insulin administration inhibits the activities of glucose 6-phosphatase, fructose 1,6-diphosphatase, pyruvate carboxylase, and phosphoenolpyruvate carboxykinase; actinomycin D pretreatment of animals also blocks this inhibition.[9,11–13]

Insulin also influences mRNA levels (Table 1). Recent studies have indicated that the production of mRNA for tyrosine aminotransferase[14] is decreased in the liver of adrenalectomized rats and that albumin,[15] fatty acid synthetase,[16] and α_{2u} globulin[17] are decreased in the liver of diabetic rats. These diminished levels of mRNA can be restored by insulin administration *in vivo*. In the adrenalectomized rat, insulin produces a generalized increase in mRNA content.[14]

Administration of insulin to diabetic rats both restores diminished pancreatic acinar cell amylase levels [18,19] and reduces increased trypsinogen levels; actinomycin D

Table 1. Influence of Insulin on
mRNA Levels in Liver and
Other Tissues[14–17,20,28]

Liver
    Albumin
    Tyrosine aminotransferase
    Fatty acid synthetase
    α_{2u}-globulin
Pancreas
    Amylase
Mammary gland
    Casein

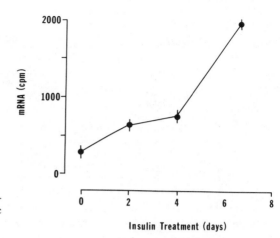

Figure 1. Effect of insulin injections on pancreatic amylase mRNA levels from diabetic rats. (Adapted from Korc et al.[20])

treatment blocks this effect.[18] Furthermore, amylase mRNA levels fall dramatically in the pancreas of diabetic rats; this drop is rapidly reversed by insulin administration[20] (Fig. 1).

In adipose tissue, insulin *in vivo* stimulates the synthesis of hexokinase II via the formation of new RNA.[13,21,22] Moreover, insulin stimulates the activity of glucose 6-P dehydrogenase, phosphofructose kinase, and pyruvate kinase[13,21,22]; these effects are blocked by the administration of actinomycin D. It has been reported that insulin *in vitro* increases lipoprotein lipase activity in 3T3-L$_1$ fibroblasts[23] in part by nuclear regulation.

In mammary glands, insulin *in vitro* activates RNA synthesis, stimulates RNA polymerase activity, the phosphorylation of histone and nonhistone proteins[24–27] and increases mRNA levels.[28]

III. BINDING SITES FOR INSULIN ON NUCLEAR ENVELOPES AND PLASMA MEMBRANES

Specific binding sites for insulin on purified liver plasma membranes were first demonstrated by Freychet et al.[29] and have now been described to many cell types.[30] The major characteristics of insulin binding to these receptors are listed in Table 2. In addition to these cell-surface binding sites, other insulin binding sites have also been described on intracellular structures, including nuclei[31–36] and nuclear membranes,[33,37,38] smooth and rough endoplasmic reticulum,[31,33] and Golgi apparatus.[39] Specific binding sites for insulin on purified rat liver nuclei free of other cellular components were first described by Horvat and co-workers[31] and then confirmed both in our laboratory[32,33] and in that of Goidl.[36] In addition, specific nuclear binding sites have been detected in thyroid nuclei.[34,35] The major site of insulin binding to the nucleus is the nuclear envelope.[37,38,40] When whole nuclei are incubated with native insulin followed by an immunofluorescence procedure, fluorescence is detected

Table 2. Characteristics of Insulin Binding to
Cellular Membranes

	Plasma	Nuclear
High-salt optimum	Yes	No
Alkaline pH optimum	Yes	No
Negative cooperativity	Yes	No
Anti-receptor antibody inhibition	Yes	No
Low-temperature stability	Yes	No
Concomitant insulin degradation	Yes	No
Regulation by exogenous insulin	Yes	Yes

only on the nuclear surface.[38] Furthermore, when nuclei are first incubated with ^{125}I-labeled insulin and then subfractionated, most of the specific hormone binding is seen with the nuclear envelope fractions.[40] In addition, when nuclei are incubated with high concentrations of detergent to remove both layers of the nuclear envelope, binding is either reduced or eliminated. Finally, insulin does not bind directly to DNA or histones, but does bind directly to purified nuclear envelopes.[38,40]

The binding of insulin to nuclear envelopes, like its binding to plasma membranes, fulfills the requirements of a hormone receptor. It is rapid, reversible, of high affinity, and hormone specific (Fig. 2). Two insulin analogs with decreased biological potencies—proinsulin and desoctapeptide insulin—were less effective in both nuclear and plasma membranes. The characteristics of insulin binding to the nuclear envelope differ, however, in a number of respects from the characteristics of insulin binding to plasma membrane (Table 2). Studies of insulin binding to liver plasma membranes have revealed two classes of binding sites.[30] In our studies of insulin binding sites to nuclear envelopes prepared by the method of Kashnig and Kasper,[41] two orders of binding sites were seen but with lower affinities (K_d 5.6 nM, 65 nM) than those seen on the plasma membrane (K_d 0.5 nM, 10 nM). Both plasma membranes and nuclear envelopes, however, have similar total insulin-binding capacities (~2 nmoles/mg protein). Horvat,[38] studying nuclear envelopes prepared by RNAase and DNAase digestion, reported only one class of binding sites for insulin having a K_d of 3 nM. In liver and other tissues, the binding of insulin to plasma membranes has three distinctive characteristics[30,38,40]: a sharp pH optimum of 8.0, enhanced binding in the presence of high concentrations of NaCl, and enhanced dissociation of labeled insulin in the presence of unlabeled insulin, which may be due to negative cooperativity. When the characteristics of insulin binding in nuclear envelopes were examined, we observed (1) a pH optimum of 7.0–7.5, (2) no enhanced binding in the presence of NaCl, and (3) that addition of unlabeled insulin did not enhance the dissociation rate of labeled insulin. Horvat also did not find any indication for negative cooperativity with insulin binding to nuclear envelopes.[38]

The serum of patients diagnosed as having severe insulin resistance and acanthosis nigricans exhibits antibodies to the plasma membrane insulin receptor; preincubation

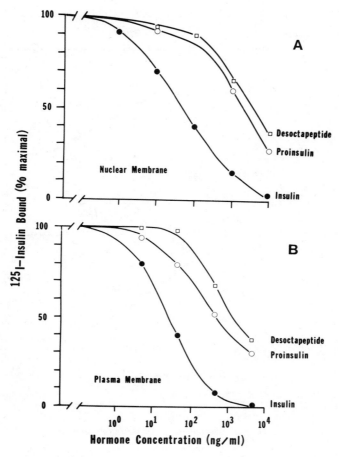

Figure 2. Inhibition of [^{125}I]insulin binding to nuclear membrane (A) and to plasma membrane (B) by native insulin, proinsulin, and desoctapeptide insulin.

of plasma membranes with these antibodies blocks the subsequent binding of insulin (Fig. 3).[37] This inhibition of binding by these antibodies can be demonstrated with insulin receptors from a variety of species and tissues. But these antibodies do not bind to receptors for other hormones, such as glucagon and growth hormone.[37] Preincubation of this antiserum with nuclei resulted in little inhibition of the subsequent binding of labeled hormone.[37] This finding suggested that the insulin-binding sites in the nuclear envelope are proteins different from the insulin binding sites on the plasma membrane. Another possibility is that they are the same binding site, but that the different milieu of the nuclear membrane significantly alters the characteristics of insulin binding. For instance, the lipid composition of the nuclear envelope, especially

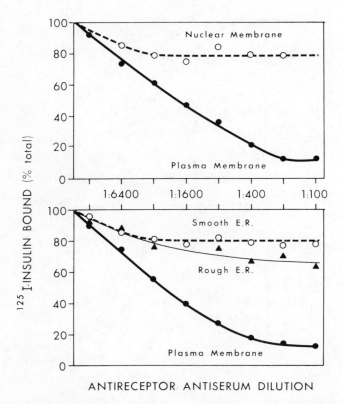

Figure 3. Effect of preincubation with an antiserum to the plasma membrane insulin receptor on the subsequent specific binding of ^{125}I-labeled insulin to isolated nuclear membranes, rough and smooth endoplasmic reticulum (ER), and plasma membranes. (Adapted from Goldfine *et al.*[38])

the cholesterol content,[42] is markedly different from that of the plasma membrane. Since the lipid environment of the plasma membrane causes alterations in insulin binding,[42] there is support for the latter hypothesis.

IV. STRUCTURE OF THE NUCLEAR ENVELOPE

The nuclear envelope is a bilayered membrane structure that separates the nucleoplasm and cytoplasm of eukaryotic cells. The outer nuclear envelope is associated with the endoplasmic reticulum, and the inner nuclear membrane is intimately associated with the peripheral heterochromatin[43,44] (Fig. 4). It is known that small molecules can freely enter into and exit from the nucleus.[44,45] There is evidence, however, that the translocation of RNA and other macromolecules to and from the nucleus may proceed via a more complicated mechanism.[44] Thus, it is likely that the nuclear envelope has more than a passive role in nuclear cytoplasmic interactions.

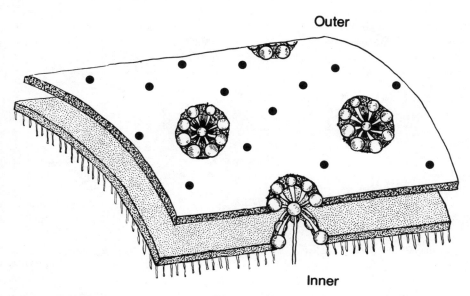

Outer

Inner

Figure 4. Schematic drawing of the nuclear envelope. Nuclear pores can be seen joining the two layers of the nuclear envelope.

Pores filled with a unique structure termed the nuclear pore complex are dispersed throughout the nuclear envelope (Fig. 4). The nuclear pore complex has a double annulus, each having eight peripheral subunits of ~250 Å in diameter.[43,44] In addition, there is a central granule between the double annuli (Fig. 4). The nuclear pore complex is attached to both the inner and outer nuclear envelope, but the nuclear pore complex is not covered on its inner surface by heterochromatin. Furthermore, it has been postulated that both the central granule and peripheral subunits are hollow tubes. Thus, the nuclear pore complex could play a role in the nuclear cytoplasmic translocation of mRNA.

V. RELATIONSHIP BETWEEN mRNA EFFLUX AND NUCLEAR MEMBRANE NTPase

In an effort to gain insight into the mechanism of mRNA transport from the nucleus, rats have been injected with radiolabeled orotic acid or uridine and the efflux of labeled RNA from isolated nuclei into a surrogate cytoplasm has been studied.[46–49] Most investigators have employed nuclei from liver. Many criteria suggest that the nature of the transported RNA under the conditions studied is mRNA.[47–49] These criteria include size, base and poly A content, activity in directing protein synthesis, incorporation into polysomes, and inclusion into specific RNP particles.

RNA transport *in vitro* involves both intranuclear RNA processing and subsequent efflux (Fig. 5). A source of high-energy phosphate is necessary for transport but not

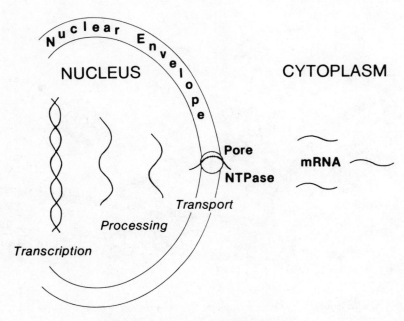

Figure 5. Major sites of mRNA processing.

for processing.[47–49] One high-energy phosphate bond is hydrolyzed to transport one nucleotide of mRNA. Studies indicate that the high-energy phosphate specificity is not highly selective, since ATP, UPT, CTP, and GTP are all effective.[47–49]

There is considerable evidence suggesting that a nuclear membrane triphosphatase (NTPase) provides the energy for the transport of mRNA. For example, the activation energy for RNA transport is 13 kcal/mole and for NTPase activity is 13.3–13.8 kcal/mole.[49] Furthermore, the affinities of ATP for both NTPase activity and facilitated RNA transport are similar. Also, cAMP stimulates both functions, whereas NaF inhibits both of them.[47–49] Finally, trypsin treatment of nuclei inactivates both functions.[48] Histocytochemical studies from our laboratory indicate that this enzyme is located throughout the nuclear envelope.[49] Others, however, have suggested that this enzyme resides in the nuclear pore complex.[48]

VI. INSULIN ACTION IN ISOLATED NUCLEI AND NUCLEAR ENVELOPES

In view of the influence of insulin on mRNA levels in liver, several studies have been carried out *in vitro* with isolated nuclei and nuclear envelopes. Schumm and Webb measured mRNA transport from liver nuclei of normal rats prelabeled 30 min *in vivo* with [^{14}C]orotic acid and found that the direct addition of insulin *in vitro* to these nuclei markedly enhanced mRNA transport.[47] In these studies, however, higher than physiological levels of insulin were needed (100 nM). We have modified their

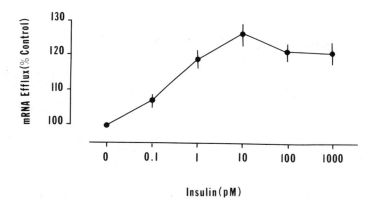

Figure 6. Effect of insulin on release of [¹⁴C]-RNA from isolated rat liver nuclei from diabetic rats. (From Purrello *et al.*[61])

methods by both eliminating liver cytosol and using diabetic rats. We now find that *in vitro* insulin as low as 1 pM can stimulate mRNA efflux (Fig. 6). A maximal effect of insulin is seen at 10 pM, and at higher insulin levels the hormonal effect is diminished.

In light of the observations that insulin may directly stimulate nuclear mRNA efflux and that nuclear membrane NTPase activity is necessary for this function, we investigated whether insulin directly influenced nuclear membrane NTPase activity. Highly purified nuclear membranes were prepared by the method of Monneron.[50] In these membranes we found basal nuclear membrane NTPase activity to be higher in liver of normal rats than in liver from hypoinsulinemic diabetic rats.[51] Moreover, the direct addition of insulin to purified nuclear envelopes of liver from diabetic rats stimulated NTPase activity (Fig. 7). As with stimulation of mRNA efflux, an effect

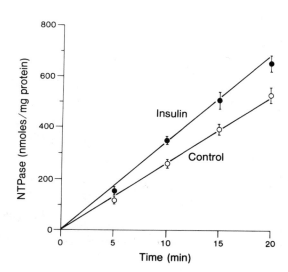

Figure 7. Time course of insulin stimulation of NTPase activity in purified nuclear envelopes from diabetic animals. (From Purrello *et al.*[57])

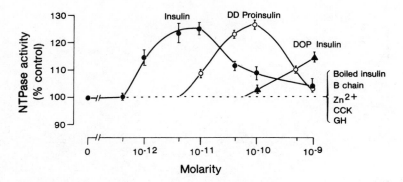

Figure 8. Dose–response curve of insulin stimulation of NTPase activity in nuclear envelopes from diabetic rats. (From Purrello *et al.*[57])

was detectable at 1 pM, and maximal effects were seen at 10–100 pM (Fig. 8). Other studies indicated that insulin increased the V_{max} of the enzyme (Fig. 9).[51] By contrast, no effects of insulin were seen with purified rat liver plasma membranes (Fig. 10).

Several groups have reported the presence of protein kinase and phosphatase activity in isolated nuclear envelopes[52–55]; it has been proposed that these reactions may have a regulatory role in nucleocytoplasmic transport.[52–55] Moreover, it has been proposed that phosphorylaton reactions regulate nuclear envelope NTPase activity.[53] Accordingly, we investigated whether insulin has direct effects on the phosphorylation of nuclear envelopes from rat liver.[61] The direct addition of insulin to highly purified nuclear envelopes but not plasma membranes decreased ^{32}P incorporation into tri-chloroacetic acid precipitable proteins (Fig. 11). As with stimulation of mRNA efflux

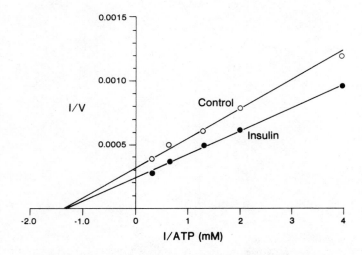

Figure 9. Effect of insulin on the V_{max} of nuclear envelope NTPase.

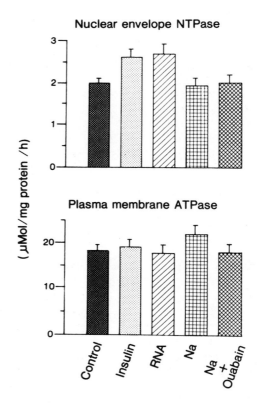

Figure 10. Effect of insulin on ATPase activity in purified plasma and nuclear membranes. (From Purrello et al.[57])

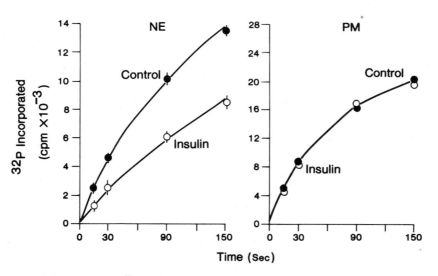

Figure 11. Effect of insulin on ^{32}P incorporation into proteins of the nuclear envelope and plasma membranes from fasted rats. (From Purrello et al.[61])

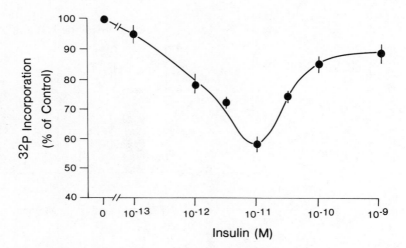

Figure 12. Dose–response curve showing insulin inhibition of ^{32}P incorporation into proteins of nuclear envelopes from fasted rats. (From Purrello *et al.*[61])

and NTPase activity, an effect was detectable at 1 pM (Fig. 12). The dose–response curve was biphasic and was a mirror image of the curves for regulation of mRNA efflux and NTPase activity (Fig. 13). Other studies suggested that insulin was acting via stimulation of phosphatase activity.[61]

In whole cells two agents are known to mimic the action of insulin by acting on the insulin receptors: polyclonal antireceptor antibodies and plant lectins.[42,56] It was therefore important to determine whether these agents, like insulin, would influence the phosphorylation of nuclear envelope.[57] When polyclonal antiserum from a patient with acanthosis nigricans and insulin resistance was tested, this agent inhibited the phosphorylation of nuclear envelope; by contrast, normal human serum was without

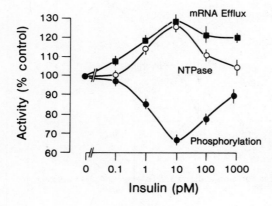

Figure 13. Effect of insulin on (1) stimulation of [^{14}C]-RNA release from isolated liver nuclei obtained from diabetic rats, (2) NTPase activity in isolated nuclear envelopes, and (3) ^{32}P incorporation into nuclear envelope proteins.

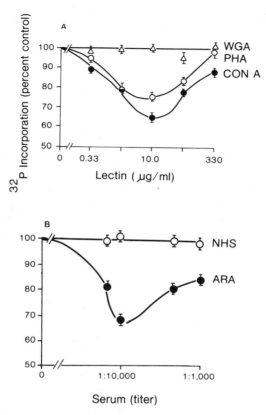

Figure 14. Effects of (A) insulin plant lectins and (B) receptor antibody on inhibition of ^{32}P incorporation into nuclear envelope proteins from fasted rats. WGA, wheat germ agglutinin; PHA, phytohemagglutinin; Con A, concanavalin A; NHS, normal human serum; ARA, antireceptor antibody. (From Purrello *et al.*[57])

effect (Fig. 14). Three plant lectins were studied: phytohemaglutinin, concanavalin A, and wheat germ agglutinin. The first two lectins, like insulin, inhibited the phosphorylation of the nuclear envelope (Fig. 14); wheat germ agglutinin, however, was not effective. When maximally inhibiting concentrations of either the polyclonal antiserum or the lectins were added with insulin, no further inhibition of phosphorylation was seen, suggesting that these agents were working in a manner similar to that of insulin.[61]

These data suggest a model of how insulin may regulate mRNA efflux. Nuclear envelope NTPase, the enzyme that regulates mRNA efflux, may be in either phosphorylated (inactive) or dephosphorylated (active) states (Fig. 15). When dephosphorylated, the enzyme has an increased affinity for both mRNA and ATP; as a result, the nucleocytoplasmic transport of mRNA is enhanced and the hydrolysis of ATP is accelerated.

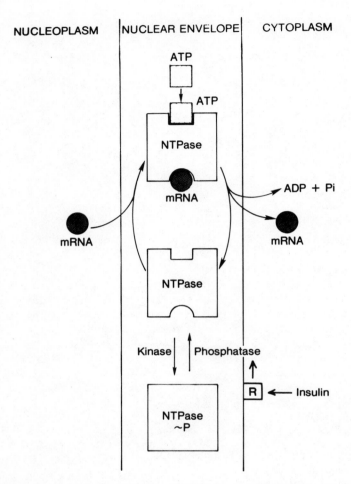

NUCLEOPLASM | NUCLEAR ENVELOPE | CYTOPLASM

Figure 15. Proposed model of insulin stimulation of mRNA release from nuclei. Insulin first stimulates phosphoprotein phosphatase activity, which in turn activates NTPase activity. NTPase then hydrolyzes ATP and promotes mRNA efflux.

VII. CONCLUSION

The nuclear envelope and its pore complex play a major role in the transport of mRNA from the nucleus. It had been documented that the nuclear envelope contains specific high-affinity binding sites for insulin, but until recently the significance of these binding sites was unknown. Current studies *in vitro* now indicate that insulin not only stimulates the release of mRNA directly from isolated nuclei, but also increases the activity of nuclear envelope NTPase, the enzyme that regulates mRNA efflux. Moreover, insulin inhibits ^{32}P incorporation into nuclear envelope proteins. These observations demonstrate that *in vitro* insulin regulates nuclear functions by acting directly at the nuclear surface.

The question arises as to whether internalized insulin regulates biological function in intact cells. Two recent observations have suggested this possibility. Draznin and Trowbridge, in their work on hepatocytes, have shown that dansylcadavarine inhibits intracellular insulin processing.[58] This agent has no effect on insulin stimulation of glycogen metabolism, but does influence insulin inhibition of protein degradation. This group has suggested that certain functions of insulin require the internalization and processing of the hormone.

Recently, we reported that insulin can be covalently conjugated to the B (binding) subunit of the toxin ricin. This molecule has the ability to bind to the insulin receptor, the ricin receptor, or both receptors. In a study of insulin-resistant cells with low numbers of insulin receptors, we showed that this conjugate can produce biological effects greater than those of insulin itself.[59] In a recent study of insulin-receptor-deficient cells, Hofmann *et al.*[60] showed that while native insulin has no effect on glycogen synthesis, insulin ricin B chain conjugate can stimulate this function at concentrations as low as 100 pM. Galactose, an inhibitor of ricin binding to its receptor, blocked this effect of the conjugate. The ricin B chain by itself had no effect. These studies with insulin ricin conjugates suggest that insulin action can be mediated via receptors other than the insulin receptor. Since the ricin B chain first binds to cells and then functions to internalize the toxin, these data raise the possibility that the internalization of insulin ricin B chain conjugate may be important for the action of the hybrid molecule.

These studies with dansylcadavarine and insulin ricin B chain conjugate suggest that insulin may exert certain actions on the cell interior. Further studies are necessary to understand the relationship between the effects of insulin on the intact cell and the direct effect of insulin on isolated intracellular organelles such as the nucleus.

ACKNOWLEDGMENTS. This research wes supported by grant AM26667 from the National Institutes of Health and by the Elise Stern Haas Research Fund, Harold Brunn Institute, Mount Zion Hospital and Medical Center.

VIII. REFERENCES

1. Gey, G.O., and Thalhimer, W.J., 1924, *JAMA* **82**:1609.
2. Rechler, M.M., Podskalny, J.M., Goldfine, I.D., and Wells, C.A., 1974, *J. Clin. Endocrinol. Metab.* **39**:512–521.
3. Smith, G.L., and Temin, H.M., 1974, *J. Cell Physiol.* **84**:181–192.
4. Chochinov, R.H., and Daughaday, W.H., 1976, *Diabetes* **25**:994–1004.
5. Bucher, N.L.R., and Weir, G.C., 1976, *Metab. Clin. Exp.* **25**:1423–1425.
6. Price, J.B., Jr., 1976, *Metab. Clin. Exp.* **25**:1427–1428.
7. Richman, R.A., Claus, T.H., Pilkis, S.J., and Friedman, D.L., 1976, *Proc. Natl. Acad. Sci. USA* **73**:3589–3593.
8. Koontz, J.W., and Iwahashi, M., 1980, *Science* **211**:947–949.

9. Steiner, D.F., 1966, *Vitam. Horm.* **24:**1–61.
10. Morgan, C.R., and Bonner, J., 1970, *Proc. Natl. Acad. Sci. USA* **65:**1077–1080.
11. Steiner, D.F., and King, J., 1964, *J. Biol. Chem.* **239:**1292–1298.
12. Weber, G., 1972, *Isr. J. Med. Sci.* **8:**325–343.
13. Krahl, M.E., 1974, *Annu. Rev. Physiol.* **36:**331–360.
14. Hill, R.E., Lee, K.-L., and Kenny, F.T., 1981, *J. Biol. Chem.* **256:**1510–1513.
15. Peavy, D.E., Taylor, J.M., and Jefferson, L.S., 1978, *Proc. Natl. Acad. Sci. USA* **75:**5879–5883.
16. Pry, T.A., and Porter, J.W., 1981, *Biochem. Biophys. Res. Commun.* **100:**1002–1009.
17. Roy, A.K., Chatterjee, B., Prasad, M.S.K., and Unakar, J.J., 1980, *J. Biol. Chem.* **255:**11614–11618.
18. Söling, H.D., and Unger, K.O., 1972, *Eur. J. Clin. Invest.* **2:**199–212.
19. Korc, M., Iwamoto, Y., Sankaran, H., Williams, J.A., and Goldfine, I.D., 1981, *Am. J. Physiol.* **240** (*Gastrointest. Liver Physiol.* 3):G56–G62.
20. Korc, M., Owerbach, D., Quinto, C., and Rutter, W.J., 1981, *Science* **213:**351–353.
21. Hansen, R.J., Pilkis, S.J., and Krahl, M.E., 1967, *Endocrinology* **81:**1397–1404.
22. Hansen, R.J., and Pilkis, S.J., 1970, *Endocrinology* **86:**57–65.
23. Spooner, P.M., Chernick, S.S., Garrison, M.M., and Scow. R.O., 1979, *J. Biol. Chem.* **254:**10021–10029.
24. Stockdale, F.E., and Topper, Y.J., 1966, *Proc. Natl. Acad. Sci. USA* **56:**1283–1289.
25. Turkington, R.W., 1968, *Endocrinology* **82:**540–546.
26. Topper, Y.J., Friedberg, S.H., and Okta, T., 1970, *Dev. Biol.* (Suppl.)**4:**101–113.
27. Terry, P.M., Banerjee, M.R., and Lui, R.M., 1977, *Proc. Natl. Acad. Sci. USA* **74:**2441–2445.
28. Bolander, F.F., Jr., Nicholas, K.R., Van Wyk, J.J., and Topper, Y.J., 1981, *Proc. Natl. Acad. Sci. USA* **78:**5682–5684.
29. Freychet, P., Roth, J., and Neville, D.M., Jr., 1971, *Proc. Natl. Acad. Sci. USA* **68:**1833–1837.
30. Goldfine, I.D., 1978, in *Receptors in Pharmacology* (J.R. Smythies and R.J. Bradley, eds.), pp. 335–377, Marcel Dekker, New York.
31. Horvat, A., Li, E., and Katsoyannis, P.G., 1975, *Biochem. Biophys. Acta* **382:**609–620.
32. Goldfine, I.D., and Smith, G.J., 1976, *Proc. Natl. Acad. Sci. USA* **73:**1427–1431.
33. Vigneri, R., Pliam, N.B., Cohen, D.C., Pezzino, V., Wong, K.Y., and Goldfine, I.D., 1978, *J. Biol. Chem.* **253:**8192–8197.
34. Brisson-Lougarre, A., and Blum, C.J., 1979, *C.R. Acad. Sci. (D) (Paris)* **289:**129–132.
35. Brisson-Lougarre, A., and Blum, C.J., 1980, *C.R. Acad. Sci. (D) (Paris)* **290:**889–892.
36. Goidl, J.A., 1979, *Biochemistry* **18:**3674–3769.
37. Goldfine, I.D., Vigneri, R., Cohen, D,, and Pliam, N.B., 1977, *Nature* **269:**698–700.
38. Horvat, A., 2978, *J. Cell Physiol.* **97:**37–47.
39. Bergeron, J.J.M., Evans, W.H., and Geschwind, I.I., 1973, *J. Cell Biol.* **59:**771–776.
40. Vigneri, R., Goldfine, I.D., Wong, K.Y., Smith, G.J., and Pezzino, V., 1978, *J. Biol. Chem.* **253:**2098–2103.
41. Kashnig, D.M., and Kasper, C.B., 1969, *J. Biol. Chem.* **244:**3786–3792.
42. Goldfine, I.D., 1978, *Life Sci.* **23:**2639–2648.
43. Franke, W.W., 1974, *Int. Rev. Cytol.* (suppl.) **4:**71–236.
44. Harris, J.R., and Agutter, P.S., 1976, in *Biochemical Analysis of Membranes* (A.H. Maddy, ed.) pp. 132–173, Wiley, New York.
45. Feldherr, C.M., 1972, *Adv. Cell Mol. Biol.* **2:**273–281.
46. Ishikawa, K., Sato-Odani, S., and Ogata, K., 1978, *Biochem. Biophys. Acta* **521:**650–654.
47. Schumm, D.E., and Webb, T.E., 1978, *J. Biol. Chem.* **253:**8513–8519.
48. Agutter, P.S., McCaldin, B., and McArdle, H.J., 1979, *Biochem. J.* **182:**811–819.
49. Clawson, G.A., James, J., Woo, C.H., Friend, D.S., Moody, D., and Smuckler, E.A., 1980, *Biochemistry* **19:**2748–2756.
50. Monneron, A., Blobel, G., and Palade, G.E., 1973, *J. Cell Biol.* **55:**104–125.
51. Purrello, F., Vigneri, R., Clawson, G.A., and Goldfine, I.D., 1982, *Science* **216:**1005–1007.
52. Agutter, P.S., Cocknee, J.B., Lavine, J.E., McCaldin, B., and Sim, R.B., 1979, *Biochem. J.* **181:**647–658.
53. Lam, K.S., and Kasper, O.B., 1979, *Biochemistry* **18:**307–311.
54. Steer, R.C., Wilson, M.J., and Ahmed, K., 1979, *Exp. Cell Res.* **119:**403–406.
55. Steer, R.C., Wilson, M.J., and Ahmed, K., 1979, *Biochem. Biophys. Res. Commun.* **89:**1082–1087.

56. Van Obberghen, E., and Kahn, C.R., 1981, *Mol. Cell. Endocrinol.* **22:**277–293.
57. Purrello, F., Burnham, D.B., and Goldfine, I.D., 1983, *Science* **221:**462–464.
58. Draznin, B., and Trowbridge, M., 1982, *J. Biol. Chem.* **257:**11988–11993.
59. Roth, R.A., Iwamoto, Y., Maddux, B., and Goldfine, I.D., 1983, *Endocrinology* **112:**2193–2199.
60. Hofmann, C.A., Lotan, R.M., Ku, W.W., and Oeltmann, T.N., 1983, *J. Biol. Chem.* **258:**11774–11779.
61. Purrello, F., Burnham, D.B., and Goldfine, I.D., 1983, *Proc. Natl. Acad. Sci. USA* **80:**1189–1193.

20

Hormonal Regulation of Phosphoenolpyruvate Carboxykinase Gene Expression

David S. Loose, Anthony Wynshaw-Boris, Herman M. Meisner, Yaacov Hod, and Richard W. Hanson

I. INTRODUCTION

Glucose homeostasis in all vertebrates involves the integration of a number of metabolic processes that are under hormonal control. A key element in this process is hepatic and renal gluconeogenesis, which provides the major source of blood glucose during starvation.[1] Tissues such as the red blood cells, brain, and kidney medulla are metabolically dependent on a continued source of glucose, a fact that underlines the importance of proper regulation of gluconeogenesis for the survival of the organism. It is not therefore surprising that a number of hormones, including insulin, glucagon, glucocorticoids, and thyroid hormone, all play a role in coordinating glucose homeostasis by controlling gluconeogenesis.

The focus of this chapter is on the hormonal regulation of a single key gluconeogenic enzyme, P-enolpyruvate carboxykinase (GTP)* (EC 4.1.1.32). This enzyme is the first committed step in gluconeogenesis and is the rate-limiting reaction in the pathway.[2] It has no known allosteric modifiers and responds to changes in metabolic flux over the gluconeogenic pathway by a rapid increase in enzyme protein.[1] A number of earlier studies [3–5] established that the increase in hepatic P-enolpyruvate carboxy-

---

* P-enolpyruvate is phosphoenolpyruvate.

David S. Loose, Anthony Wynshaw-Boris, Herman M. Meisner, Yaacov Hod, and Richard W. Hanson ● Department of Biochemistry, Case Western Reserve University School of Medicine, Cleveland, Ohio.

kinase activity noted during starvation or diabetes is due to a rapid increase in the synthesis rate of the enzyme. Feeding glucose to a rat starved for 48 hr causes a decrease in the rate of enzyme synthesis, with a half-life of 40 min.[4,5]

The major hormones that regulate the synthesis of hepatic P-enolpyruvate carboxykinase are glucagon, acting via cyclic adenosine monophosphate (cAMP); glucocorticoids, which induce the synthesis of the enzyme[3,6,7]; and insulin, which causes a rapid deinduction.[6,8,9] The mechanisms by which each of these hormones act on the enzyme are now being traced to the level of the gene, using the techniques of modern molecular biology. The rapidity of response of the gene to hormonal stimuli and its relatively high level of expression in liver and kidney[10,11] make the enzyme unique as a model for studying gene transcription. This chapter concentrates on the latest information available on the hormonal regulation of the P-enolpyruvate carboxykinase gene. More recent studies are emphasized in which direct measurements of enzyme gene expression have been made. By necessity, this review focuses on the mechanism of action of cAMP, since the effects of this compound on P-enolpyruvate carboxykinase gene expression are understood best at the present time. Wherever possible, we also discuss the important role of insulin in regulating gene expression for the enzyme.

II. BIOLOGY OF P-ENOLPYRUVATE CARBOXYKINASE

This section presents a brief review of the background information on the properties of P-enolpyruvate carboxykinase. The earlier studies on the properties of this enzyme have been extensively reviewed by Tilghman et al.[1] and by Utter and Kolenbrander[12]; the interested reader will find a more complete coverage of this aspect of the biology of P-enolpyruvate carboxykinase in these reviews.

A. Properties of the Enzyme

Figure 1 summarizes the properties of P-enolpyruvate carboxykinase from rat cytosol. Depending on the species, the enzyme is located in varying proportions in the cytosol and mitochondria.[1] In the rat, 90% of the total activity is cytosolic, whereas in the livers of most birds, the enzyme is completely mitochondrial.[1] However, most species studied, including human subjects, have both the mitochondrial and cytosolic forms of P-enolpyruvate carboxykinase.[1] The two forms of the enzyme have similar molecular weights and kinetic properties but are immunologically distinct proteins.[13] It is the cytosolic form of the enzyme that is so markedly inducible by hormones; this isozyme is the subject of the present review.

The enzyme is restricted in its tissue distribution to liver, kidney cortex, adipose tissue, and small intestine,[1] although very low activities have also been reported for lung[14] and muscle.[15] In general, the relative proportion of P-enolpyruvate carboxykinase in the cytosol and mitochondria is the same in all tissues containing the enzyme within a specific species. The chicken is an exception to this rule in that the liver contains only a mitochondrial form, with no detectable cytosolic enzyme, while the

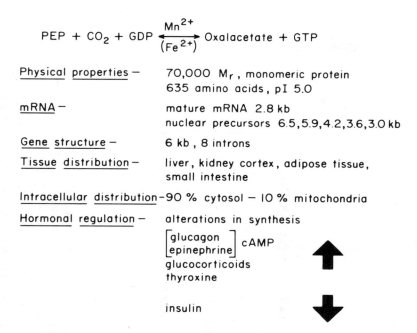

$$PEP + CO_2 + GDP \xrightleftharpoons[(Fe^{2+})]{Mn^{2+}} Oxalacetate + GTP$$

Physical properties —	70,000 M_r, monomeric protein 635 amino acids, pI 5.0
mRNA —	mature mRNA 2.8 kb nuclear precursors 6.5, 5.9, 4.2, 3.6, 3.0 kb
Gene structure —	6 kb, 8 introns
Tissue distribution —	liver, kidney cortex, adipose tissue, small intestine
Intracellular distribution —	90% cytosol — 10% mitochondria
Hormonal regulation —	alterations in synthesis

$\begin{bmatrix}\text{glucagon} \\ \text{epinephrine}\end{bmatrix}$ cAMP
glucocorticoids
thyroxine

insulin

Figure 1. Properties of P-enolpyruvate carboxykinase from rat liver cytosol.

kidney expresses both forms of the enzyme.[16] The physiological significance of both the intracellular and tissue distribution of P-enolpyruvate carboxykinase has been extensively reviewed[1,17]; space limitations do not permit us to deal with this aspect in the present chapter.

P-enolpyruvate carboxykinase is a monomeric protein of 70,000 M_r with no known allosteric modifiers. Although the amino acid sequence of the enzyme is not known, it does contain 635 amino acids, including 13 cysteines. The importance of these cysteines to the catalytic function of the enzyme has been emphasized by Colombo *et al.*[18] and Carlson *et al.*[19] who identified a pair of vicinal thiols, one of which is essential for catalytic activity but which can be readily oxidized to form a disulfide bridge.[19] The oxidized form of the enzyme is inactive. Lardy and associates[20,21] isolated a protein, termed ferroactivator, that activates P-enolpyruvate carboxykinase in the presence of Fe^{2+}. It has been suggested that Fe^{2+} may act as a physiological regulator of the enzyme and that factors that alter the intracellular flux of Fe^{2+} might control the activity of the enzyme.[22] Whether such a regulating mechanism involves hormones that regulate gluconeogenesis, such as insulin and glucagon, remains to be established.

B. Properties of the Gene and mRNA for P-Enolpyruvate Carboxykinase

The gene for cytosolic P-enolpyruvate carboxykinase from the rat has been isolated from a recombinant library, using as a probe a 2.6-kbase cDNA for the enzyme.[23]

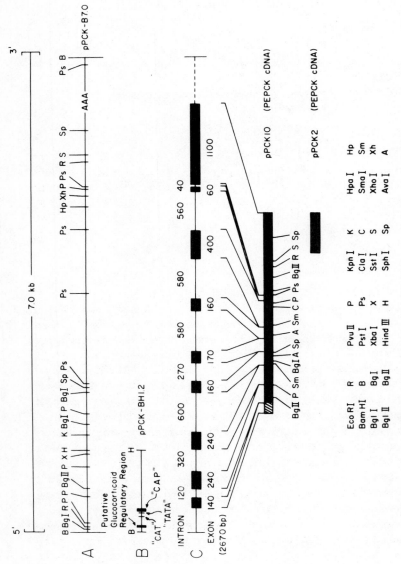

Figure 2. Structure of the gene for cytosolic P-enolpyruvate carboxykinase from the rat. (A) Detailed restriction map of the subclone pPCK-B7.0, which contains the entire P-enolpyruvate carboxykinase gene and 3′ and 5′-flanking sequences. (B) Positions of putative regulatory sequences in the genomic subclone pPCK-BH1.2. (C) Top line shows the structure of the P-enolpyruvate carboxykinase gene showing exons (■) and introns (connecting lines). Numbers indicate the average number of nucleotides in each exon and intron. The solid bars in the middle and bottom lines represent the 2.6-kbase cDNA (pPCK-10) and the 0.6-kbase cDNA (pPCK2), which have been aligned relative to the gene. The hatched segment is an area of cDNA for which no restriction sites have been identified in the genomic DNA. For details, see Yoo-Warren et al.[23]

The gene is approximately 6 kbase in length (coding for an mRNA of 2.8 kbase) and contains eight introns (Fig. 2). A control region at the 5'-end of the gene contained within a 1.2-kbase restriction fragment (Fig. 2B) was isolated and shown to contain the usual transcription start sequence and to have promoter activity as measured by its transcription in an *in vitro* assay system.[23] The largest exon in the gene is a 1.1 kbase segment at the 3'-end, which is colinear with a cDNA clone (pPCK2) containing the same region. Southern blot analysis of rat genomic DNA showed that the P-enolpyruvate carboxykinase gene is a single-copy gene, located entirely within a 7-kbase *BamH1* fragment.[23]

The mature mRNA for the enzyme is 2.8 kbase in length, whereas the 635 amino acids in rat cytosolic P-enolpyruvate carboxykinase require only 1.9 kbase of coding sequence.[24] There are also a number of nuclear RNA species, identified by hybridization techniques using a specific cDNA probe, that appear to be precursors of mature enzyme mRNA.[25] The largest of these species is 6.5 kbase in length (about the size of the P-enolpyruvate carboxykinase gene). The sequence abundance of these putative enzyme RNA precursors can be markedly altered by factors that alter the transcription rate of the gene. Preliminary experiments indicate that these nuclear RNA species are precursors of mature P-enolpyruvate carboxykinase, since they hybridize to segments of intervening sequences in the gene coding for the enzyme (M. Iliatis, D.K. Cameron, J. Short, and R.W. Hanson, unpublished results).

III. HORMONAL REGULATION OF P-ENOLPYRUVATE CARBOXYKINASE SYNTHESIS

This section briefly reviews the information currently available on the hormonal regulation of gene expression for P-enolpyruvate carboxykinase, stressing the role of glucagon (cAMP), insulin, and glucocorticoids, in this process. However, in order to fully evaluate the regulation of the P-enolpyruvate carboxykinase gene, it is necessary to first consider tissue differences in the response of the enzyme to hormones. Hepatic P-enolpyruvate carboxykinase responds rapidly to alteration in dietary carbohydrate, whereas the renal enzyme responds more slowly. For example, the administration of Bt_2cAMP to a glucose fed rat will cause a sixfold induction in the concentration of hepatic P-enolpyruvate carboxykinase mRNA within 60 min[11] but has only a marginal effect on the renal enzyme. Deinduction of the enzyme due to carbohydrate feeding is also more rapid in the liver than the kidney.[10] Alternatively, renal P-enolpyruvate carboxykinase is responsive to changes in the acid–base status of the animal, being induced by metabolic acidosis; the liver enzyme remains unchanged.[10] These differences in the pattern of enzyme induction in the two tissues reflects the physiological roles played by each in glucose homeostasis. Comparisons of the mechanisms responsible for P-enolpyruvate carboxykinase induction in liver and kidney may therefore provide some insights into the process of differentiation of cellular function in eukaryotes. We shall return to these tissue differences in the hormonal regulation of P-enolpyruvate carboxykinase in Section 4.

A. Regulation by cAMP

Early studies on the mechanism of cAMP induction of P-enolpyruvate carboxykinase from the cytosol of rat liver, using inhibitors such as actinomycin D, suggested that the cyclic nucleotide acts by stimulating transcription of the gene.[26,27] Recently, we provided direct evidence showing an effect of Bt_2cAMP on transcription of the P-enolpyruvate carboxykinase gene in isolated nuclei from rat liver.[25] In these experiments, both adrenalectomized and glucose-fed rats were used to ensure that Bt_2cAMP administration was not causing an increased release of glucocorticoids. Figure 3 shows the time course of changes in the rates of P-enolpyruvate carboxykinase gene transcription by isolated liver nuclei, as well as changes in the sequence abundance of nuclear and cytosolic RNA for the enzyme after the injection of Bt_2cAMP. The basal rate of P-enolpyruvate carboxykinase gene transcription, measured using liver nuclei from glucose-fed (adrenalectomized) rats, was about 0.05% of the total RNA synthesized. Within 20 min after the injection of Bt_2cAMP, the transcription rate increased sevenfold to 0.34%, but by 60 min the rate had declined to 0.18% (Fig. 3, hatched bars).

It is interesting to compare the kinetics of induction of P-enolpyruvate carboxykinase RNA in the nucleus and cytosol after Bt_2cAMP administration with the changes in the rate of gene transcription by isolated liver nuclei. The sequence abundance of nuclear RNA increases in parallel with the transcription rate of the P-enolpyruvate

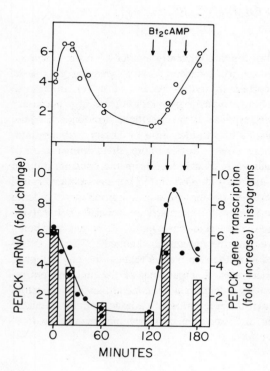

Figure 3. Time course of change in the rates of P-enolpyruvate carboxykinase gene transcription and in the sequence abundance of enzyme RNA in isolated rat liver nuclei and cytosol following glucose refeeding or after treatment with Bt_2cAMP. (●) Nuclear RNA; (○) cytosolic RNA. The procedures used to determine these parameters are given in detail in Lamers et al.[25]

carboxykinase gene (Fig. 3). There is even a similar overshoot in nuclear RNA levels 20 min after Bt$_2$cAMP injection (as was noted with the transcription rate). However, the response of cytosolic P-enolpyruvate carboxykinase mRNA to Bt$_2$cAMP was not as rapid as the rate of gene transcription, and the increase in enzyme mRNA in the cytosol was continuous.

There is also extensive literature pointing to an effect of cAMP at the level of mRNA translation. For example, Roper and Wicks[28] reported that the addition of Bt$_2$cAMP to cultured hepatoma cells accelerates the synthesis rate of elongation of tyrosine aminotransferase nascent chains six- to tenfold, while Snoek et al.[2] noted, using the same experimental system, that Bt$_2$cAMP doubled the number of enzyme nascent chains. By contrast, dexamethasone did not alter the translation of enzyme mRNA despite the fact that it elevated the activity of tyrosine aminotransferase five- to tenfold in hepatoma cells. There is indirect evidence that cAMP acts at a step in the translation of specific RNA for tyrosine aminotransferase and also for P-enolpyruvate carboxykinase.[6,30]

B. Regulation by Insulin

It is clear from a number of previous studies that insulin causes a marked reduction in the activity [31–33] synthesis rate[3,9] and mRNA levels[8] of rat cytosolic P-enolpyruvate carboxykinase. The rapid rate of deinduction of enzyme synthesis in hepatic cells in the absence of cAMP[6] has made it difficult to determine whether insulin alters the expression of the gene for P-enolpyruvate carboxykinase by a direct effect, or whether it acts indirectly via changes in the intracellular concentration of cAMP. That cAMP and insulin have mutually antagonistic effects on the synthesis of P-enolpyruvate carboxykinase was clearly established by Tilghman et al. in 1974.[3] These studies demonstrated that the administration of insulin (together with glucose) caused a rapid deinduction of enzyme synthesis, with a half-time of approximately 40 min. The same half-time was also noted for P-enolpyruvate carboxykinase in Reuber H35 cells when Bt$_2$cAMP was removed from the culture medium.[6] Other studies using Reuber H35 cells suggested a concentration-dependent interaction between the two hormones in regulating the synthesis of the enzyme.[9] What remains unresolved is the mechanism of action of the two hormones. An increasing number of experiments suggest that insulin decreases the synthesis of P-enolpyruvate carboxykinase by a mechanism independent of a simple decrease in intracellular cAMP. Recently, Beale et al.[34] used forskolin, a potent activator of adenylate cyclase, to increase the concentration of cAMP in H-4 hepatoma cells and noted that while insulin had no effect on the elevated level of cyclic nucleotide in the hepatoma cells, it markedly decreased the sequence abundance of P-enolpyruvate carboxykinase mRNA. This finding strongly suggests that insulin antagonizes the effect of cAMP on enzyme induction by a mechanism other than lowering the concentration of cAMP.

It is now established that insulin acts directly to decrease the transcription rate of the P-enolpyruvate carboxykinase gene. This was suggested in studies with the intact rat by Lamers et al.,[25] who demonstrated that refeeding glucose to starved rats caused a rapid decrease in the transcription rate of the gene for P-enolpyruvate car-

boxykinase as measured by nuclear runoff (see Fig. 2). The half-time for this deinduction of transcription is approximately 40 min, in close agreement with the change
observed in the sequence abundance of enzyme RNA found in the nuclei. Recently,
Granner et al.[35] demonstrated directly, using H4 hepatoma cells, that at concentrations
within the physiological range, insulin can specifically decrease the rate of transcription
of the gene for P-enolpyruvate carboxykinase, even in the presence of cAMP. These
workers also noted a rapid decline in the concentration of nuclear RNA transcripts for
the enzyme, which was followed by a proportionate decline in cytosolic P-enolpyruvate
carboxykinase mRNA and synthesis rate of the enzyme.

It would thus seem established that insulin and cAMP change the expression of
the gene for P-enolpyruvate carboxykinase by altering the transcription rate of the gene
for the enzyme. The mechanism by which these hormones act is unknown. (Section
5 reviews several approaches currently being used to study the transcription of the P-
enolpyruvate carboxykinase gene, both in vitro and in vivo.) We are also using gene
transfection techniques to study the sequence required for response to both cAMP and
insulin.

IV. TISSUE DIFFERENCES IN THE REGULATION OF P-
ENOLPYRUVATE CARBOXYKINASE GENE EXPRESSION

One interesting aspect of the biology of P-enolpyruvate carboxykinase is the tissue-
specific regulation of gene expression for the enzyme. The cytosolic form of the enzyme
in the chicken and rat provides excellent examples to illustrate these differences in
response to hormones. This section outlines the effects of cAMP, insulin, and glucocorticoids on the regulation of gene expression for cytosolic P-enolpyruvate carboxykinase from the liver and kidney cortex.

A. Tissue Differences in the Response of P-Enolpyruvate Carboxykinase
to cAMP

1. Chicken

P-enolpyruvate carboxykinase was first discovered in livers of chickens by Kurahashi and Utter[36] and, until the recent discovery of the cytosolic form of the enzyme
in the kidney,[16] the chicken was generally considered to have only a mitochondrial
enzyme. We have recently succeeded in cloning a cDNA to the chicken cytosolic P-
enolpyruvate carboxykinase[37] and have shown that the enzyme has limited homology
with the mitochondrial enzyme, despite the fact that the two proteins have similar
molecular weight. Also, mRNA for the cytosolic form of P-enolpyruvate carboxykinase
in the chicken kidney is inducible by starvation[37] and by the addition of Bt_2cAMP
to isolated kidney tubules in vitro. By contrast, the mitochondrial form of P-enolpyruvate carboxykinase in the liver is not inducible by starvation, and we could detect
no mRNA for the cytosolic form using a sensitive hybridization assay.[37] However,
the administration of Bt_2cAMP to the chicken causes the initial appearance of mRNA

for the cytosolic form of P-enolpyruvate carboxykinase in the liver.[37] This finding indicates that the gene for the cytosolic form of the enzyme can be expressed in the liver if the level of cAMP in the tissue is elevated by the injection of exogenous cyclic nucleotides. Studies on isolated chicken hepatocytes (M. Watford, personal communication) have also shown an induction of cytosolic P-enolpyruvate carboxykinase. The physiological role of the enzyme in chicken liver is not clear. It is possible that the normally high concentration of blood glucose in the chicken (250 mg%) that changes only marginally during starvation,[38] may contribute to the suppression of gene expression of the hepatic cytosolic enzyme, perhaps by inhibiting the secretion of glucagon. Since the proximal tubule cells of the kidney reabsorb and degrade insulin rapidly and completely in a nonsaturable manner[39] under normal physiological conditions, the insulin concentration may be inadequate to depress the expression of the gene for the cytosolic form of renal P-enolpyruvate carboxykinase.

2. Rat

In the rat, a qualititative difference between kidney and liver is observed in the effect of Bt$_2$cAMP. In liver, transcription and cytosolic mRNA levels are increased 8–10-fold,[25] while in kidney only a 1.5–3-fold elevation has been found.[40] Despite this large induction of enzyme mRNA, the synthesis rate of the enzyme and its subsequent activity are not increased to the same extent in either tissue. For example, the enzyme activity and synthesis rate in liver rises about 2-fold,[9,30] while in kidney only a 1.2–1.3-fold induction has been found.[10]

Thus, after stimulation by hormones, liver and kidney cortex differ markedly in the stoichiometric relationship between the rate of P-enolpyruvate carboxykinase gene expression and subsequent increase in sequence abundance of enzyme mRNA and the synthesis rate of P-enolpyruvate carboxykinase. This relationship is compared in Fig. 4 in the activity of P-enolpyruvate carboxykinase, its synthesis rate, the concentration of enzyme mRNA, and the transcription rate of the P-enolpyruvate carboxykinase gene in liver and kidney. The rate of transcription of the gene for cytosolic P-enolpyruvate carboxykinase in kidney, induced by starvation, is lower than noted in liver. However, the concentration of enzyme mRNA, as well as its synthesis rate and activity, are two- to threefold greater in the kidney.

This point can be illustrated by comparing the rates of transcription of the P-enolpyruvate carboxykinase gene in both tissues. In both liver and kidney, the basal transcription rate of the gene is 200–300 ppm.[25] Starvation or the administration of cAMP raises this level to 2000–3500 ppm in the liver, but only to 800–900 ppm in the kidney. By contrast, cytosolic mRNA concentrations in starved rats, as measured by a hybridization[24] or translation assay,[11] are two- to threefold higher in kidney than in liver. P-enolpyruvate carboxykinase activity, as well as its relative synthesis rate under the same conditions, is also threefold greater in kidney than in liver.[10] The 10-fold difference between the rates of transcriptional activity of the P-enolpyruvate carboxykinase gene in kidney and liver and the steady-state concentration of enzyme RNA and protein indicates that hormones may regulate the levels of P-enolpyruvate carboxykinase in the kidney at points other than transcription.

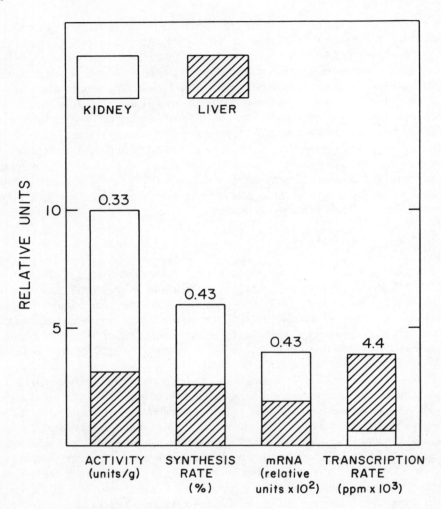

Figure 4. Comparison of aspects of gene expression for cytosolic P-enolpyruvate carboxykinase from rat liver and kidney. The activity of the enzyme, its rate of synthesis relative concentration of mRNA, and transcription rate for both liver and kidney are presented in appropriate units. Open bars show the relative activity for kidney P-enolpyruvate carboxykinase; hatched bars represent the hepatic enzyme. The number at the top of each bar is the ratio of the relative activity in liver as compared with the kidney for each process listed on the abscissa. For a more detailed treatment of these data, see Meisner *et al.*[40]

V. STUDIES OF THE MOLECULAR MECHANISMS OF HORMONAL AND TISSUE-SPECIFIC REGULATION OF P-ENOLPYRUVATE CARBOXYKINASE GENE EXPRESSION

Up to this point we have described studies that have determined the biological mechanisms by which hormones regulate the levels of P-enolpyruvate carboxykinase gene expression. There is considerable evidence that the expression of the P-enolpyruvate carboxykinase gene is regulated by hormones at the level of transcription. The tissue-specific expression of this gene also appears to occur at the level of transcription. A major emphasis in our laboratory has been to investigate the molecular mechanisms of hormonal and tissue-specific regulation of P-enolpyruvate carboxykinase gene expression.

This section describes our most recent studies of these molecular mechanisms, beginning with a discussion of our investigation into the sequence requirements for hormonal regulation and tissue-specific expression of rat cytosolic P-enolpyruvate carboxykinase. Having succeeded in identifying a region of 400 base pair (bp) within the gene required for induction by cAMP, we are in the process of defining sequences that might be responsible for the tissue-specific expression of this gene. This discussion is followed by an outline of *in vitro* studies that will identify the hormonal and tissue-specific transcriptional regulatory factors interacting with sequences in this gene to modulate its expression.

A. Sequence Requirements for Hormonal Regulation

Our approach has been to transfect either the entire P-enolpyruvate carboxykinase gene or fragments of this gene into cells by DNA-mediated gene transfer. We have used cell lines shown to be sensitive to one or more of the known regulators of hepatic P-enolpyruvate carboxykinase gene expression: glucocorticoids (induction), cAMP (induction), or insulin (deinduction). In the process of identifying a suitable cell system in which to study the hormonal regulaton of transfected genes, we were able to define, *in vivo*, the promoter for this gene. After transfection of the vectors described below into thymidine kinase-deficient rat hepatoma cells (T. Gross Lugo and R.F.K. Fournier, unpublished observations), we have also been able to identify a region at the 5'-end of the P-enolpyruvate carboxykinase gene required for cAMP inducton.

1. Vector Construction

The P-enolpyruvate carboxykinase gene constructs we have used for all these vectors are shown in Fig. 5. A BamH1 restriction fragment of 7 kbase (B7.0) was cloned intact into transfection vectors. Alternatively, a 600-bp *Bam*H1-*Bgl*II fragment at the 5' end of B7.0 (Fig. 5) was used in the construction of chimeric genes. These fragments contain the P-enolpyruvate carboxykinase promoter as well as sequences required for cAMP induction of gene expression. The parent vectors used in our transfection studies were generously provided by Dr. Richard Flavell.[41] pOPF contains the herpes simplex virus thymidine kinase gene (TK) together with the promoter, (Fig. 5A), while pTM is identical to pOPF, except that the aminoglycoside phosphotransferase gene (AGPT) has been inserted in the TK gene of pOPF after the removal of TK coding sequences. The AGPT gene confers resistance to the antibiotic, G418, which is toxic to all cells.

The entire B7.0 fragment was ligated into pOPF or pTM at a unique *Bam*H1 restriction site to give vectors 3F and 9G, respectively (Fig. 5B). These vectors can be introduced into any cell line that does not produce P-enolpyruvate carboxykinase. However, it would be advantageous to study the regulation of transfected P-enolpy-

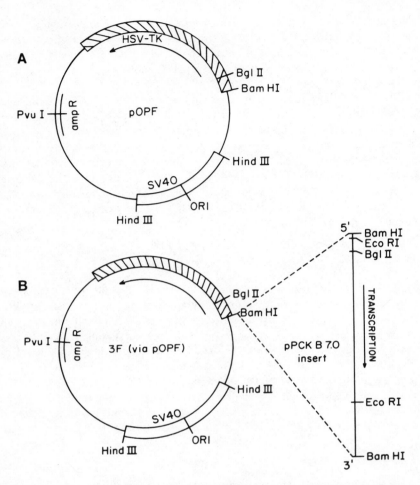

Figure 5. Vectors constructed with the P-enolpyruvate carboxykinase gene and TK or AGPT genes. (A) pOPF is a pBR327-based vector that contains the herpes simplex virus (HSV) thymidine kinase gene (TK), as well as a 1.1-kbase *Hind*III fragment of SV40 containing the origin of replication, early and late promoters, as well as the 72 base pair enhancer repeats. The arrow indicates the direction of transcription of HSV-TK. pTM is identical to pOPF, except that the structural sequences for the aminoglycoside 3′-phospho-transferase genes (AGPT) replaces the sequence for the TK structural gene. (B) 3F was constructed by ligating the *Bam*H1 7.0-kbase subcloned fragment into the unique *Bam*H1 site of pOPF. The directions of transcription of HSV-TK and P-enolpyruvate carboxykinase are indicated by arrows. AG is identical to 3F, except that the TK structural sequences are replaced by the AGPT structural sequences. (C) 6A is a P-enolpyruvate carboxykinase-TK fusion gene constructed by ligating the *Bam*H1-*Bgl*II 600-bp fragment containing the P-enolpyruvate carboxykinase promoter and 100 base pairs of 5′ untranslated sequence to the *Bgl*II site of the TK structural gene, which is missing the TK promoter. This links the P-enolpyruvate carboxykinase promoter to the TK structural gene in the proper orientation for transcription (see arrows). 12R is identical to 6A, except that the TK structural sequences are replaced by the AGPT structural sequences. (D) 6B is a P-enolpyruvate carboxykinase-TK fusion gene, as in 6A, except that the P-enolpyruvate carboxykinase promoter is opposite to the TK structural gene relative to transcription (see arrows). 12C is identical to 6B, except that the TK structural sequences are replaced by the AGPT structural sequences.

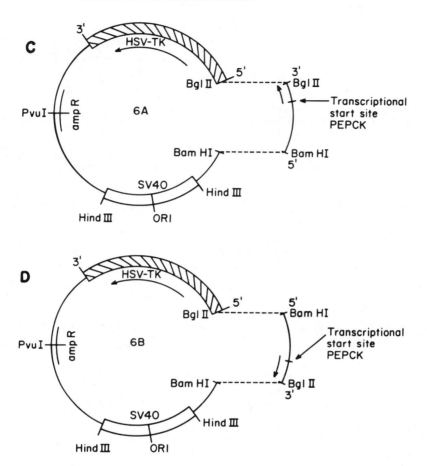

Figure 5. (continued)

ruvate carboxykinase genes in cells that express this gene in a hormonally regulated fashion. This approach permits comparison of hormonal regulation of the endogenous gene with the transfected gene. The problem with this approach for studying the expression of the P-enolpyruvate carboxykinase gene is that the product of the transfected gene in 3F or 9G is identical to the endogenous gene product (see below). Therefore, we also constructed chimeric genes containing the start site of transcription and 400 bp of 5'-flanking sequence of the P-enolpyruvate carboxykinase gene (the BamH1-BglII 600-bp fragment mentioned above) ligated to the TK or AGPT promoters, respectively. We were able to select vectors that had the P-enolpyruvate carboxykinase promoter in the proper orientation relative to transcription of the TK gene (6A, Fig. 5) or the AGPT gene (12R), as well as vectors that contained the putative promoter in the opposite orientation relative to the transcription of TK (6B, Fig. 5) or AGPT (12C).

The vectors containing the chimeric genes can be introduced in all cell lines, including those that express the P-enolpyruvate carboxykinase gene. In these cells, the regulation of transfected gene can be compared with the regulation of endogenous genes. Since they also contain selectable markers expressed via the P-enolpyruvate carboxykinase promoter, cells can be selected in appropriate media that have active P-enolpyruvate carboxykinase promoters. Expression of TK vectors in cells deficient in thymidine kinase can be monitored by selection in HAT medium[42]; expression of the AGPT vectors in all cells can be mediated by selection in G418-containing medium.

2. Identification of the Promoter of the P-Enolpyruvate Carboxykinase Gene by Gene Transfection

The promoter* of the rat P-enolpyruvate carboxykinase gene was identified by DNA-mediated gene transfer of the above vectors into a variety of cell types, including COS cells, human HeLa cells, and mouse L cells.

COS cells are a permanent cell line selected by Gluzman.[43] They were derived from African green monkey kidney cells which have incorporated into their genome simian virus 40 (SV40) virus with a defective origin of replication. COS cells produce SV40 T antigen constitutively and are permissive for SV40 infection. Thus, they provide an ideal host for DNA vectors containing an SV40 origin of replication. Such vectors will be replicated in COS cells to high copy number and COS cells thus provide a rapid means (48 hr as compared with 6 weeks in other systems) by which one can assay promoter functions.

COS cells were transfected with the TK vector outlined above by the calcium phosphate precipitation method of Graham and van der Eb,[44] as modified by Wigler et al.,[45] followed by a short glycerol shock.[46] The TK vectors contain an SV40 origin of replication and can therefore be amplified and expressed at high levels in COS cells. It is clear from Fig. 6 that COS cells transfected with 3F (containing an intact P-enolpyruvate carboxykinase gene) produce an RNA species hybridizing with the P-enolpyruvate caboxykinase cDNA (lanes 5 and 6) and identical in size to mature liver P-enolpyruvate carboxykinase mRNA (lane 7). As controls, COS cells transfected with salmon sperm DNA or pOPF (which is identical to 3F except that it does not contain the B7.0 fragment) do not produce any RNA species that hybridize to the P-enolpyruvate carboxykinase cDNA (lanes 1–4). The transfected gene in 3F produces as much P-enolpyruvate carboxykinase mRNA in COS cells as does the endogenous gene in maximally induced rat liver.

When COS cells were transfected with 9G containing the entire P-enolpyruvate carboxykinase gene, again only a mature-sized transcript hybridized to a cDNA probe to the enzyme (data not shown). The amount of this RNA species is much reduced at 48 hr, presumably because 9G does not contain an SV40 origin of replication and hence cannot be amplified in COS cells. HeLa cells were transfected with 3F, and similar amounts of mature-sized P-enolpyruvate carboxykinase mRNA were produced

---

*We define promoter as those sequences required for the proper initiation of gene transcription.

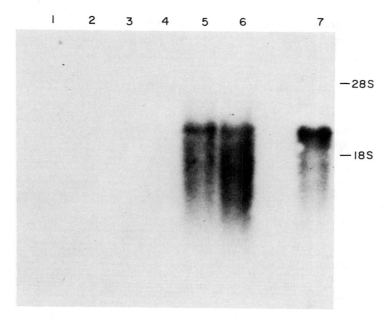

Figure 6. Analysis of RNA in COS cells transfected with various vectors. COS cells were transfected with appropriate vectors and, after 48 hr, the RNA was isolated from the cells. Twenty µg of total cellular RNA was separated on 1% agarose gels, with formaldehyde as the denaturing agent, transferred to nitrocellulose, hybridized to nick translated P-enolpyruvate carboxykinase cDNA, washed, and exposed to x-ray film at −70°C for 20 hr with intensifying screens. Lanes 1–7 contain RNA isolated from cells transfected with the following DNA: lanes 1 and 2, salmon sperm DNA; lanes 3 and 4, pOPF; lanes 5 and 6, 3F. Lane 7 contains 20 µg of total cellular RNA from maximally induced rat liver. A detailed description of the construction of each of these vectors is given in the legend to Fig. 5.

after 48 hr as were produced in COS cells transfected with 9G. The reasons for this low level of transcription in HeLa cells are unknown. HeLa cells also consistently express genes from SV40-containing vectors, although they are semipermissive with respect to SV40 infection. The above results in COS and HeLa cells indicate that the B7.0 fragment contains all the information needed by the cellular transcription and processing machinery to produce mature rat P-enolpyruvate carboxykinase mRNA, including the promoter for this gene.

The promoter region could be localized more precisely by using the chimeric gene construct outlined in Fig. 5. We took advantage of the fact that these chimeric genes produce selectable gene products and utilized the following strategy: Cells were transfected with chimeric genes or intact genes and exposed to selective medium; the ability of genes to protect cells from the killing effect of selective medium was taken as evidence of promoter activity and was scored as colony formation.

Mouse thymidine kinase-deficient L cells were transfected with varying amounts of either the TK-containing vectors (pOPF, 6A, 6B) or the AGPT-containing vectors

Table 1. Transfection Efficiency of Vectors Constructed with
the P-Enol + Pyruvate Carboxy + Kinase Gene and TK or
AGPT Genes[a,b]

Vector	μg DNA/plate	Colonies/plate
Salmon sperm DNA[c]	20	0
pOPF	0.5	970
6A	0.5	115
6A	0.5	115
6B	1.0	0
Salmon sperm DNA[b]	20	0
pTM	0.5	610
12R	0.5	85
12R	0.5	80
12C	1.0	0

[a] Mouse L cells were transfected with the TK-containing vectors (Fig. 5A). Twenty-four hr later, cells were changed to HAT medium; after 3 weeks, plates were stained and colonies counted (average of three 100-mm plates).
[b] Mouse L cells were transfected with AGPT-containing vectors (Fig. 5B) changed to 200 μg/ml G418, stained, and counted after 3 weeks (average of three 100-mm plates). A description of each vector can be found in the legend to Fig. 5.

(pTM, 12R, 12C) and selected in HAT medium or G418-containing medium, respectively (Table 1). No colonies were found in plates transfected with salmon sperm DNA in either HAT medium or G418-containing media. Conversely, many colonies were found in plates transfected with intact TK (pOPF) or AGPT (pTM) genes and selected in appropriate medium. Cells transfected with chimeric P-enolpyruvate carboxykinase TK genes (6A) or AGPT genes (12R) in which the promoter and structural genes are aligned in the correct orientation relative to transcription of the TK gene were also protected using selective medium, but at only 10% of the level of the intact genes. If cells were transfected with chimeric genes in which the P-enolpyruvate carboxykinase promoter was aligned in the opposite orientation relative to the transcription of the TK (6B) or AGPT (12C) genes, no cells were protected.

These experiments localize the P-enolpyruvate carboxykinase promoter to a fragment of DNA 600 bp in length. The directionality of the promoter was also confirmed. It is not clear to us why the chimeric genes protected only 10% as many cells as the intact genes, but further experiments are in progress to investigate this phenomenon.

3. Identification of a cAMP-Responsive Region of the P-Enolpyruvate Carboxykinase Gene

We have tested the ability of intact and chimeric P-enolpyruvate carboxykinase genes to respond to the positive modulators cAMP and dexamethasone after transfecton into COS and HeLa cells. We have been unable to detect hormonal regulation of these genes in either of these heterologous cell lines. Further studies have therefore employed a homologous rat hepatoma cell line that is TK deficient. Using this cell line, we have demonstrated cAMP regulation of the transfected chimeric genes (A. Wynshaw-Boris, T. Gross Lugo, J.M. Short, R.E.K. Fournier, and R.W. Hanson, unpublished results).

These experiments have been done in collaboration with T. Gross Lugo and R.E.K. Fournier at the University of Southern California, who developed these cells. We transfected the TK-containing vectors into FTO-2B TK-deficient rat hepatoma cells and selected for transformation in HAT medium. Cells transfected with pOPF had 10-fold more colonies per plate than did 6A- or 6H-transfected cells; 6B afforded no protection to these cells in HAT medium.

Mass cultures of transfectants (nonclonal) containing each of the TK vectors were treated with Bt$_2$cAMP (500 μM) and theophylline (1 mM) for 16 hr; cellular extracts were isolated and TK activity was measured.[47] cAMP treatment induced the thymidine kinase activity of cells transfected with the chimeric P-enolpyruvate carboxykinase-TK genes 1.9–3.3-fold over untreated cells. Cells transfected with intact TK genes were not induced by cAMP treatment. This finding compares well with the threefold induction of endogenous P-enolpyruvate carboxykinase activity in cells treated with dibutyryl cAMP and theophylline for 16 hr.

These experiments were repeated in clonally isolated cell lines. Again, we found cAMP induction of TK activity only in clones transfected with the chimeric genes. The levels of induction were two- to fourfold under the experimental conditions described above. Clones isolated from cells transfected with intact TK genes were not sensitive to cAMP induction. It is interesting to note that the basal TK activities in the cells transfected with the chimeric genes were usually greater than those in cells transfected with the intact TK genes, in spite of the fact that only 10% of the cells were protected in HAT medium by cells transfected with the chimeric genes as compared with the intact TK genes. We can think of no obvious reason for these observations, but we hope to address them in future experiments.

4. Further Studies to Identify Hormonally Responsive Regions of the Gene

We are in the process of determining whether the chimeric genes described above contain sequences responsive to glucocorticoids and insulin. We will construct chimeric genes containing more 5'-flanking sequences if these sequences are not present in the vectors described. We plan to further define the mRNA produced in these cells in terms of sequence abundance and fidelity of initiation. We would also like to determine whether the hormonally sensitive transfected genes are transcriptionally regulated. The major difficulty with these studies is that the mRNA produced from the chimeric genes is at least 20-fold less abundant than transcripts from the endogenous gene.[48] It is possible that we are missing key regulatory sequences that might be necessary for greater expression of chimeric genes. New vectors are being constructed to address these questions.

We are also constructing fine deletion mutants of the 600 bp BamH1-BglII promoter fragment in order to identify precisely the sequences necessary for hormonal regulation. We may then be able to determine whether the various sequences work independently to regulate P-enolpyruvate carboxykinase gene expression or whether there is some degree of interaction or overlap of hormonally responsive sequences.

B. Sequences Responsible for the Tissue-Specific Expression of Rat Cytosolic P-Enolpyruvate Carboxykinase

We have already described the tissue-specific pattern of expression of the genes for P-enolpyruvate carboxykinase in the rat and chicken. The study of tissue-specific regulation must be carried out in the whole animal. The sequence requirements for tissue-specific regulation can be approached as the study of hormonally regulated sequences above, i.e., via gene transfer. However, the introduction of genes into whole animals presents unique problems. We have decided on two approaches to these problems. We are introducing genes into fertilized mouse eggs via microinjection to study the tissue-specific expression of such genes when they go through the entire developmental program. We are also attempting to introduce genes into adult animals by using viral transducing particles in order to study the tissue-specific expression of genes that have not gone through the developmental program. The questions we hope to answer initially using these approaches are as follows: Are there sequences in genes that specify tissue-specific expression? Are these sequences responsive at all times in an animal's life, or must they be present during development?

As these experiments are currently in their initial stages, they are not described in detail here.

C. Approaches for Investigation of Transcriptional Regulatory Factors

1. Regulation of P-Enolpyruvate Carboxykinase Transcription in Vitro

A fundamental requirement for investigation into mechanisms of transcriptional regulation is a completely *in vitro* system that is capable of responding to exogenously added factors. Such a system will be used to identify cellular factors responsible for transcriptional regulation by cAMP and to examine the effects of glucocorticoid receptor complexes on P-enolpyruvate carboxykinase mRNA synthesis. The followng sections describe some of the current assays of gene transcription, as well as their drawbacks and their use in determining the mechanism of hormonal regulation of gene expression.

2. Isolated Nuclear Runoff Assay

Studies using nuclei isolated from rats administered Bt_2cAMP and glucocorticoids[25] T3 and epinephrine (D.S. Loose, K. Cameron, R.W. Hanson, unpublished results) have shown that these hormones change the transcriptional activity of the P-enolpyruvate carboxykinase gene. Recently, we have found that cAMP, as well as the catalytic and the regulatory subunit of protein kinase are without effect on P-enolpyruvate carboxykinase transcription when added directly to nuclei isolated from control rats. This may reflect an inadequacy in our assay for the measurement of gene transcription or protein kinase may not directly affect isolated nuclei. The nuclear runoff assay initiates transcription poorly[48] and in fact reflects only the number of active polymerase molecules bound to template DNA at the time the nuclei are isolated from the liver.

Even with this drawback, there have been some reports of *in vitro* regulation of gene transcription in isolated nuclei. Teyssot *et al.*[49] reported that fractions of a cellular extract from prolactin-treated tissues contains a factor of ~5000 M_r that specifically increased casein gene transcription when added to isolated nuclei. Furthermore, only cellular fractions from prolactin-sensitive tissues were able to elicit the *in vitro* response. Another disadvantage of using nuclei for transcription studies is that the entire genome, as opposed to a single specific gene, is available as a template. Using highly purified estrogen receptors added to isolated nuclei, Taylor and co-workers,[50] measured *in vitro* alterations in ovalbumin synthesis. While an induction in specific ovalbumin synthesis was noted, a significant part of the change could be explained by a nonspecific increase in overall RNA synthesis.

Current efforts in our laboratory using intact nuclei are centered on increasing the detection of the mRNA chains initiated *in vitro*. One approach is to employ a hybridization probe taken from the region just 3' of the starting site of transcription. This is especially powerful when combined with a transcription system employing α- or β-sulfhydryl deoxynucleotides. These modified nucleotides are incorporated with intact sulfhydryl moieties present only on the 5'-end of the nascent RNA chain. Newly initiated RNA chains are separated from other RNA species by passage through a Hg-Sepharose column and quantitated by hybridization to a P-enolpyruvate carboxykinase-specific [^{32}P]-cDNA. This technique has been successfully employed to assess changes in the initiation of transcription of mouse mammary tumor virus.[51]

3. In Vitro Transcription Using Purified DNA Templates

An advantage of using isolated intact nuclei to study transcription is integrity of nuclear structure. The major disadvantage is that the transcriptional activity of the entire complex rat genome is also being measured. An alternative method is the use of an *in vitro* transcription system described by Weil *et al.*[52] and Manley *et al.*[53] to circumvent two problems found with isolated intact nuclei since: (1) the DNA template for transcription is unique and purified, and (2) the assay readily permits quantitation of the initiation of transcription. This assay uses a crude cellular extract, generally prepared from HeLa cells, to supply soluble factors (such as the necessary RNA polymerase) and transcribes added genomic fragments. To date, we have used this assay to identify the 5'-flanking sequences of the P-enolpyruvate carboxykinase gene, which exhibits promoter activity.[23] Using a 1.2-kbase *Bam*H1-*Hind*III fragment present at the 5'-end of the P-enolpyruvate carboxykinase gene (Fig. 2), we were able to map the promoter to a position approximately 100 nucleotides 5' of a unique *Bgl*II site. Using fine deletions in this area of the gene, we are currently mapping sequences important for regulation of transcription *in vitro*. One important limitation of this system is its inability to respond to regulatory factors such as cAMP or steroid-receptor complexes. To date, only in viral systems employing SV40 T antigen has an *in vitro* transcription assay proved useful for studying regulatory mechanisms. As demonstrated by Myers *et al.*,[54] T antigen of SV40 binds to specific regions of the viral genome and modulates viral transcription in a completely *in vitro* system. Using a somewhat different approach, recent studies by Davison *et al.*[55] examined the formation of stable initiation complexes using a combination of an *in vitro* transcription assay and a DNA binding assay. Preincubation of a cellular extract with fragments of the ovalbumin or conalbumin gene before addition of a second transcriptional template permitted indirect measurements of events occurring on the first DNA segment. Factors necessary for transcription are bound to DNA during the first incubation and are unavailable for transcription of the second template. This type of system should permit identification of transcriptional regulators necessary for *in vitro* transcripton and may reveal regulatory factors. Regulating the transcription of eukaryotic genes in any *in vitro* system may be difficult due to a lack of either chromatin structure or of tissue-specific factors not supplied by the heterologous HeLa cell extracts added to the assay. Merling *et al.*[56] have shown extracts made from chicken fibroblasts function well in *in vitro* transcription of chicken collagenase genes. Thus, an entirely homologous *in vitro* system could be used to examine tissue-specific transcriptional regulators. Further studies with such homologous systems, e.g., liver or kidney cell extracts, and the P-enolpyruvate carboxykinase gene are required to determine whether this system will be useful in identifying regulatory tissue-specific factors, or both.

4. Construction of Minichromosomes for Transcriptional Studies

An additional approach in our laboratory has been the construction of minichromosomes that contain segments of the P-enolpyruvate carboxykinase gene. Using a bovine papilloma virus (BPV) vector obtained from P. Howley, National Cancer Institutes, we have inserted into the virus either the entire gene or a 2.2-kbase minigene consisting of the 5' and 3' portions of the P-enolpyruvate carboxykinase gene. Mouse fibroblasts transfected with this virus carry the recombinant vector as a stable episomal element.[57] It is important to note that these episomal minichromosomes have not lost higher-order structure. Thus, they overcome a major limitation of *in vitro* transcription systems, which usually use linearized, naked DNA. The DNA found in these episomes is completely defined and, in certain cases, synthesizes mRNA in a hormonally regulated manner.[58] In studies that are currently just under way, these minichromosomes will be used to characterize tissue-specific factors involved in the action of hormones on the transcription of P-enolpyruvate carboxykinase.

VI. CONCLUSIONS

The mechanisms by which cAMP and insulin regulate gene expression in eukaryote cells are not understood. From a variety of experiments chiefly based on indirect *in vivo* measurements of gene expression, it would appear that the ability of both hormones to alter gene transcription has been established. However, to study the intermediate steps in the action of cAMP or insulin at the level of the gene will require the development of new *in vitro* assay systems capable of responding to purified factors. We will soon know which sequences in the 5'-regulatory regions of the P-enolpyruvate carboxykinase gene are required for response to cAMP or insulin, but we will have far more difficulty in identifying the factors responsible for regulating the expression of the gene. What is required is an *in vitro* system that accurately transcribes well-characterized purified genes and that is also responsive to regulation by intermediates in the action of hormones—in short, a hormonally regulated transcription system.

This chapter has outlined the transcription systems now in general use and suggested new approaches currently being employed in our laboratory to study the effects of cAMP and insulin on the regulation of P-enolpyruvate carboxykinase gene expression. It is probable that work on the hormonal regulation of this enzyme and on other proteins that respond to cAMP and insulin will greatly benefit from the rapid rate of progress in molecular biology and that many of the questions posed in this chapter will be answered.

ACKNOWLEDGMENTS. This work was supported by grants AM-21859, AM-25541, and ES01170 from the National Institutes of Health and by a grant from the Kroc Foundation.

VII. REFERENCES

1. Tilghman, S.M., Ballard, F.J., and Hanson, R.W., 1976, in *Gluconeogenesis: Its Regulaton in Mammalian Species* (R.W. Hanson and M.A. Mehlman, eds.), pp. 47–87, Wiley, New York.
2. Rognstad, R., 1979, *J. Biol. Chem.* **254**:1875–1878.
3. Tilghman, S.M., Hanson, R.W., Reshef, L., Hopgood, M.F., and Ballard, F.J., 1974, *Proc. Natl. Acad. Sci. USA* **71**:1304–1308.
4. Hopgood, M.F., Ballard, F.J., Reshef, L., and Hanson, R.W., 1973, *Biochem. J.* **134**:445–453.
5. Kioussis, D., Reshef, L., Cohen, H., Tilghman, S., Iynedjian, P.B., Ballard, F.J., and Hanson, R.W., 1978, *J. Biol. Chem.* **253**:4327–4332.
6. Tilghman, S.M., Gunn, J.M., Fisher, L., Hanson, R.W., Reshef, L., and Ballard, F.J., 1975, *J. Biol. Chem.* **350**:3322–3329.
7. Gunn, J.M., Hanson, R.W., Meynhac, O., Reshef, L., and Ballard, F.J., 1975, *Biochem. J.* **150**:195–203.
8. Andreone, T.L., Beale, E.G., Bar, R.S., and Granner, D.K., 1982, *J. Biol. Chem.* **257**:35–38.
9. Gunn, J.B., Tilghman, S.M., Hanson, R.W., Reshef, L., and Ballard, F.J., 1975, *Biochemistry* **14**:2350–2351.
10. Iynedjian, P.B., Ballard, F.J., and Hanson, R.W., 1975, *J. Biol. Chem.* **250**: 5596–5603.
11. Iynedjian, P.B., and Hanson, R.W., 1977, *J. Biol. Chem.* **252**:655–662.
12. Utter, M.F., and Kolenbrander, H.M., 1972, in *The Enzymes* (P.D. Bayer, ed.), Vol. 6, pp. 136–154, Academic Press, New York.
13. Ballard, F.J., and Hanson, R.W., 1969, *J. Biol. Chem.* **244**:5625–5630.
14. Hanson, R.W., and Garber, A.J., 1972, *Am. J. Clin. Nutr.* **25**:1010–1021.
15. Duff, D.A., and Snell, K., 1982, *Biochem. J.* **206**:147–152.
16. Watford, M., Hod, Y., Utter, M.F., and Hanson, R.W., 1981, *J. Biol. Chem.* **256**:10023–10027.
17. Soling, H-B., and Kleineke, J., 1976, in *Gluconeogenesis: Its Regulation in Mammalian Species* (R.W. Hanson and M.A. Mehlman, eds.), pp. 369–462, Wiley–Interscience, New York.
18. Colombo, G., Carlson, G.M., and Lardy, H.A., 1978, *Biochemistry* **17**:5321–5329.
19. Carlson, G.M., Colombo, G., and Lardy, H.A., 1978, *Biochemistry* **17**:5329–5338.
20. Bentle, L.A., Snoke, R.E., and Lardy, 1976, *J. Biol. Chem.* **251**:2922–2928.
21. Bentle, L.A., and Lardy, H.A., 1977, *J. Biol. Chem.* **252**:1431–1440.
22. Merryfield, M.L., and Lardy, H.A., 1982, *J. Biol. Chem.* **257**:3628–3635.
23. Yoo-Warren, H., Monahan, J.E., Short, J., Short, H., Bruzel, A., Wynshaw-Boris, A., Meisner, H., Samols, D., and Hanson, R.W., 1983, *Proc. Natl. Acad. Sci USA* **80**:3656–3660.

24. Cimbala, M.A., Lamers, W., Nelson, K., Monahan, J.E., Yoo-Warren, H., and Hanson, R.W., 1982, *J. Biol. Chem.* **257:**7629–7636.
25. Lamers, W.H., Hanson, R.W., and Meisner, H.M., 1982, *Proc. Natl. Acad. Sci. USA* **79:**5137–5141.
26. Yeung, D., and Oliver, I.T., 1968, *Biochemistry* **7:**3231–3239.
27. Yeung, D., and Oliver, I.T., 1968, *Biochem. J.* **108:**325–331.
28. Roper, M.D., and Wicks, W.D., 1978, *Proc. Natl. Acad. Sci USA* **75:**140–144.
29. Snoek, G.T., Van de Poll, K.W., Voorma, H.O., and Van Wijk, R., 1981, *Eur. J. Biochem* **114:**27–31.
30. Wicks, W.D., 1971, *Ann. NY Acad. Sci.* **185:**152–165.
31. Strago, E., Lardy, H.A., Nordlie, R.C., and Foster, D.O., 1963, *J. Biol. Chem.* **238:**3188–3192.
32. Foster, D.O., Ray, P.D., and Lardy, H.A., 1964, *Biochemistry* **5:**555–562.
33. Sharp, E., Young, J.W., and Lardy, H.A., 1967, *Science* **158:**1572–1573.
34. Beale, E., Brotherton, A., and Granner, D., 1983, *Endocrin. Soc. Abst.*, p. 271.
35. Granner, D., Andreone, T., Sasaki, K., and Beale, E., 1983, *Nature* **305:**549–551.
36. Utter, M.F., and Kurahashi, K., 1973, *J. Am. Clin. Soc.* **75:**758–762.
37. Hod, Y., Morris, S., and Hanson, R.W., 1984, *J. Biol. Chem.* (in press).
38. Simon, J., and Rosselin, G., 1978, *Horm. Metab. Res.* **10:**93–98.
39. Chamberlain, M., and Stimmler, L., 1967, *J. Clin. Invest.* **46:**911–919.
40. Meisner, H.M., Loose, D.S., and Hanson, R.W., 1984, *Biochemistry* (in press).
41. Grosreld, F.G., Lund, T., Murray, E.J., Mellor, A.L., Dahl, H.H.M., and Flavell, R.A., 1982, *Nucleic Acids Res.* **10:**6715–6732.
42. Szybalska, E.H., and Szybalski, W., 1962, *Proc. Natl. Acad. Sci. USA* **48:**2026–2034.
43. Gluzman, Y., 1981, *Cell* **23:**175–182.
44. Graham, F.L., and van der Eb, A.J., 1973, *Virology* **52:**456–469.
45. Wigler, M., Sweet, R., Sim, G.R., Wold, B., Pellicer, A., Lacy, E., Maniatis, T., Silverstein, S., and Axel, R., 1979, *Cell* **16:**775–785.
46. Parker, B.A., and Stark, G.R., 1979, *J. Virol.* **31:**360–369.
47. Brinster, R.L., Chen, H.Y., Trumbauer, M., Senear, A.W., Warren, R., and Palmiter, R., 1981, *Cell* **27:**223–231.
48. Wynshow-Boris, A., Lugo, T.G., Short, J.M., Fournier, R.E.K., and Hanson, R.W., 1984, *J. Biol. Chem.*, **259:**12161–12170.
49. MaClean, N., and Gregory, S.P., 1981, *The Cell Nuclease* (H. Busch, ed.), Vol. 8, pp. 139–191, Academic Press, New York.
50. Teyssot, B., Houdebine, L.M., and Dijane, J., 1981, *Proc. Natl. Acad. Sci. USA* **78:**6729–6733.
51. Taylor, R.W., and Smith, R.C., 1982, *Biochemistry* **21:**1781–1787.
52. Stallcup, M.R., and Washington, L.D., 1983, *J. Biol. Chem.* **258:**2802–2807.
53. Weil, P.A., Segall, J., Harris, B., Ng, S-Y., and Roeder, R.G., 1979, *J. Biol. Chem.* **254:**6163–6173.
54. Manley, J.L., Fire, A., Cano, A., Sharp, P.A., and Gefter, M.L., 1980, *Proc. Natl. Acad. Sci. USA* **77:**3855–3859.
55. Myers, R.M., Rio, D.C., Robbins, A.K., and Tijan, R., 1981, *Cell* **25:**373–384.
56. Davison, B.L., Egly, J.M., Mulvihill, E.R., and Chambon, P., 1983, *Nature* **30:**680–686.
57. Mereling, G.T., Tyagi, T.S., deCrombrugglie, B., and Pastan, I., 1982, *J. Biol. Chem.* **257:**7254–7261.
58. Law, M-F., Lowy, D., Dvoretzfy, J., and Howley, P., 1981, *Proc. Natl. Acad. Sci. USA* **78:**2727–2731.
59. Ostrowski, M.C., Richard-Foy, H., Wolford, R.G., Berard, D.S., and Hager, G.L., 1983, *Mol. Cell Biol.* **3:**2045–2057.

21

Hormonal Regulation of the Expression of the Genes for Malic Enzyme and Fatty Acid Synthase

Alan G. Goodridge

I. INTRODUCTION

In many cell types, the primary function of the pathway for *de novo* fatty acid synthesis is to provide long-chain fatty acids for membrane lipids. In the liver, however, the maximum rates of *de novo* fatty acid synthesis can be several orders of magnitude higher than that required for membrane biosynthesis. The primary function of hepatic lipogenesis is to convert excess dietary carbohydrate or protein to fatty acids, which are stored as triglyceride in adipose tissue and used as a source of energy during periods of restricted food intake. Regulation of hepatic fatty acid synthesis is consonant with this function. Thus, synthesis of long-chain fatty acids in the liver is inhibited by starvation, whereas refeeding starved animals stimulates fatty acid synthesis to normal levels or to supranormal levels if the diet is high in carbohydrate.[1,2]

Fatty acid synthesis is therefore an "altruistic" function of the liver. This lipogenic pathway and its precise regulation by diet have evolved as mechanisms that enable animals to survive alternate periods of feast and famine. Cyclic accessibility to food may be attributable to seasonal variability in food availability or, in animals with relatively high metabolic rates, to non-feeding periods during the diurnal cycle. When food is available, some fraction of the calories is stored for use during those periods of deprivation. In many human populations, this function still plays an important role in survival. In more affluent populations, however, the combination of a plentiful and

Alan G. Goodridge ● Departments of Pharmacology and Biochemistry, Case Western Reserve University, Cleveland, Ohio.

continuous food supply with a precisely regulated hepatic lipogenic pathway contributes significantly to the problem of obesity.

Insulin was the first hormone to be implicated in the regulation of fatty acid synthesis. In *in vivo* experiments, Stetten and Boxer[3] used incorporation of the deuterium of heavy water into fatty acids to measure pathway function. In subsequent experiments, liver slices from diabetic rats were shown to incorporate ^{14}C-labeled glucose or acetate into fatty acids very slowly as compared with liver slices from normal animals.[4,5] The rate of fatty acid synthesis in the liver of diabetic animals is similar to that observed in the livers of starved animals. Administration of insulin restores normal rates of fatty acid synthesis from acetate in liver slices from diabetic animals.[5,6] The addition of insulin *in vitro* to various liver preparations also stimulates fatty acid synthesis.[7–11] Insulin levels are high when dietary carbohydrate intake is high, and vice versa. Thus, insulin is considered to be one of the signals in the blood that communicates the nutritional state of the whole animal to the liver and that regulates the rate of fatty acid biosynthesis in that organ.

Glucagon also plays an important role in the regulation of fatty acid synthesis. Glucagon inhibits fatty acid synthesis in liver slices and isolated hepatocytes[11–13] and in liver *in vivo*.[14] Glucagon also inhibits the stimulation of hepatic fatty acid synthesis that normally accompanies the refeeding of starved rats.[15] The secretion of glucagon is inversely related to the plasma glucose concentration,[16–18] consistent with a physiological role for this hormone in the dietary regulation of fatty acid synthesis.

Thus far, the effects of diet and hormones on fatty acid synthesis have been discussed in terms of regulation of carbon flux through the overall pathway. That regulation is imposed at two levels. Rapid changes in flux are initiated by altering the catalytic efficiency of pace-setting enzymes in the pathway which converts glucose to fatty acids. Both phosphorylation–dephosphorylation and allosteric mechanisms are involved in these rapid adjustments. A slower, more long-term adjustment of enzyme activity occurs when the stimulus for altered flux is maintained for a prolonged period.

Two of the lipogenic enzymes whose total activities are regulated by starvation and realimentation are malic enzyme [L-malate-NADP oxidoreductase (decarboxylating), EC1.1.1.40] and fatty acid synthase. Both enzymes are localized in the soluble fraction of the cell. Fatty acid synthase is a multifunctional protein which catalyzes the conversion of malonyl CoA and acetyl CoA to saturated long-chain fatty acids using NADPH for the reductive steps. The reaction catalyzed by malic enzyme furnishes much of the NADPH used during fatty acid synthesis.[2,19]

The treatments that modulate minute-to-minute flux through the pathway for fatty acid synthesis also modulate the total activities of the lipogenic enzymes. Thus, the total activities of malic enzyme and fatty acid synthase in liver are decreased coordinately when animals are starved or made diabetic and are increased coordinately when starved animals are refed or diabetic animals are treated with insulin[20–26] (for review, see refs. 2, 19). Glucagon inhibits enzyme induction *in vivo*[27] and in liver cells in culture.[28] Immunochemical studies have established that the changes in total enzyme activity are due to changes in the concentrations of the enzyme proteins.[29–33] Insulin and glucagon are therefore considered important regulators of the concentrations of the lipogenic enzymes in liver.

Thyroid hormone is also an important regulator of the concentrations of malic enzyme and fatty acid synthase. Injection of thyroid hormone into normal rats increases the total activities of malic enzyme and fatty acid synthase.[23,25,34,35] Conversely, the activities of these enzymes are decreased in the hypothyroid state.[23,35,36] Similar results have been reported for the chicken.[37] The physiological significance of the effects of thyroid hormone on malic enzyme and fatty acid synthase activities is difficult to assess. Despite reports that chronic starvation lowers the plasma level of triiodothyronine,[38,39] other evidence indicates that thyroid hormone does not play a major role in the nutritional regulation of the activities of malic enzyme and fatty acid synthase.[23,25,34,40] Thyroid hormone *in vivo* stimulates the total activities of several enzymes involved in lipogenic and nonlipogenic pathways.[41] Thus, thyroid hormone may play a permissive role, being required for the synthesis of several enzymes of diverse pathways.

In this introductory section, I have reviewed the teleological and physiological bases for the actions of the hormones whose basic mechanisms we wish to analyze. Studies of basic mechanisms are best carried out in cell culture where environmental conditions can be controlled. Hepatocytes were selected because a strictly regulated, high-capacity pathway for fatty acid synthesis is a specialized function of the liver. Prenatal chick embryos were chosen as the source of hepatocytes because *de novo* fatty acid synthesis, and its component enzyme activities are virtually absent in the high fat/low carbohydrate environment of the chick embryo but rapidly induced when neonatal birds are fed. In addition, chick embryo hepatocytes are easy to maintain in culture for prolonged periods.

II. SIMILARITY OF REGULATION IN CULTURE AND IN VIVO

Feeding neonatal chicks produces coordinated increases in flux through the pathway for fatty acid synthesis and in the total activities of malic enzyme and fatty acid synthase. Incorporation of [^{14}C]glucose into fatty acids increased about 1000-fold in the first few days after feeding was initiated.[42] When hepatocytes isolated from 18–19-day old embryos were incubated in culture in the presence of serum for 4 days, fatty acid synthesis from glucose increased more than 1000-fold (Fig. 1A). In intact chicks, hepatic malic enzyme activity increased 70-fold[20]; in culture in the absence of exogenous hormones, it increased about 80-fold (Fig. 1B). *In vivo*, fatty acid synthase activity increased fourfold just before hatching and another 15-fold when the neonatal chicks were fed, for a total increase of about 60-fold.[26] In culture, fatty acid synthase activity increased 20-fold (Fig. 1C). Thus, the pattern of development of lipogenesis and the lipogenic enzymes in hepatocytes in culture was virtually identical to that observed when newly hatched chicks were fed.

When exogenous triiodothyronine was added to the cells in culture, a further threefold increase was observed for fatty acid synthesis, malic enzyme, and fatty acid synthase (Fig. 1). Fatty acid synthesis, malic enzyme activity and fatty acid synthase activity are markedly stimulated by thyroid hormone in intact animals.[23,25,34–36] *In vivo*, the effect of thyroid hormone on these variables in hypothyroid animals is usually much greater than threefold.[35] This was also true in the hepatocytes in culture. Studies

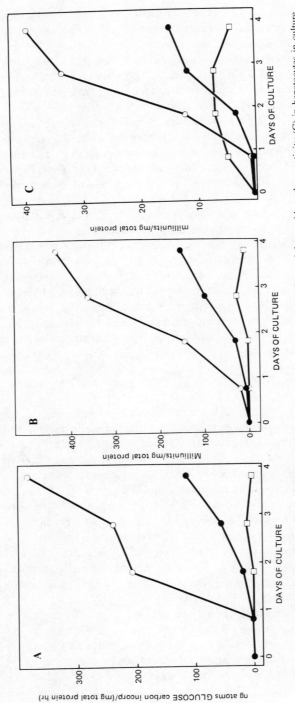

Figure 1. Incorporation of [U-^{14}C]glucose into total fatty acids (A), malic enzyme activity (B), and fatty acid synthase activity (C) in hepatocytes in culture. Cells were isolated from 18-day-old chick embryos as previously described[10] and incubated in Nutrient Mixture F-10 containing 15% horse serum and 2.5% fetal calf serum. At the designated times, the cells were harvested and the indicated measurements carried out. The results are expressed as nanogram atoms of glucose carbon incorporated into total fatty acids per milligram of total cellular protein per hour or as milliunits of malic enzyme or fatty acid synthase activity per milligram total cellular protein. (●) Cells incubated without added hormone: (○) Cells with triiodothyronine (2 μg/ml): (□) Cells with glucagon (1 μg/ml).

with specific thyroid hormone-binding proteins established that a large fraction of the stimulation of the lipogenic pathway in culture was due to thyroid hormone in the added serum.[28] In the presence of proteins that bind most of the thyroid hormone in serum, excess exogenous triiodothyronine stimulated malic enzyme activity more than 10-fold.[28] The thyroid hormone effects we observe in hepatocytes in culture clearly mimic a physiological phenomenon characteristic of liver cells in intact animals.

Addition of glucagon to the hepatocytes in culture blocked the induction of the pathway and its enzymes (Fig. 1). Several studies with intact animals indicate that glucagon plays an important role in the regulation of fatty acid synthesis and its associated enzymes.[11–18,27] Starvation or starvation followed by refeeding causes the plasma level of glucagon to rise or fall, respectively,[16–18] consistent with the finding that this hormone plays an important role in the starve–refeed response of liver enzymes involved in fatty acid synthesis. Our finding that the hepatocytes in culture mimic this important in vivo regulatory phenomenon adds further confidence to the use of these cells in culture as a model for the regulation of lipogenesis and its associated enzymes.

Curiously, glucagon stimulated the accumulation of fatty acid synthase activity on the first day of culture. During subsequent days in culture, glucagon markedly inhibited accumulation of the enzyme (Fig. 1C). This stimulatory response is still not understood, but it may be analogous to the stimulation of fatty acid synthase activity by glucagon or cyclic adenosine monophosphate, cAMP, which has been reported to occur in the livers of chick embryos just before hatching.[43] These stimulatory effects of glucagon on fatty acid synthase in vivo were elicited only in tissue from prenatal chick embryos. Loss of sensitivity of hepatic fatty acid synthase to stimulation by glucagon parallels the normal prenatal accumulation of the enzyme, suggesting a possible mechanistic similarity in these two phenomena.

When hepatocytes were incubated in serum-containing media, neither insulin nor anti-insulin serum affected the rate of fatty acid synthesis or the activity of malic enzyme.[28] This may have been due to the presence of factors in the serum that mimicked the effects of insulin. Removal of molecules of $<12,000\ M_r$ from the serum caused a significant decrease in the induction of the pathway, even in the presence of exogenously added triiodothyronine.[28] By contrast, neither insulin nor anti-insulin guinea pig serum had any effect on the induction of malic enzyme in cells incubated with media containing dialyzed serum. However, as will be noted later, when the hepatocytes were incubated in serum-free medium, insulin has a clear stimulatory effect on fatty acid synthesis, malic enzyme and fatty acid synthase.

One of the specialized functions of the liver is a very high rate of lipogenesis when food is available and contains a high percentage of carbohydrate. With the exception of some specialized lipid-synthesizing tissues, the liver is the organ with the highest capacity for fatty acid synthesis and the highest activities for the lipogenic enzymes. After 4 days in culture with serum and triiodothyronine in the medium, incorporation of [^{14}C]glucose into total fatty acids was three times higher in hepatocytes in culture than the maximum rate observed in liver slices from fed chicks (cf. ref. 42). Malic enzyme activity was more than seven times the maximum observed in intact chicks (cf. ref. 20). Fatty acid synthase activity also was significantly higher than the maximum levels observed in fed chicks (cf. ref. 26). Higher activities in hepatocytes culture may be due to our ability to add optimal concentrations of the various factors and hormones. We can, for example, completely eliminate glucagon; in intact animals, low levels of the hormone may persist even when animals are well fed. In addition, the cells in culture are continuously bathed by medium containing the regulatory hormone. *In vivo*, plasma hormone levels will fluctuate dramatically during the diurnal cycle. Very high lipogenic enzyme activities are further evidence that the hepatocytes in culture are an appropriate model for regulation of lipogenesis and its associated enzymes in intact animals.

III. HEPATOCYTES IN CULTURE WITH SERUM-FREE MEDIUM

The experiments with serum-containing media indicated that at least three factors were involved in the regulation of the lipogenesis pathway in hepatocytes in culture. Thyroid hormone and an unknown serum factor were stimulatory and glucagon inhibitory. Our next step was therefore to investigate the molecular mechanisms by which the hormones regulated the activity of malic enzyme and fatty acid synthase.

Incubating the cells in a serum-containing medium presented two difficulties. The first was technical. To inhibit the overgrowth of fibroblasts, the incubations were carried out in untreated petri plates. Under these conditions, the hepatocytes formed large aggregates that were not attached to the plates. Changing the medium or the culture conditions required cumbersome centrifugation steps. In serum-free medium, the cells continued to form aggregates but, even in untreated petri plates, the aggregates were firmly attached to the plates.[44] The second reason was experimental. In order to understand the mechanisms by which the hormones worked, it was important to eliminate all unknown components present in serum. A highly enriched Waymouth medium was selected as our chemically defined incubation medium.

When hepatocytes were incubated for 3 days in the defined medium with no added hormones, fatty acid synthesis from glucose increased about 25-fold over the rate in freshly prepared hepatocytes (cf. ref. 28 and Table 1). In the presence of insulin or triiodothyronine alone, the increase was almost 40-fold. The combination of insulin and triiodothyronine caused fatty acid synthesis to increase about 900-fold. The further addition of glucagon blocked almost 90% of the stimulation caused by insulin and triiodothyronine (Table 1). A similar result was obtained when we measured the activity of fatty acid synthase. Without hormones, activity increased fivefold. This may be the *in vitro* counterpart of the perinatal increase in enzyme activity, an increase that was independent of feeding.[26] Addition of either insulin or triiodothyronine alone caused a small further increase in activity. Addition of insulin plus triiodothyronine resulted in a fatty acid synthase activity which was 75-fold higher than that in freshly prepared cells. Glucagon inhibited 85% of the increase in activity caused by insulin plus triiodothyronine.

Malic enzyme activity also was inducible in hepatocytes incubated in chemically defined medium, but the response patterns were significantly different from those for flux through the fatty acid synthesis pathway or activity of fatty acid synthase. In the absence of hormones, there was no change or a small decline in the activity relative to that in freshly prepared cells. Addition of insulin alone caused a modest fourfold

Table 1. Effects of Hormones on the Synthesis of Fatty Acids, Malic Enzyme, and Fatty Acid Synthase and on the Mass of Protein and DNA[a]

Activity measured	No additions	Insulin	Triiodothyronine	Insulin plus triiodothyronine	Glucagon plus insulin plus triiodothyronine
Glucose→fatty acids[b]	2.5	3.9	3.8	89	13
Malic enzyme activity	7	27	304	890	26
Fatty acid synthase activity	8	14	13	115	24
Malic enzyme relative synthesis[c]	0.012	0.014	0.74	1.3	0.032
Fatty acid synthase relative synthesis	0.05	0.12	0.13	2.0	0.31
Total protein	0.97	2.11	0.73	1.79	1.89
Total DNA[d]	26	45	26	44	44

[a] Freshly isolated cells were incubated with the additions indicated. The medium was changed to one of the same composition at 24 hr of incubation. The measurements were made at 72 hr of incubation, as previously described[31,45]. Results are expressed as the averages of five to eight experiments, except as noted. Enzyme activities are expressed as nanomoles of NADP reduced or NADPH oxidized per minute per milligram soluble protein for malic enzyme and fatty acid synthase, respectively. Total protein and DNA are expressed as milligrams and micrograms per plate, respectively. Triiodothyronine and glucagon were added at 1 µg/ml, insulin at 300 ng/ml.
[b] One experiment.
[c] Two experiments.
[d] Three experiments; during the first 14 hr of incubation, no hormones were added.

increase in activity, while triiodothyronine alone resulted in a 40-fold increase. These two hormones together caused a 125-fold increase in malic enzyme activity during the 3 days in culture. The further addition of glucagon inhibited the stimulation of malic enzyme activity by 98%.

Regulation of malic enzyme thus differed from regulation of fatty acid synthase in two respects. First, there was no increase in malic enzyme activity without added hormone. In intact chicks, malic enzyme activity does not increase during the perinatal period despite a significant increase in fatty acid synthase activity.[20,26] Second, malic enzyme activity was increased greatly by thyroid hormone alone, while fatty acid synthase required the simultaneous presence of both insulin and thyroid hormone before significant induction was observed.

IV. DETERMINING THE POINT OF REGULATION

A. Catalytic Efficiency Versus Enzyme Concentration

Changes in enzyme activity measured in cell-free assays can be caused by changes in the catalytic efficiency of a constant number of enzyme molecules. Covalent modification or the presence of tightly bound effector molecules are examples of this type of regulation. Alternatively, the cell can regulate the number of enzyme molecules. We have purified both malic enzyme and fatty acid synthase from chicken liver. Using antisera prepared against the purified proteins, we established that hormone-induced changes in enzyme activity in hepatocytes in culture were due to changes in enzyme concentration, with no detectable change in catalytic efficiency of the enzymes.[31,44]

B. Synthesis versus Degradation

Enzyme concentration can be controlled by regulating either the rate of synthesis or the rate of degradation of the enzyme. For the most part, hormonal regulation of enzyme activity in the hepatocytes in culture was accompanied by a parallel regulation of enzyme synthesis (Table 1). None of the hormones had a significant effect on the half-life of malic enzyme.[45] As noted with measurements of enzyme activity, regulation of the two enzymes was not totally coordinate. Triiodothyronine alone stimulated synthesis of fatty acid synthase only slightly despite a very large stimulation of malic enzyme synthesis. In the presence of triiodothyronine, insulin caused a much larger-fold stimulation of fatty acid synthase than of malic enzyme. A third difference was a much greater inhibition of the synthesis of malic enzyme than that of fatty acid synthase when glucagon was present. Thus, the mechanisms by which these hormones regulate the expression of malic enzyme and fatty acid synthase in hepatocytes may be quite different.

Relative synthesis of malic enzyme was not stimulated by insulin alone despite a nearly fourfold increase in enzyme activity. Direct assessment of the degradation of malic enzyme suggests little effect of insulin on malic enzyme half-life ($t_{1/2}$ = 30 hr with insulin and 24 hr without insulin[45]). Accurate measurement of the rate of synthesis of malic enzyme in the absence of hormones was difficult because the rate was so low. The effect of insulin alone on malic enzyme activity may have been due to a combination of small changes in enzyme synthesis, enzyme degradation and, possibly, catalytic efficiency as well. A stimulation of the catalytic efficiency of rat malic enzyme by insulin has been reported.[46]

C. Regulation of mRNA Sequence Abundance

1. Molecular Cloning of cDNA Sequences for Malic Enzyme and Fatty Acid Synthase

The next step in the analysis of the action of these hormones on malic enzyme and fatty acid synthase activities was to determine whether enzyme synthesis was controlled by regulating the concentrations of malic enzyme and fatty acid synthase mRNAs or by regulating the efficiency with which constant concentrations of these mRNAs were translated. Both mRNAs are relatively minor mRNA species in chicken liver, so a combination of biological and recombinant DNA techniques was used to isolate the required reagents.[47,48]

Fatty acid synthase represents more than 25% of total protein in the highly specialized uropygial gland of the goose.[49] We prepared a cDNA clone bank in pBR

322, starting with total polyadenylated RNA from the goose uropygial gland. In addition to being abundant in this gland, fatty acid synthase is also very large. Highly enriched fatty acid synthase mRNA was isolated from total uropygial gland poly (A^+) RNA by sucrose-gradient centrifugation.[47] The uropygial gland cDNA clone bank was screened with a ^{32}P-labeled cDNA copied from the highly enriched fatty acid synthase mRNA. Several potential fatty acid synthase cDNA clones were isolated. Those clones containing fatty acid synthase mRNA sequences were positively identified by hybrid-selected translation.[47] Several clones, representing about 2500 nucleotides at the 3' end of the mRNA, were obtained. Although the cloned DNAs were derived from goose tissue, they cross-hybridized with chicken fatty acid synthase mRNA.

Malic enzyme mRNA is much less abundant than fatty acid synthase mRNA in the uropygial gland, liver, or any other tissue. In goose liver, the relative rate of synthesis of malic enzyme is about 0.5% of soluble protein. By *in vitro* translation, uropygial gland RNA contains about 0.2% malic enzyme mRNA. Malic enzyme activity is higher in these two goose tissues than in any other tissue of any animal species in which activity has been measured. Since the relative abundance of malic enzyme mRNA in the two tissues was quite similar, we decided to screen our uropygial gland cDNA clone bank for malic enzyme cDNA sequences.

In addition to being of relatively low abundance, the size of malic enzyme mRNA is very similar to that of the bulk of tissue mRNAs, making purification by physical techniques very difficult. We chose to screen duplicate arrays of the uropygial gland cDNA clone bank with two populations of ^{32}P-labeled cDNAs that differed primarily in the presence or absence of malic enzyme cDNA. The "plus" and "minus" cDNAs were reverse transcribed from total poly (A^+) RNA extracted from the livers of 2-day-old goslings fed for 24 hr or starved, respectively. Feeding neonatal goslings for 24 hr increased the synthesis rate for malic enzyme about 35-fold.[48] The increase in synthesis rate of malic enzyme caused by feeding neonatal chicks is due to an increase in translatable malic enzyme mRNA sequences.[50] Of the 1400 colonies screened, 50 gave significant positive signals with the "plus" cDNA and little or no signal with the "minus" cDNA. The 50 potential clones were isolated and further screened by hybrid-selected translation. DNA from one of the potential clones selected malic enzyme mRNA.[48] As with the fatty acid synthase cDNA clones, goose malic enzyme cDNA cross-hybridized with chicken malic enzyme mRNA.

2. Effect of Hormones on Sequence Abundance

Malic enzyme and fatty acid synthase messenger-like DNA sequences were subcloned into the single-stranded bacteriophage, M13. Single-stranded, ^{32}P-labeled cDNAs were synthesized using the recombinant M13 mDNAs as templates.[48,51] Sequence abundance of malic enzyme and fatty acid synthase mRNAs were measured with a dot-blot hybridization procedure and densitometric scanning of the resultant autoradiographs.[52]

The results of these measurements are expressed in Table 2 as percentages of the mRNA levels in cells treated with insulin plus triiodothyronine. Within the limits of error of the measurements of enzyme synthesis and enzyme mRNA levels, the two

Table 2. Effect of Hormones on Relative Synthesis and mRNA Sequence Abundance for Malic Enzyme and Fatty Acid Synthase[a]

	Malic enzyme		Fatty acid synthase	
Additions	Synthesis[b]	mRNA[c]	Synthesis[b]	mRNA[c]
No additions	1	4	3	8
Insulin	1	4	6	8
Triiodothyronine	56	55	7	42
Triiodothyronine plus insulin	100	100	100	100
Glucagon plus triiodothyronine plus insulin	3	3	16	35

[a] Culture conditions are described in the legend to Fig. 1. Sequence abundances for the mRNAs were measured by hybridization of RNA bound to gene screen (New England Nuclear) to single-stranded ^{32}P-labeled malic enzyme and fatty acid synthase cDNAs cloned in M13 vectors.[48] Results are expressed as a percentage of the value for cells incubated with insulin plus triiodothyronine.
[b] Data from Table 1, expressed as a percentage of the value for cells incubated with insulin plus triiodothyronine.
[c] mRNA results are the averages of two to four experiments.

variables correlate very well in almost every instance. Therefore, triiodothyronine and glucagon regulate the synthesis of malic enzyme and fatty acid synthesis primarily, if not exclusively, at a pretranslational level. The effects of insulin, however, were not exclusively at the pretranslational level. In the absence of insulin, triiodothyronine stimulated the accumulation of fatty acid synthase mRNA to almost the same extent as it did malic enzyme mRNA sequence abundance and malic enzyme synthesis. Triiodothyronine alone, however, had virtually no effect on the relative synthesis or enzyme activity of fatty acid synthase (Tables 1 and 2). These results suggest that insulin stimulated the translation of fatty acid synthase mRNA but not that of malic enzyme mRNA. Furthermore, the stimulation of the accumulation of the mRNAs for malic enzyme and fatty acid synthase by triiodothyronine was coordinate, suggesting the possibility of a common mechanistic basis.

V. THE ROLE OF INSULIN

Insulin regulated the levels of the lipogenic enzymes but, unlike triiodothyronine and glucagon, insulin also regulated the level of DNA and total protein. We compared the responses of malic enzyme activity and protein accumulation at different concentrations of insulin, with and without triiodothyronine in the medium. The insulin dose–response curves for the stimulation of malic enzyme activity were similar both with and without triiodothyronine. Fifty percent of maximal stimulation of malic enzyme activity occurred at about 10^{-9} M insulin, a concentration within the physiological range (Fig. 2A,B). In addition, hepatocytes degrade insulin, so that the average concentration during the 48 hr between the medium change and harvesting the cells may have been significantly lower than that initially added. At high concentrations of insulin, the stimulation of malic enzyme activity was decreased (Fig. 2). The manufacturer (Eli Lilly) indicated that contamination of the insulin by glucagon was 0.007%

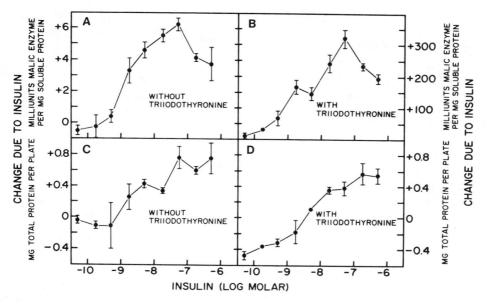

Figure 2. Dose–response curves for the effect of insulin on malic enzyme activity (A, B) and mass of total protein (C, D). Hepatocytes were prepared from 18- or 19-day-old chick embryos and incubated in chemically defined Waymouth medium MD 705/1.[45] Cells were incubated for 3 days in the presence (B, D) or absence (A, C) of triiodothyronine (1 μg/ml) and the indicated concentrations of insulin. The results are expressed as increases in malic enzyme activity or mass of protein due to insulin and are the averages of two closely agreeing independent experiments.

by weight. At 5×10^{-7} M insulin (3 μg/ml), the concentration of contaminating glucagon would be ~0.2 ng/ml, a concentration that should have caused little or no inhibition of malic enzyme induction.[44] An alternative possibility is that the inhibitory effects of high concentrations of insulin on malic enzyme induction may be due to occupancy of receptors for insulin-like growth factors. The effects of high concentrations of insulin on cells in culture are frequently attributed to the ability of insulin to occupy these receptors at high concentrations.

The dose–response curve for the stimulation of protein mass produced by insulin was shifted slightly to the right compared to that for malic enzyme activity, with 50% of the maximum response achieved at $\sim3 \times 10^{-9}$ M. In addition, at high concentrations of insulin the response of protein mass reached a plateau rather than decreasing as did malic enzyme activity (Fig. 2C,D). This latter result would be consistent with either of the explanations discussed above for the inhibition of malic enzyme activity observed at high concentrations of insulin because glucagon had no effect on protein mass and insulin-like growth factors might be expected to stimulate protein mass. The effect of insulin on protein mass could be due to increased protein synthesis or decreased protein degradation.

Insulin had another significant property. Incubation of hepatocytes with triiodo-thyronine for 24 hr caused a much larger increase in malic enzyme activity if the triiodothyronine was added after 2 days of preincubation rather than on the day the cells were prepared.[44,45] The 24-hr accumulation of malic enzyme caused by triiodothyronine in cells preincubated for 2 days was twofold higher if the medium contained insulin (2×10^{-8} M) during the 2-day preincubation period (48 vs. 23 mU/mg total protein). In these experiments, insulin was present at a concentration of 2×10^{-7} M in all incubations during the test period with triiodothyronine. The concentration dependence of this effect was similar to that for accumulation of malic enzyme activity when insulin was present throughout a 3-day incubation period. In addition to having a direct effect on malic enzyme activity, insulin primed the cells to respond more rapidly to triiodothyronine plus insulin.

The stimulatory effects of triiodothyronine and the inhibitory effects of glucagon were highly selective. Neither protein mass nor DNA mass nor relative synthesis of soluble proteins was affected by these two hormones[31] (Table 1). By contrast, insulin stimulated each of these processes (see Table 1 and Fig. 2), except for the relative synthesis of soluble protein.[31] The effects of insulin were therefore significantly less selective than those of triiodothyronine and glucagon. Thus, insulin may amplify the triiodothyronine-induced increase in malic enzyme and fatty acid synthase mRNA levels by increasing the activity of cellular components required for rapid accumulation of specific mRNAs and proteins. The ability of insulin to prime the cells for a subsequent stimulation by triiodothyronine is consistent with this suggestion. On the other hand, insulin did cause an increase in the translation of fatty acid synthase but not malic enzyme when triiodothyronine was present, indicating that it does have some selective effects on protein levels. The mechanisms whereby insulin regulates either nonselective effects on protein accumulation or selective effects on protein translation are unknown. Since glucagon appears to act via an increase in intracellular cAMP and has no effect on protein mass or DNA mass (Table 1), effects of insulin on these variables are unlikely to have been mediated by a decrease in cAMP.

VI. SUMMARY

I have described the response of fatty acid biosynthesis and two involved enzymes to insulin, triiodothyronine, and glucagon in hepatocytes from chicken embryos maintained in mass culture in a chemically defined medium. I have documented the validity of the hepatocytes in mass culture as a model for studying regulation of the expression of the genes for the lipogenic enzymes. This review has emphasized what we know about the actions of insulin on the expression of the malic enzyme and fatty acid synthase genes. However, the actions of triiodothyronine and glucagon are much larger and more selective than those of insulin and are therefore more amenable to analysis. As a consequence, much of our effort has been directed toward an understanding of the mechanisms by which those two hormones regulate the level of malic enzyme and fatty acid synthase mRNAs. Nevertheless, the experiments reviewed here indicate that insulin also plays an important role in regulating the concentration of malic enzyme and fatty acid synthase in hepatocytes.

Insulin regulates the synthesis of several proteins. Thus, in rat hepatocytes in maintenance culture, insulin stimulates the synthesis of enzymes involved in glycolysis and lipogenesis: glucokinase,[53] ATP-citrate lyase,[54] glucose 6-phosphate dehydrogenase,[55,56] L-type pyruvate kinase,[57] and malic enzyme.[58,59] The magnitudes of the effects of insulin in these instances are very similar to those we have observed for chick embryo hepatocytes. The synthesis of one enzyme, phosphoenolpyruvate carboxykinase, is inhibited by insulin in rat hepatoma cells.[60,61] This latter effect of insulin is caused by an inhibition of the accumulation of the mRNA for phosphoenolpyruvate carboxykinase, probably at the level of transcription.[61] Insulin also stimulates the synthesis of a hepatic secretory protein, albumin.[62] The effect of insulin on albumin synthesis also is mediated at the level of mRNA abundance.[62,63] Whether the synthesis rates for the rat enzymes of glycolysis and lipogenesis listed above are regulated in the same fashion is unknown. Equally unknown are the mechanisms by which insulin brings about any of these effects.

Our results show that hormones regulate the levels of malic enzyme and fatty acid synthase at a minimum of two levels: mRNA sequence abundance and translational efficiency. Insulin and glucagon appear to play important roles in dietary regulation of the enzymes of lipogenesis. Thus, the considerable changes in enzyme level caused by starvation or feeding high-carbohydrate diets may be due to regulation of specific gene expression at multiple levels by one or more hormones. In this way, small changes in the concentration of a hormone could lead to very large changes in enzyme concentration. Regulation at multiple levels also may permit both coordinate and noncoordinate regulation of the lipogenic enzymes, such as we have observed in the liver of the perinatal chick and in hepatocytes in culture. The cell culture system described is well suited to the analysis of these regulatory phenomena.

ACKNOWLEDGMENTS. I am deeply indebted to the following people who have collaborated with me on various aspects of this work: P. Silpananta, T.G. Adelman, P.W.F. Fischer, A. Garay, U.A. Siddiqui, S.M. Morris, Jr., T. Goldflam, L.K. Winberry, R.A. Jenik, D.W. Back, J.H. Nilson, M.A. McDevitt, J.E. Fisch, M.J. Glynias, and S. Deleon. Expert technical assistance was provided by I. Rusek, E. Chao and T. Uyeki. The work described in this review was supported by grants from the Medical Research Council of Canada and the National Institutes of Health of the United States.

VII. REFERENCES

1. Tepperman, H., and Tepperman, J., 1964, *Fed. Proc.* **23**:73–75.
2. Volpe, J.J., and Vagelos, P.R., 1976, *Physiol. Rev.* **56**:339–417.
3. Stetten, D., Jr., and Boxer, G.E., 1944, *J. Biol. Chem.* **156**:271–278.
4. Chernick, S.S., and Chaikoff, I.L., 1950, *J. Biol. Chem.* **186**:535–542.
5. Felts, J.M., Chaikoff, I.L., and Osborn, M.J., 1951, *J. Biol. Chem.* **193**:557–562.
6. Baker, N., Chaikoff, I.L., and Schusdek, A., 1952, *J. Biol. Chem.* **194**:435–443.
7. Winegrad, A.I., and Renold, A.E., 1958, *J. Biol. Chem.* **233**:267–272.
8. Bloch, K., and Kramer, W., 1948, *J. Biol. Chem.* **173**:811–812.
9. Topping, D.L., and Mayes, P.A., 1972, *Biochem. J.* **126**:295–311.

10. Goodridge, A.G., 1973, *J. Biol. Chem.* **248**:1924–1931.
11. Geelen, M.J.H., Beynen, A.C., Christiansen, R.Z., Lepreau-Jose, M.J., and Gibson, D.M., 1978, *FEBS Lett.* **95**:326–330.
12. Goodridge, A.G., 1973, *J. Biol. Chem.* **248**:4318–4326.
13. Haugaard, E.S., and Stadie, W.C., 1953, *J. Biol. Chem.* **200**:753–757.
14. Klain, G.J., and Weiser, P.C., 1973, *Biochem. Biophys. Res. Commun.* **55**:76–83.
15. Volpe, J.J., and Masara, J.C., 1975, *Biochim. Biophys. Acta* **380**:454–472.
16. Blackhard, W.G., Nelson, N.C., and Andrews, S.S., 1974, *Diabetes* **23**:199–202.
17. Buchanan, K.D., Vance, J.E., Dinstl, D., and Williams, R.H., 1969, *Diabetes* **18**:11–18.
18. Ohneda, A., Aguilar-Parada, E., Eisentraut, A.M., and Unger, R.M., 1969, *Diabetes* **18**:1–10.
19. Wakil, S.J., Stoops, J.K., and Joshi, V.C., 1983, *Annu. Rev. Biochem.* **52**:537–579.
20. Goodridge, A.G., 1968, *Biochem. J.* **108**:663–666.
21. Goodridge, A.G., 1968, *Biochem. J.* **108**:667–673.
22. Korchak, H.M., and Masoro, E.J., 1962, *Biochim. Biophys. Acta* **58**:354–356.
23. Tepperman, H.M., and Tepperman, J., 1964, *Am. J. Physiol.* **206**:357–361.
24. Kornacker, M.S., and Lowenstein, J.M., 1965, *Biochem. J.* **94**:209–215.
25. Wise, E.M., and Ball, E.G., 1964, *Proc. Natl. Acad. Sci. USA* **52**:1255–1263.
26. Goodridge, A.G., 1973, *J. Biol. Chem.* **248**:1932–1938.
27. Lakshmanan, M.R., Nepokroeff, C.M., and Porter, J.W., 1972, *Proc. Natl. Acad. Sci. USA* **69**:3516–3519.
28. Goodridge, A.G., Garay, A., and Silpananta, P., 1974, *J. Biol. Chem.* **249**:1469–1475.
29. Silpananta, P., and Goodridge, A.G., 1971, *J. Biol. Chem.* **246**:5754–1561.
30. Suzuki, F., Fukunishi, K., Daikuhara, Y., and Takeda, Y., 1967, *J. Biochem.* **62**:170–178.
31. Fisher, P.W.F., and Goodridge, A.G., 1978, *Arch. Biochem. Biophys.* **190**:332–344.
32. Zehner, Z.E., Joshi, V.C., and Wakil, S.J., 1977, *J. Biol. Chem.* **252**:7015–7022.
33. Craig, M.C., Nepokroeff, C.M., Lakshmanan, M.R., and Porter, J.W., 1972, *Arch. Biochem. Biophys.* **152**:619–630.
34. Diamant, S., Gorin, E., and Shafrir, E., 1972, *Eur. J. Biochem.* **26**:553–559.
35. Mariash, C.N., Kaiser, F.E., Schwartz, H.L., Towle, H.C., and Oppenheimer, J.H., 1980, *J. Clin. Invest.* **65**:1126–1134.
36. Young, J.W., 1968, *Am. J. Physiol.* **214**:378–383.
37. Chandrabose, K.A., and Bensadoun, A., 1971, *Comp. Biochem. Physiol.* **39B**:55–59.
38. Portnay, G.I., O'Brian, J.T., Bush, J., Vagenakis, A.G., Azizi, F., Arky, R.A., Ingbar, S.H., and Braverman, L.E., 1974, *J. Clin. Endocrinol. Metab.* **39**:191–194.
39. Vagenakis, A.G., Portnay, G.I., O'Brian, J.T., Rudolph, M., Arky, R.A., Ingbar, S.H., and Braverman, L.E., 1977, *J. Clin. Endocrinol. Metab.* **45**:1305–1309.
40. Marx, J.V., Richert, D.A., Westerfeld, W.W., and Ruegamer, W.R., 1971, *Biochem. Pharmacol.* **20**:3009–3020.
41. Colton, D.G., Mehlman, M.A., and Ruegamer, W.R., 1972, *Endocrinology* **90**:1521–1528.
42. Goodridge, A.G., 1968, *Biochem. J.* **108**:655–662.
43. Joshi, V.C., and Sidbury, J.B., Jr., 1976, *Arch. Biochem. Biophys.* **173**:403–414.
44. Goodridge, A.G., 1974, *Fed. Proc.* **34**:117–123.
45. Goodridge, A.G., and Adelman, T.G., 1976, *J. Biol. Chem.* **251**:3027–3032.
46. Drake, R.L., Parks, W.C., and Thompson, E.W., 1983, *J. Biol. Chem.* **258**:6008–6010.
47. Morris, S.M., Jr., Nilson, J.H., Jenik, R.A., Winberry, L.K., McDevitt, M.A., and Goodridge, A.G., 1982, *J. Biol. Chem.* **257**:3225–3229.
48. Winberry, L.K., Morris, S.M., Jr., Fisch, J.E., Glynias, M.J., Jenik, R.A., and Goodridge, A.G., 1983, *J. Biol. Chem.* **258**:1337–1342.
49. Buckner, J.S., and Kolattukudy, P.E., 1976, *Biochemistry* **15**:1948–1957.
50. Siddiqui, U.A., Goldflam, T., and Goodridge, A.G., 1981, *J. Biol. Chem.* **256**:4544–4550.
51. Messing, J., and Vieira, J., 1982, *Gene* **19**:269–276.
52. Thomas, P.S., 1980, *Proc. Natl. Acad. Sci. USA* **77**:5201–5205.
53. Spence, J.T., and Pitot, H.C., 1979, *J. Biol. Chem.* **254**:12331–12336.
54. Spence, J.T., Pitot, H.C., and Zalitis, G., 1979, *J. Biol. Chem.* **254**:12169–12173.

55. Winberry, L., Nakayama, R., Wolfe, R., and Holten, D., 1980, *Biochem. Biophys. Res. Commun.* **96:**748–755.
56. Nakamura, T., Yoshimoto, K., Aoyama, K., and Ichihara, A., *J. Biochem.* **91:**681–693.
57. Poole, G.P., Postle, A.D., and Bloxham, D.P., 1982, *Biochem. J.* **204:**81–87.
58. Wilson, E.J., and McMurray, E.C., 1981, *J. Biol. Chem.* **256:**11657–11662.
59. Mariash, C.N., McSwigan, C.R., Towle, H.C., Schwartz, H.L., and Oppenheimer, J.H., *J. Clin. Invest.* **68:**1485–1490.
60. Gunn, J.M., Tilghman, S.M., Hanson, R.W., Reshef, L., and Ballard, F.J., 1975, *Biochemistry* **14:**2350–2356.
61. Andreone, T.L., Beale, E.G., Bar, R.S., and Granner, D.K., 1982, *J. Biol. Chem.* **257:**35–38.
62. Peavy, D.E., Taylor, J.M., and Jefferson, L.S., 1978, *Proc. Natl. Acad. Sci. USA* **75:**5879–5883.
63. Liang, T.J., and Grieninger, G., 1981, *Proc. Natl. Acad. Sci. USA* **78:**6972–6976.

22

Regulation of Pyruvate Kinase Gene Expression by Hormones and Developmental Factors

Michele A. Cimbala, David Lau, and Joseph F. Daigneault

I. INTRODUCTION

These are exciting times in the development of our knowledge of the regulation of an important glycolytic enzyme—pyruvate kinase. A series of excellent laboratories are working in this area. This chapter concentrates on those agents that appear to invoke a chronic change in the total enzyme levels of pyruvate kinase. For an in-depth discussion of the acute regulation of pyruvate kinase activity via phosphorylation mechanisms, several excellent reviews are available (see refs. 1–3).

II. ISOZYMES OF PYRUVATE KINASE

Mammalian tissues contain at least three distinct isozymes of pyruvate kinase. The erythrocyte enzyme is sometimes separated into a fourth group (see refs. 3 and 4, for review). The isozymes are distinguished by their antigenic and kinetic properties and were first recognized by Tanaka.[5,6] The L isozyme is found in the hepatic parenchymal cell and as a minor isozyme of kidney cortex cells and the intestine. The mature erythrocyte contains an enzyme that is immunologically indistinguishable from the L isozyme, although kinetically they differ slightly. The L isozyme is unique in that it is the isozyme that serves as a substrate for the cyclic adenosine monophosphate

Michele A. Cimbala, David Lau, and Joseph F. Daigneault • Department of Biochemistry, University of Massachusetts Medical Center, Worcester, Massachusetts.

(cAMP)-dependent protein kinase. Antibodies against L pyruvate kinase do not cross-react with either of the other two isozymes, M_1 and M_2.

M_1 is that isozyme found in skeletal muscle, heart, and brain. Antigenically, it is related to the M_2 isozyme (also called the K isozyme), although kinetically the affinities of the two isozymes for the substrate P-enolpyruvate and for various allosteric effectors are unique for each form. The M_2 isozyme is found in adult mammalian tissues not mentioned above, but it also occurs in all fetal tissues and undifferentiated tumors. In chicken liver, the M_2 isozyme predominates and has been reported as being a substrate for phosphorylation by a cAMP-independent protein kinase.[7,8] Another report has noted that a pp60[src] kinase preparation from Rous sarcoma virus (RSV)-transformed chicken embryo fibroblasts will phosphorylate a tyrosine residue on chicken liver M_2 pyruvate kinase,[9] resulting in an inhibition of its activity. Regulation of M_2 activity by tumor factors is discussed in greater detail below.

III. REGULATION OF L PYRUVATE KINASE

A. Dietary Factors

The effect of diet on hepatic pyruvate kinase activity has been examined extensively in the rat.[5,10–23] Krebs and Eggleston[10] and later Tanaka[5] reported that the activity of rat hepatic pyruvate kinase exhibited significant changes in response to altered dietary conditions, especially the carbohydrate content.[10] When compared with the standard diet (containing 55% carbohydrate) pyruvate kinase levels rose fourfold within 2 days on an 80% carbohydrate diet. Conversely, starvation (24 hr) or a low-carbohydrate diet resulted in a 60% loss in activity. Kidney cortex pyruvate kinase showed no change with starvation and only a small stimulation by carbohydrate.[10–11] Muscle isozyme showed no change.[5,11] Historically it is interesting to note that pyruvate kinase was the first glycolytic enzyme reported to exhibit altered activity with diet.

The induction of pyruvate kinase levels by carbohydrate has the following characteristics:

1. Fructose effectively substitutes for sucrose and is a better inducer than glucose.[11–15]
2. Some metabolites can induce L pyruvate kinase activity levels, even in the diabetic rat,[12,14,15] which lacks insulin. Insulin is commonly thought to be the primary hormonal inducer of pyruvate kinase gene expression. Liver pyruvate kinase activity can, however, be induced in diabetic rats by substrates that enter glycolysis at the triose level such as fructose, sucrose, and especially glycerol.[16] The mechanism of this induction, or the hormonal factors that influence it, are unknown.
3. Although changes in enzyme induction can be easily assayed 10–24 hr after changing the diet, a relatively long lag period of 5–10 hr exists before any change is found.[1,14,15,17,18] As assayed with a specific antibody, the synthesis rate of L pyruvate kinase protein[1,15,17–19] and the translatable levels of pyruvate kinase mRNA[15,20] show a similar lag.

4. There is a species difference in the ability of carbohydrate to induce liver pyruvate kinase. Bailey et al.[13] reported little change in the activity of mouse liver pyruvate kinase with altered diet.

Recently, studies on the dietary regulation of pyruvate kinase gene expression have been extended using cloned cDNA probes to the enzyme[21-23] to assay mRNA levels directly. Hybridizable pyruvate kinase mRNA levels are also increased 10-fold, 24 hr after starting a high-carbohydrate diet. The maximum rate of induction occurred at 7–12 hr.[23] These studies confirm that changes in the assayable levels of pyruvate kinase protein reflect changes in the gene expression of this enzyme.

B. Hormones

1. Insulin

Liver pyruvate kinase activity is depressed in diabetes. The chronic induction of pyruvate kinase activity by insulin has long been recognized.[1,5,12,14,15,22,24-27] As in the dietary studies, a relatively long lag period exists before induction of pyruvate kinase levels is detectable. In an early paper, Tanaka et al.[5] reported an increase in hepatic pyruvate kinase activity only 3 hr after insulin, but Gunn and Taylor[14] reported a lag time of 12 hr. Insulin increases the synthesis rate[15,26] and the mRNA levels (both translational[15,22] and hybridizable[22]) for the enzyme. There is a lag period of 4–5 hr before the accumulation of translatable mRNA is detected.[15] However, within 24 hr, translatable mRNA accumulates ninefold, returning to normal levels. A similar time course is found for the levels of hybridizable mRNA.[22]

The mechanism through which insulin stimulates the gene expression of hepatic pyruvate kinase (or that of any gene) is unknown. This question is at the forefront of molecular endocrinology. In addition to the data discussed above that implicate a transcriptional effect of insulin, several laboratories have proposed a second, concurrent mechanism of insulin action—that of altering the in vivo specific activity of the pyruvate kinase protein molecule itself. Data that either support[18,27-33] or contradict[1,15,17] this hypothesis have been presented and to date no clear answer has been resolved. Kohl and Cottam and associates first used antibody to quantitate changes in the specific activity of hepatic pyruvate kinase.[32,33] They reported that the specific activity doubled over the first 48 hr of starvation and then leveled off. Refeeding a starved animal a high-carbohydrate diet further increased the specific activity threefold during the first 24 hr. But by 3 days the specific activity fell and returned to that found in the fed state. Similar results have been found in our laboratory using a radiosandwich assay that detects 1 ng of pyruvate kinase protein.[34] Starvation for 48 hr increased the specific activity in a linear manner from 227 ± 32 ($n = 4$) to 304 ± 26 ($n = 6$) mU/µg pyruvate kinase, although the total activity fell from 0.540 ± 0.014 ($n = 4$) to 0.292 ± 0.024 ($n = 6$) U/mg cellular protein. This value was not further increased by a longer 5-day fast. Refeeding an 80% carbohydrate diet after a 48-hr fast depressed the specific activity to control values of 186 ± 8 ($n = 3$) mU/µg pyruvate kinase within 24 hr and raised the total activity to 1.27 ± 0.08 U/mg cellular protein.[39] Other laboratories did not detect an effect of carbohydrate feeding on pyruvate kinase specific activity.[1,15,17] Several laboratories have reported that insulin or carbohydrate feeding increases rather than decreases pyruvate kinase specific activity in rat[18,27–28] and rabbit.[29,30]

In contrast to insulin, dexamethasone lowers pyruvate kinase specific activity.[31] Changes in the specific activity, whether increased or decreased, may be the result of an accumulation of antigenically reactive protein that is not enzymatically active.[18,33,34] Whether this protein represents newly synthesized but inactive pyruvate kinase monomers or whether one of the early steps in pyruvate kinase degradation involved a protease that can be hormonally regulated and limiting in the cell remains to be shown.

2. Glucocorticoids

The role of the glucocorticoids in regulating pyruvate kinase gene expression has been best characterized in cultured hepatocytes. In culture, exposure of the cells to both glucocorticoids and insulin is required to maintain *in vivo* levels of L pyruvate kinase.[35,36] When cultured in the presence of 1 μM insulin without a glucocorticoid, 50% of the assayable pyruvate kinase activity is lost in 4 days.[35] In the presence of 1 μM dexamethasone, 50% was lost in 2 days,[35] suggesting that dexamethasone alone may in fact be a deinducer. Miyanaga et al.[35] noted that a 10-min incubation in the presence of dexamethasone is sufficient to maintain pyruvate kinase activity levels in cells if they are subsequently cultured for the next 4 days in the presence of insulin.

A similar synergistic effect of dexamethasone and insulin on hepatic pyruvate kinase was found by Feliú et al. using 10^{-7} M dexamethasone.[36] The effect of the glucocorticoids may be at least partially due to the more general ability of glucocorticoids to increase the *in vitro* survival of cultured hepatocytes and to maintain their differentiated state.[36,37] A similar synergistic effect of dexamethasone and insulin has been reported for glucokinase[38] and other hepatic enzymes.[36]

3. Thyroxine

Thyroxine has been implicated as an inducer of several hepatic enzymes,[39] but its role in pyruvate kinase gene expression has not been extensively evaluated. Thyroxine administration will induce the appearance of L pyruvate kinase in newborn rat liver.[40] However, in adult rabbit liver, administration of thyroxine has no effect on L pyruvate kinase levels.[30]

C. Development

Immature fetal liver expresses the M_2 isozyme in rat[3,25,41–44] and man.[45,46] There are conflicting reports as to the precise time at which fetal rat liver begins to express L pyruvate kinase. Although many workers have noted that adult levels of L pyruvate kinase were not induced until until the rat was weaned,[25,40,48–50] some have detected relatively high levels of L pyruvate kinase before birth.[40,41,47,51] Levels of L pyruvate kinase one-half that of the adult are present from 6 to 3 days before birth.[51] From day -3 to birth, L pyruvate kinase levels fall rapidly, reaching a low value of

25% of the adult activity.[40,51] Weanling induces the reappearance of the L isozyme, which then accumulates to adult levels over the next 40 days.[25,40,50] Using a specific antibody, Poole et al. recently showed that hepatocytes which were isolated from preweaned 16-day-old rats and cultured for 4 hr in Waymouth's medium plus 5% fetal calf serum, and then maintained in Waymouth's alone, did not synthesize L pyruvate kinase.[44] The inclusion of insulin and fructose in the medium induced a de novo appearance of L pyruvate kinase after 48 hr. The induction of pyruvate kinase by fructose (2 mM) or insulin (0.1 μM) was additive. The maximal rate of pyruvate kinase synthesis obtained by the neonatal cells was 20% of the synthesis rate achieved by high-carbohydrate feeding of adult rat.[44] Fetal rat kidney does not express type L until after birth.[51]

Fetal human liver (11–15 weeks) has been reported to contain substantial type L activity.[45] Almost 80% of the total pyruvate kinase activity was estimated to be the L isozyme at 16 weeks.[45,46] Several laboratories have reported the presence of a type of pyruvate kinase that is apparently unique to the human fetus.[41,45,46,52,53] Guguen-Guillouzo et al.[46] suggest that this is a precursor form of the mature type L and that a proteolytic system that is responsible for generating type L is not yet fully active in the fetal liver. A more recent study, however, using a cDNA clone directed against L pyruvate kinase, was unable to detect hybridizable mRNA in a human fetal liver preparation (4 months) by Northern analysis.[21] Resolution of the ontogeny of human and rat L pyruvate kinase and the identification of putative type L precursors in fetal liver will no doubt soon be answered by assaying for mRNA expression with cloned pyruvate kinase probes.

D. Expression of L Pyruvate Kinase in Other Tissues

The first studies using cDNA clones directed against L pyruvate kinase have already greatly enhanced our understanding of pyruvate kinase gene expression. RNA sequences complementary to type L-cDNA are detected in liver, kidney, and the small intestine.[21,22] These results are in agreement with the early studies, which used antibodies to detect pyruvate kinase determinants. Under stringent conditions, no type L mRNA was detected in muscle, lung, spleen, or brain by Kahn's group.[21] However, under less stringent hybridization conditions, Tanaka and co-workers used clones against M_2 pyruvate kinase to identify L pyruvate kinase.[22] In all tissues in which it is found, except possibly the reticulocyte, L pyruvate kinase mRNA levels are inducible in subjects fed a high-carbohydrate diet.[23] Liver exhibits a much greater response to carbohydrate than does kidney or the small intestine. Insulin induced hepatic type L mRNA levels 10-fold in 24 hr in the diabetic rat.[22]

It is not yet clear whether hepatic pyruvate kinase and reticulocyte pyruvate kinase are transcribed from the same gene or are two closely linked genes. Genetic data strongly suggest that both enzymes are encoded by the same structural gene,[54–57] and the amino acid composition of the isozymes is very similar.[58] However, Kahn and co-workers have noted that pyruvate kinase translated by RNA extracted from rat reticulocytes or liver is synthesized in a tissue-specific manner. Reticulocyte RNA translated a peptide of 64,000 daltons, the L' form. Hepatic RNA encoded a pyruvate kinase peptide of 61,000 daltons, the L form. Mixing both RNAs resulted in the translation of both pyruvate kinase species.[52] Marie and associates[59] explained those results by proposing tissue-specific differential processing of the same primary nuclear RNA transcript. Similar results were obtained when human fetal liver RNA was examined. Two peptide species, presumably the L' and L forms, were translated by this RNA.[59]

The functional mRNA for L pyruvate kinase sediments at 22 S[53] and has been sized by Northern blot analysis at about 3400 bases. Simon et al. also report the identification of a clone that recognizes a nonfunctional 2000-base mRNA species in addition to the larger form, presumably a degradation product.[21] Since the 60,000–62,000-dalton pyruvate kinase subunit could be encoded by about 1800–1900 bases, this suggests that the 3400-base functional pyruvate kinase mRNA contains about 1500 noncoding nucleotides—a relatively large amount. Southern blot analysis suggests that both species of mRNA detected by these clones are encoded by the same gene and that there is only one copy of the L pyruvate kinase gene.[21]

IV. M_2 PYRUVATE KINASE

A. Regulation of M_2 Pyruvate Kinase

Levels of M_2 pyruvate kinase are not regulated by the same factors that alter L pyruvate kinase. The nonparenchymal cells of mammalian liver contain only the M_2 isozyme. The M_2 isozyme appears to be constitutively expressed in all fetal tissues, but factors that evoke or suppress its expression have not been defined. Insulin cannot suppress expression of the M_2 isozyme and therefore cannot alone coordinately regulate the M_2 and L pyruvate kinase genes during fetal development.

In the developing human lung, the activity of type M_2 rises sharply at about the 15th week of gestation to levels almost twice those found in the adult.[62] In the rat, high levels of pulmonary pyruvate kinase are found in the fetus, but these levels are depressed by 50% from birth until the rat is weaned, when they are reinduced to the adult level. Neither thyroxine nor cortisol was found to alter the already high levels of pulmonary M_2 pyruvate kinase levels in fetal rat lung.[62]

Expression of the type M_2 isozyme has been related to the nuclear activity of ADP-ribosyltransferase (ADPRT).[63] Induction of the hepatic M_2 isozyme was totally inhibited in cell cultures treated with inhibitors of ADPRT (10 mM 3-aminobenzamide, 25 mM nicotinamide), although the expression of α-fetoprotein (AFP) and albumin was unaffected. Levels of hepatic M_2 pyruvate kinase are also not altered by fasting, diabetes, or steroids.[64]

B. M_2 Expression in Regenerating Liver and Hepatic Tumors

The expression of type M_2 is evoked in the regenerating rat liver.[5,41,42,65,66] During liver regeneration, levels of the M_2 isozyme protein double within 48 hr. The shift from type L to type M_2 corresponds to the period of mitotic activity of the liver.

Garnett *et al.* have proposed that the synthesis of the type L or M_2 isozyme is a mutually exclusive event for the parenchymal cell,[65] an interesting hypothesis whose mechanisms would have immediate application to other coordinately regulated isozyme systems in mammals. Hybrids of the pyruvate kinase type M_2 and L protein subunits, although they can be formed *in vitro*, do not occur *in vivo* in the regenerating liver.[65] This shift in favor of M_2 isozyme synthesis continues for 9–12 days, after which the expression of the L isozyme again becomes predominate.[5,65]

It has been recognized for some time that poorly differentiated tumors express the M_2 isozyme.[5,6,41,42,67-69] Odashima *et al.* also noted that a line of Morris-hepatoma cells from a Buffalo rat (MH_1C_1) contained, surprisingly, L pyruvate kinase, while a line of normal rat liver cells (BRL) solely expressed the M_2 isozyme.[70] Induction of M_2 isozyme expression in liver can be triggered peripherally by a tumor-derived factor.[66,71] Increased levels of hepatic M_2 pyruvate kinase can be assayed in the livers of rats carrying Ehrlich ascites tumor cells[67] as a rhodamine sarcoma on their backs.[72] Type M_2 was also increased by the intraperitoneal injection of a crude tumor cell extract[67,72] or by perfusing the liver for 2 hr with blood from a tumor-bearing animal.[66,73] Using a fluorescent antibody technique, the increase in the M_2 isozyme was localized in the hepatic parenchyma cells, which usually express L pyruvate kinase.[65,66] The rapid response suggests that changes in pyruvate kinase isoenzyme synthesis may be one of the earliest assayable markers for the presence of tumor secreted factors. Much effort has gone into attempts to isolate and purify this factor,[3,67,69,72-76] but its identity has not been elucidated. The recent isolation of various tumor growth factors should greatly stimulate research in this area.

Increases in type M_2 activity are also induced in liver after feeding carcinogens such as 3'-methyl-4-dimethylaminoazobenzene to rats.[64,77] Noncarcinogens such as 2-methyldimethylaminoazobenzene failed to induce the M_2 isozyme.[64] Endo *et al.* noted that this induction precedes any measurable changes in the DNA content, histological appearances, or mitotic index[64] in the liver and that increases in pyruvate kinase activity were assayable within 2 weeks after administering the diet containing the dye.[64]

V. M_1 PYRUVATE KINASE

A. Regulation of M_1 Pyruvate Kinase

By electrophoretic analysis, M_1 pyruvate kinase appears in skeletal muscle and brain within 3 hr after birth,[6,51] the M_2 isozyme being the only form assayed in those tissues in fetal rat on day 17 of gestation. Hormones or agents that induce this differentiation have not been identified. The total levels of the skeletal muscle M_1 isozyme continue to rise throughout postnatal development until 40 days after birth, when they reach adult levels, 20-fold higher than the fetal level.[51,78] Traces of the M_2 isozyme disappear from skeletal muscle 2 days after birth.[51] Levels of M_1 pyruvate kinase in skeletal muscle are not altered by fasting, carbohydrates, insulin, or any of the other effectors known to regulate type L. Atrophic rat leg muscle contained elevated levels of the M_2 isozyme,[51] suggesting that the contractile activity of skeletal muscle may be a factor that influences the continued expression of the M_1 isozyme in that tissue. Human fetal skeletal muscle is already predominantly the M_1 isozyme at 11–16 weeks.[45]

Expression of the M_1 isozyme in rat cardiac muscle can first be detected 2 days before birth. The transition to M_1 pyruvate kinase is completed in rat heart within 4 days after birth.[51]

By Northern analysis, levels of chicken M_1-pyruvate kinase in skeletal muscle are very low 3 days after birth.[79] From 3 to 126 days after hatching, there is a dramatic rise in the sequence abundancy for this message in chicken muscle.[79] Only one mRNA species is detected on denaturing gels.

B. Gene Structure of the M_1 and M_2 Isozyme

Peptide maps have demonstrated that there is a considerable degree of structural homology between pyruvate kinase type M_1 and M_2.[80,81] Studies examining peptide maps and using monoclonal antibodies demonstrate that species variations in the structure of the M_1 isozyme are always accompanied by similiar changes in the structure of the M_2 isozyme in that species.[81,82] Because of this coordinate evolution, Hance et al. suggested that both isozymes are synthesized from the same gene.[82] However, detailed comparative studies on these isozymes from rat, mouse, and rabbit have shown that the amino acid composition of the M_1 isozyme is distinct from the M_2 form.[81–84] Each isozyme is translated by a unique mRNA species.[85,86] RNA isolated from AH-130 Yoshida ascites cells containing M_2 pyruvate kinase translates a pyruvate kinase subunit of 62,000 daltons, whereas mRNA from rat skeletal muscle translates a peptide of 61,000 daltons. Mixing the mRNA preparations generates the same two peptides species, confirming that these isozymes do not share a common mRNA.[85,86] It is not clear whether these two mature mRNAs are generated from two closely related (but different) genes or from differential processing of the same primary gene transcript.[85]

The M_2 isozyme may itself be a family of related genes, as the physical properties of various purified preparations of M_2 pyruvate kinase differ greatly within a single species.[4] Regulatory characteristics of M_2 pyruvate kinase that are unique for the isozyme in different tissues have been found. For example, in rat epididymal adipose tissue, the K_m for phosphoenolpyruvate is lowered by insulin.[87] Chicken liver pyruvate kinase can be phosphorylated by a cAMP-independent protein kinase.[7,8,88] whereas the same isozyme from rat and chicken lung is not a substrate for this phosphorylation.[89] Similarly, M_2 pyruvate kinase from rat pancreatic islets[90] and intestine[67] displays unique characteristics. It is interesting to note that adipose tissue and pancreatic islets are the only tissues that coordinately express both the M_2 isozyme of pyruvate kinase and phosphoenolpyruvate carboxykinase.[87,91] Other cells that express phosphoenolpyruvate carboxykinase, such as hepatic parenchyma and kidney cortex, express the L-isozyme gene.

Chicken muscle pyruvate kinase has recently been cloned.[79] Analysis of the cDNA clone predicts a primary structure of 529 amino acids and a molecular weight of 57,865 for this isozyme. The primary sequence has been reported.[79] The mRNA also contains 684 bases of 3′ noncoding mRNA and 80 nucleotides of 5′ noncoding nucleotides.

VI. SUMMARY

Recent advances in the molecular cloning of the isozymes of pyruvate kinase will soon reveal the ontogeny and regulation of this important glycolytic isozyme system. Specifically, the identity and number of gene copies, the intron–exon structure of those genes, and the evolutionary relationships that exist between the pyruvate kinase isozyme genes will be divulged. Studies on the hormonal regulation of L pyruvate kinase will no doubt add greatly to our current understanding of insulin action on gene expression. Research into the activation of the M_2 isozyme expression by tumor and developmental factors is also timely. The coordinate regulation of the pyruvate kinase isozyme system is not only an excellent model of hormone action, but of tissue differentiation as well. Such issues are at the forefront of molecular endocrinology research.

ACKNOWLEDGMENTS.   We are especially indebted to the laboratories of Dr. T. Tanaka, Dr. A. Kahn, and Dr. G.L. Cottom for sharing manuscripts with us prior to publication. We also thank James B. Blair for his many constructive discussions. We wish to thank Dr. Yoshitomo Oka for the English translation of reference 61. This work was supported by grant AM31402 to M.A.C. and biomedical research grant S07RR05712 from the National Institutes of Health. M.A.C. was the recipient of a Career Development Award from the American Diabetes Association and a Basil O'Conner Starter Grant from the March of Dimes.

VII. REFERENCES

1. Blair, J.B., 1980, in *Regulation of Carbohydrate Formation and Utilization in Mammals* (C.M. Veneziale, ed.), pp. 121–151, University Park Press, Baltimore.
2. Engstrom, L., 1980, *Molecular Aspects of Cellular Regulation,* (P. Cohen, ed.), Vol. 1, pp. 11–31, Elsevier Biomedical Press, New York.
3. Ibsen, K.H., 1977, *Cancer Res.* 37:341–353.
4. Hall, E.R., and Cottam, G.L., 1978, *Int. J. Biochem.* 9:785–793.
5. Tanaka, T., Harano, Y., Sue, F., and Morimura, H., 1967, *J. Biochem. (Tokyo)* 62:71–91.
6. Imamura, K., and Tanaka, T., 1972, *J. Biochem. (Tokyo)* 71:1043–1051.
7. Eigenbrodt, E., Abdel-Fattah Mostafa, M., and Schoner, W., 1977, *Hoppe Seylers Z. Physiol. Chem.* 358:1047–1055.
8. Fister, P., Eigenbrodt, E., Presek, P., Reinacher, M., and Schoner, W., 1983, *Biochem. Biophys. Res. Commun.* 115:409–414.
9. Glossmann, H., Presek, P., and Eigenbrodt, E., 1981, *Mol. Cell. Endocrinol.* 23:49–63.
10. Krebs, H.A., and Eggleston, L.V., 1965, *Biochem. J.* 94:3c–4c.
11. Bailey, E., Stirpe, F., and Taylor, C.B., 1968, *Biochem. J.* 108:427–436.
12. Weber, G., 1969, *Adv. Enzyme Regul.* 7:15–40.
13. Bailey, E., Taylor, C.B., and Bartley, W., 1968, *Nature* 217:471–472.
14. Gunn, J.M., and Taylor, C.B., 1973, *Biochem. J.* 136:455–465.

15. Noguchi, T., Inoue, H., and Tanaka, T., 1982, *Eur. J. Biochem.* **128:**583–588.
16. Takeda, Y., Inoue, H., Honjo, K., Kroaki, T., and Daikuhara, Y., 1967, *Biochim. Biophys. Acta* **136:**214–222.
17. James, M.E., and Blair, J.B., 1982, *Biochem. J.* **204:**329–338.
18. Hopkirk, T.J., and Bloxham, D.P., 1979, *Biochem. J.* **182:**383–397.
19. Cladaras, C., and Cottam, G.L., 1980, *Arch. Biochem. Biophys.* **200:**426–433.
20. Cladaras, C., and Cottam, G.L., 1980, *J. Biol. Chem.* **255:**11499–11503.
21. Simon, M.-P., Besmond, C., Cottreau, D., Weber, A., Chaumet-Riffaud, P., Dreyfus, J.C., Trepat, J.S., Marie, J., and Kahn, A., 1983, *J. Biol. Chem.* **258:**14576–14584.
22. Noguchi, T., Inoue, H., Chen, H.L., Matsubara, K., and Tanaka, T., 1983, *J. Biol. Chem.* **258:**15220–15223.
23. Weber, A., Marie, J., Cottreau, D., Simon, M.P., Besmond, C., Dreyfus, J.-C., and Kahn, A., 1984, *J. Biol. Chem.* **259:**1798–1802.
24. Weber, G., Singhal, R.L., Stamm, N.B., and Srivastava, S.K., 1965, *Fed. Proc.* **24:**745–754.
25. Weber, G., Lea, M.A., Fisher, E.A., and Stamm, N.B., 1966, *Enzyme* **7:**11–24.
26. Miyanaga, O., Nagano, M., and Cottam, G.L., 1982, *J. Biol. Chem.* **257:**10617–10623.
27. Parks, W.C., and Drake, R.L., 1982, *Biochem. J.* **208:**333–337.
28. Kohn, E.A., and Cottam, G.L., 1976, *Arch. Biochem. Biophys.* **176:**671–682.
29. Johnson, M.I., and Veneziale, C., 1980, *Biochemistry* **19:**2191–2195.
30. Veneziale, C.M., Donofrio, J.C., Hansen, J.B., Johnson, M.L., and Mazzotta, M.Y., 1981, in *The Regulation of Carbohydrate Formation and Utilization in Mammals* (C.M. Veneziale, ed.), pp. 23–44, University Park Press, Baltimore.
31. Johnson, M.L., Hansen, J.B., Donofrio, J.C., and Veneziale, C.M., 1981, *Biochim. Biophys. Acta* **675:**140–142.
32. Kohl, E.A., and Cottam, G.L., 1977, *Biochim. Biophys. Acta* **484:**49–58.
33. Hall, E.R., McCully, V., and Cottam, G.L., 1979, *Arch. Biochem. Biophys.* **195:**315–324.
34. Cimbala, M.A., Lau, D., Daigneault, J., 1985, *Biochem. J.,* in press.
35. Miyanago, O., Evans, C., and Cottam, G.L., 1983, *Biochim. Biophys. Acta* **758:**42–48.
36. Feliú, J.E., Coloma, J., Gomez-Lechon, M.-J., Garcia, M.D., and Baguena, J., 1982, *Mol. Cell. Biochem.* **45:**73–81.
37. Ichihara, A., Nakamura, T., and Tanaka, T., 1982, *Mol. Cell. Biochem.* **43:**145–160.
38. Schudt, C., 1979, *Eur. J. Biochem.* **98:**77–82.
39. Liaw, C., Seelig, S., Mariash, C.N., Oppenheimer, J.H., and Towle, H.C., 1983, *Biochemistry* **22:**213–221.
40. Greengard, O., and Jamdar, S.C., 1971, *Biochim. Biophys. Acta* **237:**476–483.
41. Schapira, F., Hatzfeld, A., and Weber, A., 1973, in *Differentiation and Control of Malignancy of Tumor Cells* (W. Nakahara, O. Tetsuo, T. Sugimura, and H. Sugano, eds.), pp. 205–220, University Park Press, Baltimore.
42. Farrina, F.A., Shatton, J.B., Morris, H.P., and Weinhouse, S., 1974, *Cancer Res.* **34:**1439–1446.
43. Singh, M., and Feigilson, M., 1981, *Arch. Biochem. Biophys.* **209:**655–667.
44. Poole, G.P., Postle, A.D., and Bloxham, D.P., 1982, *Biochem. J.* **204:**81–87.
45. Faulkner, A., and Jones, C.T., 1975, *FEBS Lett.* **53:**167–169.
46. Guguen-Guillouzo, C., Marie, J., Cottreau, D., Pasdeloup, N., and Kahn, A., 1980, *Biochem. Biophys. Res. Commun.* **93:**528–534.
47. Balinsky, D., Cayanis, E., and Bersohn, I., 1975, *Isozymes,* (C.L. Markery, ed.), Vol. III, pp. 919–933, Academic Press, New York.
48. Taylor, C.B., Bailey, E., and Bartley, W., 1967, *Biochem. J.* **105:**717–722.
49. Vernon, R.G., and Walker, D.G., 1968, *Biochem. J.* **106:**321–329.
50. Middleton, M.C., and Walker, D.G., 1972, *Biochem. J.* **127:**721–731.
51. Osterman, J., Fritz, P.J., and Wuntch, T., 1973, *J. Biol. Chem.* **248:**1011–1018.
52. Marie, J., Simon, M.-P., Dreyfus, J.-C., and Kahn, A., 1981, *Nature* **292:**70–72.
53. Marie, J., Simon, M.-P., and Kahn, A., 1982, *Biochim. Biophys. Acta* **696:**340–344.
54. Bigley, R.H., and Koler, R.D., 1968, *Ann. Hum. Genet.* **31:**383–388.

55. Nakashima, K., Miwa, S., Oda, S., Tanaka, T., Imamura, K., and Nishina, T., 1974, *Blood* **43**:537–548.
56. Shinohara, K., Miwa, K., Nakashima, K., Oda, E., Kegeoka, T., and Tsujino, G., 1976, *Am. J. Hum. Genet.* **28**:474–481.
57. Kahn, A., Marie, J., Galand, C., and Boivin, P., 1976, *Scand. J. Haematol.* **16**:250–257.
58. Harada, K., Saheki, S., Wada, K., and Tanaka, T., 1978, *Biochim. Biophys. Acta* **524**:327–339.
59. Simon, M.-P., Marie, J., Bertrand, O., and Kahn, A., 1982, *Biochim. Biophys. Acta* **709**:1–7.
60. Tanaka, T., Harano, Y., Morimura, H., and Mori, R., 1965, *Biochem. Biophys. Res. Commun.* **21**:55–60.
61. Otani, H., Imamura, K., and Tanaka, T., 1971, *J. Nutr. Sci. Vitaminol. (Tokyo)* **24**:142.
62. Greengard, O., Cayanis, E., and Bodanszky, H., 1980, *J. Dev. Physiol.* **2**:291–304.
63. Althaus, F.R., Lawrence, S.D., He, Y.-Z., Sattler, G.L., Tsukada, Y., and Pitot, H.C., 1982, *Nature* **300**:366–368.
64. Endo, H., Eguchi, M., Yanagi, S., Toris, Y., Ihehara, Y., and Kamiya, T., 1972, *Gan* **13**:235–250.
65. Garnett, M.E., Dyson, R.D., and Dost, F.N., 1974, *J. Biol. Chem.* **249**:5222–5226.
66. Suda, M., Tanaka, T., Yanagi, S., Hayashi, S., Imamura, K., and Taniuchi, K., 1972, *Gan* **13**:79–93.
67. Tanaka, T., Yanagi, S., Miyahara, M., Kaku, R., Imamura, K., Taniuchi, K., and Suda, M., 1972, *Gan* **63**:555–562.
68. Weinhouse, S., 1973, in *Differentiation and Control of Malignancy of Tumor Cells* (W. Nakahara, T. Ono, T. Sugimura, and H. Sugano, eds.), pp. 187–203, University Park Press, Baltimore.
69. Tanaka, T., Imamura, K., Ann, T., and Taniuchi, K., 1972, *Gan* **13**:219–234.
70. Odashima, S., Nishikawa, K., Nakayabu, Y., and Nagao, Y., 1981, *Biochem. Biophys. Res. Commun.* **102**:1180–1186.
71. Suda, M., Tanaka, T., Sue, F., Harano, Y., and Morimura, H., 1966, *Gan* **1**:127–141.
72. Nakamura, T., Hosoi, K., Nishikawa, K., and Horio, T., 1972, *Gan* **63**:239–250.
73. Tanaka, T., Yanagi, S., Imamura, K., Kashiwagi, A., and Nobuyuki, I., 1973, in *Differentiation and Control of Malignancy of Tumor Cells* (W. Nakahara, T. Ono, T. Sugimura, and H. Sugano, eds.), pp. 221–234, University Park Press, Baltimore.
74. Suda, M., Tanaka, T., Sue, F., Kuroda, Y., and Morimura, H., 1968, *Gan* **4**:103–112.
75. Ibsen, K.H., Basabe, J.R., and Lopez, T.P., 1975, *Cancer Res.* **35**:180–188.
76. Muroya, N., Nagao, Y., Miyasaki, K., Nishikawa, K., and Horio, T., 1976, *J. Biochem.* **79**:203–215.
77. Walker, P.R., and Potter, V.R., 1982, *Adv. Enzyme Regul.* **21**:339–364.
78. Ramponi, G., Nassi, P., and Treves, C., 1968, *Life Sci.* **7**:443–448.
79. Lonberg, N., and Gilbert, W., 1983, *Proc. Natl. Acad. Sci. USA* **80**:3661–3665.
80. Saheki, S., Saheki, K., and Tanaka, T., 1982, *Biochim. Biophys. Acta* **704**:484–493.
81. Saheki, S., Saheki, K., Tanaka, T., and Tanaka, T., 1982, *Biochim. Biophys. Acta* **704**:494–502.
82. Hance, A.J., Lee, J., and Feitelson, M., 1982, *Biochem. Biophys. Res. Commun.* **106**:492–499.
83. Harkino, R.N., Black, J.A., and Rittenberg, M.B., 1977, *Biochemistry* **16**:3831–3837.
84. Ibsen, K.H., Chiu, R.H.-C., Park, H.R., Sanders, D.A., Roy, S., Garrott, K.N., and Mueller, M.K., 1981, *Biochemistry* **20**:1497–1506.
85. Noguchi, T., and Tanaka, T., 1982, *J. Biol. Chem.* **257**:1110–1113.
86. Levin, M.J., Daegelen, D., Meienhofer, M.-C., Dreyfus, J.-C., and Kahn, A., 1982, *Biochim. Biophys. Acta* **699**:77–83.
87. Denton, R.M., Edgell, N.J., Bridges, B.J., and Poole, G.P., 1979, *Biochem. J.* **180**:523–531.
89. Eigenbrodt, E., and Schoner, W., 1977, in *Phosphorylated Proteins and Related Enzymes* (L.A. Pinna, ed.), pp. 5–16, Information Retrieval, London.
89. Schering, B., Eigenbrodt, E., Linder, D., and Schoner, W., 1982, *Biochim. Biophys. Acta* **717**:337–347.
90. Chatterton, T.A., Reynolds, C.H., Lazarus, N.R., and Pogson, C.I., 1982, *Biochem. J.* **204**:605–608.

Regulation of Membrane Components by Insulin

Hexose Transport Regulation by Insulin in the Isolated Rat Adipose Cell

Ian A. Simpson and Samuel W. Cushman

I. INTRODUCTION

Following the initial observations that insulin stimulates the metabolism of glucose in a variety of peripheral tissues,[1] it was soon recognized that this stimulatory effect was exerted at the level of glucose uptake.[2–5] Since these earlier observations, most studies pertaining to the action of insulin on glucose uptake have been carried out in adipose tissue and, more specifically, rat adipose tissue. Despite its relatively small contribution to glucose homeostasis in the whole animal,[6] adipose tissue has been preferred for such studies because of the ease with which it can be obtained, the ability to prepare isolated adipose cells relatively free of other cell types,[7] and the extreme sensitivity and responsiveness of the adipose cell to insulin. Validation of the adipose cell as a model for studies of the mechanism of insulin action on hexose transport is now being obtained by parallel studies in muscle.[8,9] However, significant and even marked differences in the magnitude and sensitivity of the response to insulin are apparent between tissues, as they are between adipose cells of different species such as guinea pig[10] and man.[11]

In the rat adipose cell, the stimulatory action of insulin occurs through a change in the maximum rate (V_{max}) of glucose transport in the absence of a change in the apparent affinity (K_m) of the transporter for glucose.[3–5,12] However, kinetic analyses alone cannot resolve whether this change in V_{max} is the result of an alteration in the intrinsic activity of individual glucose transporters or an increase in the number of transporters exposed to the extracellular medium, or a combination of both. In this

Ian A. Simpson and Samuel W. Cushman ● Cellular Metabolism and Obesity Section, National Institute of Arthritis, Diabetes, and Digestive and Kidney Diseases, National Institutes of Health, Bethesda, Maryland.

chapter, we describe our approach to resolving this problem and, together with Dr. Tetsuro Kono and colleagues in Chapter 24, we propose a novel concept for insulin action—the so-called translocation hypothesis. Both groups, independently and using entirely different methodologies, have suggested that the insulin-induced increase in glucose transport in the rat adipose cell and its rapid reversal are achieved by the translocation of glucose transporters between a large intracellular membrane pool and the plasma membrane.[13–18]

II. GLUCOSE TRANSPORT ACTIVITY IN INTACT CELLS

The rapid, reversible, and concentration-dependent insulin-induced stimulation of glucose transport activity in the isolated rat adipose cell is illustrated in Figs. 1 and 2. Before discussing the data, however, it is important to consider briefly the methodology employed in assessing glucose transport activity in this cell type. Several different glucose analogs have been used to distinguish between the transport of glucose and its subsequent metabolism. These include L-arabinose,[19] 2-deoxyglucose,[20,21] D-allose,[22] and 3-O-methylglucose,[23] of which the latter is now considered the substrate of choice. 3-O-Methylglucose has a lower K_m (higher apparent affinity) for transport than glucose ($\sim$5 mM compared with $\sim$10 mM) and, unlike the other commonly used analog, 2-deoxyglucose, is truly nonmetabolizable (for a comprehensive review, see Gliemann and Rees[24]).

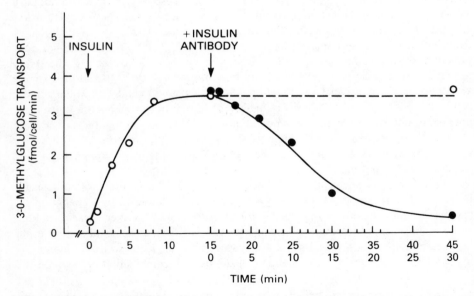

Figure 1. Time course of stimulation of glucose transport activity by insulin and its reversal by anti-insulin antiserum.

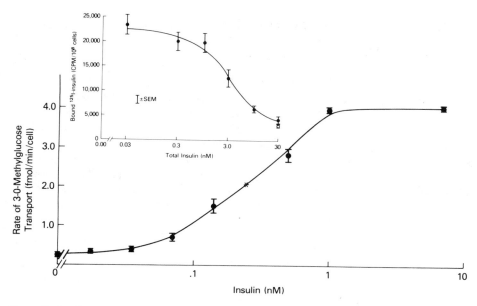

Figure 2. Insulin concentration dependence of stimulation of glucose transport acivity. (Inset) Insulin concentration dependence of insulin binding.

Typical time courses for the stimulation of 3-O-methylglucose transport by insulin and the reversal of this stimulation by anti-insulin antiserum are shown in Fig. 1.[15] The onset of the action of insulin (0.7 nM) is rapid, although lag times of up to 45 sec at 37°C have been reported.[25,26] The maximally stimulated rate, corresponding to a 10- to 40-fold increase above the basal rate, is achieved within 15 min, with a half-time of 2–3 min. A similar time course is observed in response to a large excess of insulin, indicating that hormone binding is not the rate-limiting step under these conditions. The reversal of this stimulation, initiated by addition of a 300-fold excess of anti-insulin antiserum, is equally rapid. However, under the conditions (37°C) depicted in Fig. 1, the rate of reversal cannot be distinguished from the rate of dissociation of insulin from its receptor. As will be discussed later, this is not the case at all incubation temperatures.

The concentration dependence of insulin for the stimulation of 3-O-methylglucose transport is illustrated in Fig. 2 and is to be contrasted with the typical [^{125}I]insulin binding curve illustrated in the inset.[15] The half-maximal stimulation of 3-O-meth-ylglucose transport is achieved at an insulin concentration of ~0.3 nM, whereas half-maximal binding/competition is observed at ~3.0 nM, at least a full order of magnitude higher. This observation, first made by Kono and Barham,[27] that the action of insulin on the adipose cell is fully expressed when only 5–10% of the insulin receptors are occupied—the "spare receptor" concept—still represents a major enigma in considering the mechanism of insulin action.

III. SUBCELLULAR DISTRIBUTION OF GLUCOSE TRANSPORTERS

A. Preparation and Characterization of Subcellular Membrane Fractions

In order to determine the subcellular distribution of glucose transporters, a reproducible procedure for cell fractionation has been developed in our laboratory based on the method originally described by McKeel and Jarett.[28] This procedure is shown schematically in Fig. 3.[29] Using differential ultracentrifugation, three major membrane fractions (i.e., plasma membranes and high-density and low-density microsomes) are prepared, together with a mitochondrial/nuclear fraction. While these fractions are not pure, as indicated by the distributions of marker enzyme activities (Fig. 4), each is enriched in a particular membrane species, i.e., the plasma membrane fraction with plasma membranes, the high-density microsomes with membranes of the endoplasmic

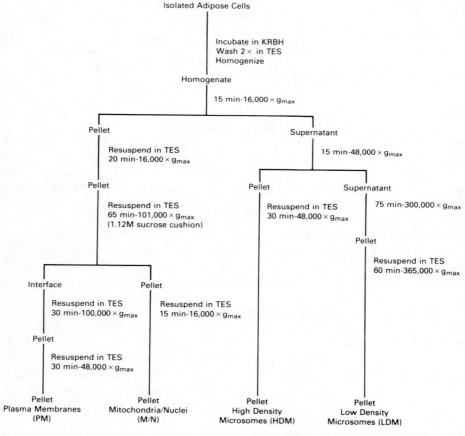

Figure 3. Subcellular fractionation procedure. KRBH, Krebs–Ringer bicarbonate/Hepes buffer; TES, Tris–EDTA–sucrose buffer.

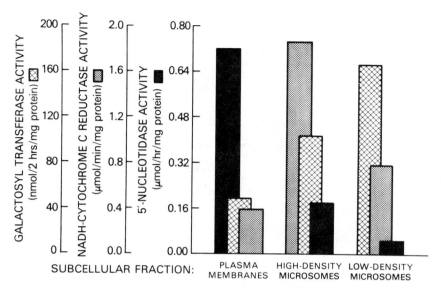

Figure 4. Distribution of marker enzyme activities among subcellular membrane fractions. 5'-Nucleotidase is a marker enzyme for plasma membranes, rotenone-insensitive NADH-cytochrome c reductase is a marker enzyme for membranes of the endoplasmic reticulum, and UDP-galactose : N-acetylglucosamine galactosyltransferase is a marker enzyme for membranes of the Golgi apparatus.

reticulum, and the low-density microsomes with membranes of the Golgi apparatus. Neither the yield nor subcellular distribution of any of these marker enzymes is affected by preincubation of the cells with insulin.[29]

B. Quantitation of Glucose Transporters by Cytochalasin B Binding

Before discussing the distribution of glucose transporters among these various membrane fractions and the effects of insulin on this distribution, we will first briefly outline the cytochalasin B binding assay used in our laboratory to quantitate the glucose transporters. Cytochalasin B, a fungal metabolite, was first shown to be a competitive inhibitor of glucose transport in the human erythrocyte.[30] It is by far the most potent of all the cytochalasin analogs isolated thus far.[31] However, like most of the cytochalasins, it also interacts specifically with other cellular proteins, the most notable of which is actin. These numerous interactions are particularly evident in a Scatchard plot of [³H]cytochalasin B binding to isolated plasma membranes.[13] Computer analysis of such curves reveals at least three distinct specific components, each of differing affinity and capacity, and a relatively high nonsaturable binding component. Therefore, in order to use [³H]cytochalasin B as a specific ligand for the glucose transporter, it was necessary to develop a method for resolving these various components. Such a method is shown in Fig. 5.[13,14] The inclusion of cytochalasin E, a cytochalasin B analog that, at a concentration of 2 μM, does not inhibit glucose transport in either

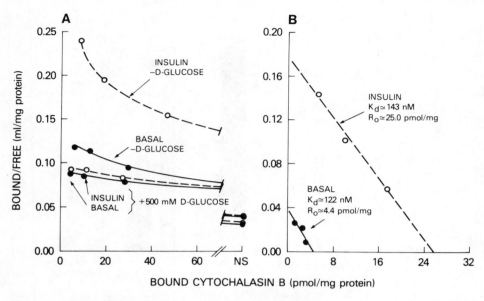

Figure 5. Cytochalasin B binding to plasma membranes prepared from basal and insulin-stimulated cells. (A) Scatchard plots. (B) Derived Scatchard plots. Derived Scatchard plots are obtained by subtracting the curves obtained in the presence of D-glucose from the respective curves obtained in the absence of D-glucose.

intact cells or isolated membranes, is found to reduce dramatically both the nonsaturable binding and the specific binding to components other than the glucose transporter.[13]

As can be seen in Fig. 5A, in the presence of cytochalasin E, the binding of [³H]cytochalasin B to plasma membranes from insulin-treated cells is clearly increased compared with that observed using plasma membranes from basal cells, yet the Scatchard plots are clearly curvilinear in both cases. However, when the binding to these membranes is performed in the presence of a saturating concentration of D-glucose (500 mM), the binding of cytochalasin B to both sets of membranes is reduced to the same final level. Furthermore, when the curves obtained in the presence of D-glucose are subtracted from the respective curves obtained in the absence of D-glucose along radial axes of constant free cytochalasin B concentration, the "derived" Scatchard plots shown in Fig. 5B are observed. Here, the specific D-glucose-inhibitable cytochalasin B-binding component is represented by a linear Scatchard plot from which both a dissociation constant (K_d) and binding capacity (R_o) can be calculated. In this particular experiment, the R_o for plasma membranes from basal cells is ~4.4 pmoles/mg of membrane protein, and from insulin-treated cells, ~25 pmoles/mg of membrane protein, with K_d values of ~122 nM and ~143 nM, respectively. Thus, insulin treatment of the intact cell increases the concentration of glucose transporters in the plasma membranes without altering their K_d for cytochalasin B binding.

C. Steady-State Response to Insulin

This same approach has subsequently been used to monitor the concentrations of glucose transporters in the other membranes fractions, altering only the centrifugation speed to ensure recovery of the respective membrane species. The distributions of glucose transporters among all three membrane fractions prepared from basal and insulin-treated cells are illustrated in Fig. 6.[29] In the membranes prepared from basal cells, the concentrations of glucose transporters in both the plasma membranes (PM) and high-density microsomes (HDM) are relatively low, with mean values from 10 experiments of ~7 pmoles/mg of membrane protein in both fractions. By contrast, substantially more glucose transporters (~82 pmoles/mg of membrane protein) are detected in the low-density microsomes (LDM), the so-called intracellular pool of glucose transporters. This distribution is dramatically changed in membranes prepared from insulin-stimulated cells. As seen in Fig. 5, the concentration of glucose transporters in the plasma membranes is increased roughly five-fold. A doubling in the concentration of glucose transporters is also observed in the high-density microsomes. However, the concentration of glucose transporters in the low-density microsomes is reduced about 60%.

The inset in Fig. 6 illustrates the K_d values for cytochalasin B binding among

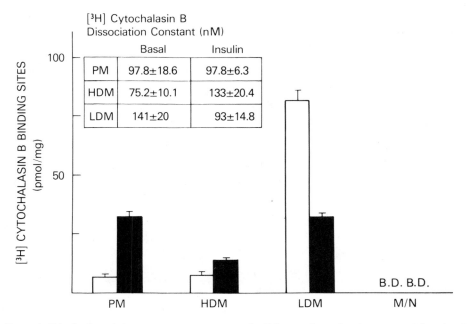

Figure 6. Distribution of glucose transporters among subcellular membrane fractions prepared from basal cells (open bars) and insulin-stimulated cells (closed bars). PM, plasma membranes; HDM, high-density microsomes; LDM, low-density microsomes; M/N, mitochondria/nuclei; B.D., below detection.

these three membrane fractions. While the K_d values are virtually identical in both plasma membrane preparations, insulin appears to exert opposite effects in the other two membrane fractions, increasing the apparent K_d in the high-density microsomes and decreasing that in the low-density microsomes.

A phenomenon similar to that described here has been simultaneously and independently reported by Kono and co-workers,[16,17] who were able to demonstrate changes in the distribution of reconstitutable glucose transport activity. Figure 7 illustrates the correlation between cytochalasin B binding to the various membrane fractions from basal and insulin-stimulated cells and the glucose transport activity reconstitutable from the same fractions.[32] Reconstitution was carried out at two dif-

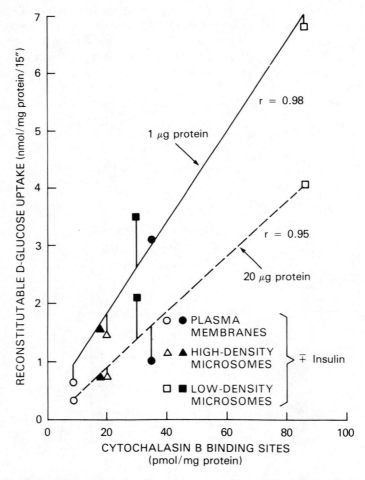

Figure 7. Comparison between reconstitutable D-glucose transport activity and the concentration of glucose transporters in subcellular membrane fractions prepared from basal and insulin-stimulated cells. (From Cushman et al.[32])

ferent protein concentrations. In both cases, the reconstitutable glucose transport activity associated with each membrane fraction correlates extremely well with the observed concentration of cytochalasin B-binding sites. These correlations strengthen the data obtained using either technique and permit direct comparisons of the respective data.

D. Time Course and Reversal of Response to Insulin

The data presented so far describe only the steady-state distribution of glucose transporters in basal and maximally insulin-stimulated cells. However, in order to establish that the translocation of glucose transporters from the low-density microsomes to the plasma membranes is the underlying mechanism for the stimulatory action by insulin, it was important to demonstrate that the movement of transporters is compatible with the rapid, reversible action of insulin on glucose transport activity in the intact cell (Fig. 1). Figure 8 illustrates the distribution of glucose transporters, expressed in picomoles per milligram of membrane protein, between the plasma membranes and low-density microsomes as a function of time of exposure of intact cells to insulin or anti-insulin antiserum.[15] The increase in the concentration of glucose transporters in the plasma membranes in response to insulin is closely mirrored by the decrease in the low-density microsomes, with half-times for both events of ~2 min. Similarly, a reciprocal movement of transporters in the opposite direction is observed when the cells are first stimulated with the minimum concentration of insulin (0.7 nM) needed

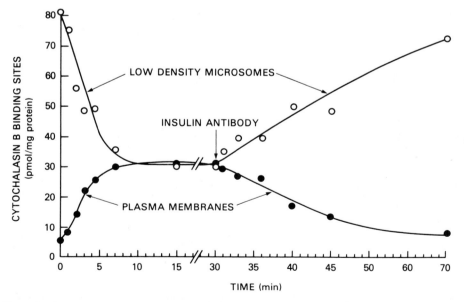

Figure 8. Time course of stimulation of glucose transporter translocation by insulin and its reversal by anti-insulin antiserum.

to achieve a fully stimulated rate of glucose transport, and this stimulation is then reversed by the addition of a 300-fold excess of anti-insulin antiserum.

A direct comparison of the changes in transport activity measured immediately before fractionation of the cells and the concentration of glucose transporters in the subsequently prepared plasma membranes is shown in Fig. 9.[15] In Fig. 9A, the appearance of glucose transporters in the plasma membranes in response to insulin appears to precede the onset of glucose transport activity. By contrast, when the action of insulin is reversed by anti-insulin antiserum, the loss of transport activity and the decrease in the concentration of transporters appear to occur concomitantly.

Figure 10 demonstrates the markedly different temperature dependencies also exhibited by the forward response to insulin and its reversal.[33,34] At 16°C, insulin action readily proceeds with a half-time of ~5 min (Fig. 10A). In fact, the stimulatory action of insulin can be detected at 4°C, although the half-times are considerably extended at temperatures below 16°C. By contrast, the reversal of insulin action in these experiments, induced by rapid insulin degradation using collagenase,[18] is accompanied by a moderately rapid decrease in insulin binding but only a very slow decrease in glucose transport activity. Thus, the half-time for reversal at 16°C is markedly different from that seen for stimulation. In order to perform these experiments, we have employed the technique developed by Kono et al.[17] for blocking either the stimulation or reversal of the translocation process with KCN. Treatment with KCN rapidly and effectively freezes the glucose transporters in their subcellular

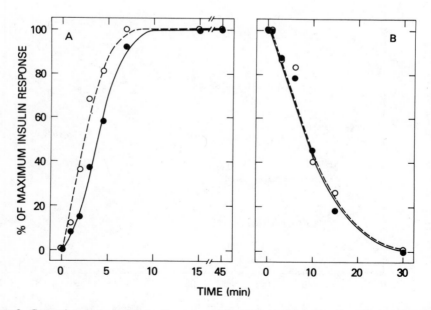

Figure 9. Comparison between (A) the time courses of stimulation of glucose transport activity (●) and translocation of glucose transporters to the plasma membrane (○) by insulin and (B) the time courses of their reversal by anti-insulin antiserum. (From Karnieli *et al.*[15])

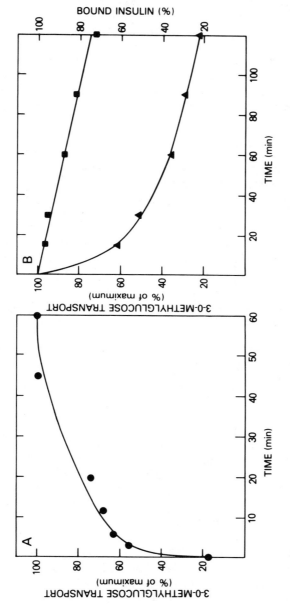

Figure 10. Time courses at 16°C of (A) stimulation of glucose transport activity by insulin and (B) its reversal (■) and the removal of bound insulin (▲) by collagenase. (From Simpson *et al.*[34])

location and permits assaying transport activity at 37°C, where uptake rates are more rapid and can be compared with control cells maintained at 37°C.

E. Stoichiometry of Response to Insulin

The question now arises as to whether these translocations of glucose transporters are stoichiometric and why the magnitude of the response to insulin assessed in the plasma membranes (i.e., four- to six-fold) is not as great as the increase in glucose transport activity measured in the intact cell (10- to 40-fold). The two questions are interrelated in that they require some assessment of the recoveries of glucose transporters in the various membrane fractions. As yet, however, no method has been developed for quantitating the number of glucose transporters in the initial cell homogenates. Both Kono's group and our own have therefore attempted to estimate the recoveries of glucose transporters by using the recoveries of marker enzyme activities.[14,18,29] This approach appears to work reasonably well for glucose transporters in the plasma membranes, although some caution should be exercised as to the choice of marker enzymes (see Simpson et al.[29]). However, as is clearly apparent from Figs. 4 and 6, the distribution of glucose transporters in either the basal or insulin-stimulated states (Fig. 6) does not correlate with the distribution of any of the marker enzymes examined (Fig. 4). More specifically, while the intracellular glucose transporters are found in a subcellular fraction enriched in membranes of the Golgi apparatus, as indicated by the marker enzyme galactosyltransferase, the distributions of glucose transporters and this marker enzyme activity in the other two fractions are clearly dissimilar. Thus, these two membrane proteins do not appear to cofractionate.

We have attempted to correct for this observation by assigning indigenous galactosyltransferase activities to both the plasma membrane and endoplasmic reticulum and then calculating the crosscontamination of each of the various membrane fractions using a computerized matrix equation to solve for the distributions of four marker enzymes simultaneously.[29] These values were used in turn to calculate corrected distributions and recoveries of glucose transporters. The results suggest that the translocation of glucose transporters induced by insulin is stoichiometric and that insulin has no acute effect on the cell's total number of glucose transporters.

In the absence of a marker enzyme activity specific for the vesicles containing the glucose transporters, however, we are unable to arrive at a reasonably accurate estimate of the extent of contamination of the plasma membranes with glucose transporters derived from the intracellular pool. Values as low as 5% would profoundly alter the apparent fold response to insulin. For example, in the basal state, a 5% contamination would represent ~5 pmoles of glucose transporters per milligram of plasma membrane protein, and in the insulin-stimulated state, ~2 pmoles/mg of plasma membrane protein. By adjusting the observed values for this level of contamination, i.e., 7 pmoles/mg − 5 pmoles/mg = 2 pmoles/mg for the plasma membranes from basal cells and 32 pmoles/mg − 2 pmoles/mg = 30 pmoles/mg for the plasma membranes from insulin-stimulated cells, an apparent 15-fold stimulation is observed, which is entirely consistent with the stimulation of glucose transport seen in the intact cell in most laboratories.

IV. THE TRANSLOCATION HYPOTHESIS

These observations taken together represent the basis for the translocation hypothesis for the stimulatory action of insulin on glucose transport in the isolated rat adipose cell, schematically represented in Fig. 11.[15] The first two steps in our working model—the binding of insulin to its receptor (step 1) and the subsequent generation of a signal (step 2)—are delineated in Chapters 3 and 6–12, respectively, and are therefore not discussed further here. However, following generation of the signal, intracellular vesicles containing the glucose transporters are envisaged initially to associate (steps 3 and 4) and subsequently fuse (step 5) with the plasma membrane in a manner analogous to the more firmly established secretory processes.

Evidence for this three-step process is provided by the time-course experiments (Figs. 1, 8, and 9) in which the appearance of glucose transporters in the plasma

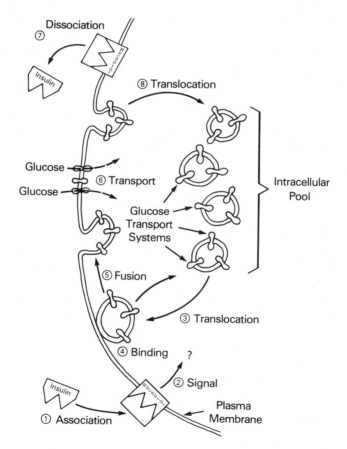

Figure 11. Schematic representation of a hypothetical mechanism of the stimulatory action of insulin on glucose transport (From Karnieli et al.[15])

membranes clearly precedes the onset of glucose transport activity. Furthermore, additional data supporting the existence of step 3 have been obtained from experiments performed at low temperatures[33,34] and in the presence of Tris,[35] in which the translocation of glucose transporters can be induced in the absence of insulin without a corresponding expression of functional glucose transport activity. The latter data also suggest that the action of insulin may be specifically localized to step 5, promoting the fusion of bound vesicles and thus exposing functional glucose transporters to the extracellular environment (step 6). The reversal of this process (step 8), initiated by the removal of insulin from its receptor (step 7), does not appear to proceed via precisely the same mechanism.

While many of the details of this working model are speculative in nature and remain to be clarified and/or established experimentally, strong support for the exocytic-like/endocytic-like features of the model is provided by our recent observations that insulin regulates the binding of insulin-like growth factor II (IGF-II) to the rat adipose cell in a similar fashion.[36] In this case, however, insulin can be directly shown to alter the cycling of the IGF-II receptor between its own large intracellular pool and the plasma membrane by monitoring the binding and internalization of ligand. The latter experimental system should be particularly useful in resolving such unanswered questions as the identity and subcellular localization of the vesicles containing the intracellular pool, the nature of any requirements for protein structure modification during the cycling process, and perhaps even the linking mechanism itself between insulin binding and the translocations of these two predominantly intracellular integral membrane proteins. It is noteworthy that insulin has also recently been shown by our laboratory to induce the rapid cycling of its own receptor between the plasma membrane and at least two distinguishable intracellular membrane compartments.[37-40]

V. COUNTERREGULATION BY CATECHOLAMINES

The counterregulatory effects of insulin on catecholamine-stimulated lipolysis have long been recognized. But it was only comparatively recently that hormones that stimulate adenylate cyclase in the adipose cell were shown to counterregulate the action of insulin. Catecholamines, adrenocorticotropic hormone (ACTH), and glucagon have now been shown to inhibit both basal and insulin-stimulated glucose transport at least in part through a cyclic adenosine monophosphate (cAMP)-mediated mechanism.[34,41-46] However, these inhibitory actions are only manifested when extracellular adenosine is removed from the incubation medium with the enzyme adenosine deaminase. Indeed, in the absence of this enzyme, isoproterenol has been shown to stimulate basal glucose transport activity.[42,43,47]

The effects of adenosine deaminase (1 U/ml) in combination with isoproterenol (1 μM) on the insulin-stimulated rate of 3-O-methylglucose transport and the subcellular distribution of glucose transporters are shown in Fig. 12. Adenosine deaminase alone induces about a 30% reduction in the insulin-stimulated rate of 3-O-methylglucose transport; in combination with isoproterenol, however, a decrease of about 70% in glucose transport activity is observed (Fig. 12A). Qualitatively, reductions in the

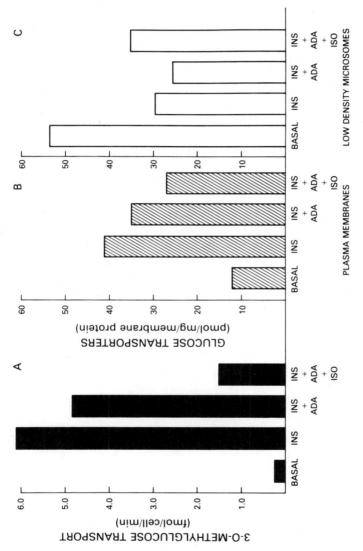

Figure 12. Effects of isoproterenol (ISO) and adenosine deaminase (ADA) on (A) glucose transport activity in and (B,C) the distribution of glucose transporters between the plasma membranes and low-density microsomes from insulin (INS)-stimulated cells.

degree of glucose transporter translocation parallel the loss of glucose transport activity (Fig. 12B,C). However, in neither case is the magnitude of the change in translocation as great as the inhibition of glucose transport activity.

The effects of adenosine deaminase alone and in combination with isoproterenol on glucose transport activity are both characterized by changes in V_{max}, in the absence of changes in K_m.[43,46] Furthermore, the combined effect of adenosine deaminase and isoproterenol can be mimicked by dibutryl-cAMP, and all the observed effects are independent of changes in cellular adenosine triphosphate (ATP) content.[46] These data therefore suggest that several cAMP-dependent levels of control of insulin-stimulated glucose transport exist, one of which modulates the translocation process (step 3, Fig. 11) and another either the intrinsic activity of the glucose transporter (step 6, Fig. 11) or the expression/fusion of transporters associated with the plasma membrane (step 5, Fig. 11).

The action of adenosine, which in packed suspensions of adipose cells can rapidly reach micromolar concentrations in the extracellular medium, is clearly complex. On the basis of our observations with adenosine deaminase, the removal of adenosine results in a roughly 30% decrease in the V_{max} for insulin-stimulated glucose transport[43,46] and is a prerequisite for the expression of either glucagon[45] or catecholamine[46] inhibition of glucose transport. These data suggest that adenosine, via its specific receptor in the plasma membrane, is exerting its insulin-promoting action by inhibiting adenylate cyclase. This is supported by the action of N^6-phenylisopropyladenosine, an adenosine analog that is not degraded by adenosine deaminase, which is capable of reversing the effects of both adenosine deaminase alone and adenosine deaminase in combination with isoproterenol at concentrations (10 nM) compatible with its known interactions with the adenosine receptor.[48]

A further complexity in the role of adenosine has been observed by Green,[45] who demonstrated that adenosine deaminase induces a rightward shift in the dose–response curve for insulin-stimulated glucose transport activity. This effect may well be related to the recent observations that isoproterenol induces a rapid loss of insulin binding.[44,49–51] Thus, both signal generation and the response to insulin appear to be under counterregulatory control.

VI. CHRONIC REGULATION IN ALTERED METABOLIC STATES

A. Insulin Resistance in the High-Fat/Low-Carbohydrate-Fed Rat

We have described several pathophysiological situations in the rat in which the stimulatory action of insulin on glucose transport in the isolated adipose cell is diminished. Examples are the streptozotocin-diabetic rat,[52] the aged obese rat,[53] the high-fat-fed rat,[54,55] and the starved rat.[56] For the purposes of this discussion, we describe only two examples: the high-fat-fed rat, as it most simply demonstrates the phenomenon of insulin resistance that appears to be common to all these models, and the more complex aged obese rat. In the former model, insulin resistance was induced by *ad libitum* feeding two groups of weanling rats for 3 weeks on diets of equicaloric density of either a high-fat/low-carbohydrate composition, 50 : 30 by calories, or a low-fat/high-carbohydrate composition, 9 : 71 by calories. The protein content of both diets was maintained at a constant 20% of total calories. The latter diet is essentially identical to the standard chow used throughout our studies. Over the 3-week investigation, the animals in both groups consumed identical quantities of food, grew at identical rates, and, at the time of sacrifice, contained epididymal adipose cells of the same size.[54]

The ability of insulin to stimulate glucose transport activity in the respective adipose cells is shown in Fig. 13A.[55] A decrease of 15–30% is seen in basal 3-O-methylglucose transport in the cells prepared from the high-fat-fed rats as compared with those from the high-carbohydrate-fed rats; however, this difference is not statistically significant ($p > 0.05$) over three experiments. The 51% reduction in the maximally insulin-stimulated rate of 3-O-methylglucose transport in these cells, on the other hand, is very reproducible. Figure 13B,C illustrates the concentrations of glucose transporters, expressed in picomoles per milligram of membrane protein, in the plasma membranes and low-density microsomes, respectively, prepared from these cells following their incubation in the absence or presence of a maximally stimulating concentration of insulin.[55] No changes in either the yields of plasma membranes or the distributions of marker enzymes between the two membrane fractions are observed. However, the yield of microsomal membrane protein from, and the intracellular water space in, the cells from the high-fat-fed rats are decreased 21% and 28%, respectively.

In the basal state, the concentrations of glucose transporters detected in the plasma membranes are essentially identical (Fig. 13B). By contrast, a 53% reduction is observed in the concentration of glucose transporters in the plasma membranes of insulin-treated cells from high-fat-fed rats as compared with the equivalent control membranes. In the low-density microsomes, on the other hand, reductions in the concentrations of glucose transporters of 48% and 37%, respectively, are observed in the membranes of basal and insulin-treated cells from high-fat-fed rats (Fig. 13C). Thus, the decrease in insulin-stimulated glucose transport activity in intact cells from high-fat-fed rats appears to be directly accounted for by a reduced number of glucose transporters in the plasma membrane. This decreased number of glucose transporters does not appear to be due to an impairment in the translocation mechanism, but rather to an absolute decrease of 59% in the number of glucose transporters residing in the low-density microsomes in the basal state.

The insulin resistance displayed in this model is typical of that also seen in the fasted rat and the streptozotocin-diabetic rat models with respect to the magnitude of the decrease in glucose transport capability. In all cases, this loss of insulin-stimulated glucose transport appears to be due primarily to a decrease in the size of the intracellular pool of glucose transporters per se and not to an impairment in either the intrinsic activity of the glucose transporter or the translocation mechanism. It is perhaps significant that each of these situations is accompanied by a reduction in the levels of circulating insulin, due either to dietary restriction or to a perturbation of insulin secretion, possibly suggesting a role for insulin itself in controlling the cellular levels of glucose transporters. Further support for this concept is provided by at least one of the models in which hyperresponsive glucose transport is seen—the chronic hyperinsulinemic rat.[57] Hyperinsulinemia, induced by the surgical insertion of osmotic

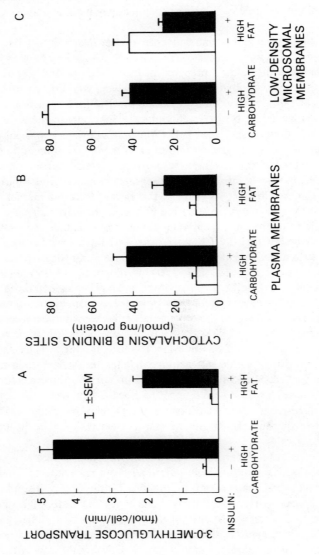

Figure 13. Effects of dietary composition on (A) glucose transport activity in and (B,C) the distribution of glucose transporters between the plasma membranes and low-density microsomes from basal and insulin-stimulated cells.

minipumps containing insulin, is accompanied by higher rates of glucose transport activity that appear to be attributable to an overall increase in the number of glucose transporters.

B. Insulin Resistance in the Aged Obese Male Rat

Hyperinsulinemia is also a characteristic feature of the aged obese rat. In this instance, however, this spontaneous hyperinsulinemia is a reflection of systemic insulin resistance and is accompanied by marked insulin resistance at the glucose transport level in the enlarged adipose cells from these animals.[53] The complexity of the latter is exemplified by the following observations:

First, the enlarged cells from 800-g rats contain ~2.5-fold more glucose transporters per cell than do the small cells from 180-g rats, although this increase is far less than would be expected relative to the enlarged cell's roughly 10-fold greater size. The enlarged cell's intracellular water space is also increased by only about two-fold.

Second, of the enlarged cell's increased number of glucose transporters, most are found in the greatly expanded plasma membrane, while the intracellular pool is relatively depleted, even in the absence of insulin. This subcellular distribution of glucose transporters is similar to that observed in small cells in the maximally insulin-stimulated state, although the concentration of glucose transporters in the enlarged cell's plasma membranes remains at the basal level, as does the intact cell's glucose transport activity per unit cellular surface area. Insulin treatment of the enlarged cell has little, if any, stimulatory effect on glucose transport activity, nor does it further influence the subcellular distribution of glucose transporters.

Third, while the enlarged cell's basal glucose transport activity is increased relative to that observed in the small cells and actually approaches the level obtained in the small cells with insulin treatment, it is not increased to the extent predicted from the calculated number of plasma membrane glucose transporters. Glucose transport activity per glucose transporter therefore appears to be reduced. Thus, while the failure of insulin to stimulate glucose transport in the enlarged cell is fully consistent with a relative depletion of glucose transporters in the intracellular pool, other factors appear to be involved in the overall lack of insulin responsiveness.[53]

VII. STRUCTURE OF THE GLUCOSE TRANSPORTER

This last section is devoted to a discussion of the structure of the rat adipose cell glucose transporter. In contrast to the human erythrocyte glucose transporter, which has been purified to homogeneity and shown to be a heavily glycosylated protein of ~55,000 M_r,[58-61] little information has been available on the structure of the rat adipose cell glucose transporter until very recently. Major reasons for this paucity of data include (1) the absence of a probe of sufficient affinity and specificity for use in monitoring the purification of the glucose transporter, and (2) the relatively low concentrations of glucose transporters in rat adipose cell membranes. For example, studies of cytochalasin B binding indicate that the concentration of glucose transporters in the plasma membranes from insulin-stimulated adipose cells is approximately one order of magnitude lower than that in the human erythrocyte plasma membrane, where glucose transporters (band 4.5) represent about 5% of the total membrane protein.

However, two probes for the rat adipose cell glucose transporter have recently been developed. The first is rabbit antisera prepared against the purified human erythrocyte glucose transporter which, under defined conditions, have been shown to crossreact with the glucose transporter of rat adipose cells.[62,63] The affinities of these antisera for the adipose cell glucose transporter are approximately 1/1000 of those for the erythrocyte glucose transporter. In addition, these antisera appear to be directed against sites that are exposed only when the protein is denatured. These limitations have prevented their use as a direct means for purifying the glucose transporter, but these antisera do recognize the Na dodecylsulfate-denatured glucose transporter when a Western blot procedure is used.

Typical results using this methodology are shown in Fig. 14, which illustrates the interaction of an antiserum prepared by Sogin and Hinkle[60] with the plasma membranes and low-density microsomes prepared from basal and insulin-stimulated rat adipose cells.[32,62] A protein (or proteins) of ~45,000–55,000 M_r appears to crossreact specifically with this antiserum. In addition, since increased interaction is observed in the plasma membranes and a corresponding decreased interaction is observed in the low-density microsomes from the insulin-stimulated as compared with basal cells, this experiment further serves as an independent confirmation of the translocation phenomenon. However, the quantitation by densitometric scanning does not precisely correspond to that observed using the cytochalasin B-binding assay, possibly suggesting the existence of differing molecular forms of the glucose transporter.[62,63]

The second probe, which has now been successfully used by several laboratories, comprises the covalent linking of cytochalasin B itself to the glucose transporter by direct photolysis.[64,65] These studies have confirmed the translocation hypothesis and further suggest the possible presence of two forms of the glucose transporter, of ~45,000 and ~54,000 M_r. We have used a similar approach in our laboratory, in which [³H]cytochalasin B is covalently bound to the glucose transporter via a photoactive bifunctional reagent.[66] As has been observed with direct photolabeling, the incorporation of [³H]cytochalasin B is relatively low (approximately 5%). However, this technique does not require the same high-intensity irradiation, and the extent of nonspecific protein crosslinking appears to be lower.

We have used this labeling technique to determine whether different molecular forms of the glucose transporter exist within the low-density microsomes, as previously suggested by the direct cytochalasin B-photolabeling studies. The results of such an investigation are shown in Fig. 15.[66] In this experiment, membrane proteins were initially covalently labeled as described above, then separated by sodium dodecylsulfate polyacrylamide gel electrophoresis (SDS-PAGE). The proteins of ~35,000–60,000 M_r were subsequently excised, eluted, and further separated by isoelectric focusing in IEF-Sephadex. Under these conditions, one major peak is observed at pH 5.5, with two smaller peaks at pH 4.5 and 4.2, respectively. An additional prominent peak can be observed at pH 6.4 when an isoelectric focusing gel with a broader pH range is

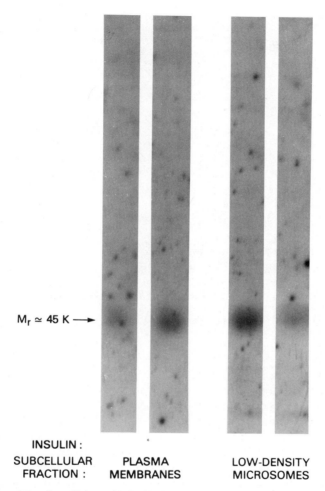

$M_r \simeq 45\ K \longrightarrow$

INSULIN :

SUBCELLULAR   PLASMA   LOW-DENSITY
FRACTION :   MEMBRANES   MICROSOMES

Figure 14. Crossreactivity of an affinity-purified rabbit IgG prepared against the purified human erythrocyte glucose transporter, with the plasma membranes and low-density microsomes from basal and insulin-stimulated cells, as assessed by the Western blot technique. (From Cushman *et al.*[32])

used (unpublished observation). By elution and subsequent Western blotting, we have further been able to demonstrate that the proteins of ~45,000 M_r isolated from these fractions crossreact with the antiserum prepared against the human erythrocyte glucose transporter.[66] Thus, we have confirmed the heterogeneity of the rat adipose cell glucose transporter and have developed methodologies that will ultimately be useful both in purifying these glucose transporters and in monitoring their structure under physiological conditions.

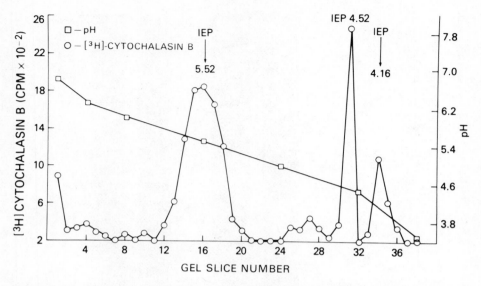

Figure 15. Isoelectric focusing of sodium dodecylsulfate polyacrylamide gel electrophoresis-purified [³H]cytochalasin B photoaffinity-labeled glucose transporters from the low-density microsomes of basal cells IEP, isoelectric point. (From Horuk *et al.*[66])

VIII. SUMMARY

Significant advances have been made during the past few years toward an understanding of the mechanism through which insulin stimulates glucose transport. However, while it now appears that the translocation concept of this mechanism is widely supported by a variety of experimental data, it is equally apparent that this mechanism and the activity of the glucose transporter itself may be subject to a variety of metabolic, hormonal, and physiological controls, the complexity of which is only recently being appreciated. Advances are being made toward purifying the rat adipose cell glucose transporter and, with the subsequent advent of specific antibodies, a further understanding of the intracellular environment of the glucose transporter and of the regulation of its movement may soon be possible. Nevertheless, a major unresolved aspect of the mechanism of insulin action lies in the initial events through which the binding of insulin to its specific receptor signals the ultimate response.

ACKNOWLEDGMENTS.   We wish to thank our many colleagues, both former and current, whose contributions to the concepts and experimental results reported here have been indispensible. These investigators include Kenneth C. Appell, James E. Foley, Peter C. Hinkle, Paul J. Hissin, Richard Horuk, Barbara B. Kahn, Eddy Karnieli, Masao Kuroda, Lester B. Salans, Ulf Smith, Lawrence J. Wardzala, and Thomas J. Wheeler. We also wish to thank Mary Jane Zarnowski, Dena R. Yver, and Steven Richards for their expert technical assistance during this work, and Louie Zalc for her patience and expertise in typing the manuscript.

IX. REFERENCES

1. Wertheimer, E., and Shapiro, B., 1948, *Physiol. Rev.* **28:**451–464.
2. Winegrad, A.I., and Renold, A.E., 1958, *J. Biol. Chem.* **233:**267–272.
3. Vaughan, M., 1961, *J. Lipid Res.* **2:**293–316.
4. Crofford, O.B., and Renold, A.E., 1965, *J. Biol. Chem.* **240:**14–21.
5. Crofford, O.B., and Renold, A.E., 1965, *J. Biol. Chem.* **240:**3237–3244.
6. Rabinowitz, D., 1970, *Annu. Rev. Med.* **21:**241–258.
7. Rodbell, M., 1964, *J. Biol. Chem.* **239:**375–380.
8. Wardzala, L.J., and Jeanrenaud, B., 1981, *J. Biol. Chem.* **256:**7090–7093.
9. Wardzala, L.J., and Jeanrenaud, B., 1983, *Biochim. Biophys. Acta* **730:**49–56.
10. Horuk, R., Rodbell, M., Cushman, S.W., and Wardzala, L.J., 1983, *J. Biol. Chem.* **258:**7425–7429.
11. Cushman, S.W., Karnieli, E., Foley, J.E., Hissin, P.J., Simpson, I.A., and Salans, L.B., 1982, *Clin. Res.* **30:**388A.
12. Vinten, J., Gliemann, J., and Østerlind, K., 1976, *J. Biol. Chem.* **251:**794–800.
13. Wardzala, L.J., Cushman, S.W., and Salans, L.B., 1978, *J. Biol. Chem.* **253:**8002–8005.
14. Cushman, S.W., and Wardzala, L.J., 1980, *J. Biol. Chem.* **255:**4758–4762.
15. Karnieli, E., Zarnowski, M.J., Hissin, P.J., Simpson, I.A., Salans, L.B., and Cushman, S.W., 1981, *J. Biol. Chem.* **256:**4772–4777.
16. Suzuki, K., and Kono, T., 1980, *Proc. Natl. Acad. Sci. USA* **77:**2542–2545.
17. Kono T., Suzuki, L., Dansey, L.E., Robinson, F.W., and Blevins, T.L., 1981, *J. Biol. Chem.* **256:**6400–6407.
18. Kono, T., Robinson, F.W., Blevins, T.L., and Ezaki, O., 1982, *J. Biol. Chem.* **257:**10942–10947.
19. Foley, J.E., Cushman, S.W., and Salans, L.B., 1978, *Am. J. Physiol.* **234:**E112–E119.
20. Livingston, J.N., and Lockwood, D.H., 1974, *Biochem. Biophys. Res. Commun.* **61:**989–996.
21. Olefsky, J.M., 1975, *J. Clin. Invest.* **56:**1499–1508.
22. Loten, E.G., Regen, D.M., and Park, C.R., 1976, *J. Cell Physiol.* **89:**651–659.
23. Gliemann J., Østerlind, K., Vinten, J., and Gammeltoft, S., 1972, *Biochim. Biophys. Acta* **286:**1–9.
24. Gliemann, J., and Rees, W.D., 1983, *Curr. Top. Membr. Transp.* **18:**339–379.
25. Häring, J.V., Kemmler, W., Renner, R., and Hepp, K.D., 1978, *FEBS Lett.* **95:**177–180.
26. Ciaraldi, T.P., and Olefsky, J.M., 1979, *Arch. Biochem. Biophys.* **193:**221–231.
27. Kono, T., and Barham, F.W., 1971, *J. Biol. Chem.* **246:**6210–6216.
28. McKeel, D.W., and Jarett, L., 1970, *J. Cell Biol.* **44:**417–432.
29. Simpson, I.A., Yver, D.R., Hissin, P.J., Wardzala, L.J., Karnieli, E., Salans, L.B., and Cushman, S.W., 1984, *Biochim. Biophys. Acta* **763:**393–407.
30. Kasahara, M., and Hinkle, P.C., 1976, *Proc. Natl. Acad. Sci. USA* **73:**396–400.
31. Wardzala, L.J., 1979, *Identification of the Glucose Transport System in Purified Rat Adipose Cell Plasma Membranes Using a Cytochalasin B Binding Assay: Effects of Insulin and Altered Physiological States*, Ph.D. dissertation, Dartmouth College, Hanover, New Hampshire.
32. Cushman, S.W., Wardzala, L.J., Simpson, I.A., Karnieli, E., Hissin, P.J., Wheeler, T.J., Hinkle, P.C., and Salans, L.B., 1984, *Fed. Proc.* **43:**2251–2255.
33. Simpson, I.A., Zarnowski, M.J., and Cushman, S.W., 1983, *Fed. Proc.* **42:**1790 (abstr.).
34. Simpson, I.A., Karnieli, E., Hissin, P.J., Smith, U., and Cushman, S.W., 1984, *Proc. Gen. Physiol. Soc. USA* **38**(in press).
35. Simpson, I.A., Martin, M.L., and Cushman, S.W., 1982, *Diabetes* **31**(Suppl. 2):2A.
36. Wardzala, L.J., Simpson, I.A., Rechler, M.M., and Cushman, S.W., 1984, *J. Biol. Chem.* **259:**8378–8383.
37. Wang, C.-C, Sonne, O., Hedo, J.A., Cushman, S.W., and Simpson, I.A., 1983, *J. Biol. Chem.* **258:**5129–5134.
38. Sonne, O., and Simpson, I.A., 1984, *Biochem. Biophys. Acta* **804:**404–413.
39. Simpson, I.A., Hedo, J.A., and Cushman, S.W., 1984, *Diabetes* **33:**13–18. (abstr.).
40. Hedo, J.A., and Simpson, I.A., 1984, *J. Biol. Chem.* **259:**11083–11089.
41. Taylor, W.M., Mak, M., and Halperin, M.L., 1976, *Proc. Natl. Acad. Sci. USA* **73:**4359–4363.
42. Kashiwagi, A., and Foley, J.E., 1982, *Biochem. Biophys. Res. Commun.* **107:**1151–1157.
43. Kashiwagi, A., Heucksteadt, T.P., and Foley, J.E., 1983, *J. Biol. Chem.* **258:**13685–13692.

44. Kirsch, D.M., Baumgarten, M., Deufel, I., Rinninger, F., Kemmler, W., and Häring, H.U., 1984, *Biochem. J.* **217**:737–745.
45. Green, A., 1983, *Biochem. J.* **212**:189–195.
46. Smith, U., Kuroda, M., and Simpson, I.A., 1984, *J. Biol. Chem.* **259**:8758–8763.
47. Ludvigsen, C., Jarett, L., and McDonald, J.M., 1980, *Endocrinology* **106**:786–790.
48. Londos, C., Cooper, D.M.F., Schlegel, W., and Rodbell, M., 1978, *Proc. Natl. Acad. Sci. USA* **75**:5362–5366.
49. Pessin, J.E., Gitomer, W., Oka, Y., Oppenheimer, C.L., and Czech, M.P., 1983, *J. Biol. Chem.* **258**:7386–7394.
50. Lönnroth, P., and Smith, U., 1983, *Biochem. Biophys. Res. Commun.* **112**:971–979.
51. Kirsch, D.M., Kemler, W., and Häring, M.U., 1983, *Biochem. Biophys. Res. Commun.* **115**:398–405.
52. Karnieli, E., Hissin, P.J., Simpson, I.A., Salans, L.B., and Cushman, S.W., 1981, *J. Clin. Invest.* **68**:811–814.
53. Hissin, P.J., Foley, J.E., Wardzala, L.J., Karnieli, E., Simpson, I.A., Salans, L.B., and Cushman, S.W., 1982, *J. Clin. Invest.* **70**:780–790.
54. Salans, L.B., Foley, J.E., Wardzala, L.J., and Cushman, S.W., 1981, *Am. J. Physiol.* **240**:E175–E183.
55. Hissin, P.J., Karnieli, E., Simpson, I.A., Salans, L.B., and Cushman, S.W., 1982, *Diabetes* **31**:589–592.
56. Kahn, B.B., and Cushman, S.W., 1984, *Diabetes* **33**(Suppl. 1):71A.
57. Kahn, B.B., Horton, E.S., and Cushman, S.W., 1984, *Clin. Res.* **32**:399A.
58. Kasahara, M., and Hinkle, P.C., 1977, *J. Biol. Chem.* **252**:7384–7390.
59. Gorga, F.R., Baldwin, S.A., and Lienhard, G.E., 1979, *Biochem. Biophys. Res. Commun.* **91**:955–961.
60. Sogin, D.C., and Hinkle, P.C., 1980, *Proc. Natl. Acad. Sci. USA* **77**:5725–5729.
61. Baldwin, S.A., and Lienhard, G.E., 1981, *Trends Biochem. Sci.* **6**:208–211.
62. Wheeler, T.J., Simpson, I.A., Sogin, D.C., Hinkle, P.C., and Cushman, S.W., 1982, *Biochem. Biophys. Res. Commun.* **105**:89–95.
63. Lienhard, G.E., Kin, H.K., Ransome, K.J., and Gorga, J.C., 1982, *Biochem. Biophys. Res. Commun.* **105**:1150–1156.
64. Carter-Su, C., Pessin, J.E., Mora, E., Gitomer, W., and Czech, M.P., 1982, *J. Biol. Chem.* **257**:5419–5425.
65. Shanahan, M.F., Olson, S.A., Weber, M.J., Lienhard, G.E., and Gorga, J.C., 1982, *Biochem. Biophys. Res. Commun.* **107**:38–43.
66. Horuk, R., Rodbell, M., Cushman, S.W., and Simpson, I.A., 1983, *FEBS Lett.* **164**:261–266.

24

Insulin-Dependent Apparent Translocation of Glucose Transport Activity: Studies by the Reconstitution Method

Tetsuro Kono

I. INTRODUCTION

Nearly a decade ago, we found that when rat epididymal adipocytes were exposed to [125I]iodoinsulin for 10 min at 37°C, 125I activity would associate with two subcellular structures that could be fractionated by sucrose density-gradient centrifugation into the plasma membrane fraction (peak 1) and a low-density microsomal fraction (peak 2).[1] The latter peak (peak 2 of 125I activity) was later identified as the peak of the internalized hormone.[2,3] During this study, however, we noticed several similarities between the characteristics of the peak 2 formation, i,e., the endocytotic internalization of the hormone-receptor complex, and those of the hormonal action on glucose transport. Thus, both reactions are very slow at a low temperature, such as 15°C,[2] both reactions require adenosine triphosphate (ATP) not only for their development[2,3] but also for their reversal,[4,5] and both reactions are completed in approximately 5–10 min at 37°C when the hormone concentration is 1 nM.[2,6] We therefore postulated as a working hypothesis that the complex of glucose and its transport carrier might be co-internalized along with the insulin-receptor complex. In theory, one can test this working hypothesis by measuring the distribution of glucose transport activity in subcellular fractions obtained from the basal and plus-insulin states of adipocytes using the aforementioned sucrose density-gradient centrifugation.

Tetsuro Kono ● Department of Physiology, School of Medicine, Vanderbilt University, Nashville, Tennessee.

II. ASSAY OF GLUCOSE TRANSPORT ACTIVITY BY RECONSTITUTION INTO LIPOSOMES

We chose to determine the glucose transport activity in the above-mentioned sub-cellular fractions by reconstituting it into egg lecithin liposomes. The basic methods of reconstitution of glucose transport activity were already worked out by Kasahara and Hinkle[7,8] for the activity from human erythrocytes, by Crane et al.[9] and Fairclough et al.[10] for that from rat intestine, and by Shanahan and Czech[11] for that from rat adipocytes. In our study, we solubilized glucose transport activity with sodium cholate, separated the macromolecular fraction by gel filtration, and incorporated the macromolecules, including the glucose transporter, into egg lecithin liposomes by sonication, freezing, and a second sonication.[12] We then measured the glucose transport activity in the reconstituted liposome system by two methods. In the first method, we incubated the reconstituted liposomes with D-[³H]glucose either in the presence or absence of cytochalasin B and calculated the results on the assumption that the agent would specifically and completely inhibit the carrier-mediated glucose transport activity.[12] In the second method, we incubated the reconstituted liposomes with a mixture of D-[³H]glucose and L-[¹⁴C]glucose and computed the data on the assumption that only D-glucose would be transported by the mediated mechanism.[12,14] The results obtained by the two methods were indistinguishable from each other.[13] In both methods, we terminated the transport reaction by adding cold mercuric chloride solution, filtered the reaction mixture with a piece of Millipore membrane, and measured the radioactivity collected on the filter.[12–14] Details of the reconstitution procedures and the transport assay in the reconstituted system have been reported elsewhere.[15] The overall efficiency of the transport assay, including all the factors involved in the solubilization, reconstitution, and transport assay itself, was approximately 25–35°.[14] The results of the assay were highly reproducible, and the observed transport activity was proportional to the amount of protein used for the reconstitution up to a certain level.[15]

III. DISTRIBUTION OF GLUCOSE TRANSPORT ACTIVITY IN MICROSOMAL SUBFRACTIONS

Figure 1 shows the distribution of glucose transport activity and several marker enzymes in microsomal subfractions (including the plasma membrane-rich fraction) obtained from the basal and plus-insulin forms of adipocytes and separated by linear sucrose density-gradient centrifugation.[14] The data show that insulin (added to cells) considerably increased the glucose transport activity in the plasma membrane-rich fraction (fractions 3–7) while decreasing the activity in the Golgi-rich fraction (fractions 13–17). These observations suggested that (contrary to our initial working hypothesis described in Introduction) insulin might induce externalization of the reserve glucose transport activity from the Golgi-rich fraction to the plasma membrane-rich fraction. The insulin-dependent increase in the glucose transport activity at the plasma membrane peak (fraction 5) was approximately sevenfold, in good agreement with the observation that the insulin-dependent stimulation of glucose transport in intact fat cells was rou-

Figure 1. Distribution of glucose transport activity in subcellular fractions separated by sucrose density-gradient centrifugation. Aliquots of rat epididymal adipocytes were exposed to either 0 or 1 nM insulin for 10 min. Their crude microsomal fractions (including the plasma membranes) were subfractionated by a linear sucrose density-gradient centrifugation (15.0–32.5% 35,000 r.p.m., 40 min). Note that insulin treatment increased the glucose transport activity in the plasma membrane-rich fraction (fractions 3–7) while decreasing the activity in the Golgi-rich fraction (fraction 13–17). (△) Protein, 10 × (mg/ml); (▲) 5'-nucleotidase, 0.25 × (nmol/min/ml); (▽) UDPGal-N-acetylglucosamine galactosyltransferase, 20 × (nmol/min/ml); (○) basal glucose transport, 0.5 × (nmol taken up at 20s/ml); (●) plus insulin glucose transport, 0.5 × (nmol taken up at 20s/ml); (x) NADH dehydrogenase, 1 × (μmol/min/ml). (From T. Kono et al.[14])

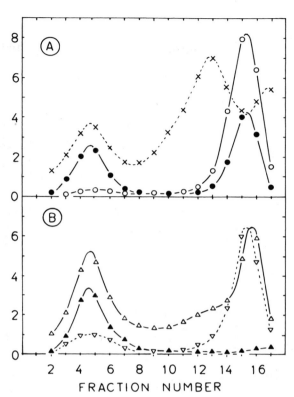

tinely five- to tenfold.[14] In our earlier reconstitution studies, however, the apparent insulin-dependent increase in the transport activity in the plasma membrane-rich fraction was only two- to threefold,[12,13] presumably because the fraction was considerably contaminated by the Golgi-rich fraction, which contained glucose transport activity affected by insulin toward the opposite direction.[14] Insulin had no detectable effect on the distribution of any of the marker enzymes tested.

IV. TIME COURSE OF INSULIN ACTION, EFFECTS OF INSULIN MIMICKERS, AND EFFECTS OF METABOLIC INHIBITORS

As shown in Fig. 2, when insulin was administered to cells at 0 time, the glucose transport activity in the plasma membrane-rich fraction was increased while the activity in the Golgi-rich fraction was decreased, and both reached steady-state levels within 5–10 min.[14] Although the change in the transport activity observed in the plasma membrane-rich fraction was less than that in the Golgi-rich fraction, this difference was thought to be caused by a difference in the recoveries of the two subcellular fractions.[14] Unfortunately, the precise degrees of recovery of the two fractions were

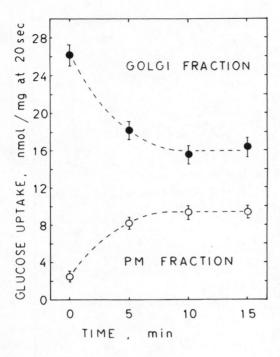

Figure 2. Time courses of insulin effects. Aliquots of adipocytes were exposed to 1 nM insulin at 0 min. The exposure was terminated at the indicated time points, the plasma membrane-rich and Golgi-rich fractions were separated, and the glucose transport activity in each fraction was determined. PM, plasma membrane. (From Kono *et al.*[14])

unknown for lack of specific markers; for example, UDP-Gal : *N*-acetylglucosamine galactosyltransferase forms a distinct peak in the plasma membrane-rich fraction in addition to a major peak in the Golgi-rich fraction (see Fig. 1). The effects of insulin on the glucose transport activity in the two subcellular fractions were reversible (data not shown); when the hormone was eliminated from the system by proteolytic mod-ification, the glucose transport activity in the plasma membrane-rich fraction was decreased, while the activity in the Golgi-rich fraction was increased, and both reached steady-state (basal) levels within 45–60 min.[14]

This effect of insulin to induce the apparent translocation of glucose transport activity was mimicked by hydrogen peroxide, vanadate, *p*-chloromercuriphenylsul-fonate, or low concentrations of trypsin.[14] All these agents are known to have in-sulinlike effects on glucose transport activity in intact adipocytes.[14] Significantly, those agents that had large effects on the glucose transport activity in the plasma membrane-rich fraction also had large effects on the apparent translocation, and *vice versa*.[14] In contrast, mechanical agitation, which stimulates glucose transport in adi-pocytes,[6] did not show any significant insulin-like effect in our preliminary recon-stitution experiments (T. Kono, unpublished observation); the reason for this is not clear.

The actions, as well as the reversal of the actions, of insulin on glucose transport activity in the plasma membrane-rich and Golgi-rich fractions were all blocked when the ATP level was reduced by treatment of cells with 1 mM 2,4-dinitrophenol or 2 mM KCN.[13] These agents are known to block the action, as well as the reversal of the action, of insulin on glucose transport in intact adipocytes.[2,4,5] The inhibitory effects of 2,4-dinitrophenol and KCN are reversible; when cells treated with these agents are washed and incubated with fresh buffer for 10 min in the presence of 2 mM glucose or pyruvate, both the ATP level and the cellular response to the hormone are almost completely restored.[2,5] As is well known, 2,4-dinitrophenol is an uncoupler of oxidative phosphorylation, while KCN inhibits cellular respiration at the cytochrome level. By contrast, neither the actions of insulin nor the reversal of the hormonal action was inhibited by 0.1 mM puromycin or 1 mM cycloheximide, which under the given conditions inhibited more than 97% of protein synthesis in adipocytes.[12] Thus, protein synthesis does not appear to be involved in the increase in glucose transport activity in either subcellular fraction, at least during our short experimental period (less than 1 hr), suggesting that the glucose transport activity is recycled or shuttled between the two subcellular fractions.[13]

V. APPARENT ABNORMALITY IN THE TEMPERATURE COEFFICIENT

Earlier data reported by Czech,[16] Vega and Kono,[6] Whitesell and Gliemann,[17] and Amatruda and Finch[18] indicated the glucose transport activity in the basal and plus-insulin forms of fat cells appears to have different temperature coefficients. More specifically, our earlier data indicated that the basal glucose transport activity at 25°C was higher than that at 37°C, while the plus-insulin activity at 25°C was lower than that at 37°C.[6] Initially, these data were thought to be incompatible with the translocation hypothesis. Later, it was found that the apparent temperature coefficient of the basal activity was distorted by an insulinlike effect of low temperature.[20] Thus, when the temperature was lowered, the glucose transport activity was slowly translocated from the Golgi-rich fraction to the plasma membrane-rich fraction by an energy-dependent process.[20] The degree of this insulinlike effect was greater as the temperature was lowered from 37° to 10°C, but the rate of translocation was rendered very slow at lower temperatures.[20] When this insulinlike action of low temperature (or temperature-dependent translocation) was blocked by the addition of 2,4-dinitrophenol, the basal and plus-insulin forms of adipocytes exhibited identical temperature coefficients for glucose transport.[20] From the Arrhenius plot of the temperature coefficient, it was calculated that the heat of activation for the transport of 3-O-methyl-D-glucose in the basal form of fat cells was 10.5 kcal/mole, and that in the plus-insulin form of cells was 10.7 kcal/mole.[20] It was concluded that the apparent abnormality observed in the temperature coefficient (see above) is not in disagreement with the translocation hypothesis. It is further suggested that the insulinlike effect of low temperature may also affect glucose transport activity in muscle, since Brown et al.[21] and Yu and Gould[22] reported that glucose transport activity in rat diaphragm and soleus muscle was stimulated when the tissue was chilled to 0°C. In addition, Kipnis and Cori[23] observed that the temperature coefficients of glucose transport were apparently different in the basal and plus-insulin forms of rat diaphragm. As for the mechanism of the insulinlike effect of low temperature, it can be speculated that a low temperature might generate an insulinlike signal, or it might simply shift the equilibrium distribution of the transport activity between the Golgi-rich and the plasma membrane-rich fractions.[20]

VI. TRANSLOCATION HYPOTHESIS OF INSULIN ACTION ON GLUCOSE TRANSPORT

The above data suggest that glucose transport activity in adipocytes is recycled between the Golgi-rich fraction and the plasma membrane-rich fraction by a mechanism that is both energy dependent and protein synthesis independent and that the distribution of the activity in the two subcellular fractions is determined by the presence or absence of insulin or certain other factors.[2,14] The mechanism of this proposed recycling, or translocation, of glucose transport activity is unknown; however, we tentatively postulate that insulin induces the incorporation of certain intracellular vesicles associated with glucose transport activity into the plasma membrane by exocytosis and that the action of insulin is reversed by endocytotic internalization of the transport activity, as schematically presented in Fig. 3. The basis of this suggestion is our initial observation that there are several similarities between the characteristics of insulin action on glucose transport and those of the endocytotic internalization of the insulin-receptor complex (see Introduction). As is well documented, endo- and exocytosis are energy-dependent processes, and their activities are substantially reduced below a threshold temperature of around 20°C,[24] although significant levels of endocytosis[25,26] and insulin action[20] are both detectable at lower temperatures, such as 10°C.

The concept of recycling of plasma membrane proteins is not new; it was described

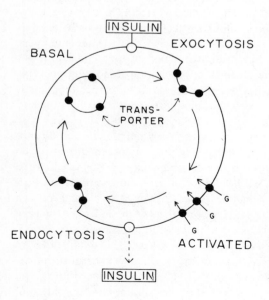

Figure 3. Proposed mechanism of insulin action on glucose transport. It is hypothesized that insulin causes exocytotic translocation of the glucose transport mechanism from certain intracellular vesicles to the plasma membrane and that elimination of the hormone triggers endocytotic internalization of the glucose transport apparatus from the plasma membrane to the intracellular site.

by Palade in 1975[27] and was considered at a recent symposium.[28] Recycling of the low-density lipoprotein (LDL) receptor[29] and that of the insulin receptor[30] was recently discussed as well. Apparently, only selected species of proteins are internalized and recycled, excluding other proteins, such as 5'-nucleotidase.[28] At least in some of the recycling processes, the Golgi apparatus appears to play an important role.[28] This view is consistent with our observation that the distribution of the intracellular glucose transport activity in the sucrose density-gradient fractions coincides with that of the Golgi marker enzyme (Fig. 1). In fact, our attempts to separate these two activities have been futile (T. Kono, unpublished results). At the same time, we have no positive evidence to support the notion that the transport activity is physically associated with the Golgi vesicles; rather, our preliminary data indicate that the action of insulin may not be inhibited by monensin (T. Kono, unpublished observations), which reportedly inhibits Golgi-mediated recycling.[31] In other words, it is yet to be determined whether the intracellular glucose transport activity is associated with the Golgi apparatus or with certain other vesicles that happened to have sedimentation characteristics identical to those of the Golgi vesicles. If the latter is the case, the vesicles might be conveniently localized just beneath the plasma membrane. Although involvement of microfilaments and microtubules is often implicated in the intracellular movement of vesicles,[24,28] our preliminary data indicate that the apparent translocation of the glucose transport activity is not detectably blocked by cytochalasin B, vinblastin, or colchicine (T. Kono, unpublished observations). However, these agents did not block the endocytotic internalization of insulin either (K. Suzuki and T. Kono, unpublished results). Therefore, the physiological significance of the above negative data is not clear. It may be noted at this point that the distribution of internalized [^{125}I]iodoinsulin in the sucrose density-gradient fractions (peak 2, see Introduction) does not coincide with that of the intracellular glucose transport activity, as we recently reported elsewhere.[43] This observation suggests that the two activities are associated with different subcellular vesicles and that the routes of their recycling may be different. Furthermore, it is yet to be ascertained how insulin induces the proposed translocation. A possible candidate for the hormonal signal is an increase in Ca^{2+} concentration, since it is generally accepted that the exocytotic discharge of neurotransmitters at the synaptic junction is preceeded by an increase in intracellular Ca^{2+} concentration. In addition, involvement of Ca^{2+} in the insulin action has been repeatedly suggested.[32] However, recent work by Hall et al.[33] and Eckel et al.[34] suggests otherwise. On the other hand, several laboratories recently reported that the actions of insulin on glycogen synthase and pyruvate dehydrogenase are mediated by a certain peptide,[35] although it is yet to be ascertained whether glucose transport is stimulated by this putative insulin mediator. As discussed in detail elsewhere,[36] insulin appears to regulate (1) glycogen synthase, pyruvate dehydrogenase, and other enzymes by stimulating dephosphorylation of the enzyme proteins (type A action), (2) phosphodiesterase by a reaction that involves phosphorylation (type B action), and (3) glucose transport by facilitating translocation of the transport activity (type C action). It is interesting to speculate whether these plural actions of insulin are mediated by a single common signal or by multiple signals.

VII. RELATED OBSERVATIONS

The results of our studies are basically in good agreement with the data independently obtained by Cushman and associates[37,38] and described in Chapter 23. Whereas we measured glucose transport activity in the microsomal subfractions using the reconstitution method, Cushman's group estimated the number of the transport carriers by measuring the D-glucose-inhibitable cytochalasin B-binding activity. Since cytochalasin B is a competitive inhibitor of glucose transport, it is assumed that the agent binds to the glucose transport carrier.[37,38] It is of interest that two groups of investigators independently obtained basically identical results using different experimental techniques. More recently, Wheeler et al.[39] and Lienhard et al.[40] found that the glucose transport protein from rat epididymal adipocytes could be titrated with the antibody against glucose transport protein from human erythrocytes. The results of their immunological studies were also consistent with the translocation hypothesis. The translocation hypothesis was further supported by Gorga and Lienhard,[41] who estimated the number of glucose transport carriers by incorporating it into a large number of liposomes.

While all these studies were carried out with adipocytes, Wardzala and Jeanrenaud[42] obtained the cytochalasin B-binding data in the rat diaphragm system, which were in agreement with the translocation hypothesis. Similarly, we recently obtained the reconstitution data in the rat heart system, which are consistent with the translocation theory.[44]

ACKNOWLEDGMENTS. Our original glucose transport studies, carried out in collaboration with Dr. Kazuo Suzuki, Lynn E. Dansey, Frances W. Robinson, Teresa L. Blevins, Dr. Osamu Ezaki, Dr. Tomoyuki Watanabe, and Melinda M. Smith, were supported by United States Public Health Service grant RO1 AM 19925 and by Juvenile Diabetes Foundation grant 80-R-340. This manuscript and most of our original articles were typed by Ms. Patsy Barrett.

VIII. REFERENCES

1. Kono, T., Robinson, F.W., and Sarver, J.A., 1975, J. Biol. Chem. 250:7826–7835.
2. Kono, T., Robinson, F.W., Sarver, J.A., Vega, F.V., and Pointer, R.H., 1977 J. Biol. Chem. 252:2226–2233.
3. Suzuki, K., and Kono, T., 1979, J. Biol. Chem. 254:9786–9794.
4. Kono, T., Vega, F.V., Raines, K.B., and Shumway, S.J., 1977, Fed. Proc. 36:341.
5. Vega, F.V., Key, R.J., Jordan, J.E., and Kono, T., 1980, Arch. Biochem. Biophys. 203:167–173.
6. Vega, F.V., and Kono, T., 1979, Arch. Biochem. Biophys. 192:120–127.
7. Kasahara, M., and Hinkle, P.C., 1976, Proc. Natl. Acad. Sci. USA 73:396–400.
8. Kasahara, M., and Hinkle, P.C., 1977, J. Biol. Chem. 252:7384–7390.
9. Crane, R.K., Malathi, P., and Preiser, H., 1976, FEBS Lett. 67:214–216.
10. Fairclough, P., Malathi, P., Preiser, H., and Crane, R.K., 1979, Biochim. Biophys. Acta 553:295–306.
11. Shanahan, M.F., and Czech, M.P., 1977, J. Biol. Chem. 252:8341–8343.
12. Suzuki, K., and Kono, T., 1980, Proc. Natl. Acad. Sci. USA 77:2542–2545.

13. Kono, T., Suzuki, K., Dansey, L.E., Robinson, F.W., and Blevins, T.L., 1981, *J. Biol. Chem.* **256:**6400–6407.
14. Kono, T., Robinson, F.W., Blevins, T.L., and Ezaki, O., 1982, *J. Biol. Chem.* **257:**10942–10947.
15. Robinson, F.W., Blevins, T.L., Suzuki, K., and Kono, T., 1982, *Anal. Biochem.* **122:**10–19.
16. Czech, M.P., 1976, *Mol. Cell. Biochem.* **11:**51–63.
17. Whitesell, R.R., and Gliemann, J., 1979, *J. Biol. Chem.* **254:**5276–5283.
18. Amatruda, J.M., and Finch, E.D., 1979, *J. Biol. Chem.* **254:**2619–2625.
19. Kono, T., 1982, in *Membranes and Transport* (A.N. Martonosi, ed.), Vol. 2, pp. 551–554, Plenum, New York.
20. Ezaki, O., and Kono, T., 1982, *J. Biol. Chem.* **257:**14306–14310.
21. Brown, D.H., Park, C.R., Daughaday, W.H., and Cornblath, M., 1952, *J. Biol. Chem.* **197:**167–174.
22. Yu, K.T., and Gould, M.K., 1981, *Diabetologia* **21:**482–488.
23. Kipnis, D.M., and Cori, C.F., 1957, *J. Biol. Chem.* **224:**681–693.
24. Silverstein, S.C., Steinman, R.M., and Cohn, Z.A., 1977, *Annu. Rev. Biochem.* **46:**669–722.
25. Steinman, R.M., Silver, J.M., and Cohn, Z.A., 1974, *J. Cell Biol.* **63:**949–969.
26. Weigel, P.H., and Oka, J.A., 1981, *J. Biol. Chem.* **256:**2615–2617.
27. Palade, G., 1975, *Science* **189:**347–358.
28. 1982, *Ciba Foundation Symposium 92: Membrane Recycling,* (D. Evered and G.M. Collins, eds.) Pitman, London.
29. Anderson, R.G.W., Brown, M.S., Beisiegel, U., and Goldstein, J.L., 1982, *J. Cell Biol.* **93:**523–531.
30. Marshall, S., Green, A., and Olefsky, J.M., 1981, *J. Biol. Chem.* **256:**11464–11470.
31. Basu, S.K., Goldstein, J.L., Anderson, R.G.W., and Brown, M.S., 1981, *Cell* **24:**493–502.
32. Czech, M.P., 1977, *Annu. Rev. Biochem.* **46:**359–384.
33. Hall, S., Keo, L., Yu, K.T., and Gould, M.K., 1982, *Diabetes* **31:**846–850.
34. Eckel, J., Pandalis, G., and Reinauer, H., 1983, *Biochem. J.* **212:**385–392.
35. Larner, J., Cheng, K., Schwartz, C., Kikuchi, K., Tamura, S., Creacy, S., Dubler, R., Galasko, G., Pullin, C., and Katz, M., 1982 *Recent Prog. Horm. Res.* **38:**511–556.
36. Kono, T., 1983, *Recent Prog. Horm. Res.* **39:**519–557.
37. Cushman, S.W., and Wardzala, L.J., 1980, *J. Biol. Chem.* **255:**4758–4762.
38. Karnieli, E., Zarnowski, M.J., Hissin, P.J., Simpson, I.A., Salans, L.B., and Cushman, S.W., 1981, *J. Biol. Chem.* **256:**4772–4777.
39. Wheeler, T.J., Simpson, I.A., Sogin, D.C., Hinkle, P.C., and Cushman, S.W., 1982, *Biochem. Biophys. Res. Commun.* **105:**89–95.
40. Lienhard, G.E., Kim, H.H., Ransome, K.J., and Gorga, J.C., 1982, *Biochem. Biophys. Res. Commun.* **105:**1150–1156.
41. Gorga, J.C., and Lienhard, G.E., 1982, *Fed. Proc.* **41:**627.
42. Wardzala, L.J., and Jeanrenaud, B., 1981, *J. Biol. Chem.* **256:**7090–7093.
43. Ezaki, O., and Kono, T., 1984, *Arch. Biochem. Biophys.* **231:**280–286.
44. Watanabe, T., Smith, M.M., Robinson, F.W., and Kono, T., 1984, *J. Biol. Chem.,* **259:**13117–13122.

25

Insulin Action on Membrane Components: The Glucose Transporter and the Type II Insulinlike Growth Factor Receptor

Yoshitomo Oka and Michael P. Czech

I. INTRODUCTION

It has been known for more than three decades that a major cellular target of insulin action is glucose transport activity.[1,2] In two major target tissues—muscle and fat—insulin action on hexose transport occurs within minutes and results in up to 10-fold stimulations (for reviews, see ref. 3 and 4). The insulin-responsive glucose transport systems are of the facilitated diffusion type, while the Na^+-dependent, active glucose transport systems appear not to be insulin sensitive. Interestingly, in most cell types other than muscle and fat containing glucose transporters that operate by a facilitated diffusion mechanism, insulin appears to have little or no effect on glucose transport. Nevertheless, the effect of insulin on glucose uptake in muscle and adipose tissue contributes significantly to the hypoglycemic effect of insulin in intact animals and man. Therefore, the underlying biochemical mechanism of insulin action on hexose transport is of clear physiological and clinical significance.

One approach taken by many investigators to the mechanism of insulin action on hexose transport has been to test the hypothesis that hexose transporter proteins are modified, either covalently or noncovalently, by the effect of insulin action. Until a few years ago, this was the predominant approach. An experimental strategy used by several laboratories involved the use of reagents with distinct chemical specificity to

Yoshitomo Oka and Michael P. Czech • Department of Biochemistry, University of Massachusetts Medical School, Worcester, Massachusetts.

probe the activity of glucose transport in insulin-sensitive cells. For example, sulfhydryl oxidants were found to mimic the stimulatory effect of insulin on adipocyte hexose transport.[5,6] However, a large number of agents are now known to mimic insulin action, and they include a wide range of specificities. Thus, mild trypsinization,[7] lectins,[8–10] low levels of neuraminidase,[11,12] and the absence of K^+[13] all mimic insulin action on glucose uptake in fat cells. The diversity of these agents has unfortunately confounded attempts to identify the specific underlying biochemical mechanisms of transport activation.

Another approach has been to develop and utilize reconstitution methodologies for the study of hexose transport activity in defined lipid environments and for use as assay systems during hexose transporter purification. Our laboratory first succeeded in reconstituting hexose transporters from an insulin-sensitive cell type—rat adipocytes.[14] Using this technique, the adipocyte glucose transporter was partially purified and could be readily dissociated from insulin receptors in detergent solution.[15] It should be pointed out, however, that upon homogenization of fat cells, the ability of insulin to activate glucose transport in membrane fractions or in reconstituted systems is lost. Thus, the activation mechanism itself cannot be studied directly in broken cell systems. On the other hand, plasma membranes derived from insulin-treated fat cells do exhibit increased hexose transport activity as compared with control membranes.[16] Attempts to fully purify glucose transporters from an insulin-sensitive tissue have to date been unsuccessful.

An alternative hypothesis to the concept that hexose transport activation results from modification of transporter proteins has derived from novel experiments reported independently from the laboratories of Kono[17] and Cushman.[18] Suzuki and Kono[17] found that the increased glucose transport activity of plasma membranes derived from insulin-treated cells was accompanied by a parallel decrease in transporter activity in a low density microsome fraction from the same insulin-treated cells. Data from Cushman's group[18] showed a similar phenomenon, with [^{3}H]cytochalasin B binding to membranes used to estimate the number of glucose transporters present in the membranes. These groups suggested that the insulin-mediated redistribution of hexose transporters from one membrane fraction to another reflected a movement of transporters from an intracellular pool to the cell-surface membrane.[19,20] The data from these laboratories and this recruitment hypothesis are discussed in detail in Chapters 23 and 24 of this volume.

The focus of studies in our laboratory over the past 2 years related to glucose transport activation by insulin has been to test the basic unproved assumption of the recruitment hypothesis. This assumption is that the low-density microsome fraction displaying high levels of glucose transport activity in control cells indeed represents an intracellular membrane fraction that is not exposed to the extracellular medium in intact cells. Although an intracellular location for low-density membrane transporters is consistent with marker enzymes in this fraction that suggest a possible Golgi origin, other possible interpretations have not been excluded. An alternative hypothesis we have tested is that the low-density membranes containing glucose transporters represent a microdomain of the cell surface and are sheared from the plasma membranes during homogenization of cells. If insulin were to act by eliciting structural changes in these membrane domains such that they were stabilized during homogenization or that lateral movement of transporters away from these membrane domains were to result, the same apparent loss of transporters from the low-density microsome fraction in response to insulin would be observed experimentally. There are known examples whereby plasma membrane microdomains indeed shear during cell lysis and fractionation.[21] With these considerations in mind, we set out to establish an experimental strategy that would rigorously test whether hexose transporters in intact cells, prior to cell lysis and membrane fractionation, reside in exposed dispositions on the cell surface or in unexposed positions in the intracellular domain.

A second model system we have employed for studying the action of insulin on membrane components such as glucose transporters involves the regulation of the type II insulinlike growth factor (IGF) receptor. It has been known for several years that $[^{125}I]$-IGF-II binding to isolated fat cells is rapidly increased by insulin action.[22,23] More recently, we were able to document that this increased binding reflects an increased activity of the type II 250,000 M_r IGF receptor structure.[24] This receptor was first identified by affinity crosslinking[25-27]; subsequent work has resulted in its complete purification[28,29] and the analysis of some of its properties (for review, see ref. 30). During the course of our studies on this receptor, we discovered that, like the glucose transport system, the IGF-II receptor exhibited a redistribution among membrane fractions when derived from insulin-treated cells.[24] The time course and dose–response relationship for this insulin-mediated apparent movement of IGF-II receptors was found to be strikingly similar to that for hexose transporters.[24] We therefore set out to test rigorously whether in intact cells the IGF-II receptor was changed from a nonexposed, intracellular domain to an exposed, cell-surface membrane location.

The aim of this chapter is to summarize our recent results related to the cellular location of glucose transporters and IGF-II receptors both before and after insulin treatment. In order to resolve this issue, it was critically important to employ intact cells and to develop novel biochemical methodologies.

II. CELLULAR LOCUS OF GLUCOSE TRANSPORTERS

Localization of hexose transporters in intact cells requires a means for biochemical identification of the transporters using specific anti-hexose transporter antibody, affinity-labeling techniques, or other such methods. Antibody preparations raised against purified red cell hexose transporters have been reported to crossreact with adipocyte transporters using the Western blot technique.[31,32] However, no data have appeared suggesting that such antibody preparations are useful in binding hexose transporters in intact membranes or isolated cells. Our approach has been to develop a photolabeling technique using [^{3}H]cytochalasin B as the labeling reagent. A number of years ago, we demonstrated that cytochalasin B was a potent inhibitor of hexose transport in insulin-sensitive rat adipocytes.[33] Both basal and insulin-stimulated hexose transport activity are abolished by this compound, as is the case for other cell types. Tritiated cytochalasin B binds with high affinity to the red cell glucose transporter at a site that appears to be on the cytoplasmic side of the red cell surface membrane.[34] The binding of [³H]cytochalasin B is highly specific for the hexose transporter and is inhibited by hexoses that have affinity for the transporter but not those that are not transported.[34] Cushman and colleagues demonstrated that [³H]cytochalasin B binds to the insulin-sensitive adipocyte hexose transport system with similar characteristic specificity.[35]

Using the above observations as the basis of our strategy, we reasoned that photoactivation of a reaction mixture of [³H]cytochalasin B and membranes containing hexose transporters might spontaneously catalyze covalent insertion of the [³H]cytochalasin B into the hexose-transporter protein. Incubation of intact red cells or red cell membranes with [³H]cytochalasin B followed by irradiation with high-intensity ultraviolet (UV) light indeed resulted in the intense labeling of a 45,000–60,000-M_r band identified on sodium dodecylsulfate polyacrylamide gels.[36] As depicted in Fig. 1, the labeling of this membrane protein fraction was markedly inhibited by 500 mM D-glucose or another transported hexose, 3-O-methylglucose, but not by 500 mM sorbitol or L-glucose. Significantly, incubation of [³H]cytochalasin B with membranes in the pres-

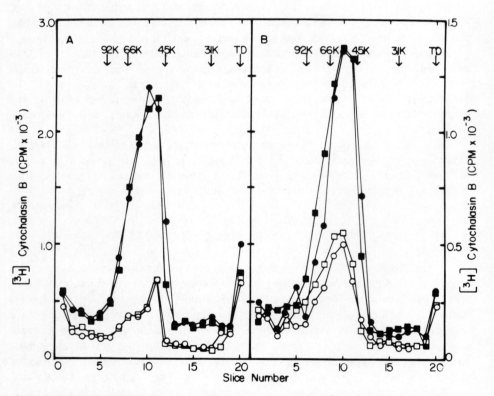

Figure 1. Photoaffinity labeling of erythrocyte membranes with 1.0 μM [³H]cytochalasin B in the presence of various monosaccharides. Erythrocyte ghosts (A) or dimethylmaleic anhydride-extracted ghosts (B) were incubated with 500 mM D-sorbitol (●), 500 mM L-glucose (■), 500 mM D-glucose (□), and 500 mM 3-O-methylglucose (○). (From Carter-Su *et al.*[36])

ence of cytochalasin E (an analog that does not inhibit hexose transport) did not result in inhibition of the photoaffinity labeling reaction. Maximal labeling of the 45,000–60,000-M_r transporter component was achieved with 2 μM [³H]cytochalasin B, while half-maximal labeling was observed at 0.7 μM [³H]cytochalasin B. These results documented an effective methodology for affinity labeling hexose transporters in erythrocytes or erythrocyte membranes.

The hexose transporter photoaffinity labeling technique is readily applied to other cell types. Plasma membrane fractions from chick embryo fibroblasts are effectively photolabeled with [³H]cytochalasin B in order to identify hexose transporter components.[37] We observe that glucose starvation of these cells results in a parallel marked increase in the hexose transport activity and photolabeling intensity of two components of 46,000 M_r and 52,000 M_r.[37] These data suggest that two transporter systems may operate in chick embryo fibroblasts. We[38] and others[39] also demonstrated successful photolabeling of hexose transporters in membranes derived from an insulin-sensitive cell type—rat adipocytes. In such membranes, the photolabeled protein migrates at 46,000 M_r on sodium dodecylsulfate polyacrylamide gels. The intensity of photolabeling is increased in plasma membranes derived from insulin-treated cells and is decreased in low-density microsomal membranes from such cells.[38] These results demonstrate the utility of this methodology for a number of diverse cell types.

In order to address the question of whether significant numbers of hexose transporters resided in an intracellular domain in intact cells, we applied the photoaffinity labeling procedure to hexose transporters in intact adipocytes.[40] Our methodology for labeling hexose transporters in intact cells is similar to that described for photoaffinity labeling transporters in isolated membranes, but several important modifications are necessary to achieve optimal labeling.[40] Tritiated cytochalasin B (10–15 Ci/mm, New England Nuclear) in 100% ethanol is dried under nitrogen and resuspended in Krebs–Ringer phosphate buffer pH 7.4. A fat cell suspension (3–3.5 × 10⁶ cells/ml) is placed onto a 50-mm-diameter plastic dish and incubated with 0.5 μM [³H]cytochalasin B and 10 μM unlabeled cytochalasin D, which has much less affinity for hexose transporters, in the presence or absence of 3-O-methylglucose. After various incubation times, the samples can be irradiated three times for 10 sec each with a 1000-W Porter-cure lamp (American Ultraviolet Co.) through a glass color filter (Farrand Optical Co., No. 7-54), which transmits only UV light in a range of 260–400 nm at a distance of 26 cm.

It is important to irradiate cells through the glass color filter in order to maintain cell viability due to potentially harmful effects of the intense light emitted by the mercury lamp. The presence of 10 μM unlabeled cytochalasin D is important in order to inhibit [³H]cytochalasin B incorporation into nontransporter proteins that bind cytochalasin B nonspecifically. Such proteins that are labeled by [³H]cytochalasin B are bovine serum albumin (BSA) and actin.[40] The presence of the cytochalasin D has little or no effect on the photolabeling reaction of hexose transporter proteins. After the photolabeling reaction of intact adipocytes, the cells are disrupted by homogenization, and specific membrane fractions are purified by standard procedures.[24,41] Analysis of hexose transporter labeling is readily performed by sodium dodecylsulfate polyacrylamide gel electrophoresis (SDS-PAGE).

Figure 2 presents the results of an experiment in which intact cells were irradiated with 0.5 μM [³H]cytochalasin B and 10 μM unlabeled cytochalasin D in the presence and absence of 100 mM 3-O-methylglucose. Clearly apparent is the labeling of a 46,000-M_r protein component in both plasma membrane and microsomes that is inhibited by the presence of 100 mM 3-O-methylglucose during the photolabeling re-

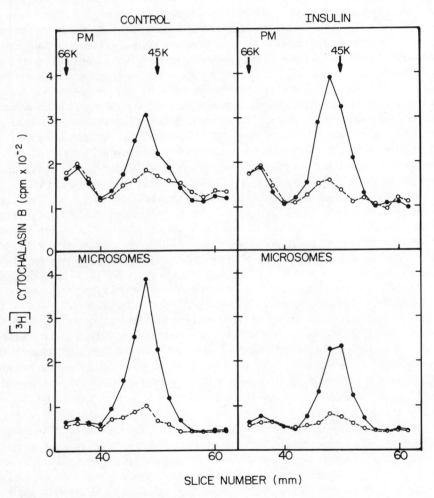

Figure 2. Effect of insulin treatment and the inhibitory effect of 3-0-methylglucose on the photoaffinity labeling in intact rat adipocytes. Isolated adipocytes in Krebs-Ringer phosphate buffer containing 3% bovine serum albumin were incubated with (right) or without (left) 2 mU/ml insulin for 15 min at 37°C, and then incubated for 30 min at 37°C with 0.5 μM [³H]cytochalasin B and 10 μM cytochalasin D in the presence (○) or absence (●) of 100 mM 3-0-methylglucose. Cells were then irradiated and homogenized. After membrane fractionation, SDS-PAGE electrophoresis was performed and incorporated [³H]cytochalasin B into protein was analyzed. PM, plasma membranes. (From Oka and Czech.[40])

action. Insulin treatment of cells before irradiation in the presence of [³H]cytochalasin B and 10 μM unlabeled cytochalasin D leads to an approximate twofold increase observed in the resultant plasma membrane fraction and a decrease of about 50% in the labeling of the 46,000-M_r protein component in the microsomal fraction. Fat cell number and integrity as measured by passive diffusion are not compromised by the irradiation procedure.[40] Significantly, treatment with insulin of intact adipocytes first irradiated in the presence of [³H]cytochalasin B leads to the expected shift in membrane distribution of the 46,000-M_r component such that its labeling is increased in the plasma membrane fraction and decreased in the low-density microsomal fraction. This important observation demonstrates that the hexose transporters in adipocytes remain responsive to insulin after the manipulations required for the affinity labeling procedure.

The experimental strategy we employed from this point involves the concept that inhibitors of the photoaffinity labeling reaction that are unable to penetrate into the intracellular space effectively should provide a basis for determining which transporters are present in an exposed versus unexposed disposition within the cell. Therefore, the photolabeling of transporters that reside in true intracellular membrane fractions would not be expected to be inhibited by a nonpermeable D-glucose analog, while the photolabeling reaction of cell surface transporters should be inhibited by such an analog. In our studies, we utilized the D-glucose analog ethylidene glucose as a poorly permeable inhibitor of the photolabeling reaction in intact cells. Ethylidene glucose has several important properties related to our objectives. It markedly inhibits D-glucose transport activity in cells[42,43] as well as the binding of [³H]cytochalasin B to hexose transporters.[44] Significantly, ethylidene glucose has been shown to be transported into cells by passive diffusion and has little or no capacity to be transported by the hexose transporters it binds.[42,43]

We reasoned that incubation of cells with ethylidene glucose for only a brief period would result in the exposure of cell-surface hexose transporters but not intracellular transporters to this glucose analog. If this were to prove the case, photolabeling of cell-surface hexose transporters by [³H]cytochalasin B would be readily inhibited under conditions in which photolabeling of intracellular hexose transporters would proceed unaffected. It should be noted that under these conditions the inhibition of [³H]cytochalasin B binding to cell-surface hexose transporters would be expected to be inhibited indirectly by the ethylidene glucose. This is because the [³H]cytochalasin B binding site appears to be on the cytoplasmic side of the cell-surface membrane,[34] whereas the binding site for ethylidene glucose appears to be on the exofacial side of the membrane.[42,43] The ability of ethylidene glucose to inhibit [³H]cytochalasin B binding thus appears to result from a noncompetitive interaction. However, this consideration does not alter the basic strategy of these experiments.

Our attempts to develop experimental conditions[40] that fulfill the expectations described above have been successful. Isolated intact adipocytes were incubated with 50 mM ethylidene glucose for 1 min at 15°C, resulting in an accumulation of only 2 mM ethylidene glucose in the intracellular compartment as measured by uptake studies. At 2 mM, this hexose has virtually no affect on the photolabeling reaction using 0.5 μM [³H]cytochalasin B, either in isolated membranes or intact cells. At 50 mM ethylidene glucose, however, the photoaffinity labeling reaction in membranes or cells is inhibited by this hexose. We also incubated 50 mM ethylidene glucose with cells for 15 min at 37°C to permit equilibration of the hexose and inhibition of the photo-labeling reaction with hexose transporters in all cellular compartments. After photo-labeling of intact cells with [³H]cytochalasin B under the various incubation conditions, the cells were homogenized and the membranes fractionated into plasma membranes and low-density microsomes and the photolabeled transporters were resolved on sodium dodecylsulfate gels.

Figure 3 shows the photolabeling of the 46,000-M_r transporter protein in fat cells from the experiment described above. Under conditions in which the ethylidene glucose

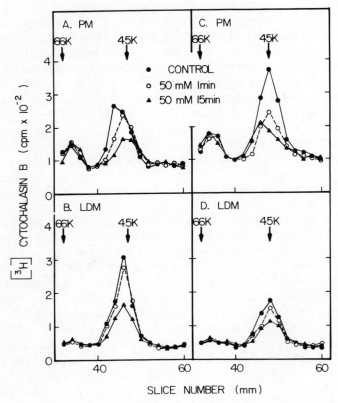

Figure 3. Effect of 1-min incubation with 50 mM ethylidine glucose on photoaffinity labeling in control cells (left) and insulin-treated cells (right). Isolated adipocytes were incubated in Krebs–Ringer phosphate buffer containing 3% bovine serum albumin with (right) or without (left) 2 mU/ml insulin for 15 min at 37°C, after which 10 μM cytochalasin D was added. Cells were then incubated with 0.5 μM [³H]cytochalasin B and 50 mM sorbitol (●) or 50 mM ethylidene glucose (○) for 1 min at 15°C and irradiated. A third group was preincubated with 50 mM ethylidene glucose for 15 min at 37°C, then incubated with 0.5 μM [³H]cytochalasin B for 1 min at 15°C and irradiated (▲). After being irradiated, cells were homogenized, and the labeled protein was analyzed in membrane fractions. The results shown are from a representative experiment among three performed with similar results. PM, plasma membranes; LDM, low-density mi-crosomes. (From Oka and Czech.[40])

concentration was low (~2 mM) in the intracellular compartment (1-min incubation at 15°C with 50 mM hexose) it can be seen that the photolabeling of hexose transporters in the low-density microsomal fraction of control cells is not affected by this treatment (Fig. 3B). In contrast, the photolabeling reaction of hexose transporters in the plasma membrane fraction of insulin-treated cells is markedly inhibited by the ethylidene glucose under these experimental conditions (Fig. 3C). When the hexose analog is equilibrated across the cell membrane at a concentration of 50 mM (15 min incubation at 37°C), the photolabeling of all cellular hexose transporters is significantly inhibited (Fig. 3A–D). The plasma membrane fraction from control cells and the low-density microsome fraction from insulin-treated cells could not be rigorously analyzed by this method because of the relatively low numbers of hexose transporters present and the carryover contamination from the corresponding fractions that have more transporters under these conditions (Fig. 3A,D). Nevertheless, the data in Fig. 3, clearly demonstrate that insulin action leads to a change in the accessibility of hexose transporters to inhibition of photolabeling by the extracellular hexose analog. We can conclude that at least one action of insulin on the hexose transport system of adipocytes is to change their disposition from a nonexposed, presumably intracellular domain, to a cell-surface location exposed to the extracellular medium.

III. REGULATION OF THE IGF-II RECEPTOR BY INSULIN

Results from our laboratory that first suggested[24] that the IGF-II receptor may be regulated by insulin in a manner similar to that for the glucose transport system are shown in Fig. 4. Intact adipocytes were incubated in the presence or absence of

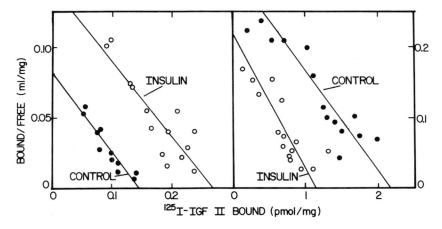

Figure 4. Scatchard analysis of [^{125}I]-IGF-II binding to isolated membrane fractions. Plasma membranes and low-density microsomal fractions were incubated with [^{125}I]-IGF-II for 90 min at 10°C. Binding was analyzed by the method of Scatchard. (Left) Plasma membranes prepared from control (●) and insulin-treated (○) adipocytes. (Right) Low-density microsomes from control (●) and insulin-treated (○) cells. The experiment was performed five times for plasma membranes and three times for low-density microsomes. (From Oppenheimer et al.[24])

insulin and then homogenized and fractionated by standard sedimentation techniques.[24,41] Scatchard analysis of [^{125}I]-IGF-II binding performed on the resulting
plasma membrane fraction demonstrated an apparent increase in IGF-II receptor number
due to insulin action (Fig. 4). In contrast, the low-density microsomal membranes
derived from insulin-treated fat cells exhibited an apparent decrease in IGF-II receptor
number compared with those from control cells (Fig. 4). When adipocytes treated with
or without insulin were homogenized and centrifuged at very high speed to sediment
all cellular membrane fractions, no difference in total apparent IGF-II receptor number
was found between the total membranes from control versus insulin-treated adipocytes.[24] Thus, the apparent membrane redistribution of the IGF-II receptor in response
to insulin action paralleled almost exactly the results obtained for the glucose transporter
as described in the previous section.

   Although the data depicted in Fig. 4 suggest a recruitment mechanism for the
increase in IGF-II binding to intact cells in response to insulin, other interpretations
are possible, as noted for early studies on the glucose transporter. In fact, Scatchard
analysis of [^{125}I]-IGF-II binding to intact adipocytes yielded plots consistent with the
interpretation that an affinity change of IGF-II receptors rather than a number change
was responsible for the effects of insulin (Fig. 5). As can be seen in Fig. 5, the slopes

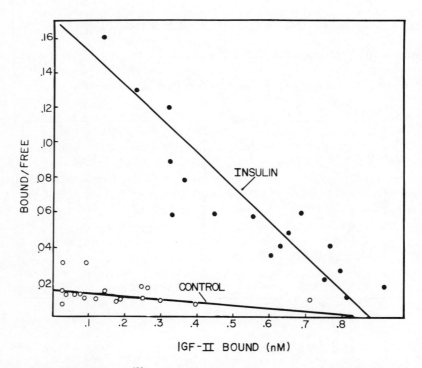

Figure 5. Scatchard analysis of [^{125}I]-IGF-II binding to intact rat adipocytes. Afer 10-min preincubation
at 37°C in the presence (●) or absence (○) of 10 nM insulin, adipocytes were incubated with [^{125}I]-IGF-II
for 30 min at 24°C. Binding was analyzed by the Scatchard method. This experiment was carried out in
duplicate with similar results. (From Oppenheimer *et al.*[24])

of the binding curve for $[^{125}I]$-IGF-II are markedly different between control and insulin-treated cells, whereas the intercepts of these binding curves on the x axis are similar. These data, which suggest an affinity change of IGF-II receptors in response to insulin action, were consistent with results reported by King *et al.*[45] Thus, the use of $[^{125}I]$-IGF-II binding to receptors was inadequate to document rigorously the mechanism of insulin action.

In light of the above considerations, we set out to develop a methodology that could rigorously analyze cell-surface IGF-II receptor numbers independent of the use of $[^{125}I]$-IGF-II binding estimates. The strategy we developed involves the use of specific anti-IGF-II receptor immunoglobulin (Ig) raised against highly purified IGF-II receptor. If insulin were to increase the affinity of the IGF-II receptor for IGF-II, it would seem unlikely that the receptor affinity for a polyclonal population of anti-receptor antibodies would also be affected. We therefore developed an assay to monitor the binding of anti-IGF-II receptor Ig to control and insulin-treated intact fat cells. Increased numbers of receptors expressed on the cell surface in response to insulin would be expected to lead to increased anti-IGF-II receptor Ig binding, whereas an insulin-mediated increase in receptor affinity would be expected to produce no change in anti-receptor Ig binding.

In order to generate anti-IGF-II receptor Ig, IGF-II receptors from rat placenta were purified to near homogeneity by an affinity–chromatography technique.[28] This technique involves solubilization of rat placenta membranes in Triton X-100 and adsorption of IGF-II receptors on immobilized rat IGF-II. The purified receptor is eluted in an acidic, high-salt solution. Figure 6 shows the polyacrylamide gel record of various stages of the purification procedure. It is evident that a highly purified receptor preparation is attained and that contamination is minimal. The specific activity of $[^{125}I]$-IGF-II binding to the receptor preparation increases about 1100-fold from the crude Triton X-100 extract to the final purified material.[28] Potent anti-IGF-II receptor antiserum was raised by injection of this purified IGF-II receptor preparation into rabbits (initial injection 50 μg, boosted with 12 μg). The resulting antisera produced immunoprecipitation lines in agar immunodiffusion dishes and showed a positive reaction at 5000-fold dilution using an enzyme-linked immunosorbent assay (ELISA).[46] Interestingly, the anti-IGF-II receptor Ig markedly inhibited $[^{125}I]$-IGF-II binding to plasma membranes or intact cells, while $[^{125}I]$-insulin binding to its receptor was not effected by the anti-IGF-II receptor Ig.[46] As little as only a few micrograms of anti-receptor Ig is observed to inhibit the interaction between IGF-II receptor and $[^{125}I]$-IGF-II. Binding to the IGF-II receptor of the anti-receptor Ig does not appear to be affected by the occupancy of the IGF-II receptor by IGF-II, however, because immunoprecipitation lines between receptor and anti-receptor Ig could still be determined in the presence of saturating concentrations of IGF-II in agar plates.[46]

The specific binding of anti-IGF-II receptor Ig to intact cells was measured by a double antibody technique. One ml of fat cell suspension was incubated with 5 μg/ml of Ig fraction from anti-receptor antiserum (prepared by precipitation with ammonium sulfate) at 21°C for 30 min in a polystyrene test tube. The Ig fraction from control rabbits represented the control condition. The cells were washed twice and further incubated with 25 μg/ml of $[^{125}I]$-goat anti-rabbit IgG for 30 min at 21°C. The $[^{125}I]$-goat anti-rabbit IgG was composed of the IgG fraction of serum from goats

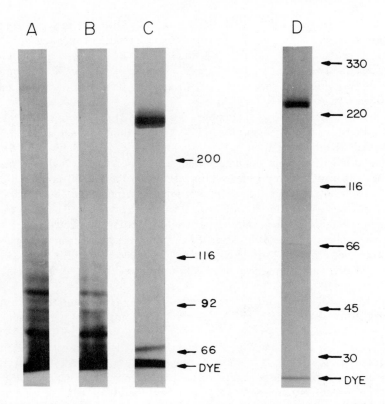

Figure 6. Sodium dodecylsulfate polyacrylamide gel electrophoresis of the IGF-II receptor after purification. Samples were boiled in the presence of 1% SDS and 50 mM dithiothreitol, then electrophoresed on polyacrylamide gels by the method of Laemmli and silver stained. (Lanes A–C), 5% polyacrylamide gel: lane A, rat placenta membranes (8 μg); lane B, Triton X-100 extract (9 μg); lane C, purified IGF-II receptor (0.9 μg). Arrows show the location of molecular-weight standards: bovine serum albumin (66,200 M_r), phosphorylase *b* (92,500 M_r), β-galactosidase (116,250 M_r), and myosin (200,000 M_r). (Lane D) 5–12% gradient polyacrylamide gel: purified IGF-II receptor (0.9 μg). Molecular-weight standards: carbonic anhydrase (30,000 M_r), ovalbumin (45,000 M_r), bovine serum albumin (66,200 M_r), β-galactosidase (116,250 M_r), ferritin half-unit (220,000 M_r), and thyroglobulin (330,000 M_r). (From Oppenheimer and Czech.[28])

raised against rabbit IgG. The cells were washed free of nonadsorbed antibodies and centrifuged with 3 ml of silicone fluid before measuring the bound [^{125}I]-goat anti-rabbit IgG. The concentrations of 5 μg/ml anti-IGF-II receptor Ig and 25 μg/ml [^{125}I]-goat anti-rabbit IgG were found to be saturating concentrations for each binding reaction. The conditions used for these experiments were such that the IGF-II receptors on the adipocytes were saturated with anti-IGF-II receptor Ig. The data in Fig. 7 show the binding of [^{125}I]-IGF-II as well as anti-receptor Ig as reflected by [^{125}I]-goat anti-rabbit IgG binding to intact rat adipocytes treated with or without insulin. As previously observed, the binding of [^{125}I]-IGF-II is increased about threefold in the insulin-treated cells. Significantly, anti-IGF-II receptor Ig binding to receptors on intact adipocytes,

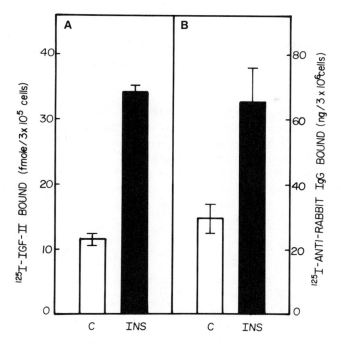

Figure 7. Effect of insulin on [^{125}I]-IGF-II binding (left) and anti-receptor Ig binding (right) to intact adipocytes. Isolated adipocytes were incubated with or without 10 nM insulin for 15 min at 37°C. For measurement of [^{125}I]-IGF-II binding (A) cells were then incubated with 1 nM [^{125}I]-IGF-II for 30 min at 21°C, and bound [^{125}I]-IGF-II was measured by the oil-flotation method. For measurement of anti-receptor Ig binding (B), control and insulin-treated cells (3 × 10^6) were incubated with 5 μg/ml of anti-receptor Ig for 30 min at 21°C. Cells were then washed and further incubated with 25 μg/ml of [^{125}I]-goat anti-rabbit serum IgG for 30 min at 21°C, and bound goat IgG was measured. Values presented are the mean ± standard deviations obtained in three separate experiments. C, control; INS, insulin. (From Oka *et al.*[46])

as estimated by [^{125}I]-goat anti-rabbit IgG binding, was also increased in the insulin-treated cells to a similar degree, as depicted in Fig. 7B. These data are consistent with the conclusion that insulin acts to increase the number of IGF-II receptors expressed on the cell surface.

It could be argued that increased anti-IGF-II receptor Ig binding to cells in response to insulin might reflect the expression of new antigenic sites on the IGF-II receptor caused by an insulin-induced change in receptor conformation. This explanation is most unlikely, however, because we found that control adipocytes could adsorb the anti-receptor Ig responsible for the insulin-induced increment of the binding. Thus, it appears that control IGF-II receptors contain the same antigenic sites that reside on receptors from insulin-treated cells. Furthermore, an insulin-induced increased affinity of IGF-II receptors for anti-receptor Ig seems very unlikely because IGF-II receptors are saturated with anti-receptor Ig under the conditions of our experiments. Interestingly, these results are consistent with those obtained from Scatchard analysis of [^{125}I]-IGF-II binding to isolated plasma membranes and low-density microsomes, but are opposite to the interpretation suggested by Scatchard analysis of [^{125}I]-IGF-II binding to intact adipocytes, which indicated an affinity change due to insulin action.[24,45] The reason for this difference is not yet known. However, the failure of Scatchard analysis of [^{125}I]-IGF-II binding in intact cells probably relates to deviations from experimental conditions in which ideal equilibrium binding is achieved. Degradation of [^{125}I]-IGF-II, for example, may interfere with the binding assay. Internalization of [^{125}I]-IGF-II in intact cells may also be a problem. In addition, IGF-II is known to mimic insulin action at high concentrations due to its binding to the insulin receptor with lower affinity. It is also possible that at high concentrations of [^{125}I]-IGF-II, this ligand mimicked the effect of insulin on the IGF-II receptor. Taken together, these considerations lead us to conclude that our results using the anti-IGF-II receptor Ig strongly support the conclusion that insulin induces a rapid exposure of new IGF-II receptors on the cell surface of intact adipocytes.

IV. CONCLUSIONS

The results from our laboratory on experiments with intact insulin-sensitive adipocytes[40,46] reviewed herein, validate the basic elements of the recruitment hypothesis originally proposed by Cushman and colleagues[18,19] and by Kono and co-workers.[17,20] In addition, our data extend this hypothesis to a second insulin-sensitive membrane component—the IGF-II receptor. Taken together, the data now available document that insulin action mediates the exposure of previously sequestered glucose transporters and IGF-II receptors in isolated adipocytes. The mechanism of this action remains unknown.

A working model which accounts for the data now available on the action of insulin on membrane components is presented in Fig. 8. This model suggests that insulin-sensitive hexose transporters and IGF-II receptors continuously undergo both endocytotic and exocytotic events such that they cycle from a cell surface, exposed disposition to an intracellular, vesicular disposition. According to this model control cells exhibit a much slower cycling rate of these membrane components when compared to insulin-treated cells. Recent experiments we have performed attempting to determine the fate of receptor bound [^{125}I]-IGF-II in the presence and absence of insulin suggest that a sorting process is also involved such that IGF-II receptor recycles and IGF-II itself is delivered to a compartment containing a degradation pathway. In these experiments, [^{125}I]-IGF-II initially associated with its receptor was shown to be rapidly degraded in the intact cells at 37°C. Significantly, we can demonstrate that the increased [^{125}I]-IGF-II degradation occurs by an IGF-II receptor-mediated process. Thus, addition of anti-IGF-II receptor antibody blocks the ability of insulin to increase [^{125}I]-IGF-II degradation. These experiments suggest that [^{125}I]-IGF-II is internalized with this receptor and is delivered to a locus in the cell that mediates degradation of the ligand. The receptor in this view is cycled back to the plasma membrane.

In the working model of Fig. 8, insulin action acts at the site of exocytosis of IGF receptors or hexose transporters rather than inhibiting an endocytotic pathway. Recent preliminary experiments in our laboratory support this site of action. In these

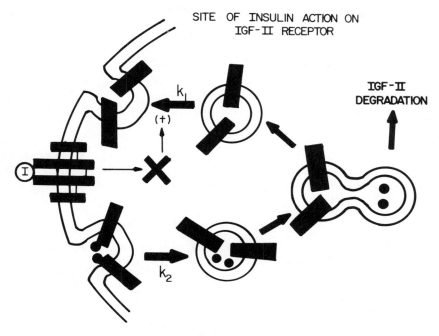

Figure 8. A working hypothesis on the action of insulin on the IGF-2 receptor. It is proposed that the IGF-2 receptor is continuously cycling in adipocytes between the cell surface membrane and internal low density vesicles. Insulin may activate the cycling rate of the IGF-2 receptor by acting at a site or sites involving the exocytosis pathway, thus increasing the number of IGF-2 receptors at the cell surface. Cycling of the IGF-2 receptor leads to uptake and degradation of IGF-2. (See text for further details.)

experiments, IGF-II receptors were specifically photolabeled in intact adipocytes at low temperature and their internalization monitored by measuring their redistribution from the plasma membrane fraction to the low density microsome fraction with time at 37°C. Internalization of photolabeled IGF-II receptors (movement to the low density microsomes) was found to be rapid and complete by about 10 min at 37°C. Significantly, insulin had no effect on the cellular movement of photolabeled IGF-II receptors from plasma membrane to low density microsomes when insulin was added after the photolabeling reaction, indicating insulin does not modulate the internalization process. This conclusion is not yet unequivocally established however because the photolabeling procedure may disrupt the normal internalization process. Taken together, these data suggest that insulin may regulate the exocytotic pathway such that increased numbers of receptors, at steady state, occur on the cell surface and that the overall cycling rate of the IGF-II receptors is stimulated. We would propose a similar mechanism for modulation of hexose transporters due to the similarity of their regulation by insulin, although no similar detailed data are yet available on internalization of hexose transporters. Further experiments are clearly necessary to test the concepts in Figure 8.

An important future objective of experiments designed to elucidate the mode of insulin action on hexose transporters and IGF-II receptors will be to determine with certainty the specific cellular domains in which these insulin-sensitive membrane components reside as well as the cellular routes whereby movement among such domains is coordinated. Defining these mechanisms will clearly require the use of both biochemical and morphological techniques.

V. REFERENCES

1. Levine, R., Goldstein, M.S., Klein, S.P., and Huddlestun, B., 1949, *J. Biol. Chem.* **179:**985–986.
2. Levine, R., Goldstein, M.S., Huddlestun, B., and Klein, S.P., 1950, *Am. J. Physiol.* **163:**70–76.
3. Morgan, H.E., and Whitfield, C.F., 1973, *Curr. Top. Membr. Transp.* **4:**255–303.
4. Czech, M.P., 1980, *Diabetes* **29:**399–409.
5. Czech, M.P., Lawrence, J.C., Jr., and Lynn, W.S., 1974, *J. Biol. Chem.* **249:**1001–1006.
6. Czech, M.P., Lawrence, J.C., and Lynn, W.S., 1974, *Proc. Natl. Acad. Sci. USA* **71:**4173–4177.
7. Kono, T., and Barham, F.W., 1971, *J. Biol. Chem.* **246:**6204–6209.
8. Czech, M.P., and Lynn, W.S., 1973, *Biochim. Biophys. Acta* **297:**368–377.
9. Czech, M.P., Lawrence, J.C., and Lynn, W.S., 1974, *J. Biol. Chem.* **249:**7499–7505.
10. Cuatrecasas, P., and Tell, G.P.E., 1973, *Proc. Natl. Acad. Sci. USA* **70:**485–489.
11. Rosenthal, J.W., and Fain, J.N., 1971, *J. Biol. Chem.* **246:**5888–5895.
12. Cuatrecasas, P., and Illiano, G., 1971, *J. Biol. Chem.* **246:**4938–4946.
13. Fain, J.N., 1968, *Endocrinology* **83:**548–554.
14. Shanahan, M.F., and Czech, M.P., 1977, *J. Biol. Chem.* **252:**8341–8343.
15. Carter-Su, C., Pilch, P.F., and Czech, M.P., 1981, *Biochemistry* **20:**216–221.
16. Martin, D.B., and Carter, J.R., 1970, *Science* **167:**873–874.
17. Suzuki, K., and Kono, T., 1980, *Proc. Natl. Acad. Sci. USA* **77:**2542–2545.
18. Cushman, S.W., and Wardzala, L.J., 1980, *J. Biol. Chem.* **255:**4758–4762.
19. Karnieli, E., Zarnowski, M.J., Hissin, P.J., Simpson, I.A., Salans, L.B., and Cushman, S.W., 1981, *J. Biol. Chem.* **256:**4772–4777.
20. Kono, T., Robinson, F.W., Blevins, T.L., and Ezaki, O., 1982, *J. Biol. Chem.* **257:**10942–10947.
21. Hoessli, D.C., and Rungger-Brandle, E., 1983, *Proc. Natl. Acad. Sci. USA* **80:**439–443.
22. Zapf, J., Schoenle, E., and Froesch, E.R., 1978, *Eur. J. Biochem.* **87:**285–296.
23. King, G.L., Kahn, C.R., Rechler, M.M., and Nissley, S.P., 1980, *J. Clin. Invest.* **66:**130–140.
24. Oppenheimer, C.L., Pessin, J.E., Massague, J., Gitomer, W., and Czech, M.P., 1983, *J. Biol. Chem.* **258:**4824–4830.
25. Massague, J., Guillette, B.J., and Czech, M.P., 1981, *J. Biol. Chem.* **256:**2122–2125.
26. Massague, J., and Czech, M.P., 1982, *J. Biol. Chem.* **257:**5038–5045.
27. Kasuga, M., Van Obberghen, E., Nissley, S.P., and Rechler, M.M., 1981, *J. Biol. Chem.* **256:**5305–5308.
28. Oppenheimer, C.L., and Czech, M.P., 1983, *J. Biol. Chem.* **258:**8539–8542.
29. August, G.P., Nissley, S.P., Kasuga, M., Lee, L., Greenstein, L., and Recher, M.M., 1983, *J. Biol. Chem.* **258:**9033–9036.
30. Czech, M.P., 1984, *Rec. Prog. Horm. Res.* **40:**347–377.
31. Wheeler, T.J., Simpson, I.A., Sogin, D.C., Hinkle, P.C., and Cushman, S.W, 1982, *Biochem. Biophys. Res. Commun.* **105:**89–85.
32. Lienhard, G.E., Kim, H.H., Ransome, K.J., and Gorga, J.C., 1982, *Biochem. Biophys. Res. Commun.* **105:**1150–1156.
33. Czech, M.P., Lynn, D.G., and Lynn, W.S., 19783, *J. Biol. Chem.* **248:**3636–3641.
34. Lin, S., and Spudich, J.A., 1974, *J. Biol. Chem.* **249:**5778–5783.
35. Wardzala, L.J., Cushman, S.W., and Salans, L.B., 1978, *J. Biol. Chem.* **253:**8002–8005.
36. Carter-Su, C., Pessin, J.E., Mora, R., Gitomer, W., and Czech, M.P., 1982, *J. Biol. Chem.* **257:**5419–5425.
37. Pessin, J.E., Tillotson, L.G., Yamada, K., Gitomer, W., Carter-Su, C., Mora, R., Isselbacher, K.J., and Czech, M.P., 1982, *Proc. Natl. Acad. Sci. USA* **79:**2286–2290.

38. Oka, Y., and Czech, M.P., in *Methods in Diabetes Research*. (S.L. Pohl and J. Larner, eds.), Wiley, New York (in press).
39. Shanahan, M.F., Olsen, S.A., Weber, J.J., Lienhard, G.E., and Gorga, J.C., 1982, *Biochem. Biophys. Res. Commun.* **107**:38–43.
40. Oka, Y., and Czech, M.P., 1984, *J. Biol. Chem.* **259**:8125–8133.
41. McKeel, D.W., and Jarett, L., 1970, *J. Cell. Biol.* **44**:417–432.
42. Baker, G.F., and Widdas, W.F., 1973, *J. Physiol. (Lond.)* **231**:129–142.
43. Holman, G.D., and Rees, W.D., 1982, *Biochem. Biophys. Acta* **685**:78–86.
44. Gorga, F.R., and Lienhard, G.E., 1981, *Biochemistry* **20**:5108–5113.
45. King, G.L., Rechler, M.M., and Kahn, C.R., 1982, *J. Biol. Chem.* **257**:10001–10006.
46. Oka, Y., Mottola, C., Oppenheimer, C.L., and Czech,M.P., 1984, *Proc. Natl. Acad. Sci. USA* **81**:4028–4032.

26

Insulin Action on the $(Na^+,K^+)ATPase$

Marilyn D. Resh

I. INTRODUCTION

Reduction of serum potassium levels was one of the earliest recognized effects of insulin administration *in vivo*.[1] There is now substantial evidence to indicate that this hypokalemic effect is due to enhanced K^+ uptake into insulin's target tissues and that intracellular concentrations of Na^+ and K^+ are altered as a result of insulin treatment. Almost all eukaryotic cells maintain high cytoplasmic K^+ concentrations and low Na^+ concentrations relative to the extracellular fluid. These cationic gradients have a profound influence on cellular volume and osmotic pressure,[2] membrane potential,[3] protein synthesis,[4] and the transport of nonelectrolytes.[5] Active transport of Na^+ and K^+ ions across the plasma membrane is mediated by the Na^+ and K^+ ion-activated adenosine triphosphatase [$(Na^+,K^+)ATPase$ or Na^+ pump],[6-8] first identified in crab nerve membranes by Skou in 1957.[9] Under physiological conditions, this membrane-bound enzyme couples the influx of two K^+ ions and the efflux of three Na^+ ions to the hydrolysis of one molecule of ATP.[6-8]

The structure and function of the $(Na^+,K^+)ATPase$ have been subjects of intensive investigation for the past 25 years. Purification of the enzyme has been achieved from a variety of sources, including mammalian kidney and brain[10,11] and eel electroplax.[12] It is generally agreed that the $(Na^+,K^+)ATPase$ consists of two subunits, a 100,000-M_r α subunit (molecular weight estimates range from 85,000 to 25,000) and a 45,000- to 60,000-M_r glycoprotein β subunit. The 100,000-M_r polypeptide is the catalytic subunit that contains an intracellular binding site for ATP[13] and an extracellular site for binding of cardiac glycoside inhibitors[14,15] and is phosphorylated by ATP and phosphate.[13,16] The function of the β subunit is not known, but it is apparently present in a 1 : 1 stoichiometry with the α subunit. Crosslinking evidence[17-19]

Marilyn D. Resh ● The Biological Laboratories, Harvard University, Cambridge, Massachusetts.

is consistent with the existence of some tetrameric $(\alpha\beta)_2$ structures for the native enzyme, or an approximate molecular weight of 280,000.

Evidence based on kinetic studies[16] and proteolytic digestion[20,21] supports the existence of two different conformational states of the enzyme, designated E_1 and E_2, which are preferentially stabilized by either Na^+ or K^+. A simplified enzymatic mechanism for the (Na^+,K^+)ATPase can be considered as follows:

$$E_1 + Na^+_i + ATP \xrightleftharpoons{Mg^{2+}} E_1 - P + ADP + Na^+_e$$

$$E_1 - P \rightleftharpoons E_2 - P$$

$$E_2-P + K^+_e \xrightleftharpoons{Mg^{2+}, H_2O} E_2 + P_i + K^+_i$$

$$E_2 \rightleftharpoons E_1$$

where subscripts i and e refer to the intracellular and extracellular location of ions, respectively. In the presence of $Na^+ + ATP + Mg^{2+}$, a covalent phosphorylated intermediate of the enzyme is formed through an aspartyl–phosphate bond,[22] with concomitant hydrolysis of ATP. Interconversion of the E_1–P form to E_2–P then occurs; this reaction can be blocked by maleimides and oligomycin.[23,24] The inhibitor ouabain binds with relatively high affinity to the E_2–P state of the enzyme.[6,7] Addition of potassium ion accelerates the hydrolysis and dephosphorylation of the acyl–phosphate bond,[6–8] generating free phosphate and enzyme in the E_2 state. Under certain conditions, the reaction sequence can be forced to operate in reverse, such that inorganic phosphate is initially incorporated into the enzyme and, with the addition of ADP, ATP is synthesized.[25]

The possibility that activation of the (Na^+,K^+)ATPase could provide a molecular explanation for the hypokalemic effect of insulin has only recently been addressed. This chapter reviews the evidence documenting insulin action on the (Na^+,K^+)ATPase in various tissue and cell types. In addition, current models invoked to elucidate the biochemical basis for this effect will be presented and discussed, with reference to insulin stimulation of another membrane-bound protein, the glucose transporter.

II. INSULIN STIMULATION OF (Na^+,K^+)ATPase IN INTACT TISSUES AND ISOLATED CELLS

Insulin-induced increases in monovalent cation transport have been reported in a variety of systems. Insulin stimulation of K^+ uptake[26–30] and Na^+ efflux[30–34] occurs in frog and rat skeletal muscle and diaphragm, as well as in rat adipose tissue,[35] duck salt gland,[36] 3T3 fibroblasts,[37] and rat hepatocytes.[38] This stimulatory action occurs in the absence of glucose in the medium and is unaffected by the addition of glucose transport inhibitors.[30] The insulin effect is inhibited by cardiac glycosides,[29,30,32–34,36–38] suggesting that the hormone affects the transport activity of the (Na^+,K^+)ATPase.

More definitive evidence linking increased K^+ flux to activation of the Na^+ pump was obtained in the isolated rat adipocyte.[39] In fat cells, insulin stimulates the uptake of Rb^+ (a K^+ analog) in a rapid, reversible fashion and increases the steady-state concentration of intracellular potassium. The hormone effect is mediated by physiological concentrations of insulin and requires insulin binding to its intact cell-surface receptors. Both basal and insulin-stimulated Rb^+ transport rates are energy dependent, ouabain inhibitable, and dependent on the concentrations of external K^+ and external Na^+. These observations establish that the adipocyte (Na^+,K^+)ATPase is an insulin-sensitive membrane-transport protein.

One could argue that increased transport activity of the Na^+ pump is a response to an insulin-induced increase in K^+ efflux from the cell. However, experiments in hepatocytes[38] and adipocytes[39] have failed to demonstrate any such change in K^+ (Rb^+) efflux, although small transient decreases in K^+ efflux have been detected following insulin treatment of rat muscle.[30,40] Furthermore, kinetic studies demonstrate that insulin increases the V_{max} of Rb^+ transport without changing the K_m of the (Na^+,K^+)ATPase for Rb^+.[39] This result can be accounted for by an increase in the number of ion pumps or by activation of existing pumps. To distinguish between these possibilities, it is necessary to quantitate the number of active Na^+ pumps both before and after insulin treatment.

III. QUANTITATION OF Na⁺ PUMPS

Initial attempts to measure the number of (Na^+,K^+)ATPases involved the use of binding of the specific inhibitor ouabain. Insulin increases [³H]ouabain binding to frog sartorius muscle by 1.7-fold, and addition of insulin to muscles pretreated with ouabain and then washed results in stimulated Na^+ efflux.[34] These results suggest that insulin unmasks previously inactive Na^+ pump sites in the muscle cell membrane. However, Clausen and Hansen[41] showed that insulin increases the rate of ouabain binding, but there is no change in the total, steady-state number of ouabain binding sites in either rat soleus muscle or adipose tissue. Closer examination of ouabain binding to isolated rat adipocytes[39] reveals that although no change is apparent in cells exposed to insulin, the stoichiometries and affinities of ouabain-binding sites are complex. Thus the presence of nonspecific binding sites[34,39] and binding to nonfunctional Na^+ pumps[39] complicates interpretation of ouabain-binding data.

Accurate determination of the number of active Na^+ pumps in insulin-treated cells was achieved[42] using an alternative technique known as "back-door phosphorylation." As discussed in the introduction, in the presence of ouabain and Mg^{2+}, an alkali-labile covalent phosphorylated intermediate of the (Na^+,K^+)ATPase is formed from inorganic [³²P]phosphate with a stoichiometry of 1 mole of phosphate per mole of enzyme.[16,43] Specific incorporation of ³²P into the 95,000-dalton catalytic subunit of the Na^+ pump is evident upon acidic pH polyacrylamide gel electrophoresis. By quantitating the maximal phosphorylation capacity of plasma membranes prepared from rat adipocytes, the number of (Na^+,K^+)ATPases is calculated to be 6×10^5 pumps/cell.[42] Moreover, there is no significant difference in the amount of ouabain-dependent phosphorylation between plasma membranes prepared from untreated or insulin-treated adipocytes.[42] One must conclude that insulin does not alter the number of active Na^+ pumps.

IV. EFFECT OF INSULIN ON MEMBRANE-BOUND (Na^+,K^+)ATPase

The reported effects of insulin on ATPase activity are multifarious. Jarett and Smith[44] found that direct addition of insulin to rat adipocyte plasma membranes produces a 12% increase in Mg^{2+}-ATPase activity, with no change in the (Na^+,K^+)ATPase, although Mg^{2+}-ATPase activity is actually slightly lower in plasma membranes prepared from insulin-treated cells.[44] Similarly, a 10–20% increase in apparent Mg^{2+}-ATPase activity following direct addition of insulin to lymphocyte plasma membranes was reported by Hadden et al.[45] In contrast, incubation of duck salt gland[36] or rat muscle[46] with insulin prior to membrane isolation results in higher (Na^+,K^+)ATPase activity, whereas no effect is observed when insulin is added directly to the membranes. Others have reported that direct addition of insulin to frog skeletal muscle membranes stimulates (Na^+,K^+)ATPase activity, but this effect is observed only at submaximal concentrations of Na^+ and ATP.[47] However, no alteration of (Na^+,K^+)ATPase activity is detected in membranes from rat skeletal muscle or rat adipocytes whether insulin is added directly to the membranes or to intact tissue prior to homogenization[42,48] or is injected into rats prior to muscle isolation.[48] One can tentatively conclude that the stimulatory effect of insulin on the (Na^+,K^+)ATPase is evident only in the intact cell.

On the basis of measurements of (Na^+,K^+)ATPase activity, K^+ transport rates, and the number of (Na^+,K^+)ATPases in the cell, several important parameters of monovalent cation transport can be determined. In plasma membranes prepared from rat adipocytes, the turnover number of the (Na^+,K^+)ATPase for ATP is 14,000 molecules/site/min at 37°C. However, according to the rate of Rb^+ (K^+) transport, the turnover number of this enzyme in the intact cell is only 1700 molecules/site/min.[42] This observation implies that under physiological conditions, the fat cell Na^+ pump operates at only 12% of its maximal rate. In soleus muscle, it is estimated that the pump works at 2.4% of maximal activity.[27] Thus constraints imposed by the cellular milieu result in inhibition or inefficient operation of the (Na^+,K^+)ATPase. One must presume that insulin stimulates the Na^+ pump by overcoming this inhibition.

V. COMPARISON WITH ACTIVATION OF THE GLUCOSE TRANSPORTER

Striking similarities are apparent when one compares activation of the (Na^+,K^+)ATPase with the well-documented effect of insulin on the transport system for D-glucose. Formation of fat cells in culture is accompanied by dramatic and simultaneous increases in the insulin sensitivity of both the glucose transporter and the Na^+ pump.[37] The insulin dose–response curves for stimulation of these two membrane transport proteins are identical in cells at various stages of growth and differentiation.[37,39] This evidence suggests a coordinate regulation of glucose- and K^+-transport systems by insulin.[49]

Further support for the hypothesis that a common mechanism is involved in activating the glucose transporter and the (Na^+,K^+)ATPase is provided by the data presented in Table I. Several agents that have insulinomimetic effects on glucose transport (hydrogen peroxide, antibody, hyperosmolarity) also stimulate ouabain-inhibitable Rb^+ uptake in rat adipocytes. On the other hand, catecholamines and Na^+ ionophores activate the (Na^+,K^+)ATPase without affecting the glucose transporter. However, these agents apparently act through a mechanism different from that of insulin, since their stimulatory effects on the Na^+ pump are additive with those exerted by maximal concentrations of insulin (Table I).

It is also interesting to compare insulin sensitivities of the glucose transporter and the (Na^+,K^+)ATPase in different tissue types (Table II). Both transport proteins show marked insulin responsiveness in two of the three primary target tissues for insulin action, i.e., fat and muscle. The situation in liver tissue is not as well defined. The glucose transport mechanism in liver is not insulin responsive, but a recent report has documented a small, but significant stimulation of the (Na^+,K^+)ATPase in isolated hepatocytes.[38] However, experiments conducted in this laboratory have not been able to reproduce this effect in hepatocytes. Thus, it is unclear, at least to this investigator, whether the (Na^+,K^+)ATPase in liver is insulin responsive. In other studies, low levels of insulin stimulation of both glucose and K^+ uptake have been documented in cultured fibroblasts. On the other hand, neither transport system is activated in lymphocytes or turkey and human erythrocytes, although these cells contain insulin receptors. It remains to be determined whether the (Na^+,K^+)ATPase in brain is insulin

Table 1. Effects of Insulinomimetic Agents on the Membrane Transport Systems for Glucose and Potassium in Rat Adipocytes[a,b]

Agent	Glucose transporter	(Na^+,K^+)ATPase
1 nM insulin	↑	↑
4 mM H_2O_2	↑	↑
H_2O_2 + insulin	↑	↑
Anti-adipocyte membrane antibody	↑	↑
200 mM Mannitol	↑	↑
1 μg/ml epinephrine	—	↑
Epinephrine + insulin	↑	↑ ↑
5 μg/ml monensin	—	↑
Monensin + insulin	↑	↑ ↑

[a] All data are unpublished results of M.D. Resh.
[b] ↑, Stimulation; ↑ ↑, stimulation, additive effect; —, no effect.

Table 2. Comparison of Insulin Effects on the Membrane-
Transport Systems for Glucose and Potassium in Various Cells
and Tissue Types[a]

Tissue or cell	Glucose transporter	$(Na^+,K^+)ATPase$
Fat	↑ ↑	↑ ↑
Muscle	↑ ↑	↑ ↑
Liver	—	↑ ,—
3T3-L1 fibroblasts	↑	↑
3T3-L1 adipocytes	↑ ↑	↑ ↑
Lymphocytes[b]	—	—
Turkey erythrocytes[c]	—	—
Human erythrocytes[d]	—	—
Brain	—	?

[a] ↑ ↑, Significant stimulation; ↑, moderate stimulation; —, no effect; ?, undetermined.
[b] Data from L.C. Cantley (personal communication).
[c] Data from M.D. Resh (unpublished results).
[d] Data from G. Guidotti (unpublished results).

insensitive, as no effect on the transport of glucose in this tissue has been demonstrated with insulin.[50]

Kinetic studies further illustrate the marked similarities between activation of these two transport proteins. When insulin is added to fat cells at supramaximal concentrations, such that binding of insulin to its cell-surface receptor is not rate limiting, a distinct lag time is observed before stimulation of transport activity occurs.[51,52] The length of this lag time as well as its temperature dependence are identical for both Rb^+ uptake and 2-deoxyglucose uptake.[53] Since the lag time presumably represents the coupling step(s) between insulin binding and stimulation of transport, it is logical to conclude that the rate-determining steps for activation of the glucose transporter and the $(Na^+,K^+)ATPase$ are kinetically indistinguishable.

In light of the many similarities between insulin activation of glucose- and K^+-uptake rates, one might argue that these two processes are catalyzed by the same enzyme. This is clearly not the case. Stimulation of Rb^+ uptake occurs in the absence of glucose in the medium and is not inhibited by cytochalasin B, a specific inhibitor of the glucose transport.[37,39] Ouabain does not affect basal or insulin-stimulated sugar transport in soleus muscle[54] or adipocytes,[37,39] although prolonged exposure to ouabain stimulates 3-O-methylglucose efflux in muscle.[54] Hexose transport is not altered when adipocytes are suspended in Na^+-free medium, whereas basal and insulin-stimulated Rb^+ uptake rates are inhibited under these conditions. Moreover, the Na^+ pump can be activated independently by the addition of the ionophore monensin (Table I), with no effect on basal or stimulated deoxyglucose uptake. Thus uptakes of glucose and K^+ are clearly mediated by two different membrane transport systems.

There is at least one obvious difference in the way in which insulin activates the

(Na$^+$,K$^+$)ATPase and the glucose transport mechanism. A convincing body of evidence indicates that the insulin-induced increase in the V_{max} of the glucose transporter is mediated through an increase in the number of plasma membrane transport units. Cushman[55] and Kono[56] and their co-workers have demonstrated that an insulin-mediated translocation of glucose transport mechanisms from an intracellular microsomal pool to the plasma membrane occurs. However, at least in the rat adipocyte, functional (Na$^+$,K$^+$)ATPases are present only in the plasma membrane, and insulin treatment does not result in a change in the number of plasma membrane Na$^+$ pumps.[42] Thus, although growth or density-dependent changes in monovalent cation transport can be effected by alteration of the number of (Na$^+$,K$^+$)ATPases at the cell surface, such a mechanism is not operative in insulin action. This is not surprising, in view of the fact that no intracellular pool of active Na$^+$ pumps exists in the adipocyte and that the plasma membrane Na$^+$ pumps operate at only 10% of their maximal capacity in the basal state. Thus, an enormous potential exists for activating already existing, but inefficiently operating, (Na$^+$,K$^+$)ATPases.

VI. ACTIVATION OF THE Na$^+$ PUMP BY GROWTH FACTORS

Since the (Na$^+$,K$^+$)ATPase pumps at submaximal capacity in the intact cell, one would expect that other compounds besides insulin might be capable of increasing transport activity. Studies on quiescent 3T3 fibroblasts by Rozengurt's laboratory first documented stimulation of ouabain-sensitive Rb$^+$ uptake following addition of serum.[57] Since then, other peptide growth factors have been shown to stimulate (Na$^+$,K$^+$)ATPase transport activity, including epidermal growth factor,[58] fibroblast-derived growth factor,[59] platelet-derived growth factor,[60] nerve growth factor,[61] and multiplication stimulating activity.[62] Moreover, vasopressin,[63] phorbol esters,[64] mellitin,[65] prolactin,[66] glucagon,[38,67] catecholamines,[68,69] and cyclic adenosine monophosphate (cAMP)[70] also induce increased Na$^+$ pump activity. Several of these factors (insulin in hepatocytes, growth factors, vasopressin, phorbol esters, and mellitin) seem to share a common mechanism for activating the Na$^+$ pump, i.e., increased Na$^+$ entry into the cell.

The strongest evidence that increased Na$^+$ influx triggers the observed stimulation of the (Na$^+$,K$^+$)ATPase comes from the work involving serum-stimulated fibroblasts[71,72] and neuroblastoma cells.[73] A serum-induced increase in sodium uptake occurs on the same time scale as stimulation of Rb$^+$ uptake.[71-73] Addition of amiloride, a Na$^+$/Na$^+$ and Na$^+$/H$^+$ exchange inhibitor, blocks serum stimulated Na$^+$ influx without affecting basal influx and prevents activation of the Na$^+$ pump by serum.[72,73] No stimulation of the pump occurs in Na$^+$-free medium.[72,73] Moreover, activation of the (Na$^+$,K$^+$)ATPase in these cells can be effected in the absence of serum by compounds that increase intracellular Na$^+$, such as the ionophores monensin[71-73] and gramicidin,[71] and the neurotoxins veratridine and scorpion toxin.[73] If one assumes that the basal activity of the (Na$^+$,K$^+$)ATPase is limited by the intracellular concentration of sodium ion, then increased Na$^+$ influx provides a molecular explanation for the resultant activation of the Na^+ pump. Since increased Na^+ influx has been implicated in stimulation of DNA synthesis[74] and induction of cell proliferation,[75,76] it is tempting to speculate that all growth factors or mitogens (including insulin) operate through this pathway.

VII. MECHANISMS FOR ACTIVATING THE $(Na^+,K^+)ATPase$

Although intracellular Na^+ concentrations are not maintained at levels optimal for Na^+ pump activity, several lines of evidence indicate that, in the rat adipocyte, insulin stimulation of the $(Na^+,K^+)ATPase$ does not involve increased Na^+ entry. No change in Na^+ influx is detected, either in ouabain-treated or untreated fat cells, after addition of insulin. No change in Na^+ efflux from preloaded, ouabain-treated cells is seen after addition of insulin. The basal rate of Rb^+ uptake is not increased when the external Na^+ concentration is raised from 145 to 195 mM, and the insulin-stimulated rate at both concentrations is identical. Although monensin activates Rb^+ uptake in adipocytes, it clearly does so by raising intracellular Na^+ concentrations. No such change is seen with insulin. Moreover, the effects of maximal concentrations of monensin and insulin are additive on Rb^+ uptake. Finally, addition of known Na^+ channel blockers or activators (tetrodotoxin, veratridine, and scorpion venom $+/-$ tetrodotoxin, tetracaine, or amiloride) has no effect on Na^+ or Rb^+ uptake in either basal or insulin-stimulated cells.[49]

In muscle, the situation is apparently different. Increased Na^+ uptake is detected when insulin is added to ouabain-treated frog or rat muscle.[30,77] This increase is not affected by tetracaine,[30] but is blocked by 2,4-dinitrophenol[30] and amiloride.[77] In frog skeletal muscle, an insulin-induced increase in intracellular pH is dependent on the Na^+ concentration gradient across the membrane and is amiloride sensitive.[76] This observation led Moore[77] to propose that insulin activates a Na^+/H^+ exchange mechanism, a possibility also invoked to explain serum-stimulated Na^+ influx.[73] Along these lines, it is equally plausible to propose that an insulin-induced increase in intracellular pH activates the $(Na^+,K^+)ATPase$. Indeed, the internal pH in a fat cell is not maintained at levels optimal for Na^+ pump activity. Artificial elevation of intracellular pH, by increasing CO_2 : bicarbonate ratios in the medium, does result in higher levels of Rb^+ uptake comparable to those observed with insulin stimulation (G. Guidotti, unpublished results). However, it has not yet been possible to establish whether this is the mechanism whereby insulin stimulates the Na^+ pump.

A plethora of other mechanisms have been proposed to account for the actions of insulin in its target tissues. Among the candidates for a second messenger of insulin action are Ca^{2+} ion,[78] intracellular peptide messengers, phospholipid turnover, and protein phosphorylation. The latter possibility is an intriguing one, as phosphorylation of the $(Na^+,K^+)ATPase$ has been shown to occur in transformed cells.[79] To date, no effect on enzyme activity has been correlated with this phosphorylation, and it remains to be determined whether insulin has any role in this reaction.

VIII. STIMULATION OF (Na+,K+)ATPase AND HYPERPOLARIZATION

The net result of insulin-induced alterations of monovalent cation uptake is an increase in the intracellular K^+/Na^+ concentration ratio.[30,32,33,39] One would therefore predict that hyperpolarization of target tissues would occur, as has been well documented in muscle and fat tissue.[80–89] However, the mechanism of insulin-induced hyperpolarization is in dispute. Zierler[80–82] has argued that insulin acts by decreasing muscle membrane permeability to Na^+, thereby inducing hyperpolarization. The increase in intracellular K^+ concentration was found to be too small and too slow to account for the observed changes in membrane potential and was thus concluded to be secondary to the hyperpolarization. It was further demonstrated that electrically induced hyperpolarization could partially mimic insulin stimulation of glucose transport.[86] Others have suggested that, in heart cells, insulin induces a net outward current, but agree with Zierler that the resulting hyperpolarization is not mediated by activation of K^+ transport.[87] On the other hand, two recent investigations have shown that cardiac glycosides completely inhibit the effect of insulin on membrane potential in muscle,[88,89] implying that the hyperpolarization is secondary to a stimulation of the (Na+,K+)ATPase. Thus, although the ability of insulin to hyperpolarize the membrane is well established, whether this is a primary effect of the hormone has not yet been resolved.

IX. A MODEL FOR ACTIVATION OF THE (Na+,K+)ATPase

Given the ample evidence demonstrating similarities between insulin activation of Rb^+ and glucose transport, it is reasonable to invoke the translocation hypothesis[55,56] to describe a molecular mechanism for insulin activation of the (Na+,K+)ATPase. If we assume that the Na^+ pump and the glucose transport mechanism respond to the same signal generated by insulin interaction with its cell-surface receptor and that no change in the number of Na^+ pumps occurs, then, on a simple level, two types of mechanisms are conceivable:

Model I:

Insulin binding to its receptor produces a signal X, which triggers vesicle translocation and fusion and also directly activates the plasma membrane-bound (Na+,K+)ATPase. Translocation of intracellular vesicles would not be a prerequisite for stimulation of the Na^+ pump, and the lag phase observed[51–53] would represent the time required to generate a sufficient quantity of signal X.

Model II:

By contrast, activation of the Na^+ pump could occur after the translocation event. The interaction of insulin and its receptor would generate a signal X, which catalyzes translocation of vesicles that contain glucose transporters and an additional protein. Following fusion with the plasma membrane, this other protein functions to activate the (Na+,K+)ATPase.

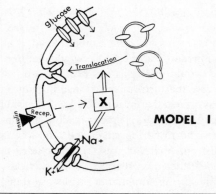

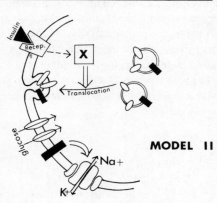

Figure 1. Two models for insulin activation of the (Na^+,K^+)ATPase. Translocation model of Karnieli *et al.*,[55] modified to include stimulation of the Na^+ pump. *Model I:* Direct activation. Insulin binding to its cell-surface receptor generates a signal X, which stimulates translocation and fusion of glucose transporter-containing vesicles to the plasma membrane. Factor X also activates the Na^+ pump. *Model II:* Activation following translocation. Insulin binding to its cell-surface receptor generates a signal X, which stimulates translocation of vesicles containing glucose transporters and an additional protein (solid rectangle). Following fusion with the plasma membrane, this other protein (rectangle) functions to activate the (Na^+,K^+)ATPase. In both models, activation of the Na^+ pump occurs only in an intact cell, and insulin treatment does not change the number of Na^+ pumps. (Reprinted with permission from Resh.[53] Copyright 1983, American Chemical Society.)

These two models are illustrated schematically in Fig. 1. Although it is not possible to distinguish definitively between these two models, several lines of evidence support the mechanism described in model II. Vesicle translocation, by analogy with exocytosis, is probably a temperature- and energy-dependent process.[90,91] Assuming that the signal that stimulates this process is rapidly generated, then time- and temperature-dependent stimulation of (Na^+,K^+)ATPase is best explained by model II. Hydrogen peroxide, an insulinomimetic compound that acts at a step distal to the insulin-binding event, stimulates glucose and Rb^+ uptake with the same lag time as insulin.[53] If peroxide is mimicking the action of factor X, then the existence of a lag time is also consistent with the translocation mechanism of model II. It should be noted that additional time requiring steps are probably involved following vesicle fusion.[55] The term *translocation* has been used to describe the phenomenon whereby increased numbers of transporters appear in the plasma membrane. Until definitive morphological evidence demonstrating vesicle translocation is available, this mechanism must be treated as a hypothesis.

X. SIGNIFICANCE OF INSULIN STIMULATION OF (Na$^+$,K$^+$)ATPase

The physiological relevance of insulin stimulation of the Na$^+$ pump is evidenced by the important role played by insulin in regulating potassium homeostasis *in vivo*.[92] Yet, on a biochemical level, one can question *why* insulin stimulates this enzyme, and whether this event is necessary for the manifestation of other biological effects of insulin. The only definitive example of a process dependent on the activation of the Na$^+$ pump is the insulin-induced hyperpolarization of the muscle membrane.[88,89] The evidence thus far reported indicates that stimulation of glucose transport is not dependent on the activity of the (Na$^+$,K$^+$)ATPase, and vice versa. It is still possible that insulin activation of the Na$^+$ pump is the means by which the hormone exerts control over intracellular metabolic machinery. However, recent experiments conducted in the laboratory of G. Guidotti have established that the antilipolytic effect of insulin (unpublished results) and the stimulation of glycogen synthetase (L. DeAntonio, unpublished results) are unaffected when the Na$^+$ pump is blocked by ouabain. Although protein synthesis rates[4] and the activities of phosphofructokinase[93] and pyruvate kinase[94] are dependent on potassium, maximal rates are attained at approximately 50 mM K$^+$.[93–95] Thus, at ambient intracellular K$^+$ concentrations (150 mM), these insulin-responsive enzymes are probably already operating at V_{max} (for K$^+$), and one would predict further increases in K$^+$ to be without significant effect.

XI. PERSPECTIVE AND CONCLUSIONS

Although the (Na$^+$,K$^+$)ATPase is biochemically an extremely well-characterized enzyme, in fact we know very little about how its activity is regulated *in vivo*. It is now established that insulin stimulates the uptake of K$^+$ into fat and muscle tissue and that this increase in K$^+$ transport is mediated by rapid, reversible activation of the (Na$^+$,K$^+$)ATPase. In muscle, concomitant stimulation of Na$^+$ efflux is also observed. The net effect of these insulin-induced alterations in ion fluxes is to increase the intracellular K$^+$/Na$^+$ ratio, ultimately leading to hyperpolarization of the target cell. Insulin treatment does not change the number of plasma membrane Na$^+$ pumps, nor is a permanent activation of (Na$^+$,K$^+$)ATPase activity evident. Thus, the (Na$^+$,K$^+$)ATPase responds only to signals effected by insulin in the intact cell.

There is compelling evidence to support the hypothesis that interaction of insulin with its receptor generates a common signal that activates two different membrane-transport proteins: the glucose transporter and the (Na$^+$,K$^+$)ATPase. Ultimately, the key to understanding the molecular mechanism of insulin action will require elucidation of the events that couple binding to the receptor to activation of the biological responses, including identification of the elusive factor X. Insulin stimulation of the (Na$^+$,K$^+$)ATPase, while providing a mechanistic explanation for the hypokalemic effect following administration of hormone *in vivo*, also provides another clue in solving the mystery of how polypeptide hormones regulate metabolic homeostasis in living cells.

ACKNOWLEDGMENT. I wish to thank Guido Guidotti, in whose laboratory my work was performed, for support and for permission to cite unpublished data from his laboratory.

XII. REFERENCES

1. Briggs, A.P., Koechig, I., Doisy, E.A., and Weber, C.J., 1924, *J. Biol. Chem.* **58:**721–730.
2. Tosteson, D.C., 1964, in *The Cellular Functions of Membrane Transport* (J.F. Hoffman, ed.), pp. 3–22, Prentice-Hall, Englewood Cliffs, N.J.
3. Katz, B., 1966, *Nerve, Muscle, and Synapse,* McGraw-Hill, New York.
4. Ledbetter, M.L.S., and Lubin, M., 1977, *Exp. Cell Res.* **105:**223–236.
5. Schwartz, A., Lindemayer, G.E., and Allen, J.C., 1972, *Curr. Top. Membr. Transp.* **3:**1–82.
6. Cantley, L.C., 1981, *Curr. Top. Bioenerg.* **11:**201–237.
7. Robinson, J.D., and Flashner, M.S., 1979, *Biochim. Biophys. Acta* **549:**145–176.
8. Glynn, I.M., and Karlish, S.J.D., 1975, *Annu. Rev. Physiol.* **37:**13–55.
9. Skou, J.C., 1957 *Biochim. Biophys. Acta* **23:**394–401.
10. Kyte, J., 1971, *J. Biol. Chem.* **246:**4157–4165.
11. Sweadner, K.J., 1979, *J. Biol. Chem.* **254:**6060–6067.
12. Dixon, J.F., and Hokin, L.E., 1974, *Arch. Biochem. Biophys.* **163:**749–758.
13. Kyte, J., 1971, *Biochem. Biophys. Res. Commun.* **43:**1259–1265.
14. Perrone, J.R., and Blostein, R.,1973, *Biochim. Biophys. Acta* **291:**680–689.
15. Ruoho, A., and Kyte, J., 1974, *Proc. Natl. Acad. Sci. USA* **71:**2352–2356.
16. Post, R.L., Toda, G., and Rogers, F.N., 1975, *J. Biol. Chem.* **250:**691–701.
17. Giotta, G.J., 1976, *J. Biol. Chem.* **251:**1247–1252.
18. Kyte, J., 1972, *J. Biol. Chem.* **247:**7642–7649.
19. Liang, S.M., and Winter, C.G., 1977, *J. Biol. Chem.* **252:**8278–8284.
20. Jorgensen, P.L., 1977, *Biochim. Biophys. Acta* **466:**97–108.
21. Castro, J., and Farley, R.A., 1979, *J. Biol. Chem.* **254:**2221–2228.
22. Post, R.L., and Kume, S., 1973, *J. Biol. Chem.* **248:**6993–7000.
23. Albers, R.W., Fahn, S., and Koval, G.J., 1963, *Proc. Natl. Acad. Sci. USA* **50:**474–481.
24. Fahn, S., Hurley, M.R., Koval, G.J., and Albers, R.W., 1966, *J. Biol. Chem.* **241:**1890–1895.
25. Taniguchi, K., and Post, R.L., 1975, *J. Biol. Chem.* **250:**3010–3018.
26. Willebrands, A.F., Groen, J., Kamminga, C.E., and Blickman, J.R., 1950, *Science* **112:**277–279.
27. Clausen, T., and Hansen, O., 1974, *Biochim. Biophys. Acta* **345:**387–404.
28. Smillie, L.B., and Manery, J.F., 1960, *Am. J. Physiol.* **198:**67–77.
29. Gourley, D.R.H., 1961, *Am. J. Physiol.* **200:**1320–1326.
30. Clausen, T., and Kohn, P.G., 1977, *J. Physiol. (Lond.)* **265:**19–42.
31. Creese, R., 1968, *J. Physiol. (Lond.)* **197:**255–278.
32. Moore, R.D., 1973, *J. Physiol. (Lond.)* **232:**23–45.
33. Lostroh, A.J., and Krahl, M.E., 1973, *Biochim. Biophys. Acta* **291:**260–268.
34. Erlij, D., and Grinstein, S., 1976, *J. Physiol. (Lond.)* **259:**13–31.
35. Gourley, D.R.H., and Bethea, M.D., 1964, *Proc. Soc. Exp. Biol. Med.* **115:**821–823.
36. Hougen, T.J., Hopkins, B.E., and Smith, T.W., 1978, *Am. J. Physiol.* **234:**C59–C63.
37. Resh, M.D., 1982, *J. Biol. Chem.* **257:**6978–6986.
38. Fehlmann, M., and Freychet, P., 1981, *J. Biol. Chem.* **256:**7449–7453.
39. Resh, M.D., Nemenoff, R.A., and Guidotti, G., 1980, *J. Biol. Chem.* **255:**10938–10945.
40. Zierler, K.L., 1960, *Am. J. Physiol.* **198:**1066–1070.
41. Clausen, T., and Hansen, O., 1977, *J. Physiol. (Lond.)* **270:**415–430.
42. Resh, M.D., 1982, *J. Biol. Chem.* **257:**11946–11952.
43. Schuurmans Stekhoven, F.M.A.H., Swarts, H.G.P., DePont, J.J.H.H.M., and Bonting, S.L., 1980, *Biochim. Biophys. Acta* **597:**100–111.

44. Jarett, L., and Smith, R.M., 1974, *J. Biol. Chem.* **249:**5195–5199.
45. Hadden, J.W., Hadden, E.M., Wilson, E.E., Good, R.A., and Coffey, R.G., 1972, *Nature New Biol.* **235:**174–177.
46. Brodal, B.P., Jebens, E., Oy, V., and Iverson, O.-J., 1974, *Nature* **249:**41–43.
47. Gavryk, W.A., Moore, R.D., and Thompson, R.C., 1975, *J. Physiol. (Lond.)* **252:**43–58.
48. Rogus, E., Price, T., and Zierler, K.L., 1969, *J. Gen. Physiol.* **54:**188–202.
49. Resh, M.D., 1982, *Regulation of Membrane Transport Proteins by Insulin,* Ph.D. thesis, Harvard University, Cambridge, Mass.
50. Lund-Anderson, H., 1979, *Physiol. Rev.,* **59:**305–352.
51. Cirialdi, T.P., and Olefsky, J.M., 1979, *Arch. Biochem. Biophys.* **193:**221–231.
52. Häring, H.U., Biermann, E., and Kemmler, W., 1981, *Am. J. Physiol.* **240:**E556–E565.
53. Resh, M.D., 1983, *Biochemistry* **22:**2781–2784.
54. Kohn, P.G., and Clausen, T., 1971, *Biochim. Biophys. Acta* **225:**277–290.
55. Karnieli, E., Zarnowski, M.J., Hissin, P.J., Simpson, I.A., Salans, L.B., and Cushman, S.W., 1981, *J. Biol. Chem.* **256:**4772–4777.
56. Kono, T., Robinson, F.W., Blevins, T.L., and Ezaki, O., 1982, *J. Biol. Chem.* **257:**10942–10947.
57. Rozengurt, E., and Heppel, L.A., 1975, *Proc. Natl. Acad. Sci. USA* **72:**4492–4495.
58. Fehlmann, M., Canivet, B., and Freychet, P., 1981, *Biochem. Biophys. Res. Commun.* **100:**254–260.
59. Bourne, H.R., and Rozengurt, E., 1976, *Proc. Natl. Acad. Sci. USA* **73:**4555–4559.
60. Mendoza, S.A., Wigglesworth, N.M., Pohjanpelto, P., and Rozengurt, E., 1980, *J. Cell Physiol.* **103:**17–27.
61. Boonstra, J., Skaper, S.D., and Varon, S., 1982, *J. Cell Physiol.* **113:**28–34.
62. Smith, G.L., 1977, *J. Cell. Biol.* **73:**761–767.
63. Mendoza, S.A., Wigglesworth, N.M., and Rosengurt, E., 1980, *J. Cell. Physiol.* **105:**153–162.
64. Dicker, P., and Rozengurt, E., 1981, *Biochem. Biophys. Res. Commun.* **100:**433–441.
65. Rozengurt, E., Gelehrter, T.D., Legg, A., and Pettican, P., 1981, *Cell* **23:**781–788.
66. Falconer, I.R., Forsyth, I.A., Wilson, B.M., and Dils, R., 1978, *Biochem. J.* **172:**509–516.
67. Ihlenfeldt, M.J.A., 1981, *J. Biol. Chem.* **256:**2213–2218.
68. Cheng, L.C., Rogus, E.M., and Zierler, K., 1977, *Biochim. Biophys. Acta* **464:**338–346.
69. Clausen, T., and Flatman, J.A., 1977, *J. Physiol. (Lond.)* **270:**383–414.
70. Paris, S., and Rozengurt, E., 1982, *J. Cell Physiol.* **112:**273–280.
71. Smith, J.B., and Rosengurt, E., 1978, *Proc. Natl. Acad. Sci. USA* **75:**5560–5564.
72. Smith, J.B., and Rozengurt, E., 1978, *J. Cell. Physiol.* **97:**441–450.
73. Moolenaar, W.H., Mummery, C.L., van der Saag, P.T., and deLaat, S.W., 1981, *Cell* **23:**789–798.
74. Rozengurt, E., and Mendoza, S., 1980, *Ann. NY Acad. Sci.* **339:**175–190.
75. Koch, K.S., and Leffert, H.L., 1979, *Cell* **18:**153–163.
76. Toback, F.G., 1980, *Proc. Natl. Acad. Sci. USA* **77:**6654–6656.
77. Moore, R.D., 1981, *Biophys. J.* **33:**203–210.
78. Sorensen, S.S., Christensen, F., and Clausen, T., 1980, *Biochim. Biophys. Acta* **602:**433–445.
79. Yeh, L.-A., Ling, L., English, L., and Cantley, L., 1983, *J. Biol. Chem.* **258:**6567–6574.
80. Zierler, K.L., 1951, *Science* **126:**1067–1068.
81. Zierler, K.L., 1959, *Am. J. Physiol.* **197:**515–523.
82. Zierler, K.L., and Rogus, E.M., 1981, *Biochim. Biophys. Acta* **640:**687–692.
83. Beigelman, P.M., and Hollander, P.B., 1964, *Proc. Soc. Exp. Biol. Med.* **116:**31–35.
84. Cheng, K., Groarke, J., Osotimehin, B., Haspel, H.C., and Sonenberg, M., 1981, *J. Biol. Chem.* **256:**649–655.
85. Davis, R.J., Brand, M.D., and Martin, B.R., 1981, *Biochem. J.* **196:**133–147.
86. Zierler, K., and Rogus, E.M., 1980, *Am. J. Physiol.* **239:**E21–E29.
87. Lantz, R.C., Elsas, L.J., and DeHaan, R.L., 1980, *Proc. Natl. Acad. Sci. USA* **77:**3062–3066.
88. Flatman, J.A., and Clausen, T., 1979, *Nature* **281:**580–581.
89. Moore, R.D., and Rabovsky, J.L., 1979, *Am. J. Physiol.* **236:**C249–C254.
90. Kono, T., Suzuki, K., Dansey, L.E., Robinson, F.W., and Blevins, T.L., 1981, *J. Biol. Chem.* **256:**6400–6407.

91. Ezaki, O., and Kono, T., 1982, *J. Biol. Chem.* **257**:14306–14310.
92. Cox, M.C., Sterns, R.H., and Singer, I., 1978, *N. Engl. J. Med.* **299**:525–532.
93. Paethau, V., and Lardy, H.A., 1967, *J. Biol. Chem.* **242**:2035–2042.
94. Kachman, J.F., and Boyer, P.D., 1953, *J. Biol. Chem.* **200**:669–682.
95. Evan, H.J., and Sorger, G.J., 1966, *Annu. Rev. Plant Physiol.* **17**:47–76.

Index